# Forest Fire and Climate Change

Hukum Singh · Narendra Kumar
Editors

# Forest Fire and Climate Change

## Insights into Science

Springer

*Editors*
Hukum Singh
Forest Research Institute
Dehradun, Uttarakhand, India

Narendra Kumar
College of Agriculture
Central Agriculture University (Imphal)
Kyrdemkulai, Meghalaya, India

ISBN 978-3-031-89966-9 ISBN 978-3-031-89967-6 (eBook)
https://doi.org/10.1007/978-3-031-89967-6



This Springer imprint is published by the registered company Springer Nature Switzerland AG
The registered company address is: Gewerbestrasse 11, 6330 Cham, Switzerland

# Preface

This book delves deep into the complex relationship between forest fires and climate change by thoroughly exploring the scientific foundations, ecological consequences, and global implications of forest fires in the climate change scenario. The book begins by elucidating the fundamentals of climate change, offering readers a clear understanding of the mechanisms driving global warming. It then seamlessly transitions into an in-depth analysis of how rising temperatures, prolonged droughts, and altered precipitation patterns exacerbate forest fires. The book demonstrates how natural forest fires become destructive, uncontrollable forces threatening ecosystems, communities, and economies through case studies and data-driven insights. It emphasizes solutions and mitigation strategies for forest fires under the climate change scenario. It explores innovative approaches to managing and preventing forest fires, from controlled burns to advanced fire prediction models. It also underscores the importance of global cooperation in reducing greenhouse gas emissions to mitigate the impacts of climate change on forest fires. This book serves as a vital resource for policymakers, scientists, and anyone interested in the intersection of climate change and forest fires. It will empower readers with the knowledge to address this alarming global issue and highlight the urgency of protectings and mitigating the devastating effects of forest fires and global climate change.

Dehradun, India — Hukum Singh
Kyrdemkulai, India — Narendra Kumar

# Contents

# About the Editors

**Dr. Hukum Singh** is a forest physiology and climate change scientist associated with the Forest Research Institute in Dehradun, India. His research covers eco-physiology, plant responses to abiotic stresses, climate-smart forestry, forest fire, climate change adaptation and mitigation, urban pollution, and more. Dr. Singh has authored over 100 publications, including research publications, review articles, books and book chapters, and conference proceedings with journals and publishers of international repute.

**Dr. Narendra Kumar** earned his Ph.D. in plant physiology from G. B. Pant University of Agriculture and Technology, Pantnagar, India. His research work focused on the plants response to abiotic stresses. Later, he served as an assistant professor in various reputed academic institutes, especially Surajmal Aggarwal (Pvt.) Kanya Mahavidyalaya, Kichha US Nagar, Dev Bhoomi Group of Institute, Dehradun, India; Uttaranchal University, Dehradun, India; Doon (PG) College of Agriculture Science and Technology, Dehradun, India. He also worked as a research associate at Forest Research Institute, Dehradun. His research areas focus primarily on Plant Physiology, ecophysiology, plant response to abiotic stresses, climate change adaptation and mitigation, functional traits, and climate change modeling. He has supervised many M.Sc. students, and many more are under supervision. He has published various papers/chapters/articles in national and international peer-reviewed reputed journals with high impact factors. He has edited so many books with an international publisher. He has served as an editor for so many journals on current subjects. He has also served as an associate guest editor for *Frontiers of Plant Science*'s for the special "Rhizophereic Manipulation for Sustainable Agriculture" issue. He was also engaged as a project scientist at the Plant Physiology Discipline, Forest Genetics and Tree Improvement Division, Forest Research Institute in Dehradun, India. Currently, he is working as an assistant professor at the College of Agriculture, Kyrdemkulai, Central Agriculture University (I), Meghalaya.

# About the Editors

# Forest Fire's Drivers: Climatic, Biological and Anthropological

**Himshikha Gusain, Jitendra Singh Butola, and NG Manitombi Devi**

**Abstract** Forest fires are complex phenomena that generally impact forest ecosystems positively and negatively. Forest fire incidents have caused significant ecological and economic disruptions worldwide in recent years. Several factors have likely contributed to the increase in these incidents. The major driving forces behind forest fires can be categorized into anthropogenic, biological, and climatic drivers (ABCs). These ABC factors are increasingly transitioning forest resources to a higher-risk state by exacerbating the danger of fires in various regions. This essay describes the major ABC drivers, emphasizing fire hazards. Among these drivers, anthropogenic factors have become the major causative agents in the past century, primarily due to the rapid expansion of industries, tourism, increased population density, and livelihood demands. Increased emissions bridge anthropogenic-driven climate and biological changes, selection of inappropriate species, and natural climatic variability, all of which contribute to the potential and intensity of fires. The driving mechanisms and impacts of these causative forces are elaborated, followed by possible measures to mitigate the destructive effects of fires. Despite large uncertainties, anthropogenic drivers can spark fire incidents, while climate and biological drivers may lead to increased fire activity. Effective, independent, and mutual stewardship is suggested to minimize the impact of ABCs on forest fires. An evidence-based approach to technological adoption, combined with the inclusion of knowledge from forest fringe communities in natural and urban landscaping, will help achieve the desired goal of fire management in fire-prone areas. Furthermore, trans-disciplinary research efforts are recommended to examine anthropogenic-driven climate change and its indirect influence on modified biological factors causing forest fires at the regional level. Reviewing and revising existing forest fire management policies, increasing forest resilience and fire plans, and minimizing biological fuel can effectively deal with the incidence and expansion of forest fires in high-risk zones. Hence, reducing the severity of fire hazards and impacts requires nations to come together on a single platform to formulate ecosystem-based guidelines for forest fire management and address the associated challenges.

---

H. Gusain (✉) · J. S. Butola · NG Manitombi Devi
Department of Forestry and Natural Resources, HNB Garhwal University, Srinagar Garhwal, India
e-mail: bhhimshikha654@gmail.com

 
H. Singh and N. Kumar (eds.), *Forest Fire and Climate Change*,
https://doi.org/10.1007/978-3-031-89967-6_1

**Keywords** Anthropological · Biological · Climatic forest fires · Factors · Fire's drivers

## 1 Introduction

A forest or forest fire denotes any unplanned or unwanted fire appearing in forest areas or wild lands without development. Even if such fires are anthropogenic, they can accidentally spread without fulfilling any significant purpose (Schmerbeck et al. 2015). Forest fires occur naturally in various regions and can positively affect the forests. For example, in natural ecosystems, especially in those forests that are spread at higher latitudes, fire plays a vital role in contributing to maintaining a healthy forest ecosystem by burning additional mulch matter or dry biomass (Barwa and Singh 2024), removing unwanted weed communities, releasing important nutrients such as carbon into the soil (Singh 2024b; Kaushik and Singh 2024), aiding seed dispersal, by breaking dormancy in the seeds of certain species of the forests (Singh 2024c; Sett et al. 2024), and by providing suitable temperature range required for seedling emergence (Kumar et al. 2021a, 2021b). However, these fires have intensified and expanded in recent years, seemingly without any specific cause. Certain nations are grappling with abundant forest fires in the present era, whereas others are facing a shortage of such incidents. Hence, due to evolving climatic conditions causing biomass to dry and organic build-up to accumulate rapidly, extensive forest fires are no longer advantageous for the overall well-being of forest ecosystems (Singh 2024d).

Certain forest fires are excessively intense in environments where such events are typically rare, unexpected, or heightened in frequency and severity. A usual method to describe forest fires occurring in forest vegetation is that their observed height should be more than 1.8 m or 6 feet. The incidence of such fires is increasing day by day across the globe. Colossal forest fire significantly influences ecosystem patterns and processes, such as vegetation distribution, structure, the carbon cycle, and climate. Forest fires' escalating occurrence and severity result in significant carbon dioxide emissions, amplifying global warming and climate change impact. As per the Fifth Assessment Report by the Intergovernmental Panel on Climate Change (IPCC), forest fires contribute annual carbon emissions ranging from 2.5 billion to 4.0 billion tons of $CO_2$, as indicated in the data from the Ministry of Environment, Forest and Climate Change (MoEFCC 2018). As global incidents of fire increase, mega forest fire incidents have become more frequent, intense, and hazardous, destroying enormous portions of the forests. The eruptive frequency and intensity of these fires have increased irrespective of seasonal variation, posing significant severe threats to the ecosystem, natural biodiversity, and the well-being of human and wildlife communities across the globe.

Despite the longstanding coexistence of humans and fire, our ability to manage it remains imperfect (Bowman et al. 2009). Every year, millions of people are displaced and losing their livelihoods due to the widespread occurrence of these fires. In India,

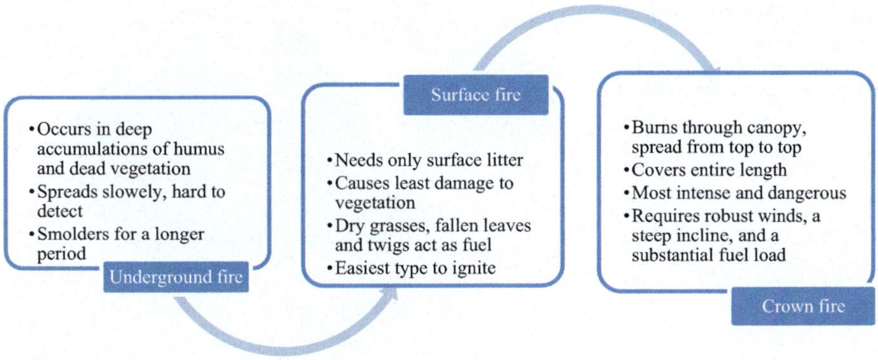


**Fig. 1** Three basic phases of forest fires

tree cover loss stands at 3.59 lakh hectares due to fires and an additional 2.15 million hectares from various other factors, leading to ecological and economic disruptions. In a large-scale context, the forest area has been reduced to half compared to the beginning of the twentieth century. In contrast to 2001, the current annual tree cover loss from fires has increased by 3 million hectares (7.4 million acres), surpassing the size of Belgium (Kimbrough 2022). Recently, in 2021, a staggering 9 million hectares of forests were lost globally, equivalent to the size of Portugal in Europe. Russia, Canada, and the United States had a combined loss of 7.8 million hectares. Italy experienced the destruction of 723,924 ha over the past 14 years, with an alarming 159,437 ha devastated by forest fires in 2021 alone. These disturbing figures paint a bleak picture of a world that continues to exploit its natural resources intensively, as highlighted by Carta et al. (2023). The rampant forest fires are responsible for over 33% of the total tree cover loss, marked as the most devastating for forests and wildlife in recorded history. In context to India, forest inventory records reveal that 54.40% of forests undergo sporadic fires, 7.49% experience slightly frequent fires, and 2.40% of forests have elevated prevalence stages. Gajendiran et al. (2024) reported that 35.71% of Indian forests are still considered vulnerable, posing a substantial risk of significant loss due to fire. Future challenges may arise due to climate change altering fire regimes, but assessing this risk is challenging as fires are still inadequately represented. Depending on state and extent, forest fires can be divided into three phases: underground (grass) forest fire, surface forest fire, and tree top (crown) fires (Fig. 1).

## 2 Forest Fire's Drivers

Forest fires are complex events likely exploded by various drivers, causing eruption and expansion. Recently, there has been a rapid increase in the destruction of global forests, attributed to intense cultivation, expanding settlements, and the ignition and

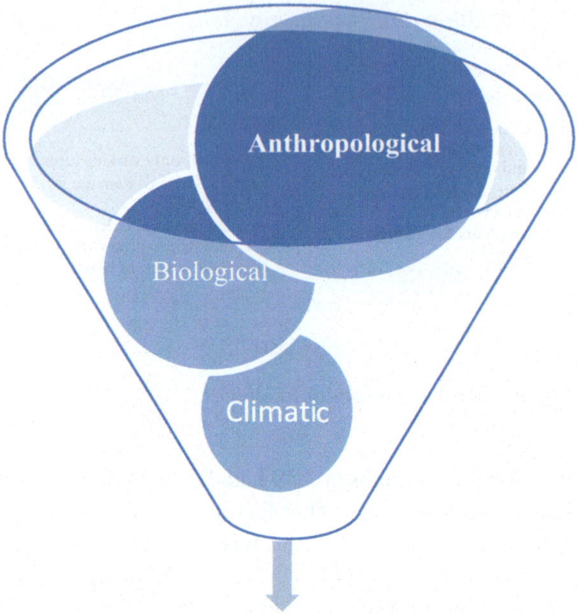


**Fig. 2** The key drivers of forest fires are Anthropological, Biological, and Climatic Drivers (ABCs)

spread of fires resulting from various fire drivers and their cumulative impact. The ignition and spread of forest fires in natural ecosystems can result from multiple factors influencing fire behaviour. For example, globally, only 4% of these fires are attributed to natural causes like lightning and windstorms, while anthropogenic factors are predominantly responsible for most fire incidences. The key drivers of forest fires can be broadly categorized into Anthropological, Biological, and Climatic drivers (ABCs), as depicted in Fig. 2. A brief description of these drivers is given below:

## *2.1 Types of Forest Fire Drivers: Anthropological, Biological, and Climatic Drivers (ABCs)*

### 2.1.1 Anthropological Drivers

Although natural causes such as lightning can spark fires, human activities often serve as the primary catalysts for numerous forest fires. Numerous studies highlight concerns regarding the historical, ongoing, and planned exploitation and conversion of forests. Sloan et al. (2017) have focused on the interconnected relationship between fire activity and forest degradation from timber extraction and agricultural practices. Several key anthropological factors that contribute to forest fires are outlined below.

***Urbanization and Population Density***

Urbanization patterns erode the distinctions between urban areas and adjacent forest lands, leading to a growing ambiguity in regions worldwide. In the last few years, the Wildland-Urban Interface (WUI) has emerged as the forefront of forest fire disasters (NDMA 2020), where the boundaries between urban and forested environments have increasingly shrunken. Rapid urbanization with increased population density has also sparked ever-increasing livelihood demands to the next level, making forests more vulnerable.

***Forest Degradations and Land Conversions***

Over the past two centuries, anthropogenic influences on fire activity have heightened significantly, attributed to the rapid conversion of lands and the degradation of forests spurred by population growth, industrial expansion, and tourism. The large-scale degradation and transformation of forests for industrial purposes contribute reciprocally to the escalation of fire incidents. Additionally, the interplay of these activities with climate anomalies, such as El Niño, has triggered pockets of drought, amplifying both the extent and severity of fire activity, particularly in warm regions.

***Deforestation***

Clearing vast areas of forests primarily for new establishments, infrastructure development, agriculture or industrial extension, etc., reduces the natural firebreaks and increases the availability of dry vegetation that can fuel fires. Deforestation led by urbanization can alter the landscape usage and pattern, making it more prone to fires (Singh et al. 2024, 2018; Yadav et al. 2019a; Singh 2021).

***Human Settlement Patterns***

Cultural preferences for certain types of settlements, such as rural communities in fire-prone areas and communities inside or in forest fringe areas, can increase the likelihood of human-caused fires. Such settlements increase the chances of accidental ignition and make fire suppression more challenging. The risk of unintentional ignition may increase when human activities occur close to forested areas. Furthermore, increased infrastructure demand for population settlement has ignored the forests and climatic-related risks.

***Recreational Tourism and Negligence***

A rising trend in the tourism sector, especially in nature-oriented areas such as ecotourism, recreational tourism, etc., has led to tourist influx in forests and surrounding areas. Ever-increased tourism activities have generated many folds to combat fire in these places. One of the most important causes of fire ignition in tourism spots inside forest areas is the irresponsible and negligent behavior of tourists, such as smoking or improper disposal of flammable material in wild trails, no accountability for certain activities such as unmanaged or unauthorized campfires, use of combustible fuel, etc. inside the forest areas. If not conducted responsibly, the fires ignited due to these activities may immediately threaten humans and wildlife within no time.

***Electricity, Equipment, and Machinery***

Sometimes, electrical equipment such as transformers, electric poles, power lines, lighthouses, etc., inside or near forest fringe areas can ignite surrounding vegetation cover. Additionally, using electronics and machinery such as chainsaws in various logging and clearing operations can spark and sometimes cause deliberate fires if not controlled adequately.

***Traditional Land Management Practices***

Indigenous or local communities have traditional land management practices that involve controlled burning in forest patches. Such practices are still prevalent in various indigenous communities across the world. For example, the Yurok, Karuk, and Hoopa tribes of North America manage fires on their traditional lands (Nature 2020). Also, local people often perform certain rituals related to sacred practices in forests, during which any ignorance may lead to the initiation of fires. Fire incidences in the northeastern Khasi-Garo hills of Meghalaya, India, and Brazil's Amazon forests are examples of such activities.

***Poor and Unrestrained Agricultural Practices***

Specific agricultural methods, like slash-and-burn techniques, may inadvertently result in forest fires. This commonly occurs in numerous forests and bordering regions, where the intentional burning of agricultural and forest crops or their remnants occurs. In some localities like northern plain regions of India, locals use open burning to clear land for agriculture or development. If not properly managed, these practices can spread fire beyond desirable areas. Moreover, Slash-and-burn practices are progressively causing more harm to the world's forests. Such malpractices play a notable role in exacerbating the twin challenges of climate change and biodiversity loss. This can escalate into uncontrolled forest fires without appropriate precautions or adverse weather conditions. The Amazon experienced over 80,000 forest fire incidents in the current year, marking a 75% surge from the previous year. Many of these fires have been linked to individuals and businesses engaged in land clearance, primarily for beef and soy farming (Huffpost 2019).

***Dwelling Forest-Derived Ecosystem Services (FDE)***

Forests offer various ecosystems (FDES). According to Schmerbeck et al. (2015), forest fires support communities' livelihoods near tropical forests. Consequently, the local inhabitants frequently engage in the controlled burning of tropical landscapes to access specific ecosystem services.

***Illegal Felling***

Sometimes, harmful elements such as poachers or people involved in illegal felling or theft deliberately ignite fires to threaten forest officials or divert their attention, causing damage to natural resources.

***Cultural Practices and Remedies***

As Roos (2021) described, cultural burning involves the deliberate ignition of more minor, controlled fires by indigenous communities to achieve specific cultural goals. This may include enhancing the well-being of plant and animal life, which is crucial for sustenance, clothing, ceremonial purposes, and other cultural necessities.

***Rising Demand for Timber and NTFPs***

The traditional practice of managing fire cycles is no longer followed due to increasing demands for timber and non-timber forest products. Additionally, the excessive extraction of non-timber forest resources has resulted in the degradation of extensive areas.

***Human-Induced Climate Change***

The impacts of climate warming on both natural and human systems are becoming more evident worldwide (Harvey 2016). Indirectly, anthropological or human-induced climate change exacerbates natural factors contributing to forest fires. This fosters frequent and intense fires, including prolonged droughts, elevated temperatures, and dry vegetation cover. Many intended fossil fuel and material burning activities release gases and particulate matter into the air, which contribute to the warming of regional temperature, creating more favorable conditions for fire. Furthermore, increased global emissions, selection of wrong species, and natural climatic variability have increased fire potential and intensity (Sharma et al. 2018; Jangra et al. 2024; Dhyani et al. 2021; Singh 2024e).

***Poor Fire Management***

Inadequate firefighting resources, absence of controlled burning, and insufficient firebreaks and fire-line management are among the factors that can lead to the swift propagation of fires within forest areas.

***Fire Mitigating Policies***

Some fire suppression policies may inadvertently lead to an accumulation of fuel, increasing the intensity of future fires. Sometimes, due to negligence or sudden exposure to factors such as wind or electric sparks, the fire breaks and controlled burning before fire seasons convert to uncontrolled fire spread, causing drastic damage to the forest ecosystem. Accumulating additional fuels along fire lines can lead to more intense forest fires. Human landscape management, fire suppression, overgrazing, invasive plants, and climate change can contribute to transitions in fire regimes (Bowman et al. 2009).

### 2.1.2 Biological Drivers

Biological drivers of forest fires mainly include plants, wildlife, and related activities. Understanding the biological drivers of forest fire is essential for long-run preparedness against fire-management strategies. Furthermore, it is crucial to recognize the role of biological factors that drive natural and anthropogenic-induced fire incidences. A few biological factors contributing to forest fires are discussed below:

***Litter Accumulation***

In natural forests, many broad-leaf tree species, such as *Shorea robusta* and *Tectona grandis,* and conifer species, such as Pine spp. produce vast amounts of dry litter biomass, fueling forest fires on favorable temperature bands. The situation becomes more vulnerable if the forest patch is only dominated by certain deciduous species.

***Microbial Influence on Combustibility***

Microorganisms are essential for breaking down plant material. The litter decomposition process leads to the buildup of combustible organic substances, which may cause fire in favorable conditions. Additionally, certain bacteria can produce volatile compounds that can contribute to the ignition and spread of fires (Doerr and Santi 2016).

***Fuel Moisture Content***

In prolonged hot and summer seasons or when the rainfall becomes too late, the accumulated litter material gradually reduces its moisture content. It becomes more flammable, increasing the risk of forest fire. Furthermore, the aspect of a slope can also influence the moisture retention capacities of accumulated dry biomass. This can be understood by the study of Babu et al. (2023), who mentioned that forest fires have primarily occurred in the northeastern regions of the landscape, which are dominated by deciduous forests and plantations. In contrast, western areas are less susceptible to fires.

***Composition and Density of Forest Vegetation***

The type and amount of vegetation can also impact fire behavior. Overgrown deciduous vegetation can fuel fires, providing more material for ignition and spread. The situation can worsen if the area is more vulnerable due to recurring fire incidences or in tropical dry zones.

***Tree Resin and Other Flammable Compounds***

Plants have evolved various adaptation mechanisms in fire-prone environments while certain species accumulate flammable compounds in their tissues, such as oils and waxes. These compounds act as accelerants, promoting the spread of fires. For example, eucalyptus trees release volatile oils, creating a highly flammable environment in certain ecosystems (Bond and Scott 2010). In natural mixed forests, many species produce highly flammable gums or resin material (Pine spp.). Furthermore, improper practice of resin collection also increases the chances of fire spread, making it more hazardous for important timber species.

***Fire-Prone Nature of Species***

Species prone to fire are crucial in fire incidents, as they frequently rely on fire for various aspects of their life. Additionally, certain species like *Pinus roxburghii* generate substantial quantities of dead plant matter and additional biological material, potentially functioning as a fuel source. Southern India is particularly susceptible to the impact of various forest-prone species, including dry and moist deciduous forests, Eucalyptus, and Teak plantations. Due to specific characteristics, the abundance of such species triggers more fire incidences. Therefore, species composition plays a vital role in shaping the fire regime of that particular forest ecosystem.

***Insect Infestations***

The presence of insects that weaken or kill trees can enhance the fire fuel supply. Research findings vary on the relationship between forest insect outbreaks and the frequency of forest fires. Despite the existing correlation between forest insects and fire disturbances, there remains a contentious aspect in the published literature, as indicated by Bai et al. (2022). According to Hicke et al. (2012), some studies suggest increased fire frequency post-insect outbreaks. In contrast, research by Berg et al. (2006) and Bond et al. (2009) indicates that the occurrence or reduction of forest fires is unaffected by the outbreak of forest insects or woodland bugs.

***Animal Activities***

Certain animals may contribute to ignition through activities like burrowing or carrying flammable materials from one side to another side of the forest. However, the extent of forest fires due to animal activity is much lower.

***Invasive Species***

Invasive plants such as grasses and weeds have rapid growth, which can form continuous dense and dry stands, resulting in increasing fuel load as dry and flammable biomass. Such species contribute to a higher risk of ignition and intensification of fires. For example, invasive vine species such as *Cuscuta* spp. may create a ladder fuel effect where fire can reach from the ground to the crown portion of vegetation, resulting in more severe damage to forest stands. Additionally, invasive species such as wild *Mentha* or *Lantana* can increase the risk due to the release of flammable, volatile compounds into the air, creating more chances of accidental or lightning-induced fire.

### 2.1.3 Climatic Drivers

Various climatic factors contribute to the initiation and acceleration of forest fires by creating conditions favorable for ignition, intensification, and rapid spread. As a natural cause, lightning can ignite fires in trees, while strong winds can further propagate them. Seasonal drought indices are somehow related to fires resulting from annual precipitation fluctuations. Evidence of this is seen in a three-fold rise in the annual occurrence of extreme fire danger days in fire-prone regions, as reported by Sloan et al. in 2017. Understanding these climatic drivers is crucial for the prediction and management of fires. Here are some critical climatic drivers:

***Temperature***

Elevated temperatures can raise the probability of forest fires by desiccating the native vegetation and increasing the evaporation rate, leading to arid conditions in forest soils and the air. A brisk upward change in temperature regime can lead to prolonged dryness, drought, and reduced atmospheric humidity, creating favorable conditions for fires to initiate and spread. It also leads to spontaneous combustion of accumulated dry matter under certain microclimatic conditions, which ignites forest fires. The significance of temperature-related factors in disturbance states rises with latitude, whereas water availability follows the opposite trend (Seidl et al. 2017). During hot, dry, and windy conditions, combustible materials typically have low moisture content and high temperatures, significantly elevating the likelihood of forest fires (Oliveira et al. 2012). Fire ignition and transition periods can also occur due to prolonged periods of drought, leading to dry conditions and making forests more susceptible to ignition.

***Humidity and Dryness***

A low level of atmospheric humidity may render vegetation drier, which further becomes more susceptible to crown fire, the most hazardous fire that is difficult to deal with. At the same time, increased dryness can reduce the moisture content of accumulated humus and dry litter material, which act as fuel to forest fires.

***Wind***

Strong and gusty winds often carry embers with increased velocity over a long distance. This can enhance the rate of combustion of ignited fires. Likewise, it may spread fires quickly in surrounding areas, making it more difficult to control. The situation becomes more dangerous if the terrain is too steep or when fire-fighting efforts become more challenging with dense vegetation cover.

***Precipitation and Lighting***

Insufficient rainfall leads to drought conditions, drying out forests, and heated surface soils, which create favorable conditions that ignite fires. Besides, intense lightning can be considered one of the primary natural causes of forest fires. It results from substantial spatial sparks from thunderstorms, which can ignite fire in unexpected forest locations. The chances of fire ignition due to lightning are higher in dense forest cover, overmature woodlots, or whipped/wolf tree patches. While forest fires caused by lightning are infrequent, they pose a significant challenge to forest fire management due to their spontaneous nature. The situations created are often unpredictable and challenging to control quickly.

Numerous research endeavors have sought to evaluate the continual changes in the interplay between biophysical conditions and human factors concerning the presence or absence, as well as the frequency (number of fires per unit area and time), of fires,

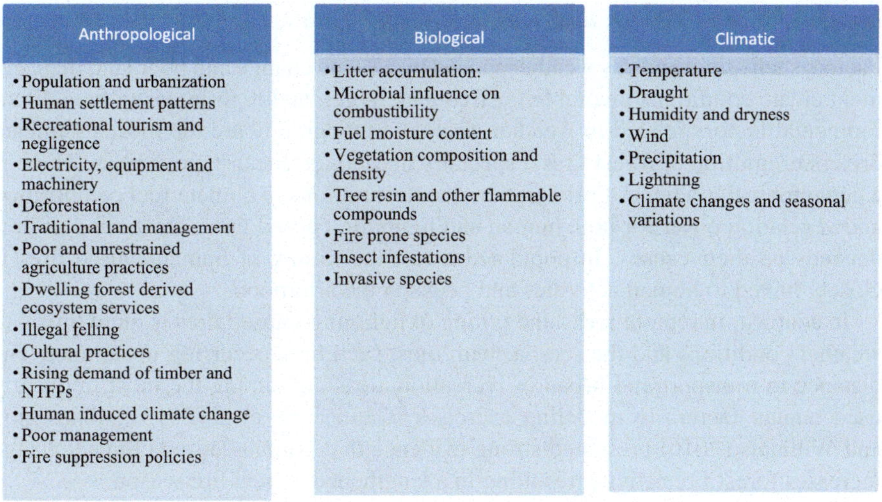


**Fig. 3** Summary of forest fire drivers and their causative factors

examining these dynamics from both a static and temporal perspective (NDMA 2020). Figure 3 (Summary) indicates that the significant influence of anthropological factors on forest fires cannot be overstated.

Concerning the factors discussed above (summarized in Fig. 3), it is clear that forest fires driven by both natural (climatic, biological) and human-induced causative agents pose significant challenges. Hence, an in-depth understanding of causative agents and their interaction is essential for effective devising measures (Singh et al. 2021; Yadav et al. 2019b; Singh et al. 2023; Sharma and Singh 2021).

## 3 Relationship Between the Mechanisms of Forest Fire Drivers

Fire's prevalence on Earth today is shaped by climate, vegetation, and land use factors at both regional and international levels, as indicated by Bowman et al. (2009). As a global phenomenon, forest fire occurrence results from the interaction of climate–weather patterns, geo-environmental variables, vegetation, fuel characteristics, and human interference (Flannigan et al. 2016). Knowing the drivers and identifying factors that affect fire distribution in the landscape is crucial for accurately mapping fire susceptibility. Here's an overview of the key drivers and their interconnections:

### *Ignition Process, Timing, and Seasonality*

Factors such as hot and dry weather caused by climate change and poor land management create conditions favorable for frequent, larger, and high-intensity fires. Environmental factors related to weather, fuel, and topography are the most significant drivers of ignition of forest fires, especially in Mediterranean-type regions. There is a geographical gradient of fire ignition due to variations in climate fuel composition and vegetation cover for both human and lightning-caused fires. The timing of fires depends on their causes. In populated areas, the timing of human-caused fires is closely linked to human activities and peaks in the afternoon.

In contrast, in remote areas, the timing of lightning-caused fires is more linked to weather conditions and the season, with most such fires occurring in summer. The distance to transport and urban or recreation areas are among the most frequently used human factors in modeling exercises (Ganteaume et al. 2013). Abatzoglou and Williams (2016) presented strong evidence that human-caused climate change increases forest fire activity, resulting in a lengthened annual fire season.

### *Spatial and Temporal Distribution*

It is a general perception that spatial distribution of forest fire can be instigated by environmental conditions, which are an outcome of locality factors such as wind, temperature, etc., in certain seasons, e.g., when humidity decreases with temperature or strong winds prevail along the slopes. The reasons for forest fires are numerous, and their distribution differs amongst countries. However, it can also vary spatially and temporally in different countries. Also, different fire management systems in cross-border areas may lead to different forest fire driving factors (Guo et al. 2017). Therefore, a comprehensive understanding of the relationships between the spatial distribution of fire occurrence and its driving factors is critical for predicting fires followed by effective forest fire management (Guo et al. 2020).

### *Spatial Scale Effects of Drivers*

Li et al. (2022) provided a reference for selecting drivers and evaluating their spatial scale effects to construct predictive regional forest fire models. It shows that meteorological factors such as relative humidity, air temperature, air pressure, sunshine hours, daily precipitation, wind speed, topographic factors such as elevation, slope, and aspect, anthropogenic factors such as population density and road network, as well as vegetation type are significantly correlated with forest fires. Constructing a prediction model to investigate the driving mechanism of fire drivers can help reveal the mechanism of fire occurrence and risk, thus contributing to the prevention and control of forest fires. A comparative evaluation of the driving elements in different nations can assist in offering prevention and control solutions.

### *Factor Interaction in the Distribution of Fire*

Previous studies have shown that the occurrence and spread of forest fires result from the interaction of many factors (Kharitonova and Kharitonova 2021). Other issues include determining what conditions forest fires represent an adequately significant threat to necessitate action and what factors regulate the frequency and the effect of forest fires in this instance (Pourtaghi et al. 2015). For example, a combination of warmer and dry weather followed by high temperatures and human activities in areas with dense vegetation can create a high-risk scenario. It can lead to increased fire frequency, contributing to more dry fuel and exacerbating the fire risk. In fire-prone areas, fire hazards can be accelerated by natural and human-induced factors, each playing a unique role in initiating and propagating these destructive events (Raturi et al. 2022; Sett et al. 2022; Singh et al. 2020; Apurva et al. 2017). The findings of various studies have revealed the linkage of severity and magnitude of forest fires to the infrastructure (e.g., roads, pathway networks, and settlements), topography (Oliveira et al. 2012), increased dry spells, high temperatures, reduced precipitation, and low humidity (Babu et al. 2023). A combination of multiple factors can accelerate more severe and uncontrollable fires.

***Early Prediction, Detection, and Monitoring***

Forest fire drivers hold a significant role in predicting forest fires in two main ways. First, the choice of forest fire driver impacts the model prediction results, meaning that different drivers may produce speculative results that are mechanistically difficult to interpret uniformly. The second aspect is that the expression of forest fire driver features affects the model prediction results. For example, expressing driver features at different spatial scales may result in predictions with various effects (Li et al. 2022). Therefore, identifying fire drivers and their mechanisms is vital for forest fire prediction. Understanding the spatiotemporal distribution of fires and future predictions is equally important for management strategies. Integrated fire management strategies should address ignition and fuel factors to mitigate the risk and impact of forest fires.

***Regional Socio-economics***

Depending on the socio-economic context of the region concerned, factors such as the unemployment rate, developmental process, societal fabrication, settlement style, and pattern, or variables linked to agricultural or commercial activity can explain the ignition of intentional and accidental fires. The frequency of fire events increases with human presence. However, it differs from region to region as humans' societal fabrication and cultural preferences change with distance, making one region more prone to fires while another is less concerned. Wenuyan et al., (2020) showed that climate variables are the main driving factors in all regions of China except the northwest region, and topographic variables and socio-economic variables are the only driving factors in a few regions. Including socio-economics in fire detection models can help improve the representation of fire ignition and management systems.

## 4 Possible Impact of Causative Forces on Forest Fire Occurrence and Spread

Forest firesare complex and generally act as positive and negative factors in ecosystems, mainly undeveloped forests. Typically, it is seen that ABC drivers work together, accelerating the impact of each other on fire ignition and spreading across the spatial and temporal levels. Nonetheless, the functioning of these drivers becomes more important in landscapes that are more prone to fire incidences and that promptly keep changing their patterns, mainly due to anthropogenic activities. Understanding fire-driving factors and their impacts is an important aspect that should be considered in forest fire management. Forest fires can cause critical damage in numerous ways. Carbon emissions from forest fires release large amounts of carbon and have important implications for the global and regional carbon cycle and atmospheric carbon concentrations (Kumar et al. 2020a, 2020b). Climate change is increasing fire size and severity, driving larger patches of high-severity fire. Many regions are predicted to experience increased fire severity where conditions are extremely hot and changes in fire regimes are evident. Increased temperatures, drought conditions, fuels, and weather are essential drivers of fire severity (Prakash et al. 2022; Nirmal and Singh 2021; Singh and Kumar 2022; Savita et al. 2017). Recent increases in fire severity are attributed to changes in climatic water deficit (CMD), vapor pressure deficit (VPD), evapotranspiration (ET), and fuels. A few causative factors of fire drivers (ABCs) and possible impact on forest fire activity have been briefly summarized in Table 1.

While alleviating the negative impact of causative forces on fire management and forest ecosystems, it is essential to employ efficient forest management techniques, raise public awareness about fire prevention, and allocate resources for interdisciplinary research to devise strategies to minimize the environmental consequences of fires. This will help maintain the health of forest ecosystems and contribute to universal efforts to combat climate change.

The combination of weather conditions and the composition of vegetation species could also impact the severity of fires. Future increases in fire severity are likely to influence forest resilience and increase the probability of forest-type conversions in many ecosystems (Wasserman and Mueller 2023). The capacity of forest fuel treatments to moderate the behavior and severity of subsequent forest fires depends on weather and fuel conditions at the burning time (Kumar et al. 2024). However, in-depth evaluations of how treatments perform are limited because encounters between forest fires and areas with extensive pre-fire data are rare (Brodie et al. 2024). At the same time, frequent forest fires may further lead to soil erosion, increased land desertification, and other adverse effects, ultimately destroying the ecological balance and affecting the human environment. Therefore, preventing the occurrence of forest fires effectively and controlling their hazards is a challenge for forest management and governmental disaster prevention and mitigation departments worldwide (Li et al. 2022). These factors can also adversely impact fire management efforts in areas more prone to fires in certain seasons.

**Table 1** Causative factors of fire drivers (ABCs) and possible impact on forest fire activity

| Fire drivers | Causative activity/factors | Possible impacts |
|---|---|---|
| Climatic | Change in temperature, precipitation, wind speed, relative humidity | Timing and intensity of monsoons, thus altering fire weather |
| | Warmer temperature | Spike in fire activity |
| | Climate water deficit, vapour pressure deficit, evapotranspiration | Increase in fire severity |
| | Thunders and lightning strikes | Fire ignition |
| | Smoke and increased carbon emissions | Increased temperature leads to more fire hazards |
| Biological | The abundance of live and dead plants, biomass, and litter accumulation | Fire properties<br>Fueling to forest fire<br>Loss of moisture, fire spread |
| | Outbreak of populations of certain species, e.g., native bark beetles | Outbreak and subsequent fire occurrence |
| | Dry forest type | Fueling of fire incidence |
| | Dynamics (vegetation composition and structure) | Post-fire ecology |
| | Forest degradation due to diseases | Increase in more fire-prone conditions |
| Anthropological | Human-caused climate change | Increase in forest fire activity, lengthen annual fire season<br>Increase in forest area burnt |
| | Demographic pressure | Exacerbate the danger of forest fire |
| | Human settlements near forests | Fire vulnerability |
| | Fossil fuel emissions | Human-caused climate change |
| | Poor/improper forest management (lack of prescribed fire treatments and management regime) | Increase in fire activity |
| | Fragmentation due to deforestation and illegal logging | More probability of fire ignition and spread |
| | Accidental and unplanned management fires | Increase in probability of forest type conversions leading to more fire incidences |
| | Rapid expansion of industries and developmental pressure | Accelerating contribution to forest fire |

# 5 Policy Intervention, Mitigation, and Management Strategies

As the hazard of forest fires increases globally, adopting policies and preventive measures to control and manage them are some of the issues of public and political attention. Knowledge of the causes of forest fires and the main driving factors of ignition is an indispensable step toward effective fire prevention policies and mitigation and management strategies. Effective policies play a crucial role in tackling forest fires for several reasons, a few of which are described below:

***Prevention and Mitigation***

Policies can focus on preventive measures and strategies to reduce forest fire risk. This may include regulations on land use, zoning, and building codes that discourage construction in high-risk fire zones. Additionally, policies can promote the creation of firebreaks, controlled burns, and other measures to reduce fuel loads and limit the spread of fires.

***Participatory Conservation Approach as Mitigation Strategy***

Mitigation strategies often involve a combination of fire-resistant landscaping, controlled burns, public awareness campaigns, and policies addressing natural and human-induced factors. Participatory conservation is an activity to mitigate and adapt to forest fires through farmer field school (FFS) activity, which forest farmer groups (FFG), non-governmental organizations (NGOs), and academics can attend.

***Integrated Learning Methods***

Acquiring knowledge ofignition's reasons and the principle user elements is essential to formulating robust fire preventive policies. For this, statistical and machine-learning methods are currently used to analyze the drivers of forest fires and their mechanisms (Tuyen et al. 2021). These methods identify forest fire drivers and their influence through an integrated learning approach for complex spatial relationships between forest fires and drivers. These methods also provide spatial predictions of forest fires. Using this, the likelihood of regional forest fires and their dangers can be realized on time. Statistical methods include commonly used frequency ratios (Abdo et al. 2022), the weight of evidence (Hong et al. 2017), and multi-criteria decision analysis (Hong et al. 2019). The basis of this type of method is to construct a relationship between historical fire data and the drivers and use this relationship in combination with domain knowledge.

***Research and Development (R&D)***

Combined with domain knowledge systems, the policies can support R&D initiatives focused on improving firefighting technologies, understanding fire behavior, and developing innovative approaches to fire management. Continuous learning and adaptation based on scientific findings can enhance the effectiveness of firefighting efforts.

***Public Awareness***

Public awareness campaigns and programs to inform communities about the risks of forest fires, the outcome of negligence, and the preventive measures they can take. This may involve communication strategies to encourage responsible behavior, such as fire-safety practices and reporting mechanisms for potential fire hazards.

***Early Detection and Rapid Response***

Constructing predictive spatial models to assess forest fires and associated risks holds significance in averting forest fire occurrence (Achu et al. 2021). Policies can support developing and implementing early detection systems, such as surveillance technologies and monitoring networks. A swift response to a detected fire is essential in preventing it from escalating into a more significant and destructive event. Policies can outline the procedures and resources required for quick and efficient responses.

***Strategicforest Fire Management***

Forest ecosystems have evolved with fire over millions of years, and many plant and animal species have adapted their life cycles to benefit from or withstand periodic forest fires. Some plant species have developed adaptations that allow their seeds to be stimulated by fire. These species often have hard seed coats that require the intense heat of fire to break and allow germination. Planting fire-adapted species in fire-prone areas can help to minimize the associated risk of vegetation loss. Under specific fire-weather conditions in particular ecosystems, restoration, and fuel treatments can decrease the extent of high-severity burns, minimizing the conversion of forests to non-forest conditions. This approach enhances forest resistance and resilience. Implementing such strategic measures for fire management in fire-prone areas can reduce the associated losses.

***Selective Removal of Vegetation***

To reduce the accumulation of extensive dry litter, selective vegetation removal is a proactive strategy for fire management. This method establishes breaks in the landscape, making it easier to contain fires and diminish their intensity and speed of propagation. Implementing thinning and prescribed fire measures has proven successful in minimizing the risk of crown fires, decreasing fuel levels, and fostering forest resilience (Wasserman and Mueller 2023). The main drawback of this method is that it necessitates the extraction of substantial quantities of biomass fuels across extensive regions, a task feasible only through mechanized methods like employing forestry mulching machines or huge human resources, often in areas where the target site is not practicable to reach.

***Resource Allocation and Procurement***

Effective policies ensure adequate resources, including personnel, equipment, and funding, are allocated to firefighting efforts. This may involve establishing and maintaining well-equipped fire departments, training personnel in firefighting techniques, and ensuring the availability of necessary resources such as water, aircraft, and firefighting equipment.

***International Cooperation***

Forest fires often cross borders, and policies can facilitate international collaboration for sharing resources, expertise, and technology. Agreements and partnerships between countries can enhance the collective capacity to address large-scale forest fires that may be beyond the capabilities of individual nations.

## 6 Conclusion

Acknowledging the intricate relationship between anthropogenic and climatic drivers with fire-adapted and fire-prone flora and fauna is vital for fostering sustainable management. Despite significant uncertainties, anthropogenic drivers can spark fire incidents, while climate and biological drivers can lead to outbursts of fire hazards in the future. Effective techniques and tools must be implemented on a broader scale to manage fire challenges. Minimizing the devastating impact on invaluable natural resources requires the strict implementation of prevention, control, and mitigation approaches to address fire occurrence and spread. The focus should be on the combined use of integrated learning tools and research, fire-resistant landscaping, seasonal mapping of fire-prone areas, training programs, public awareness campaigns, and international cooperation. Designing and implementing effective strategies with increased public responsiveness and inclusiveness is crucial. Global collaboration will be essential for promoting ecological-based fire mapping and establishing joint prevention and cooperation mechanisms to address large-scale transboundary forest fires in the coming years. This effort aims to slow down the decline of forests and wildlife. Additionally, multifaceted and trans-disciplinary research is needed to examine anthropogenic-driven climate change and its indirect influence on modified biological factors causing forest fire incidents at all levels.

## References

Abatzoglou JT, Williams AP (2016) Impact of anthropogenic climate change on wildfire across western US forests. Proc Natl Acad Sci USA 113:11770–11775

Abdo HG, Almohamad H, Dughairi A, Al-Mutiry M (2022) GIS-based frequency ratio and analytic hierarchy process for forest fire susceptibility mapping in the western region of Syria. Sustain 14:7

Achu A et al (2021) Machine-learning modelling of fire susceptibility in a forest-agriculture mosaic landscape of southern India. Eco Inform 64:101348

Apurva K, Singh H, Kumar H (2017) Elucidating morphological and growth changes in Arjun Tree (*Terminalia arjuna*) grown under elevated temperature condition. Trends Biosci 10(1):332–336

Babu KN, Gaur R, Ayushi K, Ayyappan N, Parthasarathy N (2023) Environmental drivers and spatial prediction of forest fires in the Western Ghats biodiversity hotspot, India: an ensemble machine learning approach. For Ecol Manage 540:121057

Bai M, Wang X, Yao Q, Fang K (2022) ENSO modulates interaction between forest insect and fire disturbances in China. Natur Haz Res 2(2):138–146. https://doi.org/10.1016/j.nhres.2022.04.001

Barwa S, Singh, H (2024) Climate-induced changes in forest ecosystems of Indian Himalayan region: a meta-analysis. In: Singh H (eds) Urban forests, climate change and environmental pollution. Springer, Cham. https://doi.org/10.1007/978-3-031-67837-0_39

Berg EE, Henry JD, Fastie CL, De Voider AD, Matsuoka SM (2006) Spruce beetle outbreaks on the Kenai Peninsula, Alaska, and Kluane National Park and Reserve. Yukon Territory: relationship to summer temperatures and regional differences in disturbance regimes. For Ecol Manag 227:219–232

Bond WJ, Scott AC (2010) Fire and the spread of flowering plants in the Cretaceous. New Phytol 188(4):1137–1150

Bond ML, Lee DE, Bradley CM, Hanson CT (2009) Influence of pre-fire tree mortality on fire severity in conifer forests of the San Bernardino Mountains. Calif for Sci J 2:41–47

Bowman D, Balch J, Artaxo P, Bond W, Carlson J, Cochrane M, D'Antonio C, Defries R, Doyle J, Harrison S, Johnston F, Keeley J, Krawchuk M, Kull C, Marston J, Moritz M, Prentice I, Roos C, Scott A, Pyne S (2009) Fire in the earth system. Science (New York, N.Y.)324:481–484. https://doi.org/10.1126/science.1163886.

Brodie EG, Knapp EE, Brook WR, Drury SA, Ritchie MW (2024) Forest thinning and prescribed burning treatments reduce wildfire severity and buffer the impacts of severe fire weather. Fire Ecol 20:17. https://doi.org/10.1186/s42408-023-00241-z

Carta F, Zidda C, Putzu M, Loru D, Anedda M, Giusto D (2023) Advancement in forest fire prevention: a comprehensive survey. Sens 23(14):6635

Dhyani K, Kuniyal HB, Singh H, Sobha (2021) Growth and physiological potential of *Terminalia arjuna* under elevated $CO_2$ levels in Open top chamber condition. J Appl Nat Sci 13(3):1121–1126. https://doi.org/10.31018/jans.v13i3.2490

Doerr SH, Santí C (2016) Global trends in wildfire and its impacts: perceptions versus realities in a changing world. Philos Trans R Soc B 371(1696):20150345

Flannigan MD, Wotton BM, Marshall DA, Johnston J, Jurko N, Cantin AS (2016) Fuel moisture sensitivity to temperature and precipitation: climate change implications. Clim Chan 134:58–71

Gajendiran K, Kandasamy S, Narayanan M (2024) Influence of wild life on the forest ecosystem and climate change—a comprehensive study. Environl Res 240(2):117537

Ganteaume A, Camia A, Jappiot M, San-Miguel-Avanz J, Long-Fournel M, Lampin C (2013) A review of the main driving factors of forest fire ignition over Europe. Environ Manag 51:651–662. https://doi.org/10.1007/s00267-012-9961-z

Guo U, Su Z, Wang G, Sun L, Tigabu M, Yang X, Hu H (2017) Understanding fire drivers and relative impacts in different Chinese forest ecosystems. Sci Total Environ 605–606:411–425

Guo XY, Zhang HY, Wang YQ, Zhao JJ, Zhang ZX (2020) The driving factors and their interactions of fire occurrence in Greater Khingan Mountains. China J Mt Sci 17:2674–2690. https://doi.org/10.1007/s11629-020-6036-0

Harvey B (2016) Human caused climate change is now a key driver of forest fire in the western United State. PNAS 113(42):11649–11650

Hicke JA, Johnson MC, Hayes JL, Preisler HK (2012) Effects of bark beetle-caused tree mortality on wildfire. For Ecol Manag 271:81–90

Hong H, Naghibi SA, Mostafa DM, Pourghasemi HR, Chen WA (2017) Comparative assessment between linear and quadratic discriminant analyses (LDA-QDA) with frequency ratio and weights-of-evidence models for forest fire susceptibility mapping in China. Arab J Geosci 10:7

Hong H, Jaafari A, Zenner EK (2019) Predicting spatial patterns of wildfire susceptibility in the Huichang County, China: an integrated model to analysis of landscape indicators. Ecol Ind 101:878–891

Jangra A, Kumar K, Maikhuri S, Bhandari MS, Pandey S, Singh H and Barthwal S (2024) Unveiling stress-adapted endophytic bacteria: characterizing plant growth-promoting traits and assessing cross-inoculation effects on *Populus deltoides* under abiotic stress. Plant Physiol Biochem 108610. https://doi.org/10.1016/j.plaphy.2024.108610

Kaushik P, Singh H (2024) Climate change-induced forest ecosystem functioning and alteration in water-atmosphere-soil–plant (WASP) relationship. In: Singh H (eds) Urban forests, climate change and environmental pollution. Springer, Cham. https://doi.org/10.1007/978-3-031-67837-0_38

Kharitonova AO, Kharitonova TI (2021) The effect of landscape pattern on the 2010 wildfire spread in the Mordovia State Nature Reserve. Russ Nat Conser Res 6:29–41

Kimbrough L (2022) Forest fires are getting worse, 20 years of Data confirm Monfabay. Accessed 12 Feb 2024

Kumar A, Dwivedi GK, Tewari S, Paul J, Anand R, Kumar N, Kumar P, Singh H, Kaushal R (2020a). Carbon mineralization and inorganic nitrogen pools under *Terminalia chebula* Retz.-Based Agroforestry System in Himalayan Foothills, India. Forest Sci 66(5):634–643.

Kumar A, Kumar P, Singh H, Kumar N (2020b) Adaptation and mitigation potential of roadside trees with bio-extraction of heavy metals under vehicular emissions and their impact on physiological traits during seasonal regimes. Urb For Urb Green 58:126900

Kumar A, Kumar P, Singh H, Bisht S, Kumar N (2021a) Relationship of physiological plant functional traits with soil carbon stock in temperate forest of Garhwal Himalaya. Curr Sci 120(8):1368–1373

Kumar A, Kumar P, Singh H, Kumar N (2021b) Modulation of plant functional traits under essential plant nutrients during seasonal regime in natural forests of Garhwal Himalayas. Plant Soil 465:197–212

Kumar N, Singh H, Kumar A, Singh AK, Pandey PK, Kumar A (2024) Forest-fire-induced land degradation. In: Mishra G, Giri K, Singh S, Kumar M (eds) Sustainable land management in India. Springer Nature, Singapore. https://doi.org/10.1007/978-981-97-5223-2_4

Li W, Xu Q, Yi J, Liu J (2022) Predictive model of spatial scale of forest fire driving factors: a case study of Yunnan Province. China. Sci Rep 12:19029

Nirmal, Bakshi M, Singh H (2021) Effect of $CO_2$ elevation on Shisham growth at nursery stage. Indian J Hill Farming 34:272–276

Oliveira S, Oehler F, San-Miguel- Ayanz J, Camia A, Pereira JMC (2012) Modeling spatial patterns of fire occurrence in Mediterranean Europe using Multiple Regression and Random Forest. For Ecol Manag 275(1):117–129

Pourtaghi ZS, Pourghasemi HR, Rossi M (2015) Forest fire susceptibility mapping in the Minudasht forests. Golestan Province. Iran. Environ Earth Sci 73:1515–1533

Prakash V, Hunney K, Singh H (2022) Effect of elevated carbon dioxide on growth and development of *Santalum album* L. seedlings inoculated with plant growth promoting microorganisms in Open Top Chambers. Int J Environ Agric Biotechnol 7(5):149–159

Raturi A, Singh H, Kumar P and Shukla N (2022) Characterizing the post-monsoon $CO_2$, $CH_4$, $N_2O$, and $H_2O$ vapor effluxes from a tropical wetland in the Himalayan foothill. Environ Monit Assess 194(50). https://doi.org/10.1007/s10661-021-09721-8

Roos D (2020) Native Americans used fire to protect and cultivate land Indigenous people routinely burned land to drive, prey, clear underbrush and provide pastures. https://www.history.com/news/native-american-wildfires

Savita SSK, Singh H, Singh O (2017) Forest fire: a review. Indian Forester 143(8):719–728

Schmerbeck J, Kohli A, Seeland K (2015) Ecosystem services and forest fires in India-context and policy implications from a case study in Andhra Pradesh. For Polic Econom 50(1):337–346

Seidl R, Thom D, Kautz M, Martín-Benito D, Peltoniemi VG, Wild J, Ascoli D, Petr M, Honkaniemi J, Lexer M, Trotsiuk V, Mairota P, Svoboda M, Fabrika M, Nagel T, Reyer C (2017) Forest disturbances under climate change. Nat Clim Chan 7:395–402

Sett T, Nikam BR, Singh H, Purohit S (2022) Effect of drought stress on forest evapotranspiration—a case study on Indian forests. EGU General Assembly. https://doi.org/10.5194/egusphere-egu22-510

Sett T, Nikam BR, Singh H, Purohit S (2024) Hydrological variability in Indian forest ecosystem: analysis of drought resilience, recovery and water use efficiency in moist and dry deciduous forests. In: Singh H (eds) Urban forests, climate change and environmental pollution. Springer, Cham. https://doi.org/10.1007/978-3-031-67837-0_37

Sharma R, Singh H, Kaushik M, Nautiyal R, Singh O (2018) Adaptive physiological response, carbon partitioning and biomass production of *Withania somnifera* (L.) Dunal grown under elevated $CO_2$ regimes. 3 Biotech 8:267. https://doi.org/10.1007/s13205-018-1292-1

Sharma R, Singh H (2021) Alteration in biochemical constituents and nutrients partitioning of *Asparagus racemosus* in response to elevated atmospheric $CO_2$ concentration. Environ Sci Pollut Res 29:6812–6821. https://doi.org/10.1007/s11356-021-16050-3

Singh H, Sharma R, Savita SMP, Kumar M, Verma A, Ansari MW, Sharma SK (2018) Adaptive physiological response of *Parthenium hysterophorus* to elevated atmospheric $CO_2$ concentration. Indian Forester 144:1–14

Singh S, Singh H, Sharma SK, Nautiyal R (2021) Seasonal variation in biochemical responses of bamboo clones in the sub-tropical climate of Indian Himalayan foothills. Heliyon 7(4):e06859. https://doi.org/10.1016/j.heliyon.2021.e06859

Singh H, Kumar N, Singh R, Kumar M (2023) Assessing the climate change impact on the habitat suitability of the range-restricted bird species (*Catreus wallichii*) in the Indian Himalayan ecosystem. Environ Sci Pollut Res. https://doi.org/10.1007/s11356-023-30789-x

Singh M, Singh H, Kumar A, Kumar M, Barthwal S, Thakur A (2024) Soil nitrogen availability determines the $CO_2$ fertilization effect on tree species (*Neolamarckia cadamba*): growth and physiological evidence. Environ Sustain 7:53–60. https://doi.org/10.1007/s42398-023-00300-w

Singh H and Kumar M (2022) Climate change and its impact on Indian Himalayan forests: current status and research needs. In: Rani S, Kumar R (eds) Climate change impacts, responses and sustainability in the Indian Himalaya. Springer, Cham, pp 223–242. https://doi.org/10.1007/978-3-030-92782-0_11

Singh H, Kumar N, Kumar M, Singh R (2020) Modelling habitat suitability of western tragopan (*Tragopan melanocephalus*) a range-restricted vulnerable bird species of the Himalayan region, in response to climate change. Clim Risk Manag 29:100241. https://doi.org/10.1016/j.crm.2020.100241

Singh H (2021) Process-based carbon sequestration study in relation to the energy-water-carbon flux in a forest ecosystem. In: Kumar P, Singh RK, Kumar M, Rani M, Sharma (eds) Climate impacts on sustainable natural resource management. CRC Press Taylor & Francis Group, United States, pp 336–361

Singh H (2024) Climate change and forests: biological perspectives. In: Singh H (eds) Forests and climate change. Springer Nature, Singapore. https://doi.org/10.1007/978-981-97-3905-9_1

Singh H (Ed) (2024b) Forests and climate change: biological perspectives on impact, adaptation, and mitigation strategies, Springer Nature Singapore. https://doi.org/10.1007/978-981-97-3905-9

Singh H (2024c) Climate change and environmental pollution: biological consequences on urban forests. In: Singh H (eds) Urban Forests, climate change and environmental pollution. Springer Nature, Cham, Switzerland. https://doi.org/10.1007/978-3-031-67837-0_1

Singh H (Ed) (2024d) Urban forests, climate change and environmental pollution: physio-biochemical and molecular perspectives to enhance urban resilience. Springer Nature, Cham, Switzerland. https://doi.org/10.1007/978-3-031-67837-0

Singh H (2024) Exploring adaptive modulation in plant functional traits and its impact on the productivity of *Acacia auriculiformis* under $CO_2$-enriched environment. Ind Crop Prod 210(2024):118186. https://doi.org/10.1016/j.indcrop.2024.118186

Sloan S, Locatelli B, Woosrer M, Gaveau D (2017) Fire activity in Borneo driven by industrial land conversions and drought during EL Nino periods, 1982–2010. Glob Environ Chang 47:95–109

Tuyen T, Jaafari A, Yen H, Nguyen-Thoi T, Tran P, Nguyen H, Lê H, Phuo T, Nguyen S, Prakash I, Pham B (2021) Mapping forest fire susceptibility using spatially explicit ensemble models based on the locally weighted learning algorithm. Ecol Inform 63. https://doi.org/10.1016/j.ecoinf.2021.101292

Wasserman TN, Mueller SE (2023) Climate influences on future fire severity: a synthesis of climate-fire interactions and impacts on fire regimes, high-severity fire, and forests in the western United States. Fire Ecol 19:43. https://doi.org/10.1186/s42408-023-00200-8

Wenuyan M, Feng Z, Cheng Z, Chen S, Wang F (2020) Identifying forest fire driving factors and related impacts in china using random forest algorithm. Forests 11(5):507. https://doi.org/10.3390/f11050507

Yadav SK, Singh H, Nautiyal R, Ginwal HS, Ansari SA, Barthwal S (2019) Modulation of morpho-physiological responses in *Populus deltoides* by elevated carbon dioxide and temperature. Forest Science 66:105–118. https://doi.org/10.1093/forsci/fxz048

Yadav SK, Singh H, Ginwal HS, Barthwal S (2019b) Elevated $CO_2$ enhanced growth and physiological process of *Populus deltoides* Bartr. ex Marsh. Indian Forester 145(1):23–27. https://moef.gov.in/annual-reports, https://www.ndma.gov.in/Resources/AnnualReports, https://www.nature.com/, https://www.huffpost.com/

# The Forest Fire and Climate Change Interplay

**Garima Kumari, Narendra Kumar, Hukum Singh, Apurva Malik, Shubham Singh, and Sarita Bisht**

**Abstract** A dynamic and interconnected system with broad ramifications for ecosystems, biodiversity, and the global climate is represented by the interaction between forest fires and climate change. A complex relationship occurs between forest fires and climate change despite forests' critical role in climate control. The intricate interaction between these two forces is examined in this chapter. The impact of climate change on fire regimes is explored, with changes in weather patterns, rising temperatures, and increased dryness being some of the factors. Forest fires are a significant contributor to greenhouse gas emissions. Forest fires release carbon dioxide and other gases like methane and nitrous oxide, exacerbating climate change. This study explores the interactions among increasing global temperatures, shifting precipitation patterns, and increased frequency and severity of forest fires. Forest fires are more common and intense with greater precipitation because they increase fuel loads and lengthen the fire season. Climate change makes conditions more conducive to long-lasting droughts, which reduces the resilience and health of forests and increases the frequency of forest fires. Positive reinforcement mechanisms set off by forest fires harm ecosystems and increase the susceptibility of forests to subsequent flames. The effects of modified fire regimes on the environment include the possibility of invasive species establishment, increased danger of erosion, and disturbances to fauna and flora. Comprehensive solutions are needed to mitigate the connection between forest fires and climate change. Worldwide initiatives to cut greenhouse gas emissions and sustainable forest management techniques have become essential to solving these issues. In light of changing climatic conditions, the study also highlights the significance of adaptation tactics and the necessity of resilient ecosystems and communities.

**Keywords** Climate change · Forest fire · Greenhouse gas · Forest management · Adaptations

G. Kumari (✉) · H. Singh · A. Malik · S. Singh · S. Bisht
ICFRE-Forest Research Institute, Dehradun, India
e-mail: garima82.gk@gmail.com

N. Kumar
College of Agriculture, Central Agriculture University (I), Kyrdemkulai, Meghalaya, India

 
H. Singh and N. Kumar (eds.), *Forest Fire and Climate Change*,
https://doi.org/10.1007/978-3-031-89967-6_2

# 1 Introduction

Forest fires have been a natural part of many ecosystems for millennia, shaping landscapes and influencing species distribution. However, the interplay between forest fires and climate change has garnered increasing attention due to its profound impact on ecosystems and human societies (Kumar et al. 2024; Singh 2024a). The escalating severity and frequency of forest fires are closely linked to the shifts in global climate patterns. As the climate warms, many regions experience drier and hotter conditions, contributing to forest susceptibility to fire (Singh 2024b; Barwa and Singh 2024). These changes increase the risk of ignition and foster conditions that can exacerbate the spread and intensity of fires (McKenzie et al. 2004; Stephens et al. 2013).

Furthermore, climate change can result in altered precipitation patterns, causing prolonged droughts and further vegetation drying, thereby providing ample fuel for forest fires. Extreme fire weather, characterized by high temperatures, low humidity, and strong winds, is also expected to rise, leading to an even greater challenge in forest fire management and control (McKenzie et al. 2004; Kaushik and Singh 2024).

The relationship between forest fires and climate change is bidirectional; forest fires themselves are significant contributors to climate change by releasing vast amounts of carbon dioxide, a potent greenhouse gas, thus exacerbating the warming of the atmosphere (Prakash et al. 2022; Nirmal and Singh 2021; Singh and Kumar 2022; Savita et al. 2017). This creates a feedback loop wherein climate change intensifies the conditions conducive to fires, which release emissions that contribute to further climate change (Stephens et al. 2013; Sett et al 2024).

The reality of managing forests and fires in a changing climate invites a paradigm shift in how fires are perceived, with policymakers being challenged not to categorize all fires as destructive simply because they are large or intense. Recognizing the ecological nuances and the historical context of fire regimes is crucial in adapting management strategies suitable for future conditions (Stephens et al. 2013). A thorough understanding of this intricate interaction between forest fires and climate change is essential to prepare for and respond to the increased fire activity anticipated in a warming world (McKenzie et al. 2004; Singh 2024c).

Various adaptation and mitigation strategies must be considered to fully understand and address the synergy between forest fires and climate change. Effective management practices, better prediction and modelling of fire behaviour, community engagement, and comprehensive policy initiatives are crucial for minimizing forest fire risks and mitigating their impacts on ecosystems and human well-being (Herawati et al. 2015).

## 2 Climate Change and Its Impact on Forests

Climate change profoundly impacts forest ecosystems worldwide, including shifting species distributions, altered growth patterns, and increased vulnerability to pests, diseases, and fire (Singh 2024d). As global temperatures rise due to higher levels of greenhouse gases in the atmosphere, forests face many stressors that can drastically alter their composition, structure, and functionality (Kumar et al. 2020a, 2020b).

One of the primary effects of climate change on forests is the alteration in temperature and precipitation regimes. Higher temperatures can lead to heat stress in trees, and changes in rainfall patterns can lead to drought or excessive moisture, affecting forest health. Drought conditions can weaken trees, making them more susceptible to pests and pathogens, while heavy rainfall can result in flooding and erosion (Cramer et al. 2004; Prinz 2009). Also, variability in precipitation and an increase in evapotranspiration can affect the amount of water available to forests, impacting growth and survival (Dale et al. 2000; Sett et al. 2024). Forest fires' increased incidence and severity are among climate change's most visible and immediate effects on forests. Warmer and drier conditions and changes in weather patterns create environments more prone to fires. These destroy vast forest areas and release stored carbon into the atmosphere, further contributing to climate change (Cramer et al. 2004; McKenzie et al. 2004).

Forests play a crucial role in the global carbon cycle by sequestering carbon dioxide from the atmosphere; however, the impacts of climate change, such as deforestation and disturbances like fires, are compromising this function (Singh 2024e). The release of carbon due to these activities adds to the atmospheric concentration of greenhouse gases, reinforcing the cycle of climate change (Cramer et al. 2004). In addition to changes in carbon sequestration, climate change can also affect biodiversity within forest ecosystems. As temperatures rise and precipitation patterns change, the distribution and abundance of plant and animal species may shift, threatening species sensitive to these changes. Some species may decline due to unsuitable conditions, whereas others could increase, potentially upsetting the ecological balance (McKenzie et al. 2004; Kumar et al. 2021a, 2021b).

Overall, the effects of climate change on forests necessitate urgent action in the form of mitigation to reduce greenhouse gas emissions and adaptation strategies for forest management. This includes preserving existing forests, restoring degraded forests, and improving forests' resiliency to withstand future climate change impacts (Prinz 2009).

## 3 Forest Fires: Causes and Characteristics

Fires are the most destructive natural disturbances that impact the forest's successional dynamics, mosaicity, structure, and quantitative and qualitative features. They can also result in significant losses to the environment, the economy, and society

(Jonsson and Wardle 2010). Addressing the causes of forest fires requires a comprehensive approach that includes public education on fire prevention, policy and regulation to manage human activities that pose risks, and incorporating climate adaptation strategies into forest management to reduce vulnerability to fire outbreaks.

Forest fires can be triggered by various factors, often categorized into natural and human-induced causes. Understanding these causes is essential for effective fire management and prevention strategies (Herawati et al. 2015; Földi and Kuti 2016).

## *3.1 Causes*

### 3.1.1 Natural

- Lightning is a significant cause of fire and can occur unpredictably in remote areas. It can ignite fires in insusceptible areas with abundant, dry fuels and other combustible material.
- Volcanic eruptions can also ignite fires by ejection of hot materials that can set surrounding vegetation on fire.

### 3.1.2 Human-Induced

These can be accidental or deliberate

- Fire left unattended can escalate rapidly and spread into surrounding forests.
- Agricultural practices like slash and burn can unintentionally spread and lead to uncontrollable forest fires.
- Arson, the intentional setting of fires.
- Smoldering tobacco and discarding lit cigarettes can ignite dry organic matter.
- Equipment use and malfunctions, such as from chainsaws, vehicles, or power lines, can produce sparks, leading to fires.

Each of these factors, whether natural or human-induced, contributes to the broader context in which forest fires occur. To reduce the risk of forest fires, it is necessary to adopt multifaceted approaches, including public education, stringent policies to manage risky activities, and climate adaptation strategies in forest management (Herawati et al. 2015). Furthermore, research suggests the importance of understanding the local conditions that contribute to forest fire risk to effectively tailor prevention and management strategies to different ecosystems (Fig. 1).

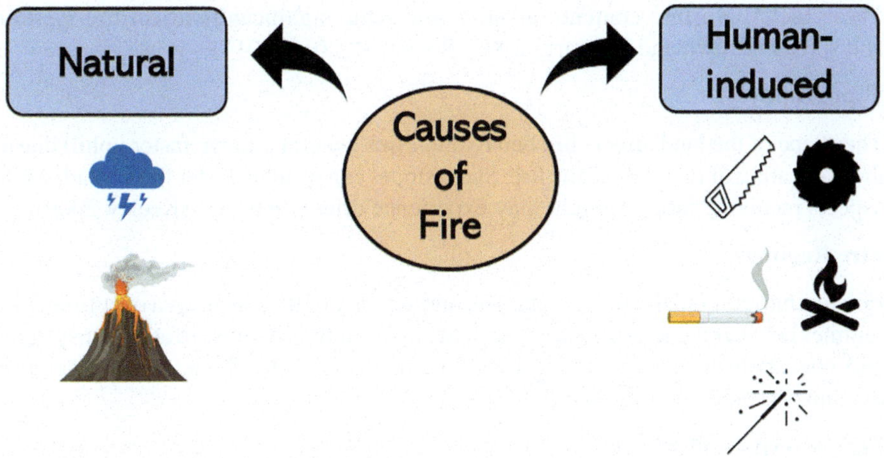


**Fig. 1** Diagram showing causes of forest fire

## 3.2 *Characteristics*

Forest fires, also known as forest fires, have characteristics that can vary widely depending on a range of factors, including the environment, weather conditions, and the type of vegetation involved. Some of the critical characteristics of forest fires include.

**Fire Behaviour**

This consists of the rate of spread, flame length, and intensity and is influenced by the fuel type, topography, and weather conditions. High-intensity fires release substantial heat quickly and can cause significant ecological changes. The higher canopy of trees is where crown fires occur, which are usually more severe and challenging to control. The forest floor is where surface fires ignite, devouring leaves, twigs, and underbrushes. The speed at which wind, slope, fuel types, fuel load, and moisture content influence a forest fire spreads. Flame length is the measurement of flames from the ground to the flame's tip, indicating the heat output (Stephens et al. 2013).

**Weather**

Weather conditions greatly influence fire behaviour. High temperatures and low humidity can dry out fuels, making them more prone to burning, while strong winds can escalate a fire's intensity and spread. Conversely, high humidity, low wind, and cooler temperatures can help to moderate fire behaviour (Westerling et al. 2006).

**Fuel**

Fuels are the combustible materials in forests, such as grass, dead trees, underbrush, timber, and organic matter on the forest floor on which the flammability depends.

These fuels' moisture content, quantity, and setup significantly affect fire ignition and spread (Brown and Pulwarty 2006; Boer et al. 2016, 2017).

**Topography**

The shape of the land affects fire behaviour. Fires generally move faster uphill due to the preheating of fuels above the fire. Steep slopes can result in faster fire spread, while aspects receiving more sunlight may experience drier conditions (Keeley 2009).

**Fire Regimes**

Fire regimes refer to the frequency, size, and severity of fires in an area over time. Fire regimes help to characterize the ecological effects of fires within different ecosystems and contribute to understanding long-term patterns and influences on vegetation dynamics (Pausas and Keeley 2014).

**Fire Severity and Effects**

The impact of a forest fire on the ecosystem can be variable. The fire intensity, duration, and the resilience of the ecosystem to fire alter severity. Low-severity fires may clear out underbrush without significantly harming mature trees. In contrast, high-severity fires can result in widespread tree mortality and changes to the soil and the forest's microclimate (Wang and Kemball 2005; Keeley 2009; Stephens et al. 2009).

## 4 Impact of Climate Change on Forest Fires

The interplay between forest fires and climate change is a pressing global concern, epitomizing the intricate relationship between natural, human-induced environmental alterations and their consequences on natural ecosystems (Raturi et al. 2022;Sett et al. 2022; Singh et al. 2020; Apurva et al. 2017). As vital components of the Earth's biosphere, forests play multifaceted roles in regulating climate patterns, maintaining biodiversity, and providing crucial ecosystem services. Globally, there is a greater risk of forest fires due to climate change. It affects weather patterns and dries forests, increasing their vulnerability to fires. Increased droughts, warmer temperatures, altered precipitation patterns, and stronger winds cause a higher fire danger.

Some of the critical points are discussed below.

**Increased Fire Frequency and Severity**

In many parts of the world, climate change, specifically rising temperatures and changed precipitation patterns, has increased the frequency and intensity of forest fires. Climate change contributes to higher fire danger levels by creating conditions conducive to more frequent and intense fires. Factors such as higher temperatures, lower relative humidity, and changes in wind patterns can elevate the risk of fire ignition and spread (Westerling et al. 2006; Abatzoglou and Williams 2016). In addition, increased thunderstorm activity and lightning-related fires may also result from warmer temperatures.

**Impact on Ecosystem and Biodiversity**

Climate change can alter natural fire regimes. Forest fires' increasing frequency and intensity due to climate change can profoundly impact ecosystems and biodiversity. Severe fires can cause habitat destruction, loss of species, and changes in vegetation composition, leading to long-term ecological consequences (Turner and Romme 1994; Moritz et al. 2014).

**Changing Vegetation Pattern**

Climate change can alter the composition of forests, with some species that are less fire-resistant proliferating, leading to more fuel for fires. The drier conditions brought about by temperature rise and altered precipitation patterns make vegetation more flammable and prone to ignition (Sharma et al. 2018; Jangra et al. 2024; Dhyani et al. 2021; Singh 2024a, b, c, d, e). Larger and more severe fires are caused by forest fires, which are more likely to occur in forests under drought stress (Abatzoglou and Williams 2016).

**Extended Fire Seasons**

In many places worldwide, prolonged fire seasons result from climate change, with fire starting earlier in the spring and continuing later into the fall. The fire risk and exposure period is increased overall by prolonging the fire season (Jolly et al. 2015).

## 5 How Do Forest Fires Contribute to Climate Change?

Forest fires, known as vegetation fires, affect the environment and the earth's climate system. These include the release of greenhouse gases and aerosols into the atmosphere, modifications to the regime of surface heat and evapotranspiration, changes to the course of the primary ecological processes (soil respiration and productivity), changes in the albedo of the burned areas following the fire due to the deposition of black carbon on snow and sea ice, and many more. The long-term post fire augmentation of albedo significantly reduces the net effect, but the combined effects of the agents above and others during the fire year and the early stages of ecosystem restoration result in a marked greenhouse effect (Randerson et al. 2006).

The atmospheric emission of aerosols and greenhouse gases from forest fires is the leading cause of climate change. When a forest burns, the carbon stored in wood, leaves, and organic matter is rapidly oxidized to carbon dioxide, a potent greenhouse gas. In addition to $CO_2$, forest fires also emit methane and nitrous oxide ($N_2O$), both of which are even more powerful than $CO_2$ in terms of their greenhouse effect over certain timescales (Van Der Werf et al. 2004, 2010).

Burning forests also sends particulate matter and black carbon (soot) into the atmosphere, affecting both the local and global climate. Black carbon particles can absorb sunlight and heat the atmosphere, and when they settle on ice and snow, they can reduce the albedo (reflectivity) and accelerate melting. This melting can lead to a positive feedback loop by exposing darker surfaces that absorb more heat, contributing further to global warming (Bonan 2008; Bond et al. 2013).

Furthermore, forest fires can change the land's reflectivity, resulting in the absorption of more solar energy and additional warming, locally and globally. The loss of forest cover also means a reduction in the forest's capacity to sequester $CO_2$ from the atmosphere, thus reducing a vital carbon sink and contributing to higher atmospheric $CO_2$ levels over time (Singh et al. 2024, 2018; Yadav et al. 2019a; Singh 2021).Furthermore, forest fires alter the properties of the land and atmosphere in ways that can worsen climate warming in addition to releasing significant volumes of greenhouse gases into the environment (Cramer et al. 2004; McKenzie et al. 2004; Stephens et al. 2013; Abatzoglou and Williams 2016).Mitigating the contributions of forest fires to climate change involves strategies for reducing the frequency and severity of fires, such as improved forest management practices, controlled burns to reduce fuel loads, and broader efforts to address climate change globally.

## 6 Consequences of Forest Fires on the Ecosystem

### Flora

Forest fires can have a severe impact on vegetation. Fires can consume the above-ground parts of plants, leading to a loss of vegetative cover. The heat from fires can destroy plant roots, impacting soil stability and hindering vegetation regrowth due to soil erosion and nutrient loss (Isaac and Kimaro 2011). Forest fires are a necessary part of some plant species' reproductive cycle. For example, certain trees require fire heat to open their cones and release seeds. The kind of plants that regenerate after a fire can vary, which may cause changes in the composition of the surrounding environment. Forest fires can thus drastically alter the immediate landscape and, depending on their intensity and frequency, can shape the long-term composition and resilience of the affected vegetation (Singh 2021; Yadav et al. 2019b; Singh et al. 2023; Sharma and Singh 2021).

### Fauna

Forest fires can have devastating effects on faunal species. It can wipe out local wildlife and alter the ecological niche on which the animals rely for food, shelter and breeding, making it unsuitable for the species to inhabit. Animals can perish in the fires if they are unable to escape, leading to direct mortality due to burns and smoke inhalation. This can be a particularly acute problem for less mobile species and those with small home ranges (Rowan et al. 2020). As vegetation is destroyed, the entire food chain can be affected, starting with herbivores that rely on plants for food and continuing up to apex predators. The stress from fires and the subsequent changes in their environment can force animals into unfamiliar territories, potentially leading to conflict with other animals or even human encounters as they search for new habitats. For populations already under stress or for species with limited distribution, fires can further threaten their genetic diversity and long-term survival prospects (Földi and Kuti 2016; Rowan et al. 2020).

**Atmosphere**

Forest fires release vast amounts of carbon dioxide, a greenhouse gas, contributing to the acceleration of global warming. The destruction of trees and woodlands reduces the photosynthetic capacity of the area, which in turn decreases oxygen production. Incomplete burning leads to the emission of carbon monoxide, which poses serious health risks to animals and firefighters. The burning process generates various toxic compounds, including volatile oils and derivatives of benzene, from the decomposition of wooden materials. Incomplete combustion also produces many organic microparticles in the air, contributing to atmospheric pollution (Andreae 2019). These particulates can be harmful near the fire and at longer distances as air movements caused by the fire carry them. Forest fires degrade air quality by introducing pollutants and greenhouse gases into the atmosphere, thus having both local and far-reaching environmental and health implications (López-Díaz et al. 2011).

**Water**

Forest fires can impact water quality in both direct and indirect ways. Ash, scale, and tiny plant fragments are among the toxic substances created by fires that can directly contaminate the surface of adjacent water bodies. Furthermore, these contaminants can be dispersed by wind over greater distances, contaminating waters that are farther away. When firefighting involves water and chemicals, these substances can run into neighboring surface waters, potentially polluting rivers and lakes. While some fire-retardant chemicals may be neutral in the soil, they can become activated in water through chemical reactions, leading to dangerous pollution across larger areas. Large quantities of water used for firefighting can also upset the water balance of an area, disturb the biological equilibrium of natural waters, and result in the reduction or loss of animal and plant life due to rapid changes in living conditions (Udawatta et al. 2011).

**Soil**

Forest fires can have a significant impact on soil. The destruction of vegetation by fire increases soil erosion as the protective flora is removed, allowing for more rapid topsoil degradation. The fire's heat can also destroy the plants' roots, weakening the soil structure and making it more susceptible to being washed away, especially in mountainous areas. The chemical composition of soil can also be altered by the materials used in fire-fighting, such as fire retardants, which can cause acidification or alkalization of the soil (Bronick and Lal 2005). This change in pH can lead to the mobilization of heavy metals in underground water that is otherwise insoluble, leading to potential food chain contamination (Földi and Kuti 2016).

**Air Movement**

Forest fires can significantly affect the movement of air. Fire heat creates strong updrafts that can alter local air circulation patterns. As a result, these updrafts can transport smoke, particles, and pollutants from the fire to higher altitudes and even over long distances. Due to incomplete combustion, forest fires can release large amounts of organic microparticles that contaminate the air. These particles can carry pollutants and accompany air movements up to 50–100 km away, dispersing pollution into remote places (Földi and Kuti 2016).

## 7 Forest Fires and Ecosystem Resilience

Forest fires are natural phenomena that dramatically and devastate landscapes, human structures, and the environment. According to Bowman et al. (2013), forest fires play a significant role in environmental transformations, impacting ecosystems, dissemination of seeds, biodiversity, nutrient cycling, carbon reserves, and climate.

These impacts are often overlooked, but it is crucial to understand and study the role of forest fires on Earth, especially in light of climate change and the increasing difficulty in controlling fires. Climate change is a critical factor that can influence forest fires' occurrence and behavior. Climate change can directly alter factors that influence the occurrence of fire ignitions, fuel availability, and fuel flammability. However, due to fuel accumulation, land-use practices, and climate change, forest fire frequency, intensity, and size are altering. This change begs the critical question of how adaptable ecosystems are to these new fire regimes. The ability of a system to reorganize and absorb disruption while maintaining its fundamental properties and functions is referred to as ecosystem resilience (Moritz et al. 2010; Dwomoh and Wimberly 2017). In a forest fire, a resilient ecosystem can rebuild itself, preserving its natural functions and offering a home for indigenous species. Comprehending the intricate correlation between forest fires and ecosystem resilience is crucial for implementing efficacious fire management tactics (Kelsey 2019).

## 8 Conclusion

Since trees take a long time to mature and adjust to changing environmental conditions, forests are particularly vulnerable to climate change. The forests should adapt not only to the fluctuations in the mean climatic indices but also to the increased unpredictability of climate conditions and the escalating dangers associated with extreme weather events, such as protracted droughts, hurricanes, or floods. The historically formed adaptable qualities of trees and forest ecosystems, as well as the social and economic factors that affect the viability of actions intended to achieve the targeted adaptation, are all included in the adaptive potential of forests. The short-term meteorological factors that impact the likelihood of a forest fire starting and spreading are related to lightning, wind, fuel, and moisture. Over an extended period, climate variables influence forest fires by altering the amount of organic material that accumulates, both alive and dead, and how that material decomposes, which affects the fuel load. Additionally, they frequently directly impact soil moisture and fuel drying. Humans can also influence fires' activity by suppressing and starting them. While prescribed fires can be beneficial, managing them requires careful planning and implementation to minimize negative impacts on air quality and human health. The relationship between climate change and forest fires concerns the health of natural ecosystems and poses increased risks to human health, property, and livelihoods. Forest fire smoke, for instance, carries health risks that can affect people far beyond the fire's perimeter, impacting air quality and causing respiratory issues, among other health problems. Climate change and forest fires are intrinsically linked, with changing climate patterns contributing to more extreme and destructive fire events. Understanding and acknowledging this interplay is crucial for developing effective fire management policies and for global efforts to mitigate the impacts of climate change on forest fires and their widespread effects on ecosystems and human societies.

## References

Abatzoglou JT, Williams AP (2016) Impact of anthropogenic climate change on wildfire across western US forests. PNAS 113(42):11770–11775

Andreae MO (2019) Emission of trace gases and aerosols from biomass burning–an updated assessment. Atm Chem Phys 19(13):8523–8546

Apurva K, Singh H, Kumar H (2017) Elucidating morphological and growth changes in Arjun Tree (*Terminalia arjuna*) grown under elevated temperature condition. Trends Biosci 10(1):332–336

Barwa S, Singh H (2024) Climate-induced changes in forest ecosystems of Indian Himalayan region: a meta-analysis. In: Singh H (eds) Urban forests, climate change and environmental pollution. Springer, Cham. https://doi.org/10.1007/978-3-031-67837-0_39

Boer MM, Bowman DM, Murphy BP, Cary GJ, Cochrane MA, Fensham RJ, Krawchuk MA, Price OF, De Dios VR, Williams RJ (2016) Future changes in climatic water balance determine potential for transformational shifts in Australian fire regimes. Environ Res Lett 11(6):065002

Boer MM, Nolan RH, Resco De Dios V, Clarke H, Price OF, Bradstock RA (2017) Changing weather extremes call for early warning of potential for catastrophic fire. Earth's Fut 5(12):1196–1202

Bonan GB (2008) Forests and climate change: forcings, feedbacks, and the climate benefits of forests. Sci 320(5882):1444–1449

Bond TC, Doherty SJ, Fahey DW, Forster PM, Berntsen T, DeAngelo BJ, Flanner MG, Ghan S, Kärcher B, Koch D (2013) Bounding the role of black carbon in the climate system: a scientific assessment. J Geophys Res Atmos 118(11):5380–5552

Bowman D, Murphy B, Boer M, Bradstock R, Cary G, Cochrane M, Fensham R, Krawchuk M, Price O, Williams R (2013) Forest fire management, climate change, and the risk of catastrophic carbon losses. Fron Ecol Environ 11(2):66–68

Bronick CJ, Lal R (2005) Soil structure and management: a review. Geoder 124(1–2):3–22

Brown TJ, Pulwarty RS (2006) A fire place for climate: the role and use of climate information in fire management and policy. For Ecol Manage 234(1):S208

Cramer W, Bondeau A, Schaphoff S, Lucht W, Smith B, Sitch S (2004) Tropical forests and the global carbon cycle: impacts of atmospheric carbon dioxide, climate change and rate of deforestation. Philos Trans R Soc Lond B Biol Sci 359(1443):331–343

Dale VH, Joyce LA, McNulty S, Neilson RP (2000) The interplay between climate change, forests, and disturbances. Sci Total Environ 262(3):201–204

Dhyani K, Kuniyal HB, Singh H, Sobha (2021) Growth and physiological potential of *Terminalia arjuna* under elevated $CO_2$ levels in open top chamber condition. J Appl Nat Sci 13(3):1121–1126. https://doi.org/10.31018/jans.v13i3.2490

Dwomoh FK, Wimberly MC (2017) Fire regimes and forest resilience: alternative vegetation states in the West African tropics. Land Ecol 32:1849–1865

Földi L, Kuti R (2016) Characteristics of forest fires and their impact on the environment. Acad App Res Mili Pub Manag Sci 15(1):5–17

Herawati H, González-Olabarria JR, Wijaya A, Martius C, Purnomo H, Andriani R (2015) Tools for assessing the impacts of climate variability and change on wildfire regimes in forests. For 6(5):1476–1499

Isaac ME, Kimaro AA (2011) Diagnosis of nutrient imbalances with vector analysis in agroforestry systems. J Environ Qual 40(3):860–866

Jangra A, Kumar K, Maikhuri S, Bhandari MS, Pandey S, Singh H, Barthwal S (2024) Unveiling stress-adapted endophytic bacteria: characterizing plant growth-promoting traits and assessing cross-inoculation effects on Populusdeltoides under abiotic stress. Plant Physiol Biochem 108610. https://doi.org/10.1016/j.plaphy.2024.108610

Jolly WM, Cochrane MA, Freeborn PH, Holden ZA, Brown TJ, Williamson GJ, Bowman DM (2015) Climate-induced variations in global wildfire danger from 1979 to 2013. Nat Comm 6(1):7537

Jonsson M, Wardle DA (2010) Structural equation modelling reveals plant-community drivers of carbon storage in boreal forest ecosystems. BiolLett 6(1):116–119

Kaushik P, Singh H (2024) Climate change-induced forest ecosystem functioning and alteration in water-atmosphere-soil–plant (WASP) relationship. In: Singh H (eds) Urban forests, climate change and environmental pollution. Springer, Cham. https://doi.org/10.1007/978-3-031-67837-0_38

Keeley JE (2009) Fire intensity, fire severity and burn severity: a brief review and suggested usage. Int J Wildland Fire 18(1):116–126

Kelsey R (2019) Wildfires and forest resilience: the case for ecological forestry in the Sierra Nevada. Unpublished report of The Nature Conservancy Sacramento, California 3

Kumar A, Dwivedi GK, Tewari S, Paul J, Anand R, Kumar N, Kumar P, Singh H, Kaushal R (2020a) Carbon mineralization and inorganic nitrogen pools under *Terminalia chebula* Retz.-based agroforestry system in Himalayan Foothills, India. Forest Sci 66(5):634–643

Kumar A, Kumar P, Singh H, Kumar N (2020b) Adaptation and mitigation potential of roadside trees with bio-extraction of heavy metals under vehicular emissions and their impact on physiological traits during seasonal regimes. Urb For Urb Green 58:126900

Kumar A, Kumar P, Singh H, Bisht S, Kumar N (2021a) Relationship of physiological plant functional traits with soil carbon stock in temperate forest of Garhwal Himalaya. Curr Sci 120(8):1368–1373

Kumar A, Kumar P, Singh H, Kumar N (2021b) Modulation of plant functional traits under essential plant nutrients during seasonal regime in natural forests of Garhwal Himalayas. Plant Soil 465:197–212

Kumar N, Singh H, Kumar A, Singh AK, Pandey PK, Kumar A (2024) Forest-fire-induced land degradation. In: Mishra G, Giri K, Singh S, Kumar M (eds) Sustainable land management in India. Springer Nature, Singapore. https://doi.org/10.1007/978-981-97-5223-2_4

López-Díaz M, Rolo V, Moreno G (2011) Trees' role in nitrogen leaching after organic, mineral fertilization: a greenhouse experiment. J Environ Qual 40(3):853–859

McKenzie D, Gedalof Z, Peterson DL, Mote P (2004) Climatic change, wildfire, and conservation. ConserBiol 18(4):890–902

Moritz MA, Hessburg PF, Povak NA (2010) Native fire regimes and landscape resilience. In: The landscape ecology of fire. Springer, pp 51–86

Moritz MA, Batllori E, Bradstock RA, Gill AM, Handmer J, Hessburg PF, Leonard J, McCaffrey S, Odion DC, Schoennagel T (2014) Learning to coexist with wildfire. Nat 515(7525):58–66

Nirmal, Bakshi M, Singh H (2021) Effect of $CO_2$ elevation on Shisham growth at nursery stage. Indian J Hill Farming 34:272–276

Pausas JG, Keeley JE (2014) Evolutionary ecology of resprouting and seeding in fire-prone ecosystems. New Phytol 204(1):55–65

Prakash V, Hunney K, Singh H (2022) Effect of elevated carbon dioxide on growth and development of *Santalum album* L. seedlings inoculated with plant growth promoting microorganisms in open top chambers. Int J Environ Agric Biotechnol 7(5):149–159

Prinz D (2009) Contributor and victim-Indonesia's role in global climate change with special reference to Kalimantan. JurnalSains&TeknologiLingkungan 1(2):139–153

Randerson JT, Liu H, Flanner MG, Chambers SD, Jin Y, Hess PG, Pfister G, Mack M, Treseder K, Welp L (2006) The impact of boreal forest fire on climate warming. Sci 314(5802):1130–1132

Raturi A, Singh H, Kumar P, Shukla N (2022) Characterizing the post-monsoon $CO_2$, $CH_4$, $N_2O$, and $H_2O$ vapor effluxes from a tropical wetland in the Himalayan foothill. Environ Monit Assess 194(50). https://doi.org/10.1007/s10661-021-09721-8

Rowan J, Beaudrot L, Franklin J, Reed KE, Smail IE, Zamora A, Kamilar JM (2020) Geographically divergent evolutionary and ecological legacies shape mammal biodiversity in the global tropics and subtropics. PNAS 117(3):1559–1565

Savita, Sharma SK, Singh H, Singh O (2017) Forest fire: a review. Indian Forester 143(8):719–728

Sett T, Nikam BR, Singh H, Purohit S (2022) Effect of drought stress on forest evapotranspiration-a case study on Indian forests. EGU General Assembly. https://doi.org/10.5194/egusphere-egu 22-510

Sett T, Nikam BR, Singh H, Purohit S (2024) Hydrological variability in Indian forest ecosystem: analysis of drought resilience, recovery and water use efficiency in moist and dry deciduous forests. In: Singh H (eds) Urban forests, climate change and environmental pollution. Springer, Cham. https://doi.org/10.1007/978-3-031-67837-0_37

Sharma R, Singh H, Kaushik M, Nautiyal R, Singh O (2018) Adaptive physiological response, carbon partitioning and biomass production of *Withania somnifera* (L.) Dunal grown under elevated $CO_2$ regimes. 3 Biotech 8:267. https://doi.org/10.1007/s13205-018-1292-1

Sharma R, Singh H (2021) Alteration in biochemical constituents and nutrients partitioning of *Asparagus racemosus* in response to elevated atmospheric $CO_2$ concentration. Environ Sci Pollut Res 29:6812–6821. https://doi.org/10.1007/s11356-021-16050-3

Singh H, Sharma R, Savita SMP, Kumar M, Verma A, Ansari MW, Sharma SK (2018) Adaptive physiological response of *Parthenium hysterophorus* to elevated atmospheric $CO_2$ concentration. Indian Forester 144:1–14

Singh S, Singh H, Sharma SK, Nautiyal R (2021) Seasonal variation in biochemical responses of bamboo clones in the sub-tropical climate of Indian Himalayan foothills. Heliyon 7(4):e06859. https://doi.org/10.1016/j.heliyon.2021.e06859

Singh H, Kumar N, Singh R, Kumar M (2023) Assessing the climate change impact on the habitat suitability of the range-restricted bird species (*Catreus wallichii*) in the Indian Himalayan ecosystem. Environ Sci Pollut Res. https://doi.org/10.1007/s11356-023-30789-x

Singh M, Singh H, Kumar A, Kumar M, Barthwal S, Thakur A (2024) Soil nitrogen availability determines the $CO_2$ fertilization effect on tree species (*Neolamarckia cadamba*): growth and physiological evidence. Environ Sustain 7:53–60. https://doi.org/10.1007/s42398-023-00300-w

Singh H, Kumar M (2022) Climate change and its impact on Indian Himalayan forests: current status and research needs. In: Rani S, Kumar R (eds) Climate change. Springer Climate. Springer, Cham, pp 223–242. https://doi.org/10.1007/978-3-030-92782-0_11

Singh H, Kumar N, Kumar M, Singh R (2020) Modelling habitat suitability of western tragopan (*Tragopan melanocephalus*) a range-restricted vulnerable bird species of the Himalayan region, in response to climate change. Clim Risk Manag 29:100241. https://doi.org/10.1016/j.crm.2020.100241

Singh H (2021) Process-based carbon sequestration study in relation to the energy-water-carbon flux in a forest ecosystem. In: Kumar P, Singh RK, Kumar M, Rani M, Sharma (eds) Climate impacts on sustainable natural resource management. CRC Press Taylor & Francis Group, United States, pp 336–361

Singh H (2024a) Climate change and forests: biological perspectives. In: Singh H (eds) Forests and climate change. Springer Nature, Singapore. https://doi.org/10.1007/978-981-97-3905-9_1

Singh H (ed) (2024b) Forests and climate change: biological perspectives on impact, adaptation, and mitigation strategies, Springer Nature Singapore. https://doi.org/10.1007/978-981-97-3905-9

Singh H (2024c) Climate change and environmental pollution: biological consequences on urban forests. In: Singh H (eds) Urban forests, climate change and environmental pollution. Springer Nature, Cham, Switzerland. https://doi.org/10.1007/978-3-031-67837-0_1

Singh H (ed) (2024d) Urban forests, climate change and environmental pollution: physio-biochemical and molecular perspectives to enhance urban resilience. Springer Nature, Cham, Switzerland. https://doi.org/10.1007/978-3-031-67837-0

Singh H (2024e) Exploring adaptive modulation in plant functional traits and its impact on the productivity of *Acacia auriculiformis* under $CO_2$-enriched environment. Ind Crops Prod 210(2024):118186. https://doi.org/10.1016/j.indcrop.2024.118186

Stephens SL, Moghaddas JJ, Edminster C, Fiedler CE, Haase S, Harrington M, Keeley JE, Knapp EE, McIver JD, Metlen K (2009) Fire treatment effects on vegetation structure, fuels, and potential fire severity in western US forests. EcolAppl 19(2):305–320

Stephens SL, Agee JK, Fule PZ, North MP, Romme WH, Swetnam TW, Turner MG (2013) Managing forests and fire in changing climates. Sci 342(6154):41–42

Turner MG, Romme WH (1994) Landscape dynamics in crown fire ecosystems. Landscape Ecol 9:59–77

Udawatta RP, Garrett HE, Kallenbach R (2011) Agroforestry buffers for nonpoint source pollution reductions from agricultural watersheds. J Environ Qual 40(3):800–806

Van Der Werf GR, Randerson JT, Collatz GJ, Giglio L, Kasibhatla PS, Arellano AF Jr, Olsen SC, Kasischke ES (2004) Continental-scale partitioning of fire emissions during the 1997 to 2001 El Nino/La Nina period. Sci 303(5654):73–76

Van der Werf GR, Randerson JT, Giglio L, Collatz G, Mu M, Kasibhatla PS, Morton DC, DeFries R, van Jin Y, van Leeuwen TT (2010) Global fire emissions and the contribution of deforestation, savanna, forest, agricultural, and peat fires (1997–2009). Atm Chem Phys 10(23):11707–11735

Wang GG, Kemball KJ (2005) Effects of fire severity on early development of understory vegetation. Canad J for Res 35(2):254–262

Westerling AL, Hidalgo HG, Cayan DR, Swetnam TW (2006) Warming and earlier spring increase western US forest wildfire activity. Sci 313(5789):940–943

Yadav SK, Singh H, Ginwal HS, Barthwal S (2019b) Elevated $CO_2$ enhanced growth and physiological process of *Populus deltoides* Bartr. ex Marsh. Indian Forester 145(1):23–27

Yadav SK, Singh H, Nautiyal R, Ginwal HS, Ansari SA and Barthwal S (2019a) Modulation of morpho-physiological responses in *Populus deltoides* by elevated carbon dioxide and temperature. Forest Sci 66:105–118. https://doi.org/10.1093/forsci/fxz048

# Post-fire Resilience: An Analysis of Vegetation Structure, Diversity, and Regeneration Potential in Fire Prone Forest Types of Uttarakhand

**Himanshu Bargali, Neha Tiwari, Dinesh Bhatt, and R. C. Sundriyal**

**Abstract** Forest fires are common in Uttarakhand, Western Himalaya, causing substantial harm to forests and ecosystems. This study aimed to assess vegetation recovery across three high-risk forest types in Uttarakhand. We conducted vegetation surveys before and after the fire season to analyze vegetation structure, diversity, and regeneration potential at each site. Non-metric Multidimensional Scaling (NMDS) was employed to evaluate species diversity. The results indicated that post-fire regeneration varied significantly between sites. The high-elevation mixed forests showed the most remarkable recovery of vegetation cover and diversity. In contrast, the lower-elevation Sal and Chir-pine forests exhibited slower recovery rates and reduced diversity. Fire severity was also a key factor influencing vegetation recovery, with areas experiencing more intense fires recovering more slowly. These findings highlight the need for ecosystem-specific management strategies to enhance post-fire recovery. Management approaches should be tailored to each ecosystem, considering forest type, elevation, and fire severity factors. The study underscores the importance of preventive measures, including controlled burning, to reduce the risk of forest fires and improve ecosystem resilience in Uttarakhand and other fire-prone regions.

**Keywords** Forest fire · Diversity · Regeneration potential · Vegetation structure · Uttarakhand

H. Bargali (✉) · D. Bhatt
Gurukula Kangri (Deemed to be University), Haridwar, India
e-mail: himanshubargali44@gmail.com

H. Bargali
Centre for Biodiversity Conservation and Management, G.B. Pant National Institute of Himalayan Environment, Kosi-Katarmal, India

N. Tiwari
Uttarakhand Open University, 263139 Haldwani, Uttarakhand, India

R. C. Sundriyal
Department of Forestry, H.N.B. Garhwal, Srinagar, India

 
H. Singh and N. Kumar (eds.), *Forest Fire and Climate Change*,
https://doi.org/10.1007/978-3-031-89967-6_3

# 1 Introduction

Forests provide many tangible and intangible benefits (Joshi and Negi, 2011). Consequently, conserving forests is crucial for environmental protection due to their essential ecological and climate services (Humpenoder et al. 2014; Svatia et al. 2017; Singh and Kumar 2022)). With its rich biodiversity and unique ecology, the Uttarakhand Himalaya is a critical repository of biological and cultural diversity that must be preserved for future generations (Bargali et al. 2022). However, forest fires in Uttarakhand significantly contribute to forest degradation.

Forest fires are a natural and frequent occurrence with beneficial and harmful impacts (Bargali et al. 2017; Kumar et al. 2024). While they can promote biodiversity and stimulate plant growth, fires destroy habitats, alter soil chemistry, and reduce forest biomass (Barwa and Singh 2024; Singh 2024d). Forest fires' role in forest ecosystems is vital for effective conservation and management, particularly in fire-prone regions like the Uttarakhand Himalaya, where fire has been traditionally viewed negatively rather than as a natural process (Babu, 2019; Bargali et al. 2020; Kaushik and Singh 2024). Historical Bahuguna and Upadhyay (2002) indicated that fire frequency increased with human activity (Bargali and Singh, 2022) , making fire a primary driver in forest ecosystems with significant consequences (Stenseth and Dunlop 2009; Sett et al. 2024). The impacts of forest fires include human and economic losses, reduction of forest resources, soil erosion, greenhouse gas emissions, biodiversity loss, etc. (Savita et al. 2017; Sati and Bandooni 2018; Singh 2024a).

In Uttarakhand, rising temperatures may increase forest fire frequency, most of which are anthropogenic (Semwal and Mehta 1996). Fire behavior, influenced by the interaction of fuels, weather, and topography, is critical in shaping forest ecosystems (Singh 2024b). Frequent fires disrupt forest regeneration by altering the seed bank and promoting fire-tolerant species (Cochrane and Barber 2009; Brando et al. 2012).

Anthropogenic fires are a significant concern, increasing the frequency and spread of forest fires (Bargali et al. 2017; Gadgil 1992). Chir-pine (*Pinus roxburghii*) forests are particularly fire-prone, with fires accelerating due to human activities (Negi 2019; Bargali et al. 2020). Analyzing the forests' structure, diversity, and regeneration potential is crucial for understanding their resilience to fire and ensuring their conservation (Singh 2024c).

## *1.1 The Current State of Knowledge*

Table 1 summarizes critical research studies on forest fire impacts on various components of the forest ecosystem of Uttarakhand, India. The table is divided into five themes, each highlighting specific objectives and key findings.

**Theme 1: Causes and Characteristics of Forest Fires**

**Table 1** Forest fire research performed in the Uttarakhand state

| S. No. | Objective of the study | Key findings | References |
|---|---|---|---|
| *Theme 1: Causes and characteristics of forest fires in Uttarakhand* | | | |
| 1 | To identify the causes, consequences, and remedial measures of forest fires in Uttarakhand | The main causes of forest fires in Uttarakhand are human activities, such as burning agricultural residue and forests for shifting cultivation, and climate change | Negi (2019) |
| 2 | To provide an overview of forest fires in Uttarakhand Himalaya | Forest fires are common and significantly impact the local environment and people, including loss of biodiversity, soil erosion, and disruption of tourism and transportation | Bhandari et al. (2012) |
| 3 | To examine the forest fire behavior, including frequency, size, and seasonal patterns in the pre-monsoonal environment | Forest fires are frequent and occur mainly during the pre-monsoon season, with a higher frequency in areas with high population density and dry forests | Singh et al. (2016) |
| 4 | To investigate the climate and recovery of trees after forest fires in the central Himalayas | Forest fires in the central Himalayas are influenced by climate factors such as temperature and precipitation, and the recovery of trees depends on fire severity and the forest's ability to regenerate | Sharma and Rikhari (1997) |
| *Theme 2: Forest fires impact on forest ecosystem and its components* | | | |
| 1 | To assess forest fire's impact on forest biodiversity in the Indian Himalayas | Forest fires significantly negatively impact forest biodiversity, including loss of plant and animal species, changes in vegetation composition, and soil erosion | Parashar and Biswas (2003) |
| 2 | To study the effect of forest fires on the ground flora dynamics of forest ecosystems | Forest fires significantly impact the ground flora, including changes in the species composition and abundance and a reduction in plant species diversity | Rikhari et al. (1999) |

(continued)

**Table 1** (continued)

| S. No. | Objective of the study | Key findings | References |
|---|---|---|---|
| 3 | To investigate the frequency of forest fires in the Nainital district and its impacts on forest biodiversity | Forest fires harm forest biodiversity, reducing plant and animal species' density and richness and changing vegetation structure | Bargali et al. (2017) |
| 4 | To assess the impact of fire frequency on forest types in Uttarakhand | Forest fires have a significant negative impact on the forest types, with a greater effect on the moist deciduous forests than on other forest types | Bargali et al. (2022) |
| *Theme 3: Ecological and environmental effects of forest fires* | | | |
| 1 | To study the forest fire ecology in Chir pine (*Pinus roxburghii* Sarg.) forests of Garhwal Himalaya | Significant ecological effects on Chir pine forests, including changes in the species composition, reduced soil organic matter, and alteration in the nutrient cycle | Semwal and Mehta (1996) |
| 2 | To assess the forest status under the fire regime and the ecological impacts on the Upper Kosi watershed in the Almora district | Forest fires have significant ecological impacts, including changes in the vegetation structure, loss of soil fertility, and reduced soil moisture content | Deepa Bisht et al. (2020) |
| 3 | To study the effect of prescribed fire on bacterial abundance and their enzymatic activity in burnt and unburnt soil of the Chilla Forest in Rajaji National Park | Prescribed burning led to changes in the bacterial abundance and enzymatic activity in burnt and unburnt soil in the Chilla Forest of Rajaji National Park. The bacterial abundance was higher in the burnt soil, indicating that prescribed burning may enhance soil microbial activity | Dhiman et al. (2018) |
| *Theme 4: Fire frequency and patterns in different forest types* | | | |

(continued)

**Table 1** (continued)

| S. No. | Objective of the study | Key findings | References |
|---|---|---|---|
| 1 | Forest fire frequency in Nainital district | Anthropogenic factors, such as human population density and land use changes, significantly influence forest fire frequency in Nainital district. The highest fire frequency occurs during the pre-monsoonal period when fuel availability and dry conditions are high | Bargali et al. (2017) |
| 2 | Forest fire ecology in Chir pine (*Pinus roxburghii Sarg.*) forest of Garhwal Himalaya | Reported 8–10 years fire return interval with stand-replacing fires occurring every 25–30 years<br>Fire severity is influenced by fuel load, stand structure, and weather conditions | Semwal and Mehta (1996) |
| *Theme 5: Modelling risk of forest fire and analysis of satellite-derived data* | | | |
| 1 | The forest fire danger index is based on modifying the Nesterov Index and fuel load and anthropogenic activities with AQUA and TERRA of MODIS and TRMM satellite datasets | These indices, fuel load, and anthropogenic activities can help predict fire risk in Uttarakhand<br>The datasets from MODIS AQUA, TERRA, and TRMM satellite can be used | Babu et al. (2016) |
| 2 | Modeling of forest fire risk in the Himalayas of Uttarakhand using TERRA satellite datasets | The TERRA satellite datasets could be used for modeling forest fire risk | Babu et al. (2016) |

This theme presents four objectives based on studies by Negi (2019), Bhandari et al. (2012), Singh et al. (2016), and Sharma and Rikhari (1997), focusing on understanding the causes and characteristics of forest fires in Uttarakhand.

**Theme 2: Impacts on Biodiversity and Ecosystem Services**

This emphasizes the impact of forest fires on the various components of forest ecosystem and services.

**Theme 3: Ecological and Environmental Effects**

This theme explores three objectives with key findings from studies by Semwal and Mehta (1996), Deepa Bisht et al. (2020), and Dhiman et al. (2018), focusing on the broader ecological and environmental consequences of forest fires.

**Theme 4: Fire Frequency and Patterns in Forest Types**

Two objectives are analyzed, based on findings from Bargali et al. (2018) and Semwal and Mehta (1996), examining fire frequency and patterns across forest types.

**Theme 5: Forest Fire Risk Modeling and Satellite Data Analysis**

This theme details two objectives focusing on forest fire risk assessment using a modeling approach and satellite-based data analysis.

The studies indicate that human activities and climate change predominantly drive forest fires in Uttarakhand. These fires have significant environmental impacts, including biodiversity loss, soil erosion, and disruptions to tourism and transportation. The research highlights the urgent need for effective forest fire risk modeling and management strategies to mitigate these impacts.

Limited research on the effects of forest fires on the Himalayan ecosystem of Uttarakhand highlights the need for further studies to comprehensively understand how fires impact vegetation structure, diversity, and regeneration in these ecosystems.

## 2 Materials and Methods

### *2.1 Study Area*

Uttarakhand is a mountainous state that serves as a transition zone between three distinct biogeographic provinces: the Tibetan Plateau (north), the Upper Gangetic Plains (south), and the Eastern Himalayan (west). The state is divided geographically into the Himalayas, Terai, and Shiwalik regions. With 45.6% of its area under forest cover, Uttarakhand is one of the most species-rich regions in the Western Himalayas. The forest cover is categorized into dense forests (20.79%), moderately dense forests (52.53%), and open forests (26.66%). The state's protected areas include 6 National Parks, 7Wildlife Sanctuaries, and 4 Conservation Reserves, collectively occupying 53,483 km2 (14.41%) of Uttarakhand's total area.

Uttarakhand accounted for 21,487 forest fires out of 345,989 reported in India between November 2020 and June 2021 (ISFR, 2021). These fires, primarily from November to May, are driven by natural and human factors, including the buildup of flammable materials such as dry leaves, pine needles, and twigs.

The study sites for investigating forest the impact on regeneration and diversity are located in Uttarakhand, in the Western Himalayan region between 78° 0′ 0″ S to 81° 0′ 0″ S and 29° 0′ 0″ N to 31° 0′ 0″ N (Fig. 1). The investigation focused on three major forest types in Uttarakhand: Sal Forest (Motichur, Haridwar), Pine Forest (Mauna, Nainital), and Mixed Forest (Dhaulchina, Almora). Table 2 summarizes the broad characteristics of these forest stands, distributed between 300 and 1600 m.a.s.l.

### *2.2 Mapping of Forest Fire Frequency*

To analyze forest fire frequency in Uttarakhand over a decade (2011–2020), satellite imagery from Landsat 5, 7, and 8 was utilized. MODIS hotspot data (MOD14) was

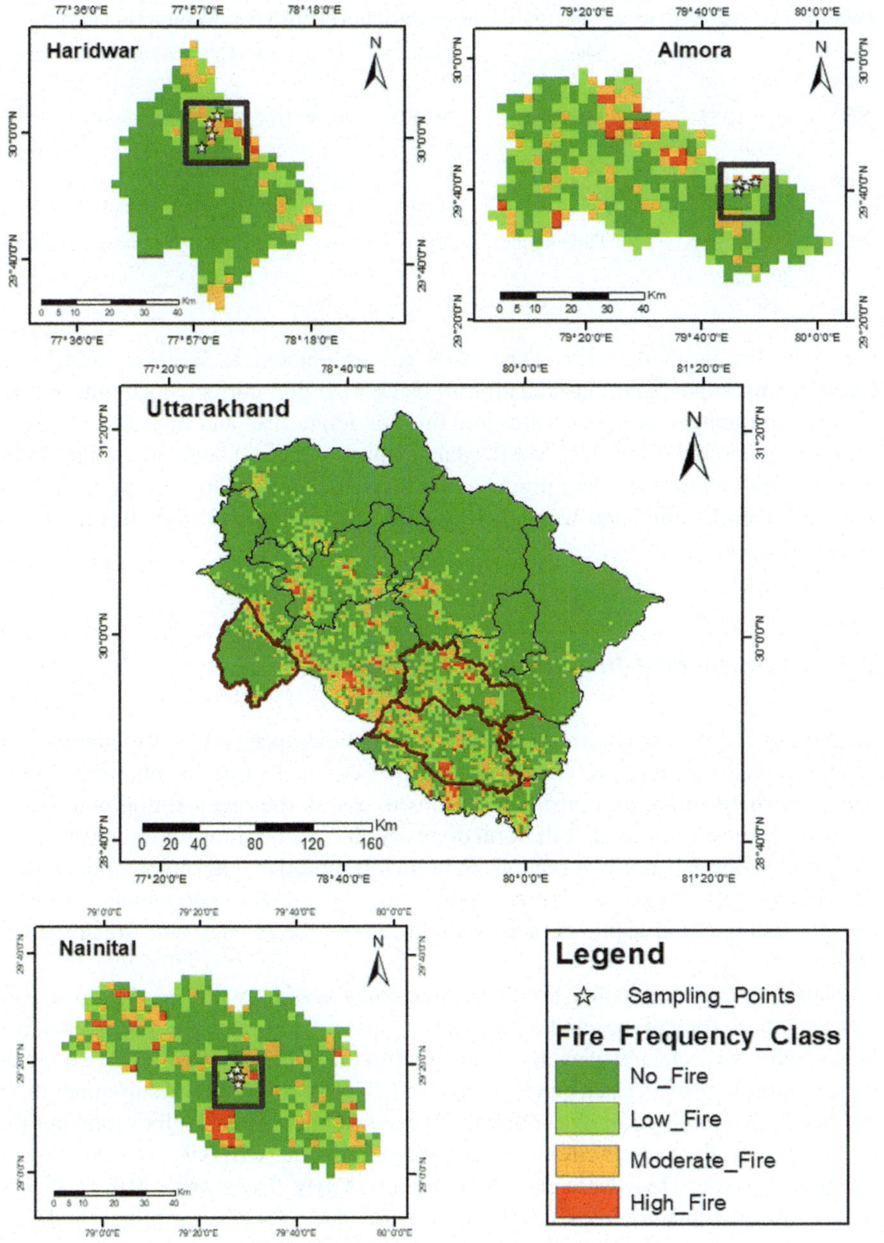


**Fig. 1** Fire frequency map, sampling points, and dominant species of the forest types at the studied sites

**Table 2** Key characteristics of the forest types studied in Uttarakhand, Western Himalaya

| Vegetation type | Elevation (m) | Site | Latitude (N) | Longitude (E) | Main canopy tree species |
|---|---|---|---|---|---|
| Sal | 355 | Motichur | 29° 59′ 0.064′ | 078° 10.217′ | *S. robusta* |
| Pine | 1596 | Mauna | 29° 41.049′ | 079° 48.173′ | *P roxburghii* |
| Mixed | 1575 | Dhaulchina | 29° 31.184′ | 079° 39.147′ | *Q leucotrichophora* |

used with fire point shapefiles. The study was conducted in Sal, Pine, and Mixed forests using supervised classification to categorize geo-corrected satellite images into fire frequency classes: no fire, low fire, moderate fire, and high fire. A fishnet grid was generated using ARCMAP's data management tool, and the spatial joining of fire points with raster data produced the fire frequency map. Fire points around 16,206 across Uttarakhand were classified into the abovementioned fire frequency classes (Fig. 1).

## 2.3 *Pre-and-Post-fire Vegetation Sampling*

To identify the plant communities present, a reconnaissance survey was conducted in each selected forest type and corresponding fire frequency class. Vegetation sampling was performed in pre and post-fire seasons to assess the regeneration potential of the chosen sites, capture the full floral diversity, and account for seasonal variations. This study adopted stratified random sampling to establish 240 sample plots, each 1 hectare in size, for the quantitative assessment of vegetation parameters, including trees, seedlings, and saplings, across the different forest sites and fire frequency classes.

Sampling was conducted in all fire frequency classes within each forest type, covering three forest types and 12 frequency classes. For tree layer assessment, a 1-hectare plot was selected at each site, and within this plot, ten 10 m × 10 m quadrats were randomly placed. Each quadrat was divided into two 5 m × 5 m sub-quadrats for tree seedling enumeration. We computed species richness as each plot's total number of tree species. Each tree's circumference at breast height (CBH) was measured using a measuring tape. Trees were classified based on CBH. Trees with CBH $\geq$ 31.5 cm were considered mature trees, CBH between 10.5 and 31.4 cm as saplings, and CBH less than 10.5 cm and height $<$ 30 cm were considered seedlings. Further, biophysical data, i.e., tree basal area, density, and frequency, were calculated using the protocol described by Misra (1968).

Vouchers of plant specimens that could not be identified were collected from the sampled sites, and a herbarium was maintained for further identification at the GBPNIHE, Kosi-Katarmal, Almora. This comprehensive sampling approach allowed for assessing plant species richness, floristic diversity, and other vegetation parameters across pre and post-fire seasons.

### 2.4 CBH Class and Regeneration Status

We used the quadrat method to analyze species distribution within the selected forest types during pre- and post-fire periods. This method involved categorizing species into girth classes based on their girth measurements, following the approach of Deb and Sundriyal (2008). Girth classes were divided into seven groups: A (0–10 cm), B (10–30 cm), C (31–60 cm), D (61–120 cm), E (120–150 cm), F (150–180 cm), and G (180–210 cm) to assess the population structure of tree species.

### 2.5 Data Analysis

We assessed species richness and density across various forest types, calculating the Shannon–Wiener diversity index (H′). We also quantified the species number and the density of growth forms (herbs, shrubs, and trees) for each forest type. We employed Principal Component Analysis (PCA) to identify critical components of pre- and post-fire variables and environmental factors. Additionally, we used Non-Metric Multidimensional Scaling (NMDS) to compare forest diversity before and after the fires.

## 3 Results

### 3.1 Site Characteristics and Vegetation Structure

This study reported 109 plant species from 56 families across all stands. Herbaceous species represented the highest proportion of the observed species (52%), followed by tree species (27%) and shrub species (21%). Notably, the total species richness was highest for trees, followed by shrubs and herbs, with 13, 20, and 32 species, respectively, during the pre-fire period, and lowest with 10, 11, and 25 species, respectively, during the post-fire period. The distribution of species across different forest types varied. In the Sal, Chir-pine, and Mixed Forest types, there were 2, 4, and 7 tree species, 2, 9, and 9 shrub species, and 5, 10, and 17 herbaceous species, respectively, during the pre-fire period. Similarly, after the fire, species distribution was recorded as 2, 3, and 5 tree species, 2, 4, and 5 shrub species, and 4, 8, and 13 herbaceous species, respectively, across Sal, Chir-pine, and Mixed Forest types.

## 3.2 Stand Density and Total Basal Area (TBA)

The study recorded trees' density and basal area across selected forest types before and after a fire event. The pre-fire tree density varied between 240 and 560 $ha^{-1}$, with the highest being in Sal (560 $ha^{-1}$), followed by mixed f (437 $ha^{-1}$), and the lowest in Pine Forest (437 $ha^{-1}$). During post-fire, the tree density ranged from 238 to 559 $ha^{-1}$, with the highest density observed in Sal Forest (559 $ha^{-1}$), followed by mixed forest (430 $ha^{-1}$), and the lowest density in Pine Forest (238 $ha^{-1}$). The sapling density ranged from 140 to 1540 $ha^{-1}$ during pre-fire and from 52 to 380 $ha^{-1}$ post-fire. The seedling density varied from 250 to 1150 $ha^{-1}$ during pre-fire and from 78 to 370 $ha^{-1}$ post-fire. The total basal area was maximum for Pine Forest during pre-fire (44.84 $m^2$ $ha^{-1}$) and post-fire (37.52 $m^2$ $ha^{-1}$), followed by Sal Forest (38.22 $m^2$ $ha^{-1}$ during pre-fire and 36.24 $m^2$ $ha^{-1}$ during post-fire), and minimum for Mixed Forest (22.72 $m^2$ $ha^{-1}$ during pre-fire and 17.82 $m^2$ $ha^{-1}$ during post-fire). The density of dominant species ranged from 320 to 520 $ha^{-1}$ during pre-fire and from 310 to 512 $ha^{-1}$ during post-fire. The density of co-dominant species ranged from 60 to 130 ha – 1 during pre-fire and from 55 to 120 $ha^{-1}$ during post-fire. During pre- and post-fire, the highest density of co-dominant species was reported in Sal, Pine, and Mixed Forests, respectively. The changes in tree density during post-fire were – 2.5% in Pine Forest, 2.7% in Sal Forest, and -1.6% in Mixed Forest, compared to pre-fire. The sapling density decreased by 63.6% in Mixed Forest, 35.5% in Pine Forest, and 74.5% in Sal Forest during post-fire, compared to pre-fire. The seedling density decreased by 69.6% in Pine Forest, 32.2% in Sal Forest, and 67.8% in Mixed Forest during post-fire, compared to pre-fire. The total basal area decreased by 16.3% in Mixed Forest, 19.5% in Sal Forest, and 16.3% in Pine Forest during post-fire, compared to pre-fire. The density of dominant species decreased by 1.9% in Sal Forest, 1.9% in Mixed Forest, and 1.5% in Pine Forest during post-fire, compared to pre-fire. The density of co-dominant species decreased by 15.4% in Pine Forest, 8.3% in Mixed Forest, and increased by 7.7% in Sal Forest during post-fire, compared to pre-fire. Table 3 provides details on the site characteristics and vegetation structure for each of the selected sites.

**Table 3** Biophysical characteristics of studied the forest types

| Parameters | Sal forest | | Pine Forest | | Mixed Forest | |
|---|---|---|---|---|---|---|
| | Pre-fire | Post-fire | Pre-fire | Post-fire | Pre-fire | Post-fire |
| Tree density (ind./ha) | 560 | 559 | 240 | 238 | 437 | 430 |
| Sapling density (ind./ha) | 1540 | 380 | 260 | 98 | 140 | 52 |
| Seedling density (ind./ha) | 1150 | 370 | 250 | 780 | 650 | 310 |
| Total basal area ($m^2$/ha) | 38.22 | 36.24 | 44.84 | 37.52 | 22.72 | 17.82 |
| Density of dominant species (ind./ha) | 420 | 510 | 520 | 512 | 320 | 310 |
| Density of co-dominance species (ind./ha) | 110 | 120 | 60 | 55 | 120 | 110 |

## 3.3 Species Diversity and Evenness

The study assessed species diversity, concentration of dominance, and species evenness in the tree, shrub, and herb layers of selected forest types during pre- and post-fire. During the pre-fire period, species diversity ranged from 1.01 to 2.13 for the tree, 2.23 to 3.01 for the shrub, and 1.61 to 3.65 for the herb. The mixed forest exhibited the highest species diversity in the tree and shrub layers. The dominance varied between 0.18 and 0.33 for the tree, 0.09 to 0.25 for the shrub, and 0.11 to 0.38 for the herb. Species evenness ranged from 0.51 to 0.87 for the tree, 0.67 to 1.22 for the shrub, and 0.43 to 0.80 for the herb. During the post-fire period, species diversity ranged from 0.90 to 2.06 for the tree, 2.10 to 2.69 for the shrub, and 0.25 to 2.02 for the herb. Like the pre-fire period, the mixed forest exhibited the highest species diversity in the tree and shrub layers. The dominance varied between 0.12 and 0.30 for the tree, 0.08–0.25 for the shrub, and 0.06–0.18 for the herb. Species evenness ranged from 0.48 to 0.79 for the tree, 0.63–1.39 for the shrub, and 0.32–0.76 for the herb.

Overall, the study did not report significant differences in species diversity, evenness, and concentration of dominance in the studied forest types did not differ significantly between pre- and post-fire conditions, suggesting that these forest ecosystems may have high resilience to fire (Table 4).

**Table 4** Indices for diversity, dominance, and evenness for tree, shrub, and herb strata during pre-And-post-fire in selected forest types

| Parameters | Sal | | Pine | | Mixed | |
|---|---|---|---|---|---|---|
| | Pre-fire | Post-fire | Pre-fire | Post-fire | Pre-fire | Post-fire |
| *Shannon index* | | | | | | |
| Tree | 1.41 | 1.40 | 1.01 | 0.90 | 2.13 | 2.06 |
| Shrub | 2.24 | 2.11 | 2.23 | 2.10 | 3.01 | 2.69 |
| Herbs | 1.61 | 0.25 | 3.65 | 1.12 | 2.35 | 2.02 |
| *Concentration of dominance (CD)* | | | | | | |
| Tree | 0.32 | 0.30 | 0.33 | 0.26 | 0.18 | 0.12 |
| Shrub | 0.25 | 0.18 | 0.10 | 0.06 | 0.09 | 0.07 |
| Herbs | 0.38 | 0.32 | 0.11 | 0.09 | 0.28 | 0.26 |
| *Species evenness (SE)* | | | | | | |
| Tree | 0.51 | 0.48 | 0.76 | 0.69 | 0.87 | 0.79 |
| Shrub | 0.67 | 0.63 | 1.01 | 0.99 | 1.22 | 1.13 |
| Herbs | 0.80 | 0.76 | 0.43 | 0.32 | 0.77 | 0.68 |

### 3.4 CBH Class and Regeneration Status

The study highlighted variations in the density of lower girth class individuals, including seedlings and saplings, across different forest types before and after the fire. The Sal Forest exhibited a consistently higher density of individuals of lower girth class in both pre-and post-fire periods. At the same time, the Pine Forest maintained an even distribution across all girth classes. In contrast, the Mixed Forest was dominated by young trees (saplings). At the stand level, the distribution of seedlings, saplings, and mature trees differed among the forest types before and after the fire. For example, during the pre-fire period, the Sal Forest had distributions of 35.38% for seedlings, 47.38% for saplings, and 17.23% for trees, while the Mixed Forest had distributions of 51.58%, 11.11%, and 37.3%, respectively. In the post-fire period, the Sal Forest had distributions of 33.48% for seedlings, 43.69% for saplings, and 22.82% for trees, while the Mixed Forest had distributions of 48.24%, 10.4%, and 41.34%, respectively (Fig. 2). These findings suggest that different forest types respond differently to fires. Post-fire recovery may depend on the pre-existing forest structure and composition.

### 3.5 Principal Component Analysis (PCA) Analysis

Table 5 shows the results of a principal component analysis (PCA). The PCA analysis resulted in 5 principal components, with the first PC accounting for 59.96% of the variance and the fifth PC accounting for 1.20%. In terms of the variables, the results indicate that the first principal component (PC1) is strongly influenced by all of the variables, with the highest loadings for soil moisture, water holding capacity, organic carbon, and nitrogen. PC1 likely represents overall soil fertility or quality.

The second principal component (PC2) is heavily influenced by the pine species before and after the fire and the mixed species after the fire. This may represent changes in vegetation following the fire. PC1 and PC2 explain 85.936% of the total variance in the data. PC1 is positively associated with the following components: Water Holding Capacity, Organic Carbon, Nitrogen, Soil Moisture, and Altitude. This suggests that PC1 captures the variation in soil and environmental factors associated with the soil's moisture-holding capacity, organic carbon, nitrogen content, and altitude. These factors likely represent the overall health and fertility of the soil. PC2, on the other hand, has a high positive association with the following components: Shrub Diversity, Shrub Richness, Tree Richness, Tree Diversity (H′), and Sapling Density. This indicates that PC2 captures the variation in the diversity and richness of the vegetation, mainly shrubs and trees. The remaining principal components have more complex loadings but represent other aspects of vegetation composition and soil characteristics (Fig. 3).

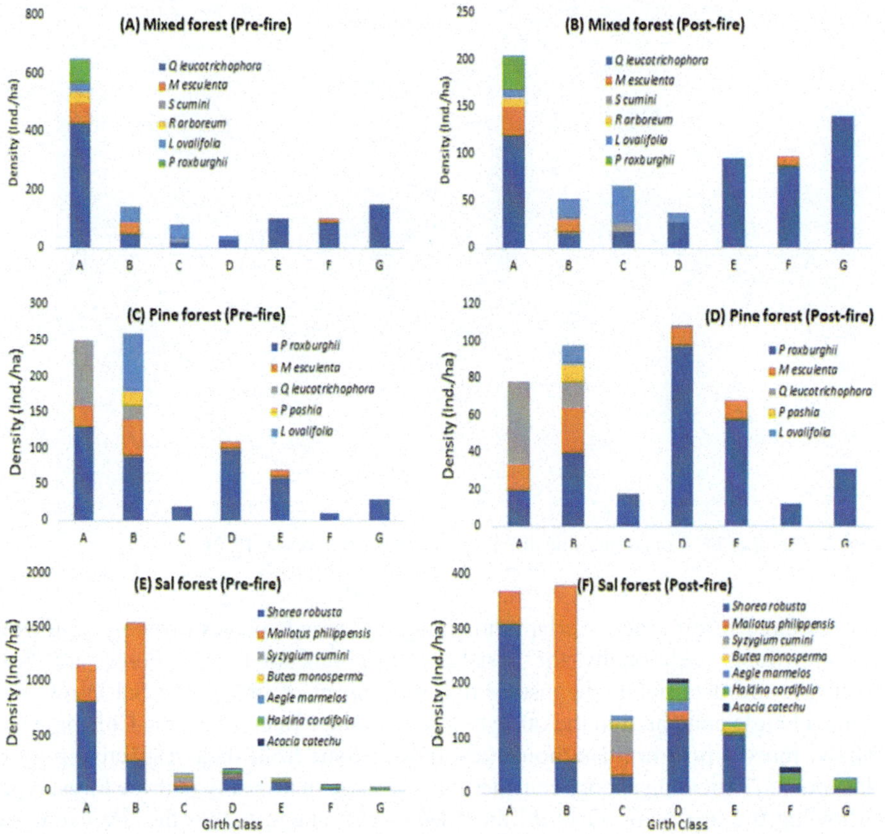


**Fig. 2** Girth class structure for the selected forest types in Uttarakhand presented for both pre-and-post-fire seasons. The girth classes are categorized as follows: **a** (0–10 cm), **b** (10–30 cm), **c** (31–60 cm), **d** (61–120 cm), **e** (120–150 cm), **f** (150–180 cm), **g** (180–210 cm), and **h** (>210 cm)

**Table 5** Eigenvalues and principal components of pre-and-post-fire variables, along with variances in percentage

| PC | Eigenvalue | % Variance |
|---|---|---|
| 1 | 13.7911 | 59.961 |
| 2 | 5.97428 | 25.975 |
| 3 | 2.03708 | 8.8569 |
| 4 | 0.922628 | 4.0114 |
| 5 | 0.274942 | 1.1954 |

## *3.6 Non-metric Multidimensional Scaling (NMDS)*

The NMDS plot shows that the pre-fire diversity of studied forest types, i.e., Sal, Chir-pine, and Mixed forest, were separated. However, the post-fire diversity data

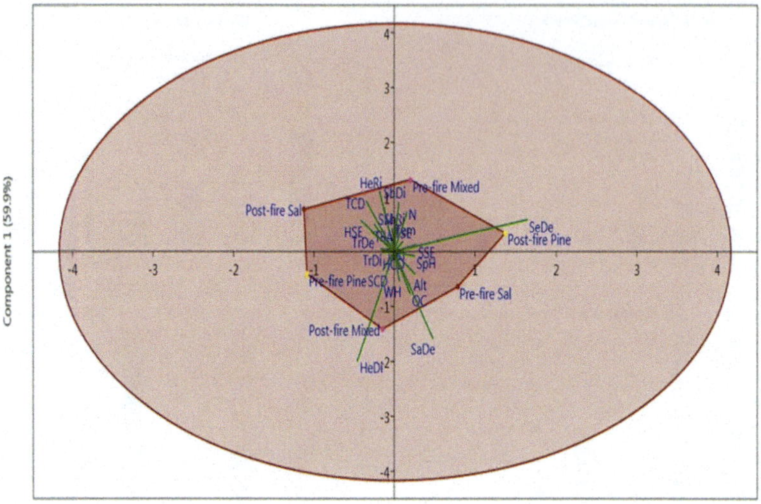


**Fig. 3** Biplot of PCA of twenty-one different variables (PC1 versus PC2)

for the three forest types overlap, indicating decreased species diversity following the fire event. Additionally, the post-fire data points for the Sal Forest are close to its pre-fire data point, suggesting that the impact of fire on the Sal Forest was relatively less compared to the other two forest types. In contrast, the Chir-pine and Mixed forest's post-fire data points are farther apart from their respective pre-fire data points, indicating a more considerable decrease in diversity in these forest types following the fire event. Overall, the NMDS plot suggests that the fire event has harmed the diversity of all three forest types, but the severity of the impact varied across the forest types. The Sal forest was relatively less affected, while the Chir-pine and Mixed forest were more severely impacted (Fig. 4).

## 3.7 *Discussion*

The results showed that fire significantly affected each forest type differently. The Sal forests experienced the most negligible negative impact. Before the fire, species richness was highest among trees, followed by shrubs and herbs, but all three categories saw a reduction in species richness post-fire. The fire also reduced species diversity, concentration of dominance, and species evenness, leading to a decline in the density of dominant and co-dominant species across all forest types. Additionally, the total basal area decreased, potentially impairing the ecosystem services provided by the forests. The density of saplings and seedlings also dropped after the fire, indicating challenges in forest regeneration.

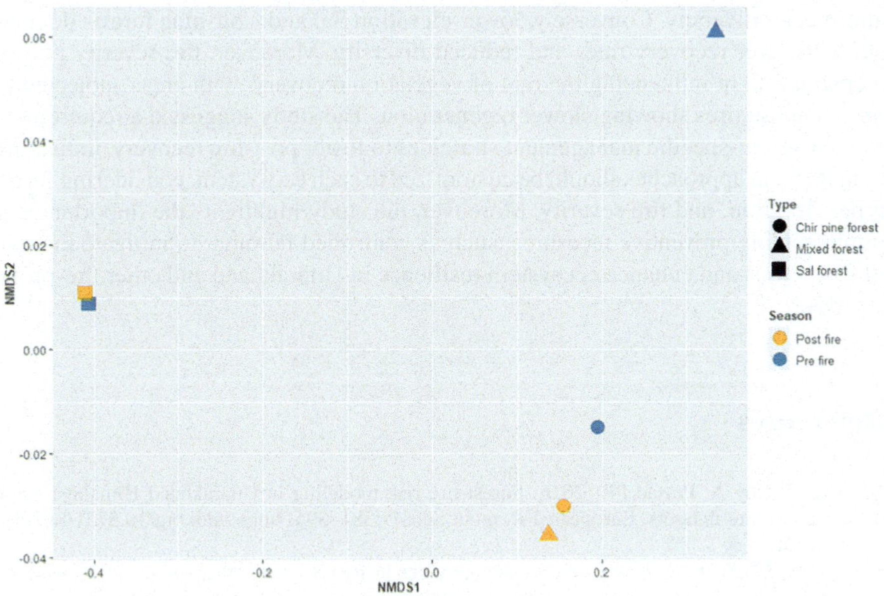


**Fig. 4** NMDS plot revealing pre-fire diversity among three forest types

Studies have reported decreased species diversity while increasing the dominance of fire-tolerant species following fire events. The current findings suggest that fire may aid the regeneration of seedlings and saplings in mixed forest species. This highlights the importance of conservation measures to restore forest ecosystems and prevent further degradation.

The findings support this view that fire can contribute to and maintain species richness in forests with localized and patchy disturbance histories. Kodandapani (2001) observed that forest fire significantly enhances forest ecosystem productivity, releasing nutrients trapped in older vegetation and increasing seedling regeneration. Naidu and Sribasuki (1994) noted that young plants are more susceptible to fire damage than mature ones. Pre-fire species diversity (H′) ranged from 1.01 to 2.13 for the tree, 2.23–3.01 for the shrub, and 1.61–3.65 for the herb across the studied forest types, consistent with findings by Joshi et al. (2022) for similar forest types in Uttarakhand.

## 4 Conclusion

The study concludes that forest fires are frequent in the Uttarakhand region of the Western Himalayas, leading to significant damage to forests and ecosystems. The extent of post-fire regeneration varies considerably among different forest types, and high-elevation mixed forests display the most substantial recovery of vegetation cover and species diversity. Conversely, lower-elevation Sal and Chir-pine forests demonstrated slower recovery rates and reduced diversity. Moreover, fire severity played a crucial role in influencing the rate of vegetation recovery, with areas subjected to more intense fires showing slower regeneration. The study suggested a requirement for ecosystem-specific management strategies to foster post-fire recovery, indicating management approaches should be customized to each ecosystem, considering forest type, elevation, and fire severity. Moreover, the study highlights the importance of implementing preventive measures, such as controlled burning, to mitigate the risk of forest fires and enhance ecosystem resilience in Uttarakhand and other fire-prone regions.

## References


Babu KVS, Roy A, Prasad PR (2016) Forest fire risk modeling in Uttarakhand Himalaya using Terra satellite datasets. European J Remote Sen 49:381–395. https://doi.org/10.5721/EuJRS20164921

Babu KVS (2019) Developing forest fire danger index using geo-spatial techniques. PhD thesis, pp 1–131

Bahuguna VK, Upadhay A (2002) Forest fires in India: policy initiatives for community participation. Int For Rev 4:122–127. http://www.jstor.org/stable/43740075

Bargali H, Gupta S, Malik DS, Matta G (2017) Estimation of fire frequency in Nainital District of Utarakhand state by using satellite images. J Remote Sens GIS 6(4):1–5

Bargali H, Singh P, Bhatt D (2020) Role of Chir Pine (*Pinusroxburghii* Sarg.) in the forest fire of Uttarakhand Himalaya. ENVIS Bull Himalayan Ecol 28:82–85

Bargali H, Calderon LPP, Sundriyal RC, Bhatt D (2022) Impact of forest fire frequency on floristic diversity in the forests of Uttarakhand, western Himalaya. Trees, For People 9:100300

Bargali H, Kumar A, Singh P (2022) Plant studies in Uttarakhand, Western Himalaya: a comprehensive review. Trees Forests People 100203. https://doi.org/10.1016/j.tfp.2022.100203

Barwa S, Singh H (2024) Climate-induced changes in forest ecosystems of Indian Himalayan region: a meta-analysis. In: Singh H (eds) urban forests, climate change and environmental pollution. Springer, Cham. https://doi.org/10.1007/978-3-031-67837-0_39

Bhandari BS, Mehta JP, Semwal RL (2012) Forest fire in Uttarakhand Himalaya: an overview. In: Negi GCS, Dhyani PP (eds) Glimpses of forestry research in the Indian Himalayan region. G.B. Pant Institute of Himalayan Environment Development, Almora, pp 167–176

Bhatt R, Negi SS, Bhatt JP (2019) Fire affects regeneration potential of moist temperate forests in the western Himalaya. SciRep 9(1):12130. https://doi.org/10.1038/s41598-019-48592-6

Brando PM, Nepstad DC, Balch JK, Bolker B, Christman MC, CoeM PFE (2012) Fire-induced tree mortality in a neo-tropical forest: the roles of bark traits, tree size, wood density and fire behavior. Glo Change Biol 18:630–641. https://doi.org/10.1111/j.1365-2486.2011.02533.x

Cochrane MA, Barber CP (2009) Climate change, human land use and future fires in the Amazon. GloChange Biol 15:601–612. https://doi.org/10.1111/j.1365-2486.2008.01786.x

Deepa Bisht D, Singh P, Sundriyal RC (2020) Forest status under fire regime: an ecological assessment in Upper-Kosi watershed in Almora district, Uttarakhand, India. Indian J For 43(3):206–211

Dhiman P, Mehta JP, Singh P, andSharesthBaldotra SS (2018) Effect of prescribe fire on bacterial abundance and their enzymatic activity in burnt and unburnt soil of Chilla Forest, Raja Ji National Park, Uttarakhand, India. Plant Arch 18(1):1125–1128

Deb P, Sundriyal RC (2008) Tree regeneration and seedling survival patterns in old-growth lowland tropical rainforest in Namdapha National Park, north-east India. For Ecol Manag 255:3995–4006. https://doi.org/10.1016/j.foreco.2008.03.046
Gadgil M (1992) Conserving biodiversity as if people matter: a case study from India. Ambio 21:266–270
Humpenoder F, Popp A, Dietrich JP, Klein D, Lotze-Campen H, Bonsch M, Muller C (2014) Investigating afforestation and bioenergy CCS as climate change mitigation strategies. Environ Res Let 9:064029. https://doi.org/10.1088/1748-9326/9/6/064029
Jhariya MK, Bargali SS, Swamy SL, Kittur B (2012) Vegetational structure, diversity and fuel load in fire affected areas of tropical dry deciduous forests in Chhattisgarh. Int JPlant Res 25:210–224
Joshi G, Negi GCS (2011) Quantification and valuation of forest ecosystem services in the western Himalayan region of India. Int J Biodivers SciEcosys Service Manage 7:2–11
Joshi VC, Bisht D, Sundriyal RC, Pant H (2022) Species richness, diversity, structure, and distribution patterns across dominating forest communities of low and mid-hills in the Central Himalaya. Geol Ecol Landscape 1–12. https://doi.org/10.1080/24749508.2021.2022424
Kaushik P, Singh H (2024) Climate change-induced forest ecosystem functioning and alteration in water-atmosphere-soil–plant (WASP) relationship. In: Singh H (eds) Urban forests, climate change and environmental pollution. Springer, Cham. https://doi.org/10.1007/978-3-031-67837-0_38
Kodandapani N (2001) Forest fires origins and ecological paradoxes. Resonance 34–41
Kumar N, Singh H, Kumar A, Singh AK, Pandey PK, Kumar A (2024) Forest-fire-induced land degradation. In: Mishra G, Giri K, Singh S, Kumar M (eds) Sustainable land management in India. Springer Nature, Singapore. https://doi.org/10.1007/978-981-97-5223-2_4
Misra R (1968) Ecology workbook. Oxford and IBH Publishing Co., New Delhi
Naidu CV, Srivasuki KP (1994) Effect of forest fire on tree spices on different areas of Aeshachalam Hills. J Trop For 10(3):31–37
Negi GCS (2019) Forest fire in Uttarakhand: causes, consequences and remedial measures. IJ Ecol Environ Sci 45:31–37
Parashar A, Biswas S (2003) The impact of forest fire on forest biodiversity in the Indian Himalayas (Uttaranchal). In: XII World Forestry Congress, vol 358 B-1
Sati VP, Bandooni SK (2018) Forests of Uttarakhand: diversity, distribution, use pattern and conservation. ENVIS Bull Himalayan Ecol 26:21–27
Savita, Sharma SK, Singh H, Singh O (2017) Forest Fire: A Review. Indian Forester, 143(8): 719–728. https://doi.org/10.36808/if/2017/v143i8/118959
Semwal RL, Mehta JP (1996) Ecology of forest fires in Chir pine (*Pinus roxburghii* Sarg.) forest of Garhwal Himalaya. Curr Sci 70:426–427
Sett T, Nikam BR, Singh H, Purohit S (2024) Hydrological variability in Indian forest ecosystem: analysis of drought resilience, recovery and water use efficiency in moist and dry deciduous forests. In: Singh H (eds) Urban forests, climate change and environmental pollution. Springer, Cham. https://doi.org/10.1007/978-3-031-67837-0_37
Sharma S, Rikhari HC (1997) Forest fire in the central Himalaya: climate and recovery of trees. Int J Biometeorol 40:63–70
Singh SP (2002) Balancing the approaches of environmental conservation by considering ecosystem services as well as biodiversity. Curr Sci 82:1331–1335
Singh H, Kumar M (2022) Climate change and its impact on Indian Himalayan forests: current status and research needs. In: Rani S, Kumar R (eds) Climate change. Springer Climate. Springer, Cham, pp 223–242. https://doi.org/10.1007/978-3-030-92782-0_11
Singh RD, Gumber S, Tewari P, Singh SP (2016) Nature of forest fires in Uttarakhand: frequency, size and seasonal patterns in relation to pre-monsoonal environment. Curr Sci 111:398–403. https://doi.org/10.18520/cs/v111/i2/398-403
Singh H (2024a) Climate change and forests: biological perspectives. In: Singh H (eds) Forests and climate change. Springer Nature, Singapore. https://doi.org/10.1007/978-981-97-3905-9_1

Singh H (Ed) (2024b) Forests and climate change: biological perspectives on impact, adaptation, and mitigation strategies, Springer Nature Singapore. https://doi.org/10.1007/978-981-97-3905-9

Singh H (2024c) Climate change and environmental pollution: biological consequences on urban forests. In: Singh H (eds) Urban forests, climate change and environmental pollution. Springer Nature, Cham, Switzerland. https://doi.org/10.1007/978-3-031-67837-0_1

Singh H (Ed) (2024d) Urban forests, climate change and environmental pollution: physio-biochemical and molecular perspectives to enhance urban resilience. Springer Nature, Cham, Switzerland. https://doi.org/10.1007/978-3-031-67837-0

Skre O, Wielgolanski FE, Moe B (1998) Biomass and chemical composition of common forest plants in response to fire in western Norway. J Vege Sci 9:501–510.https://doi.org/10.2307/3237265

Stenseth NC, Dunlop ES (2009) Evolution: unnatural selection. Nature 457:803–804. https://doi.org/10.1038/457803a

# The Carbon Conundrum: Forests' Dual Role as Carbon Absorber and Emitter During Forest Fire

**Vasudha Agnihotri**

**Abstract** Forests seem like a carbon conundrum as they function as carbon absorbers and emitters. When forests are disturbed or destroyed due to events like forest fires, deforestation, or land conversion for human purposes such as agricultural activities or infrastructure development, they can release large amounts of carbon dioxide back into the atmosphere. Under normal conditions, forests act as carbon sinks by absorbing $CO_2$ from the atmosphere during photosynthesis and storing it in their biomass, including trunks, branches, leaves, and roots. Even when trees die, their degradation occurs slowly, resulting in a gradual release of carbon. Additionally, carbon is stored in forest soil for centuries, depending on soil type, climate, and microbial activity. Conditions such as forest age, tree types, and changes in temperature and precipitation patterns can also impact the ability of forests to sequester carbon. Human activities, such as afforestation (planting trees on previously non-forested land), reforestation (replanting trees in deforested areas), sustainable forest management, and reducing emissions from deforestation and forest degradation (REDD+), can all influence the role of forests in the carbon cycle. While forests play a crucial role in sequestering and storing carbon, their net effect on the climate depends on various factors, including land use changes, management practices, and interactions with the changing climate. Balancing conservation efforts with the sustainable use of forests is key to maximizing their potential as carbon sinks while meeting other societal needs. This chapter contains information about the role of forests as carbon emitters and absorbers, maintaining the carbon exchange in equilibrium.

**Keywords** Forest · Carbon absorption · Carbon conundrum · Biomass · Soil carbon

---

V. Agnihotri (✉)
G. B. Pant National Institute of Himalayan Environment-Himachal Pradesh Regional Centre, Mohal, Kullu, Himachal Pradesh, India
e-mail: vasudha@gbpihed.nic.in

 
H. Singh and N. Kumar (eds.), *Forest Fire and Climate Change*,
https://doi.org/10.1007/978-3-031-89967-6_4

# 1 Introduction

Carbon is a fundamental building block of life, forming the basis of organic molecules essential for all living organisms. The carbon cycle is a crucial natural process that regulates the distribution and movement of carbon throughout the Earth's atmosphere, oceans, land, and living organisms (Kaushik and Singh 2024). This cycle consists of various interconnected processes such as photosynthesis, respiration, combustion, weathering and erosion, carbon sequestration, etc., that continuously exchange carbon between different reservoirs (Singh et al. 2024a, b, c; Yadav et al. 2019a; Singh 2021; Singh et al. 2018). The importance of the carbon cycle in the Earth's climate system cannot be overstated. Carbon dioxide and other greenhouse gases trap heat in the Earth's atmosphere, leading to the greenhouse effect and warming the planet. Human activities, mainly burning fossil fuels and deforestation, have significantly altered the carbon cycle by releasing large amounts of $CO_2$ into the atmosphere Sett et al (2024). This has increased atmospheric $CO_2$ concentrations and enhanced the greenhouse effect, resulting in global warming and climate change (Raturi et al. 2022; Sett et al. 2022; Singh et al. 2020a; Apurva et al. 2017). Understanding and managing the carbon cycle is essential for mitigating the impacts of climate change Barwa and Singh (2024). Efforts to reduce carbon emissions, protect carbon sinks like forests and wetlands, and promote carbon sequestration technologies are crucial for stabilizing the Earth's climate and safeguarding ecosystems and human societies (Singh 2024a).

The "carbon conundrum" refers to the complex challenge of managing carbon emissions and carbon sequestration to mitigate climate change (Nibleus and Lundin 2010; Kumar et al. 2020a, b).Forest fires play a significant role in the carbon cycle. When forests burn, they release carbon stored in vegetation and soil into the atmosphere as $CO_2$ (Kumar et al. 2024; Singh 2024b). This process contributes to carbon emissions and can exacerbate climate change. However, forests also act as carbon sinks, absorbing $CO_2$ from the atmosphere through photosynthesis and storing it in biomass and soil organic matter. The balance between carbon emissions from forest fires and carbon absorption by forests is crucial for understanding their impact on the carbon cycle and climate change (Singh 2024c). Factors such as the frequency, intensity, and extent of forest fires and the ability of forests to recover and regrow after fires all influence this balance (Kumar et al. 2021a, b).

Forest fires are a natural phenomenon that occurs in forested areas worldwide. These fires can be ignited by lightning, human activity, or volcanic eruptions, and they play a significant role in shaping ecosystems and landscapes. Climate and environmental factors like drought, high temperature, wind and topography, etc., also help spread forest fires (Singh 2024c). While some fires benefit ecosystem health, others can devastate communities, wildlife, and the environment. One of the most significant impacts of forest fires is their contribution to carbon emissions. When trees and vegetation burn, carbon dioxide ($CO_2$) and other greenhouse gases are released into the atmosphere. Forest fires can release large amounts of quickly, making them a significant source of carbon emissions, particularly during intense fire seasons

(Singh et al. 2021c; Yadav et al. 2019b; Singh et al. 2023; Sharma and Singh 2021). In addition to $CO_2$, forest fires release other pollutants such as particulate matter, nitrogen oxides, and volatile organic compounds (Singh 2024d). These pollutants can harm air quality and human health, leading to respiratory problems and other health issues, especially for those living near the fire-affected areas.

The severity and frequency of forest fires have been increasing in recent years due to a combination of factors, including climate change, land-use practices, and human activities such as fire suppression and land clearing. As a result, there is growing concern about the impact of wildfires on carbon emissions and their contribution to climate change. Efforts to mitigate the effects of forest fires on carbon emissions include better forest management practices, such as prescribed burning and forest thinning, to reduce fuel loads and decrease the likelihood of catastrophic wildfires. Additionally, a growing focus is on restoring degraded forests and implementing strategies to adapt to changing fire regimes in a warming climate. Forest fires play a complex role in carbon emissions, both as a source of $CO_2$ and a driver of ecosystem dynamics (Singh 2024e). Understanding the relationship between forests, carbon emissions due to forest fires, and carbon sequestration or absorption by the forest is crucial for developing effective strategies to mitigate climate change and protect ecosystems and communities from the impacts of forest fires (Prakash et al. 2022; Nirmal et al. 2021; Singh and Kumar 2022; Savita et al. 2017).

## 2 Forest Fire and Carbon Emission

### *2.1 The Carbon Emission Phase*

During a forest fire, carbon emissions occur in several phases, each with its dynamics and environmental impacts (Fig. 1). When a fire ignites in a forest, the initial combustion phase begins. In this phase, the fire consumes dry organic matter such as leaves, branches, and vegetation. During this stage, carbon stored in the biomass is released into the atmosphere primarily in the form of carbon dioxide ($CO_2$) through combustion (Kumar et al. 2022). As the fire spreads and intensifies, it enters the active burning phase. This phase is characterized by high temperatures and vigorous combustion of surface fuels and standing vegetation. The rapid combustion releases large amounts of carbon stored in trees and other organic matter into the atmosphere.

Along with $CO_2$, other greenhouse gases such as methane ($CH_4$) and nitrous oxide ($N_2O$) may also be emitted, albeit in smaller quantities. After the active burning phase, some parts of the fire may enter a smoldering phase. Smoldering occurs when the fire burns at lower temperatures with limited oxygen supply, leading to incomplete combustion. In this phase, significant amounts of carbon can be released as particulate matter, black carbon, and volatile organic compounds (VOCs), contributing to air pollution and having adverse health effects (Malik et al. 2024). Garcia-Hurtado et al. (2013) assessed and quantified fire emissions from Mediterranean shrublands and

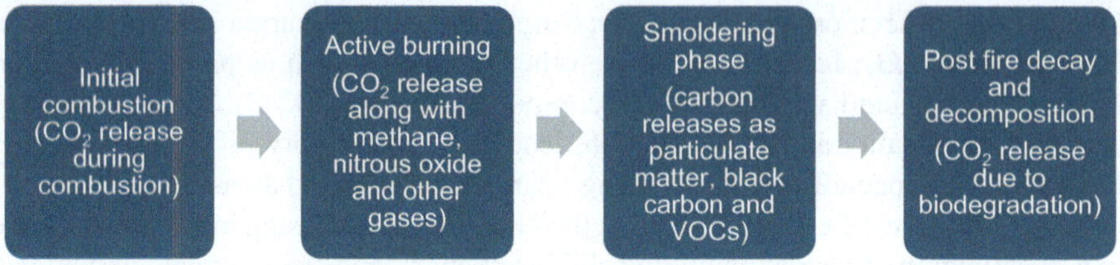


**Fig. 1** Carbon emission phases during the forest fire

observed the release of gaseous pollutants and particulate matter. While analyzing different fire phases, 25% carbon emission was observed from the flaming phase, while 75% was observed during the smoldering phase.

After the fire is extinguished, the remaining biomass undergoes decay and decomposition. During this phase, microorganisms break down the charred organic matter, releasing additional $CO_2$ into the atmosphere as part of the natural carbon cycle. Campbell et al. (2016) estimated the impact of forest fires on carbon dynamics, explicitly focusing on the rate at which fire-killed biomass decays and emits carbon to the atmosphere while studying the fire incidences of southwestern Oregon, USA (Biscuit Fire of 2002) over ten years. The decomposition rate was higher for leaves and fine roots and lower for large-diameter wood in forest fire remains. Biomass equivalent to 4.7 TgC was not combusted but killed, while 85% of the biomass remained even after ten years and continuously contributed to carbon emissions. It's important to note that the magnitude of carbon emissions from forest fires can vary depending on factors such as the size and intensity of the fire, the type of vegetation burned, weather conditions, and the effectiveness of fire management strategies. Large and intense forest fires can release substantial amounts of carbon dioxide into the atmosphere, contributing to climate change and impacting regional and global ecosystems (Singh et al. 2024a, b, c).

## 2.2 *Impact of Forest Fires on Atmospheric Carbon Emissions*

When trees and other vegetation burn during forest fires, the carbon stored in these plants is released into the atmosphere as $CO_2$. This leads to a sudden increase in atmospheric $CO_2$ levels near the fire. Large forest fires, especially those that burn for extended periods or across extensive areas, can emit millions of metric tons of $CO_2$. When forests burn, not only is $CO_2$ released into the atmosphere, but the loss of vegetation reduces the capacity of the ecosystem to sequester carbon in the future. Climate change can influence the frequency and severity of forest fires, creating a feedback loop that amplifies the release of $CO_2$ into the atmosphere. Drier conditions, prolonged droughts, and higher temperatures associated with climate change can create more favorable conditions for forest fires to occur and spread. Overall, forest fires play a significant role in the carbon cycle. They can influence atmospheric

$CO_2$ levels in the short term through immediate emissions and, in the long term, through their effects on ecosystem carbon storage and interactions with climate change. Understanding these dynamics is crucial for accurately assessing the impacts of forest fires on the global carbon budget and climate system.

Quantifying carbon emissions from forest fires involves several factors and methods, as it can be a complex process due to the variability in fire intensity, duration, fuel types, and environmental conditions. The first step is to estimate the amount of biomass (organic matter) burned during the fire. This involves assessing the area burned, fuel load (amount of vegetation available to burn), and biomass density in the affected area. Remote sensing techniques such as satellite imagery, aerial photography, or ground surveys can be used to map the burned area and estimate biomass. Wang et al. (2023) used the NSSI-NDVI vegetation index triangular space method, which enables the calculation of flammable non-photosynthetic vegetation (NPV), combustible photosynthetic vegetation (PV), and incombustible bare soil (BS) fractional cover in forest areas. Their impact was estimated before and after forest fires, which verified their accuracy in calculating post-fire vegetation-damaged areas. Once the biomass burned is estimated, the next step is to determine the carbon content of the biomass. Different types of vegetation (trees, shrubs, grasses) have varying carbon densities, so the carbon content needs to be estimated based on the type of vegetation present in the burned area.

Not all carbon in the biomass is emitted as carbon dioxide ($CO_2$) during combustion. The combustion efficiency, which depends on factors such as fire intensity, moisture content of the vegetation, and combustion conditions, determines how much carbon is released as $CO_2$ versus other combustion byproducts like carbon monoxide (CO) or particulate matter. Typically, a combustion efficiency factor is applied to estimate $CO_2$ emissions.Since carbon is often reported in units of carbon (C), while $CO_2$ emissions are typically noted in units of $CO_2$, a conversion factor of 3.67 is used (1 tonne of carbon equals 3.67 tonnes of $CO_2$).Forest fires can also emit other greenhouse gases such as methane ($CH_4$), Nitrogen dioxide ($NO_2$), and nitrous oxide ($N_2O$) (Borgohain et al. 2023). These emissions must be quantified separately and converted to their $CO_2$ equivalent using their respective global warming potentials. While studying bushfires in Australia, Wan et al. (2023) assessed the spatiotemporal variations of carbon mono oxide (CO) and nitrogen dioxide gas ($NO_2$) in different vegetation types, including temperate and savanna forests, using Tropospheric Monitoring Instrument (TROPOMI). The role of flaming and smoldering was observed in atmospheric composition and air quality after forest fires.

Liang et al. (2022) assessed organic compounds emitted from conifer-dominated forest fires in the western US. Sampling was done using Aerodyne Mobile Laboratory (AML), equipped with a comprehensive suite of real-time instruments. The presence of resin acids and diterpenoids, which might be released from heated, unburnt vegetation, was also reported in the same study. These compounds have also been reported to contribute to forming organic aerosol. Sheoran et al. (2023) observed biomass burning as a source of primary brown carbon (BrC) during the winter season (December-February), and the primary source is supposed to be a large-scale forest fire, biomass burning for heating, energy production, and cooking purposes (Laskin et al. 2015).

Quantifying carbon emissions from forest fires involves inherent uncertainties due to variability in input data, assumptions made in calculations, and limitations of measurement techniques. Uncertainty analysis helps assess emission estimates' reliability and identify areas for further research or data refinement. Li et al. (2019) merged high temporal resolution Geostationary Operational Environmental Satellite-Fire Radiative Power (GOES FRP) with high spatial resolution Moderate Resolution Imaging Spectroradiometer Fire Radiative Power (MODIS FRP) data for the estimation of biomass burning emissions (BBE) in the conterminous United States. The seasonal variations in GOES-MODIS BBE align well with existing inventories and the magnitude differs from existing inventories by a factor of less than seven. Quantifying carbon emissions from forest fires requires a combination of field measurements, remote sensing techniques, empirical models, and emission factors, and careful consideration of the factors influencing fire behavior and carbon release. Collaboration between scientists, land managers, and policymakers is essential to improve the accuracy and reliability of carbon emission estimates from forest fires.

## 2.3 Emerging Technologies and Approaches for Monitoring Forest Fire Carbon Emissions

Emerging technologies and approaches for monitoring forest fire carbon emissions are continually evolving, offering more efficient and accurate methods for assessing the impacts of forest fires on carbon stocks and greenhouse gas emissions. Here are some notable technologies and approaches:

**Satellite Remote Sensing**: Satellite-based sensors provide valuable data for monitoring forest fires and estimating carbon emissions. Advanced satellite instruments, such as those onboard NASA's Moderate Resolution Imaging Spectroradiometer (MODIS) and the European Space Agency's Sentinel satellites, can detect active fires, map burned areas, and estimate the extent of fire-induced carbon emissions.

**Unmanned Aerial Vehicles (UAVs) or Drones**: UAVs equipped with specialized sensors, such as multispectral or thermal cameras, can collect high-resolution imagery of fire-affected areas. Drones offer flexibility in accessing remote or inaccessible regions and can capture detailed information on fire severity, vegetation damage, and carbon release.

**LiDAR (Light Detection and Ranging)**: LiDAR technology uses laser pulses to generate three-dimensional forest structure and biomass maps. Combined with field data and post-fire surveys, LiDAR can help quantify changes in aboveground biomass and carbon stocks caused by forest fires with high precision.

**Ground-Based Monitoring Systems**: Deploying ground-based sensors and monitoring stations in fire-prone areas allows for real-time measurement of atmospheric carbon dioxide ($CO_2$) and other greenhouse gases released during forest fires. These monitoring systems provide localized data to complement satellite observations and validate emission estimates.

**Airborne Measurements**: Aircraft equipped with gas analyzers and sampling equipment can conduct airborne surveys to measure carbon dioxide, methane, and other emissions during and after forest fires. Airborne measurements provide detailed insights into emission rates, plume dispersion, and the composition of forest fire smoke.

**Integrated Modeling Frameworks**: Advanced modeling frameworks integrate satellite data, ground-based observations, and ecosystem models to simulate fire behavior, carbon emissions, and the long-term impacts of forest fires on carbon cycling. These models help forecast fire-induced carbon losses, assess ecosystem resilience, and inform land management strategies.

**Machine Learning and Artificial Intelligence (AI)**: Machine learning algorithms and AI techniques can analyze large datasets from satellite imagery, remote sensing, and field observations to detect fire events, map burned areas, and estimate carbon emissions more accurately. These technologies enable automated processing and rapid analysis of complex environmental data.

**Crowdsourcing and Citizen Science**: Crowdsourcing platforms and citizen science initiatives engage volunteers in collecting ground observations, photographs, and geospatial data related to forest fires. These participatory approaches enhance spatial and temporal coverage of fire monitoring efforts and support community-based resilience planning.

By leveraging these emerging technologies and approaches, researchers, policymakers, and land managers can improve the monitoring and assessment of forest fire carbon emissions, leading to more effective mitigation and adaptation strategies in response to forest fires and climate change impacts.

## 3 The Carbon Absorption Phase

### *3.1 Role of Forests as Carbon Sinks*

Forests are crucial in mitigating climate change by acting as carbon sinks. A carbon sink is any natural or artificial reservoir that absorbs more carbon (carbon dioxide) from the atmosphere than it releases. Forests are among Earth's most significant carbon sinks, performing several functions contributing to carbon sequestration (Fig. 2). It stores around 45% of the organic carbon in soil and biomass and absorbs

2Gt C annually. Trees and other vegetation absorb carbon dioxide during photosynthesis, using sunlight to convert it into organic compounds such as sugars and cellulose. This process removes carbon dioxide from the atmosphere and stores it in biomass. Plants store carbon in their tissues, including roots, stems, branches, leaves, and other organic matter. As forests grow, they accumulate biomass and sequester more carbon over time (Talberth and Carlson 2024). DellaSala et al. (2022) described the relationship between tree age and carbon stock and found it directly proportional. Carbon is released slowly during the decay of tree parts, some part of which is emitted into the atmosphere while the rest of the part is added to soil in organic form, which is locked up in the environment for a longer duration (Luyssaert et al. 2008; Singh et al. 2024a, b, c).

Forest soils also store significant amounts of carbon in organic matter, such as decomposing plant material and soil organic carbon (Chiti et al. 2024). Soil absorbs approximately. 2500 Gt C is affected by vegetation, type of soil, and climatic conditions, e.g., carbon stores better in soil in wet and colder conditions. Out of total global carbon storage (330 tonnes per hectare), maximum carbon storage is reported to be done by tropical forests (120 tonnes per hectare). In the case of total global carbon storage by soil, out of 1808 tonnes per hectare C, a maximum of 36% C storage is reported to be done by wetland soil (Fig. 3). Food and Agricultural Organization (FAO) report on the Global Forest Resource Assessment, 2020 shows a carbon stock of 662 Gt in the total forests, 300 Gt in soil organic carbon, 68 Gt in litter and dead wood, and 295 Gt in the living biomass. As per Biennial Update Report 2021, India's total Green House Gas (GHG) emissions, including Land Use Change and Forestry

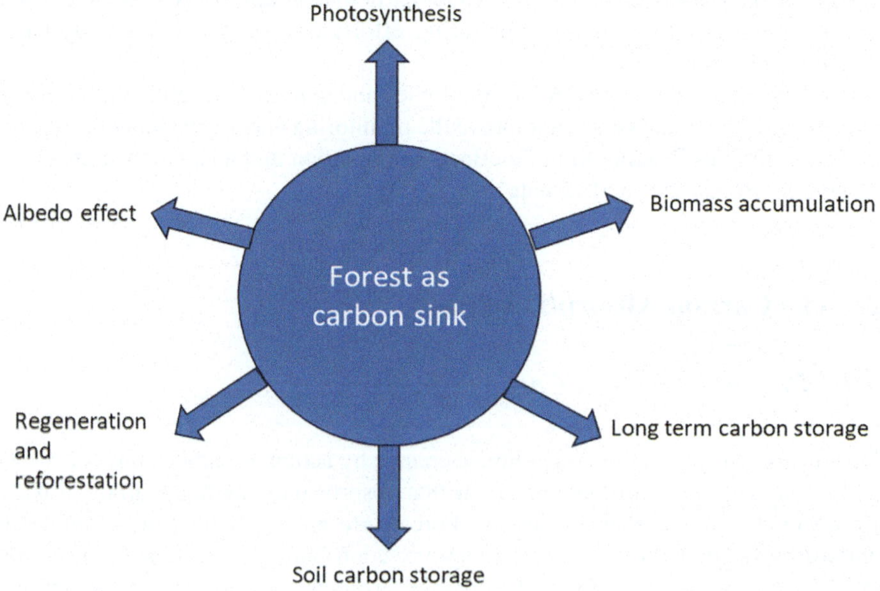


**Fig. 2** Carbon absorption phase

(LULUCF) was 2531.07 million tonnes of $CO_2$ during 2016, and the net sink of LULUCF was around 307,820 Gt $CO_2$eq (MoEFCC 2021). At the national level, forest carbon stock was estimated for different components, including above-ground biomass (AGB), below-ground biomass (BGB), dead wood, litter, and soil, and a total of approximately 7204.0 million tonnes of carbon stock were reported for the year 2021. Compared to 2019, a net increase of approximately 79.4 million tonnes of carbon stock was observed. Annually, an increase of 39.7 million tonnes of carbon stock was observed, equivalent to 145.6 million tonnes of $CO_2$ (Fig. 4).

Healthy forest ecosystems with rich biodiversity tend to have more robust soil carbon storage capacities. Some carbon stored in forests can remain sequestered for long periods, especially in old-growth forests with mature trees. Every year, around 8.1 billion t $CO_2$ is released back into the atmosphere as trees die naturally or are harvested sustainably, and it is used in long-lived wood products like furniture or buildings. The carbon they contain remains stored, thus preventing it from re-entering the atmosphere. Forest fires also contribute to this process, and their increased frequency in the last few years, along with deforestation, has increased the risk of carbon release in the atmosphere. Reforestation efforts and regenerating degraded or deforested lands contribute to carbon sequestration by establishing new forests or restoring existing ones. Proper forest management practices, including afforestation (planting trees where there were none before) and reforestation (replanting trees in previously forested areas), can enhance carbon sequestration potentials. Forests also influence the Earth's albedo or its reflectivity. Their canopy cover reflects sunlight differently than other surfaces like bare soil or urban areas, affecting local and regional climate patterns. Cooler temperatures resulting from increased forest cover can indirectly influence carbon sequestration rates by enhancing the health and growth of trees. Given the urgent need to mitigate climate change and reduce greenhouse gas emissions, preserving existing forests, restoring degraded lands, and expanding forested areas through sustainable practices are essential strategies for enhancing the role of forests as carbon sinks. Protecting forests from deforestation and degradation is critical to maintaining their carbon sequestration capacities and preserving their biodiversity and ecosystem services.

### *3.2 Factors Influencing the Carbon Absorption Capacity of Forests*

The carbon absorption capacity of forests, referred to as carbon sequestration, is influenced by many factors, ranging from biological characteristics to environmental conditions. Different types of forests (e.g., tropical rainforests, temperate deciduous forests, boreal forests) have varying capacities for carbon sequestration. Additionally, the composition of a species within a forest can affect its carbon storage potential, as some species are more efficient at capturing and storing carbon than others. Figure 5 shows the carbon stock efficiency of different forest types. Himalayan dry and moist

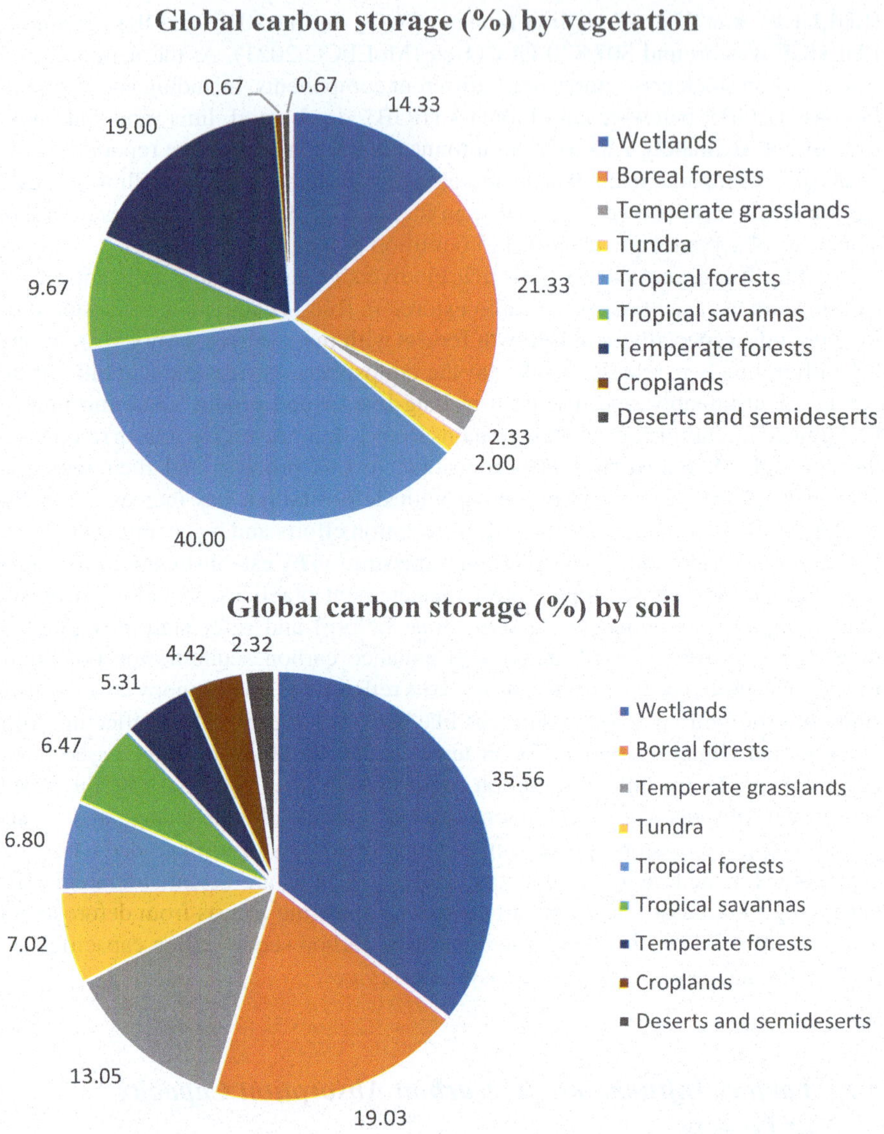


**Fig. 3** Average carbon stored (tonnes per hectare) at a ground depth of one meter by vegetation and soil

forests have shown a higher carbon stock potential, possibly due to a slow degradation rate. Young forests typically sequester carbon at higher rates than mature or old-growth forests, as they actively grow and accumulate biomass. However, older forests may store more carbon in their giant trees and soil organic matter. The Net Ecosystem Production (NEP), which is the difference between $CO_2$ assimilation and loss through

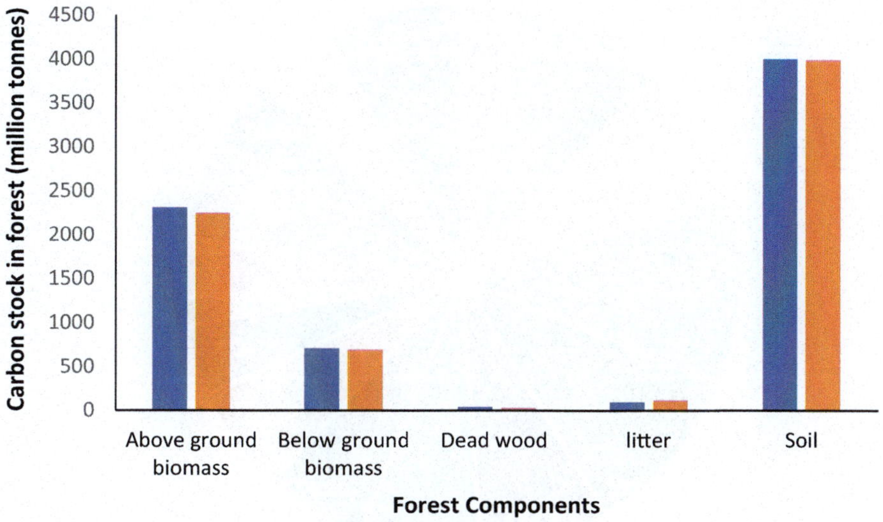


**Fig. 4** Forest carbon stock in the form of different forest components. *Source* India state of forest report 2021

soil and plant respiration, was found to be positive while analyzing the $CO_2$ data of different aged forests (15–800 years old forests) (Steffen et al. 1998; Luyssaert et al. 2008). In acarbon-neutral situation in the forest, the ratio of heterotrophic respiration and net primary production (NPP) is one. NPP is calculated by subtracting heterotrophic respiration in terms of $CO_2$ from plant organic matter (Gross Primary production). Old forests are considered carbon neutral, but Luyssaert et al. (2008) observed $< 1$ NPP for 200-year-old forests and noted no increase in NPP with forest age.

The capacity of century-old trees to accumulate carbon through climatic, edaphic, and biological factors also affects the carbon accumulation forces. The density and vertical structure of vegetation within a forest influence its carbon sequestration capacity. Forests with dense canopies and abundant understory vegetation tend to sequester more carbon than open or sparse forests. Soil properties such as texture, organic matter content, andmicrobial activity are significant in carbon storage. Soils rich in organic matter can store substantial amounts of carbon, while disturbances like erosion or compaction can reduce soil carbon stocks. Climatic factors such as temperature, precipitation, and sunlight influence the rates of photosynthesis, biomass production, and decomposition in forests.

Warmer temperatures and longer growing seasons generally enhance carbon uptake, while water availability can be a limiting factor in arid regions. Natural disturbances such as forest fires, insect outbreaks, and storms can release stored carbon back into the atmosphere, temporarily reducing a forest's carbon sequestration capacity. Conversely, sustainable forest management practices like afforestation, reforestation, and reduced-impact logging can enhance carbon storage and

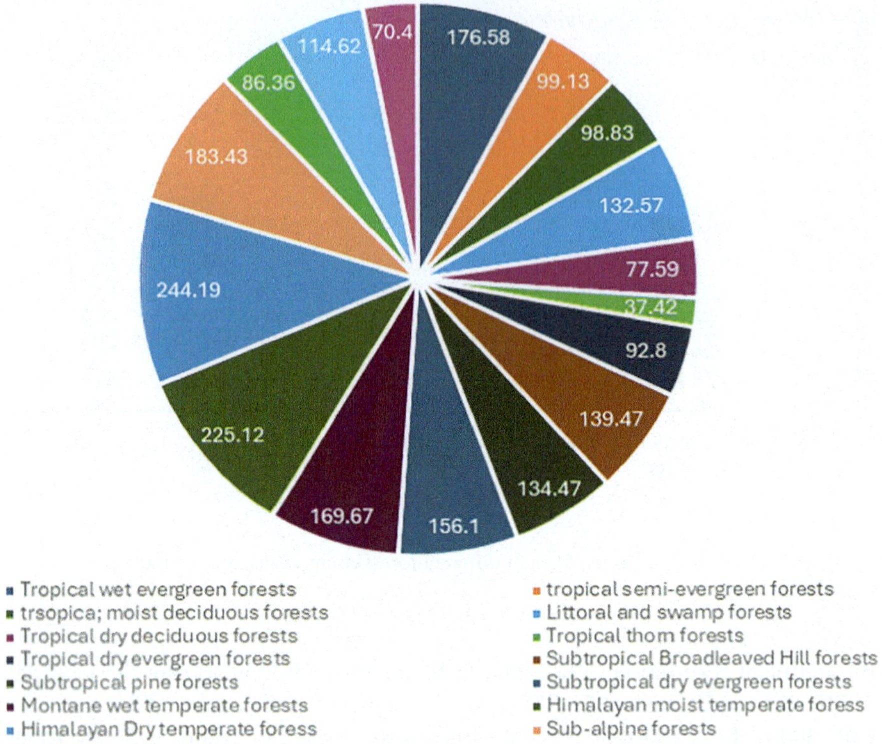


**Fig. 5** Carbon stock (tonnes/ hectare) in different forest types in India. *Source* India state of forest report 2021

promote ecosystem resilience. The allocation of carbon within different parts of a tree (e.g., leaves, stems, roots) and the organic matter turnover rates influence the long-term carbon storage in forests. Slow-growing, long-lived trees tend to allocate more carbon to durable woody tissues, while fast-growing species may allocate more carbon to rapid biomass accumulation. Biodiversity loss can impact ecosystem functioning and resilience, affecting carbon cycling processes such as nutrient uptake, decomposition, and carbon storage. Diverse forests with various species and functional traits may exhibit greater stability and carbon sequestration potential. Overall, the carbon absorption capacity of forests results from complex interactions among biological, ecological, and environmental factors. Understanding these factors is essential for effective forest management and conservation strategies to maximize carbon sequestration and mitigate climate change.

## *3.3 How Forests Recover and Regenerate Post-fire*

Forest recovery and regeneration post-fire is a complex and dynamic process influenced by various factors, including fire severity, ecosystem type, climate, soil conditions, and the presence of fire-adapted species. In the immediate aftermath of a fire, the forest ecosystem experiences significant changes. The intensity and severity of the fire determine the extent of damage to vegetation, soil, and other ecosystem components. Fire-adapted plant species, such as conifers or shrubs with fire-resistant traits like thick bark or serotinous cones, may survive or quickly resprout from surviving root systems or seeds. Fire can stimulate the germination of certain plant species' seeds, especially those with hard seed coats or seeds that require heat to break dormancy. Seed dispersal mechanisms, including wind, water, and animals, are crucial in distributing seeds across burned areas. Seedlings from surviving trees and shrubs, as well as from new seed sources, begin to establish in the burned area.

Pioneer species, often fast-growing and opportunistic plants, are among the first to colonize burned areas. These species play a crucial role in stabilizing soils, providing organic matter, and creating microclimates conducive to further plant establishment. Examples of pioneer species include grasses and certain shrubs adapted to post-fire environments. Over time, as vegetation begins to recover, successional processes occur, leading to species composition and ecosystem structure changes. Early successional species may give way to shade-tolerant species, including trees, as the forest canopy matures. Succession may proceed through several stages, from herbaceous plants and shrubs to young trees and mature forest stands. Fire can have both positive and negative effects on soil properties. While intense fires can lead to soil erosion, loss of organic matter, and changes in nutrient availability, they can also release nutrients stored in vegetation and promote the decomposition of dead organic matter. Soil recovery processes, including nutrient cycling, microbial activity, and organic matter accumulation, contribute to the restoration of soil fertility and ecosystem productivity over time.

Forest recovery post-fire is an ongoing process that can take years to decades, depending on the severity of the fire and environmental conditions. Resilient forests can withstand and recover from disturbances like fire, maintaining their structure, function, and biodiversity over time. Active management practices, such as prescribed burning, reforestation, and invasive species control, can facilitate forest recovery and promote ecosystem resilience following forest fires. Overall, forest recovery post-fire is a natural and adaptive process driven by ecological interactions and succession, highlighting the resilience of forest ecosystems in the face of disturbance. However, the increasing frequency and intensity of forest fires due to climate change underscore the importance of proactive management and conservation efforts to support forest recovery and enhance ecosystem resilience in the long term.

## 4 Case Studies and Research Findings on the Net Carbon Effects of Forest Fires

Several case studies and research findings shed light on the net carbon effects of forest fires:

**Boreal Forests of North America**: Research conducted in boreal forests, such as those in Canada and Alaska, has shown that while severe forest fires can result in significant short-term carbon emissions, the long-term carbon balance depends on ecosystem recovery. Studies have found that boreal forests can recover relatively quickly following fires, with new vegetation sequestering carbon from the atmosphere. However, the frequency and severity of forest fires in these regions are expected to increase due to climate change, potentially altering the carbon balance. Veraverbeke et al. (2021) reviewed the fire events of arctic boreal forests. They found that, on average, 142 Tg C per year is emitted by these forests, which represents 7% of the global carbon emissions due to forest fires, although this burnt area represents only 1.6% of the total global burnt area (Walker et al. 2019; McCarty et al. 2020). This carbon emission is sourced from the carbon stored in permafrost and peatland soils (Hugelius et al. 2020).

**Australian Bushfires**: The intense bushfires that ravaged Australia in recent years have sparked considerable research into their carbon implications. Studies have shown that the 2019–2020 bushfires released unprecedented carbon dioxide into the atmosphere, significantly contributing to Australia's greenhouse gas emissions for that year (Senf et al. 2023). However, the long-term carbon effects are complex, as they depend on factors such as the regrowth of vegetation, soil carbon dynamics, and future fire regimes in the affected areas (Garcia-Hurtado et al. 2013).

**Amazon Rainforest Fires**: Forest fires in the Amazon rainforest have attracted global attention due to their potential impact on carbon emissions and biodiversity. While some studies suggest that severe fires in the Amazon can lead to substantial carbon emissions, others highlight the resilience of the rainforest ecosystem and its ability to recover over time. However, ongoing deforestation and land-use changes in the Amazon threaten the integrity of the forest and its carbon storage capacity. Basso et al. (2023) highlighted no significant trend in carbon uptake by Amazonian forests between 2010 and 2018, indicating that the carbon balance remains stable despite variations in environmental conditions. The observations of Aragão et al. 2018, during 2003–2015 differed. Gross carbon emissions through forest fires were approximately. $989 \pm 504$ Tg $CO_2$ year $- 1$, although deforestation was a 76% decline. In a 12-year duration, around 36% increase in fire incidences was observed. The impact was observed more in drought situations.

**California Forest fires**: The frequent and severe forest fires in California have been the subject of numerous studies examining their carbon dynamics. Research has shown that while forest fires in California release significant amounts of carbon dioxide and other pollutants into the atmosphere, the long-term carbon balance depends on factors such as forest regeneration, soil carbon storage, and future fire regimes. Climate change-induced drought and warmer temperatures are expected to exacerbate forest fire risk in the state, potentially affecting its carbon balance (Jerrett et al. 2022).

**Global Synthesis Studies**: Several global synthesis studies have attempted to assess the overall impact of forest fires on the carbon cycle. These studies often use satellite data, ecosystem modeling, and field measurements to estimate carbon emissions from forest fires and their subsequent effects on carbon sequestration. While the exact net carbon effects vary depending on the region and specific circumstances, these studies highlight the importance of considering both short-term emissions and long-term ecosystem dynamics when evaluating the carbon balance of forest fires.

Overall, research on the net carbon effects of forest fires underscores the complexity of the issue and the need for interdisciplinary approaches to understanding and managing fire-prone ecosystems in the context of climate change.

## 5 Mitigation and Adaptation Strategies

Implementing measures to reduce greenhouse gas emissions, including those from forest fires, is essential for mitigating climate change. This can involve strategies such as improving forest management practices, reducing deforestation and land-use change, and promoting renewable energy sources to minimize the drivers of climate change. Protecting and managing forest carbon stocks is crucial for mitigating the impact of forest fires on the climate. This includes conserving intact forests with high carbon density, restoring degraded forests to enhance their carbon sequestration capacity, and implementing sustainable forestry practices to maintain or increase carbon stocks. Implementing effective fire management and prevention strategies can help reduce the severity and frequency of forest fires, thereby mitigating their carbon emissions (Sharma et al. 2018; Jangra et al. 2024; Dhyani et al. 2021; Singh 2024e). This can involve prescribed burning, fuel reduction treatments, early detection and suppression of forest fires, and community education and outreach on fire safety.

Enhancing the resilience of forest ecosystems to climate change and forest fires is essential for adaptation. This can include restoring degraded ecosystems, promoting biodiversity and ecosystem diversity, and incorporating climate-smart forest management practices that consider future fire risk and changing environmental conditions. Improving community preparedness and response to forest fires can help reduce forest fires' socio-economic and ecological impacts. It invests in fire-resistant infrastructure, implements evacuation plans, provides adequate firefighting resources, and engages communities in forest fire preparedness and mitigation efforts. Adopting an integrated landscape management approach that considers the interactions between forests, land use, and climate change is critical for adaptation. This includes strategies such as ecosystem-based adaptation, landscape planning and zoning, and collaborative governance mechanisms that involve multiple stakeholders in decision-making processes. Establishing monitoring and early warning systems for forest fires can improve preparedness and response efforts. This consists of using remote sensing technologies, satellite imagery, weather forecasting, and predictive modeling to detect and anticipate forest fire risks, allowing for timely intervention and mitigation measures. Overall, addressing the implications of forest fires for climate change mitigation and adaptation requires a multifaceted approach that integrates forest management, land-use planning, community engagement, and policy interventions. By implementing effective strategies to reduce the risk and impact of forest fires while enhancing the resilience of forest ecosystems and communities, societies can better adapt to the challenges posed by climate change and protect the vital services provided by forests.

## 6 International Cooperation and Frameworks for Addressing Forest Fire Emissions and Carbon Sequestration

International cooperation and frameworks are crucial in addressing forest fire emissions and carbon sequestration. Here are some key initiatives and frameworks:

**United Nations Framework Convention on Climate Change (UNFCCC)**: The UNFCCC is a critical international treaty addressing climate change. Parties to the UNFCCC collaborate on various initiatives, including reducing emissions from deforestation and forest degradation (REDD+). REDD+ aims to incentivize developing countries to reduce deforestation and forest degradation emissions, conserve forests, and enhance forest carbon stocks. It provides a framework for financial and technical support from developed to developing countries.

**Paris Agreement**: Under the UNFCCC, the Paris Agreement aims to strengthen the global response to climate change. It includes provisions for countries to enhance their efforts to reduce emissions and enhance sinks, including forests. Parties to the agreement commit to nationally determined contributions (NDCs), which may include forest conservation, restoration, and sustainable management targets.

**Intergovernmental Panel on Climate Change (IPCC)**: The IPCC provides scientific assessments on climate change, including the role of forests in carbon sequestration and emissions. Its reports inform international policy discussions and guide effective forest management practices for climate mitigation and adaptation.

**Global Forest Observations Initiative (GFOI)**: GFOI is a partnership of countries and international organizations working to improve forest monitoring for climate mitigation and adaptation. It supports the development of national forest monitoring systems, including the measurement, reporting, and verification (MRV) of forest carbon stocks and emissions.

**Collaborative Partnerships and Initiatives**: Various collaborative partnerships and initiatives bring together governments, non-governmental organizations (NGOs), indigenous peoples, and other stakeholders to address forest fire emissions and carbon sequestration. Examples include the Collaborative Partnership on Forests (CPF), the Global Environment Facility (GEF), and the World Bank's Forest Carbon Partnership Facility (FCPF).

**Regional Agreements and Partnerships**: Regional agreements and partnerships, such as the European Union's Forest Strategy and the Asia–Pacific Network for Sustainable Forest Management and Rehabilitation (APFNet), also contribute to international efforts on forest fire emissions and carbon sequestration. These initiatives promote regional cooperation, knowledge sharing, and capacity building.

**Bilateral and Multilateral Agreements**: Countries may also engage in bilateral and multilateral agreements to address forest fire emissions and carbon sequestration. These agreements can involve collaboration on research, technology transfer, capacity building, and financial support for forest conservation and management efforts.

Overall, international cooperation and frameworks provide essential mechanisms for addressing forest fire emissions and carbon sequestration at a global scale. By working together, countries can enhance their efforts to conserve and sustainably manage forests for climate mitigation and adaptation.

## 7 Research Gaps and Priorities in Understanding the Carbon Dynamics of Forest Fires

While significant progress has been made in understanding the carbon dynamics of forest fires, several research gaps and priorities still exist. Addressing these gaps is crucial for improving our ability to quantify and manage carbon emissions from forest fires effectively. Some key research gaps and priorities include:

**Quantifying Carbon Losses and Emissions**: There is a need for more accurate methods to quantify carbon losses from different vegetation types during forest fires, including aboveground biomass, belowground carbon, and soil organic matter. Improving our understanding of the factors influencing carbon combustion and emissions, such as fire severity, fuel moisture, and combustion completeness, is essential.

**Assessing Post-Fire Carbon Dynamics**: Research on the long-term effects of forest fires on carbon sequestration, ecosystem recovery, and soil carbon storage is lacking. Understanding how fire regimes influence vegetation regeneration, carbon uptake by regenerating forests, and soil carbon stabilization processes is critical for predicting ecosystem resilience to recurrent fires and climate change.

**Spatial and Temporal Variability**: Limited knowledge about the spatial and temporal variability of carbon emissions within fire-affected landscapes exists. Investigating how fire behavior, vegetation characteristics, topography, and climate variability influence carbon dynamics at different scales can improve the accuracy of carbon accounting and modeling efforts.

**Effects of Climate Change and Disturbance Interactions**: Climate change alters fire regimes and ecosystem dynamics, leading to complex interactions between forest fires and carbon cycling. Research is needed to assess how changing climate conditions, such as prolonged droughts and increased temperatures, interact with other disturbances (e.g., insect outbreaks and land-use changes) to affect carbon dynamics and the resilience of forest ecosystems.

**Role of Pyrogenic Carbon**: Pyrogenic carbon, or charcoal, produced during forest fires, represents a significant but poorly understood component of the global carbon cycle. Investigating the formation, transport, and long-term fate of pyrogenic carbon in soils and sediments can improve our understanding of its role in carbon storage, nutrient cycling, and ecosystem functioning.

**Integration of Observational and Modeling Approaches**: There is a need to integrate observational data, remote sensing techniques, and ecosystem models to improve our ability to predict fire-induced carbon emissions and their impacts on carbon budgets. Developing integrated frameworks that combine field measurements, satellite observations, and process-based modeling can enhance our understanding of the complex interactions driving carbon dynamics in fire-prone ecosystems.

**Impacts on Human Health and Air Quality**: Research on the health effects of forest fire smoke and airborne particulate matter is essential for assessing the broader societal impacts of forest fires. Understanding forest fire emissions' composition, dispersion, and health risks can inform public health policies and emergency response strategies.

**Community Engagement and Indigenous Knowledge**: Incorporating Indigenous knowledge and community perspectives into forest fire carbon dynamics research is critical for developing culturally appropriate and socially inclusive management strategies. Collaborative approaches that involve local stakeholders in monitoring, adaptation planning, and traditional fire management practices can enhance the resilience of forest ecosystems and human communities.

Addressing these research gaps and priorities will require interdisciplinary collaboration, long-term monitoring efforts, and sustained funding support. By advancing our understanding of the carbon dynamics of forest fires, we can develop more effective strategies for mitigating carbon emissions, conserving ecosystem carbon stocks, and enhancing the resilience of forest ecosystems in a changing climate.

## 8 Long-Term Strategies for Promoting Forest Resilience and Carbon Balance in a Changing Climate

Promoting forest resilience and carbon balance in a changing climate requires long-term strategies that address both the drivers of forest degradation and the impacts of climate change. Here are some key long-term strategies:

**Sustainable Forest Management**: Implementing sustainable forest management practices, such as selective logging, reducing fuel loads, and protecting old-growth forests, helps maintain forest health and resilience to climate change. Sustainable management also enhances carbon sequestration and storage capacity in forests.

**Forest Restoration and Reforestation**: Investing in forest restoration and reforestation initiatives can help rehabilitate degraded landscapes, enhance biodiversity, and increase carbon sequestration. Restoring native vegetation and establishing resilient tree species can improve ecosystem services and promote climate adaptation.

**Fire Management and Prevention**: Implementing proactive fire management strategies, such as prescribed burning, creating firebreaks, and enhancing early detection systems, reduces the risk of high-severity forest fires and associated carbon emissions. Integrating indigenous fire management practices and community-based approaches can strengthen ecosystem resilience to fire.

**Climate-Resilient Tree Species**: Promoting the use of climate-resilient tree species and diverse forest ecosystems increases the adaptive capacity of forests to changing climatic conditions. Selecting species well-suited to local climatic conditions and tolerant of drought, pests, and diseases enhances forest resilience and carbon storage potential.

**Ecosystem-Based Adaptation**: Implementing ecosystem-based adaptation strategies, such as conserving wetlands, protecting riparian zones, and restoring natural hydrological processes, helps buffer forests from climate impacts and enhances carbon storage in soils and vegetation.

**Reducing Deforestation and Land-Use Change**: Addressing drivers of deforestation and land-use change, such as agricultural expansion, infrastructure development, and illegal logging, is critical for maintaining forest carbon stocks and biodiversity. Implementing policies and incentives that promote sustainable land use and conservation can help reduce emissions from deforestation and forest degradation (REDD+).

**Integrated Landscape Management**: Adopting an integrated landscape approach that considers the interactions between forests, agriculture, water resources, and human settlements promotes synergies among land uses and enhances ecosystem resilience. Incorporating ecosystem services valuation and participatory decision-making processes ensures that conservation and development goals are aligned with carbon balance objectives.

**Research, Monitoring, and Capacity Building**: Investing in research, monitoring, and capacity building enhances our understanding of forest carbon dynamics, climate impacts, and effective management strategies. Developing robust monitoring systems, leveraging remote sensing technologies, and strengthening institutional capacity support evidence-based decision-making and adaptive management approaches.

**International Cooperation and Financing**: Strengthening international cooperation, partnerships, and financial mechanisms is essential for strengthening forest resilience and carbon balance efforts. Mobilizing resources for forest conservation, sustainable management, and climate adaptation through initiatives like REDD+, climate finance, and public–private partnerships facilitates collective action and shared responsibility for global forest protection.

**Community Engagement and Indigenous Rights**: Recognizing and respecting the rights of Indigenous peoples and local communities to land, resources, and traditional knowledge is crucial for sustainable forest management and carbon balance. Engaging communities in decision-making processes, supporting traditional land management practices, and fostering equitable benefit-sharing contribute to forest resilience and climate justice.

By adopting these long-term strategies and fostering multi-stakeholder collaboration, countries can enhance the resilience of forests to climate change, conserve biodiversity, and contribute to global efforts to mitigate climate change by maintaining and enhancing forest carbon stocks.

## 9 Conclusion

Addressing the carbon conundrum of forest fires is critical for effective climate change mitigation. Forest fires release substantial amounts of carbon dioxide ($CO_2$) and other greenhouse gases into the atmosphere, contributing to global warming. Forest fires are significant sources of carbon emissions in regions with frequent and intense fire activity. Tackling these emissions is essential for achieving climate stabilization goals and minimizing the impacts of climate change on ecosystems and societies. Forest fires can create feedback loops that amplify climate change. As temperatures rise, fire risk increases, leading to more frequent and severe forest fires, which release additional carbon stored in vegetation and soils, further exacerbating climate change. Breaking these feedback loops is crucial to reducing the risk of runaway climate change. Forest fires also result in the loss of carbon stocks stored in vegetation and soils, diminishing the capacity of forests to sequester carbon and mitigate climate change.

Additionally, forest fires degrade ecosystems, reduce biodiversity, and disrupt ecosystem services such as water purification, soil stabilization, and habitat provision. The cumulative impacts of recurrent forest fires on carbon budgets can be substantial, especially in fire-prone regions with large forested areas. Even small fire frequency or severity increases can significantly affect carbon stocks and the global carbon cycle. Addressing the carbon emissions from forest fires is essential for accurately accounting for carbon fluxes and balancing carbon budgets at local, regional, and global scales.

Forests play a crucial role in regulating the Earth's climate by sequestering carbon through photosynthesis and storing it in biomass and soils. Protecting forests from the impacts of forest fires and managing fire regimes sustainably is essential for preserving their carbon sequestration capacity and maintaining their role as carbon sinks. Addressing the carbon conundrum of forest fires is integral to maintaining the health and integrity of forest ecosystems and their contributions to climate change mitigation efforts. Implementing effective fire management strategies and promoting sustainable land management practices may reduce the carbon conundrum of forest fires and enhance the capacity of forests to mitigate climate change. We must prioritize forest fire management and carbon sequestration efforts as components of our climate change mitigation strategies.

## References

Apurva K, Singh H, Kumar H (2017) Elucidating morphological and growth changes in Arjun tree (*Terminalia arjuna*) grown under elevated temperature condition. Trends Biosci 10(1):332–336

Aragão LE, Anderson LO, Fonseca MG, Rosan TM, Vedovato LB, Wagner FH, Silva CVJ, Silva CHL, Arai E, Aguiar AP, Barlow J, Berenguer E, Deeter MN, Domingues LG, Gatti L, Gloor M, Malhi Y, Marengo JA, Miller JB, Phillips OL, Saatchi S (2018) 21st century drought-related fires counteract the decline of Amazon deforestation carbon emissions. Nat Comm 9(1):536

Barwa S, Singh H (2024) Climate-induced changes in forest ecosystems of Indian Himalayan Region: a meta-analysis. In: Singh H (ed) Urban forests, climate change and environmental pollution. Springer, Cham. https://doi.org/10.1007/978-3-031-67837-0_39

Basso LS, Wilson C, Chipperfield MP, Tejada G, Cassol HL, Arai E, Williams M, Smallman TL, Peters W, Naus S, Miller JB, Gloor M (2023) Atmospheric $CO_2$ inversion reveals the Amazon as a minor carbon source caused by fire emissions, with forest uptake offsetting about half of these emissions. Atm Chem Phys 23(17):9685–9723

Borgohain A, Gogoi M, Barman N et al (2023) Impact of biomass burning on black carbon and $NO_2$ over north eastern region of india using multi-satellite observations. J Indian Soc Remote Sens 51:1605–1617. https://doi.org/10.1007/s12524-023-01692-2

Campbell JL, Fontaine JB, Donato DC (2016) Carbon emissions from decomposition of fire-killed trees following a large wildfire in Oregon, United States. J Geophys Res Biogeosci 121:718–730. https://doi.org/10.1002/2015JG003165

Chiti T, Benilli N, Mastrolonardo G, Certini G (2024)The potential for an old-growth forest to store carbon in the topsoil: a case study at SassoFratino, Italy. J For Res 35(1):10. https://doi.org/10.1007/s11676-023-01660-z

DellaSala DA, Mackey B, Norman P, Campbell C, Comer PJ, Kormos CF, Keith H, Rogers B (2022) Mature and old-growth forests contribute to large-scale conservation targets in the conterminous United States. Front For Glob Change 5:979528

Dhyani K, Kuniyal HB, Singh H, Sobha (2021) Growth and physiological potential of *Terminalia arjuna* under elevated $CO_2$ levels in Open top chamber condition. J Appl Nat Sci 13(3):1121–1126. https://doi.org/10.31018/jans.v13i3.2490

Garcia-Hurtado E, Pey J, Baeza MJ, Carrara A, Llovet J, Querol X, Alastuey A, Vallejo VR (2013) Carbon emissions in Mediterranean shrubland wildfires: an experimental approach. Atm Environ 69:86–93

Hugelius G, Loisel J, Chadburn S, Jackson RB, Jones M, MacDonald G, Marushchak M, Olefeldt D, Packalen M, Siewert MB, Treat C, Turetsky M, Voigt C, Yu Z (2020) Large stocks of peatland carbon and nitrogen are vulnerable to permafrost thaw. PNAS 117(34):20438–20446

Jangra A, Kumar K, Maikhuri S, Bhandari MS, Pandey S, Singh H, Barthwal S (2024) Unveiling stress-adapted endophytic bacteria: characterizing plant growth-promoting traits and assessing cross-inoculation effects on Populusdeltoides under abiotic stress. Plant Physiol Biochem, 108610. https://doi.org/10.1016/j.plaphy.2024.108610

Jerrett M, Jina AS, Marlier ME (2022) Up in smoke: California's greenhouse gas reductions could be wiped out by 2020 wildfires. Environ Pollu 310:119888

Kaushik P, Singh H (2024) Climate change-induced forest ecosystem functioning and alteration in water-atmosphere-soil–plant (WASP) relationship. In: Singh H (ed) Urban forests, climate change and environmental pollution. Springer, Cham. https://doi.org/10.1007/978-3-031-67837-0_38

Kumar A, Singh H, Kumari G, Bisht S, Malik A, Kumar N, Singh M, Raturi A, Barthwal S, Thakur A, Kaushal R (2022) Adaptive resilience of roadside trees to vehicular emissions via leaf enzymatic, physiological, and anatomical trait modulations. Environ Poll 313:120191

Kumar A, Dwivedi GK, Tewari S, Paul J, Anand R, Kumar N, Kumar P, Singh H, Kaushal R (2020a) Carbon mineralization and inorganic nitrogen pools under terminaliachebula retz.-based agroforestry system in Himalayan Foothills, India. Forest Sci 66 (5):634–643

Kumar A, Kumar P, Singh H, Kumar N (2020b) Adaptation and mitigation potential of roadside trees with bio-extraction of heavy metals under vehicular emissions and their impact on physiological traits during seasonal regimes. Urb For Urb Green 58:126900

Kumar A, Kumar P, Singh H, Bisht S, Kumar N (2021a) Relationship of physiological plant functional traits with soil carbon stock in temperate forest of Garhwal Himalaya. Curr Sci 120(8):1368–1373

Kumar A, Kumar P, Singh H, Kumar N (2021b) Modulation of plant functional traits under essential plant nutrients during seasonal regime in natural forests of Garhwal Himalayas. Plant Soil 465:197–212

Kumar N, Singh H, Kumar A, Singh AK, Pandey PK, Kumar A (2024) Forest-fire-induced land degradation. In: Mishra G, Giri K, Singh S, Kumar M (eds) Sustainable land management in India. Springer, Singapore. https://doi.org/10.1007/978-981-97-5223-2_4

Laskin A, Laskin J, Nizkorodov SA (2015) Chemistry of atmospheric brown carbon. Chem Rev 115(10):4335–4382. https://doi.org/10.1021/cr5006167

Li F, Zhang X, Roy DP, Kondragunta S (2019) Estimation of biomass-burning emissions by fusing the fire radiative power retrievals from polar-orbiting and geostationary satellites across the conterminous United States. Atm Environ 211:274–287

Liang Y, Stamatis C, Fortner EC, Wernis RA, Van Rooy P, Majluf F, Yacovitch TI, Daube C, Herndon SC, Kreisberg NM, Barsanti KC, Goldstein AH (2022) Emissions of organic compounds from western US wildfires and their near-fire transformations. Atm Chem Phys 22(15):9877–9893

Luyssaert S, Schulze ED, Börner A, Knohl A, Hessenmöller D, Law BE, Ciais P, Grace J (2008) Old-growth forests as global carbon sinks. Nature 455(7210):213–215

Malik A, Kumar K, Kumari G, Kumar N, Singh H (2024) Application of Functional Traits in Modelling Productivity and Resilience Under Climate Change. Springer, Singapore, pp 77–96

McCarty JL, Smith TEL, Turetsky MR (2020) Arctic fires re-emerging. Nat Geosci 13:658–660. https://doi.org/10.1038/s41561-020-00645-5

MoEFCC (2021) India: third biennial update report to the United Nations framework convention on climate change. Ministry of Environment, Forest and ClimateChange, Government of India

Nibleus K, Lundin R (2010) Climate change and mitigation. Ambio 39:11–17

Nirmal, Bakshi M, Singh H (2021) Effect of $CO_2$ elevation on Shisham growth at nursery stage. Ind J Hill Farm 34:272–276

Prakash V, Hunney K, Singh H (2022) Effect of elevated carbon dioxide on growth and development of *Santalum album* L. seedlings inoculated with plant growth promoting microorganisms in Open Top Chambers. Int J Environ Agric Biotechnol 7(5):149–159

Raturi A, Singh H, Kumar P, Shukla N (2022) Characterizing the post-monsoon $CO_2$, $CH_4$, $N_2O$, and $H_2O$ vapor effluxes from a tropical wetland in the Himalayan foothill. Environ Monit Assess 194(50). https://doi.org/10.1007/s10661-021-09721-8

Savita, Sharma SK, Singh H, Singh O (2017) Forest fire: a review. Ind Forester 143(8):719–728

Senf F, Heinold B, Kubin A, Müller J, Schrödner R, Tegen I (2023) How the extreme 2019–2020 Australian wildfires affected global circulation and adjustments. Atm Chem Phys 23(15):8939–8958

Sett T, Nikam BR, Singh H, Purohit S (2022) Effect of drought stress on forest evapotranspiration—a case study on Indian forests. EGU General Assembly 2022. https://doi.org/10.5194/egusphere-egu22-510

Sett T, Nikam BR, Singh H, Purohit S (2024) Hydrological variability in Indian forest ecosystem: analysis of drought resilience, recovery and water use efficiency in moist and dry deciduous forests. In: Singh H (eds) Urban forests, climate change and environmental pollution. Springer, Cham. https://doi.org/10.1007/978-3-031-67837-0_37

Sharma R, Singh H, Kaushik M, Nautiyal R, Singh O (2018) Adaptive physiological response, carbon partitioning and biomass production of *Withania somnifera* (L.) Dunal grown under elevated $CO_2$ regimes. 3 Biotech 8:267. https://doi.org/10.1007/s13205-018-1292-1

Sharma R, Singh H (2021) Alteration in biochemical constituents and nutrients partitioning of *Asparagus racemosus* in response to elevated atmospheric CO2 concentration. Environ Sci Pollut Res 29:6812–6821. https://doi.org/10.1007/s11356-021-16050-3

Sheoran R, Dumka UC, Hyvärinen AP, Sharma VP, Tiwari RK, Lihavainen H, Virkkula A, Hooda RK (2023) Assessment of carbonaceous aerosols at Mukteshwar: a high-altitude (~2200 m amsl) background site in the foothills of the Central Himalayas. Sci Total Environ 866:161334

Singh H, Sharma R, Savita SMP, Kumar M, Verma A, Ansari MW, Sharma SK (2018) Adaptive physiological response of partheniumhysterophorus to elevated atmospheric $CO_2$ concentration. Ind Forester 144:1–14

Singh H, Kumar M (2022) Climate change and its impact on Indian Himalayan Forests: current status and research needs. In: Rani S, Kumar R (eds) climate change. Springer Climate. Springer, Cham, pp 223–242. https://doi.org/10.1007/978-3-030-92782-0_11

Singh H, Kumar N, Kumar M, Singh R (2020a) Modelling habitat suitability of western tragopan (*Tragopan melanocephalus*) a range-restricted vulnerable bird species of the Himalayan region, in response to climate change. Clim Risk Manag 29:100241. https://doi.org/10.1016/j.crm.2020.100241

Singh S, Singh H, Sharma SK, Nautiyal R (2021c) Seasonal variation in biochemical responses of bamboo clones in the sub-tropical climate of Indian Himalayan foothills. Heliyon 7(4):e06859. https://doi.org/10.1016/j.heliyon.2021.e06859

Singh H, Kumar N, Singh R, Kumar M (2023) Assessing the climate change impact on the habitat suitability of the range-restricted bird species (*Catreus wallichii*) in the Indian Himalayan ecosystem. Environ Sci Pollut Res. https://doi.org/10.1007/s11356-023-30789-x

Singh H, Kumar N, Kumar A (2024a) enhancing resource use efficiency in crops through plant functional traits. Springer, Singapore, pp 97–117

Singh M, Jyoti, Kumar N, Singh H (2024b) Plant functional traits assisted crop adaptation to abiotic and biotic stress. Springer, Singapore, pp 239–255

Singh M, Singh H, Kumar A, Kumar M, Barthwal S, Thakur A (2024c) Soil nitrogen availability determines the $CO_2$ fertilization effect on tree species (*Neolamarcki acadamba*): growth and physiological evidence. Environ Sustain 7:53–60. https://doi.org/10.1007/s42398-023-00300-w

Singh H (2021) Process-based carbon sequestration study in relation to the energy-water-carbon flux in a forest ecosystem. In: Kumar P, Singh RK, Kumar M, Rani M, Sharma (eds) Climate impacts on sustainable natural resource management. CRC Press, Taylor & Francis Group, United States, pp 336–361

Singh H (2024a) Climate change and forests: biological perspectives. In: Singh H (ed) Forests and climate change. Springer, Singapore. https://doi.org/10.1007/978-981-97-3905-9_1

Singh H (ed) (2024b) Forests and climate change: biological perspectives on impact, adaptation, and mitigation strategies. Springer, Singapore. https://doi.org/10.1007/978-981-97-3905-9

Singh H (2024c) Climate change and environmental pollution: biological consequences on urban forests. In: Singh H (eds) Urban forests, climate change and environmental pollution. Springer, Cham, Switzerland. https://doi.org/10.1007/978-3-031-67837-0_1

Singh H (ed) (2024d) Urban forests, climate change and environmental pollution: physio-biochemical and molecular perspectives to enhance urban resilience. Springer, Cham, Switzerland. https://doi.org/10.1007/978-3-031-67837-0

Singh H (2024e) Exploring adaptive modulation in plant functional traits and its impact on the productivity of *Acacia auriculiformis* under $CO_2$-enriched environment. Ind Crops Prod 210(2024):118186. https://doi.org/10.1016/j.indcrop.2024.118186

Steffen W, Noble I, Canadell J, Apps M, Schulze ED, Jarvis PG (1998) The terrestrial carbon cycle: implications for the Kyoto Protocol. In IGBP Terrestrial Carbon Working Group, Science 280(5368):1393–1394

Talberth J, Carlson E (2024) Forest carbon tax and reward: regulating greenhouse gas emissions from industrial logging and deforestation in the US. Environ Dev Sustain, 1–22

Veraverbeke S, Delcourt CJ, Kukavskaya E, Mack M, Walker X, Hessilt T, Rogers B, Scholten RC (2021) Direct and longer-term carbon emissions from arctic-boreal fires: a short review of recent advances. Curr Opin Environ Sci Heal 23:100277

Walker XJ, Baltzer JL, Cumming SG, Day NJ, Ebert C, Goetz S, Johnstone JF, Potter S, Rogers BM, Schuur EA, Turetsky MR (2019) Increasing wildfires threaten historic carbon sink of boreal forest soils. Nat 572(7770):520–523

Wan N, Xiong X, Kluitenberg GJ, Hutchinson JMS, Aiken R, Zhao H, Lin X (2023) Estimation of biomass burning emission of NO2 and CO from 2019–2020 Australia fires based on satellite observations. Atmos Chem Phys 23:711–724. https://doi.org/10.5194/acp-23-711-2023

Wang X, Yan J, Tian Q, Li X, Tian J, Zhu C, Li Q (2023) Estimation of forest fire burned area by distinguishing non-photosynthetic and photosynthetic vegetation using triangular space method. Rem Sens 15(12):3115

Yadav SK, Singh H, Nautiyal R, Ginwal HS, Ansari SA, Barthwal S (2019a) Modulation of morpho-physiological responses in *Populus deltoides* by elevated carbon dioxide and temperature. For Sci 66:105–118. https://doi.org/10.1093/forsci/fxz048

Yadav SK, Singh H, Ginwal HS, Barthwal S (2019b) Elevated $CO_2$ enhanced growth and physiological process of *Populus deltoides* Bartr. ex Marsh. Ind Forester 145(1):23–27

# Forest Fire Effects on Structure, Composition, and Functioning of Ecosystem

**Shivam Kumar Sharma and Krishan Kumar**

**Abstract** Forest fires are natural, vital components of many ecosystems, intricately shaping their structure and function globally. They bring immediate devastation but also foster renewal, driving diverse successional stages. Post-fire, habitat shifts, altered species distribution, and ecosystem heterogeneity emerge, impacting community dynamics. Certain species adapt to fire, while others struggle, reshaping trophic interactions and community composition. Ecosystem functions undergo significant shifts; nutrient cycling, carbon storage, and water dynamics are affected. Though fires deplete nutrients initially, they enhance availability over time, influencing productivity and resilience. Altered hydrological patterns, post-fire, can disrupt water quality and quantity downstream. Understanding these complexities is vital for effective forest management. Integrating fire ecology into conservation strategies promotes resilient forests capable of withstanding and adapting to fire, mitigating risks while sustaining biodiversity and ecosystem services. This chapter underscores the intricate relationship between fires and ecosystems, emphasizing the need to integrate fire ecology principles into conservation efforts for healthy, resilient forests amidst changing environmental conditions.

**Keywords** Forest fires · Biodiversity · Ecological succession · Ecosystem dynamics

## 1 Introduction

Forest fires, also known as wildfires, are uncontrolled blazes in forests, grasslands, or other wild lands (Underwood and Held 2012). These fires can originate from natural causes such as lightning strikes, human activities like campfires, discarded cigarettes, or intentional arson. They threaten ecosystems, wildlife, human lives, and property. Forest fires are of different types based on their cause, behavior, and impact (Miyanishi 2001). One classification method categorizes them as surface fires, which

S. K. Sharma · K. Kumar (✉)
School of Environmental Sciences, Jawaharlal Nehru University, New Delhi, India
e-mail: krishan_kumar@mail.jnu.ac.in



H. Singh and N. Kumar (eds.), *Forest Fire and Climate Change*,
https://doi.org/10.1007/978-3-031-89967-6_5

burn through ground-level vegetation; crown fires, which spread through the forest canopy; and ground fires, which smolder in the soil's organic layer (Kumar et al. 2024). Fires can be characterized by their intensity, speed of spread, size, and duration. Understanding the causes and classifications of forest fires is crucial for effective prevention, management, and mitigation strategies. Prevention efforts involve public education, fire bans during dry seasons, and monitoring human activities in fire-prone areas. Management techniques include controlled burns, which reduce fuel buildup and decrease the risk of catastrophic fires (Vasudeva 2018). Mitigation involves early detection systems, efficient firefighting strategies, and restoring burned areas to restore ecosystems. Comprehensively defining and classifying forest fires may help stakeholders develop more targeted approaches to protect natural environments, mitigate risks, and safeguard communities against the devastating impacts of wildfires.

Natural and human-caused forest fires influence ecosystems worldwide. Many complex interplay between natural and anthropogenic factors are responsible for igniting and spreading forest fires. Understanding these elements is crucial for devising effective fire management and mitigation strategies (Singh 2024a). Natural factors, such as lightning strikes, climatic conditions, and vegetation dynamics, significantly influence the occurrence and behavior of forest fires (Adámek et al. 2018). Lightning, as a primary natural ignition source, interacts with prevailing weather patterns, including temperature, humidity, wind, and drought, shaping the likelihood and intensity of fires (Veraverbeke et al. 2017). The composition, structure, and flammability of vegetation play pivotal roles in fire initiation and propagation within ecosystems (Singh et al. 2021c; Yadav et al. 2019b; Singh et al. 2023; Sharma and Singh 2021). Human activities constitute a significant catalyst for forest fires, accounting for a substantial proportion of ignitions globally. Anthropogenic factors include negligence, arson, land-use practices, agricultural actions, infrastructure development, and global climate change (Kelkar and Kumthekar 2019).

The encroachment of human settlements into wildland-urban interfaces escalates the potential for fire occurrence, amplifying risks to both natural habitats and human communities (Singh 2024b). The complex interplay between natural factors and human activities highlights the necessity for employing diverse strategies in managing forest fires. Strategies encompassing fire prevention, preparedness, suppression, and post-fire ecosystem restoration are imperative. Public awareness campaigns, strict enforcement of fire regulations, land-use planning, and early warning systems are pivotal in mitigating human-induced fire incidents. Advancements in technology, including remote sensing, GIS (Geographic Information Systems), and predictive modeling, offer valuable fire monitoring, forecasting, and management tools (Nayak and Zlatanova 2008). Integrating traditional ecological knowledge with modern scientific techniques enhances our ability to predict fire behavior, assess risks, and formulate adaptive strategies in fire-prone regions (Vázquez-Varela et al. 2022).

Understanding the effects of forest fires on ecosystems is paramount due to their significant impact on the structure, composition, and functioning of natural environments. These fires influence ecological processes, biodiversity, and landscape dynamics, making their study essential for effective ecosystem management and conservation (He et al. 2019; Kumar et al. 2021b; Singh 2024c). The impacts of forest fires on ecosystem structure encompass alterations in vegetation patterns, changes in species composition, and modifications to the physical environment (Kumar et al. 2020a, b, 2021a). Fires can remove vegetation cover, change canopy structures, and affect plant communities' diversity and distribution. The severity and frequency of fires often dictate the regeneration capacity of ecosystems, influencing the establishment of different plant species and altering overall biodiversity (Raturi et al. 2022; Sett et al. 2022; Singh et al. 2020a; Apurva et al. 2017).

Additionally, the composition of ecosystems is profoundly influenced by fire events. Some plant species have evolved adaptations to thrive post-fire, while others may struggle to recover, leading to species dominance and ecological balance shifts (Popkin 2021). Understanding these shifts is crucial for predicting long-term ecosystem changes and identifying species most vulnerable to fire-induced alterations (Singh 2024d). Forest fires impact ecosystem functioning and various ecological processes by altering nutrient cycling, soil properties, and habitat availability (McLauchlan et al. 2020). While fires can release nutrients locked in vegetation, they can also result in soil erosion and nutrient loss, affecting soil fertility and water quality (Verma and Jayakumar, 2012). Furthermore, fire-mediated changes in vegetation structure and composition can directly impact wildlife habitats, influencing species interactions and ecosystem resilience.

## 2 Ecological Impacts of Forest Fires

Forest fires are natural ecological phenomena that shape ecosystems' structure, composition, and functioning throughout the globe. These fires' immediate and long-term impacts on flora and fauna are multifaceted, influencing biodiversity, ecosystem dynamics, and habitat availability (Singh 2024d). This chapter synthesizes forest fires' complex effects on flora and fauna, highlighting their interdependent relationships within ecosystems. Forest fires' immediate effects on flora involve destroying vegetation, altering the physical structure, and reducing resource availability (Malik et al. 2024; Singh 2024e). Although certain plant species display resilience and have adapted to fire, others encounter significant obstacles, resulting in temporary habitat loss and fragmentation (Midgley and Bond 2013). These fires also have diverse immediate impacts on fauna, causing displacement, mortality, and altered behavioral patterns. Species-specific responses vary, with some adapted to fire-dependent ecosystems, while others face severe consequences, especially in highly affected regions (Barwa and Singh 2024).

However, forest fires also initiate ecological processes that drive long-term impacts. Post-fire environments often witness ecological succession, promoting habitat regeneration and providing opportunities for pioneer species to thrive (Zymaroieva et al. 2023; Kaushik and Singh 2024). This succession gradually restores ecosystem functionality and promotes biodiversity recovery. Fire-induced changes in vegetation structure can create diverse habitats, benefitting specific fauna by offering new niches and foraging opportunities. Nonetheless, altered landscapes and vegetation compositions post-fire can challenge the re-establishment of specific flora and fauna species, leading to potential shifts in community composition. The heightened frequency and intensity of fires affected by climate change present further obstacles, potentially worsening the immediate and enduring effects on plant and animal life (Nitschke and Innes 2007; Sett et al. 2024). Understanding the intricate relationships between forest fires, flora, and fauna is crucial for effective conservation and management strategies. Utilizing adaptive strategies that account for the ever-changing dynamics of ecosystems following fires is crucial in lessening adverse effects on biodiversity and fostering resilience amidst evolving fire patterns.

Forest fires have been acknowledged as a natural ecological phenomenon that influences forest ecosystems' biodiversity and species composition (Brockerhoff et al. 2017). These fires exert multifaceted impacts, significantly altering the intricate balance of species diversity and composition. The aftermath of a forest fire presents a dynamic landscape where the extent and severity of the blaze play pivotal roles in determining the scale of ecological transformations. Biodiversity, a cornerstone of ecosystem health, experiences immediate and long-term consequences following a fire event (Eisenhauer et al. 2019). While certain species succumb to the direct effects of flames, others display remarkable adaptations to fire, often thriving in post-fire environments. The complex interaction among species capable of withstanding fire, those adapted to it, and those vulnerable to it influence the path of ecosystem recovery. Additionally, fire severity influences habitat heterogeneity, fostering niches for diverse species to recolonize and regenerate (Dove and Hart 2017).

Species composition undergoes a discernible shift because of forest fires. Fire's selective pressures can favor species that exhibit resilience to heat, possess fire-adapted traits such as serotiny or resprouting abilities, or rely on fire-dependent reproductive strategies for regeneration. Conversely, species less equipped to cope with fire-induced changes may experience declines or localized extinctions. These alterations in species composition may lead to novel species assemblages, impacting trophic interactions and ecological relationships (Jackson et al. 2017). While fires contribute to nutrient cycling and stimulate rejuvenation, severe blazes can disrupt critical ecosystem processes, altering nutrient availability, water quality, and carbon sequestration capacities. The immediate post-fire period may witness fluctuations in primary productivity, affecting food webs and ecosystem stability (Hudiburg et al. 2023).

Forest fires are crucial in shaping ecosystem dynamics by influencing various facets of forest structure, composition, and functioning. While often perceived as destructive, fires are natural phenomena that can bring short-term disturbances and long-term ecological benefits to forested landscapes (Keane et al. 2008). One of the primary impacts of forest fires lies in altering the ecosystem's structure. Fires can clear out accumulated biomass, such as dead vegetation and woody debris, which helps in nutrient recycling and provides space for new samplings. This disruption alters plant density, canopy coverage, and spatial organization, consequently impacting the forest's overall structure and biodiversity. Additionally, it influences the composition of plant communities. Certain species, often called fire-adapted species, have evolved adaptations to thrive post-fire, while others may struggle or perish (Johansson et al. 2021). The balance between these responses dictates the post-fire flora composition, leading to shifts in dominance and diversity patterns. The functioning of an ecosystem, including nutrient cycling, carbon storage, and water dynamics, is intricately tied to fire regimes. Fires release nutrients stored in vegetation back into the soil, stimulating new growth and enhancing the overall productivity of the ecosystem. It also affects how forests interact with the atmosphere, local climates, and global carbon cycles. Although the immediate aftermath of a forest fire may appear devastating, such occurrences are essential components of natural forest cycles. They promote regeneration, stimulate biodiversity, and strengthen the ecosystem's resilience against upcoming disturbances (Löf et al. 2019). Altered fire regimes due to anthropogenic activities disrupt these natural cycles, leading to ecological imbalances and increased vulnerability to severe wildfires. Comprehending the significance of forest fires in ecosystem dynamics is essential for making well-informed decisions regarding land management. Embracing controlled burns, restoring natural fire regimes, and implementing strategies that consider the ecological importance of fire can help maintain healthy, resilient forests capable of adapting to changing environmental conditions.

## 3 Structural Changes in Forests After Fires

Forest fires are crucial in shaping ecosystem dynamics by significantly impacting vegetation structure and driving successional processes. While seemingly destructive, these events are integral to many ecosystems worldwide, influencing vegetation patterns, biodiversity, and landscape-level processes. The alteration of the physical environment marks immediate post-fire impacts on vegetation structure. The degree of vegetation loss, ranging from complete canopy reduction in high-intensity fires to moderate burning in low-intensity fires, is determined by the intensity and severity of the fire, resulting in different levels of organic matter remaining on the forest floor. Which resets the ecological clock, initiating a series of environmental responses that dictate the subsequent stages of vegetation succession (González-Pérez et al. 2004).

Succession following a forest fire unfolds through distinct stages. Initially, pioneer species, often adapted to fire-prone environments, rapidly colonize the burned area. These early successional species are typically fast-growing and fire-resistant or possess traits that aid post-fire establishment. They stabilize the soil, facilitate nutrient cycling, and create suitable microhabitats for colonizing other species. As time progresses, the composition of the plant community undergoes dynamic changes. The post-fire environment becomes increasingly hospitable to a broader array of species, fostering diversity and complexity in vegetation structure. Shrubs, grasses, and herbaceous plants initially dominate, gradually giving way to shrublands and young forests (Kruger, 1984). Over time, the community matures, re-establishing the pre-fire climax species and restoring the ecosystem to its former state. It is essential to recognize that while the occurrence of a fire is significant, it is not the sole determinant of its effects on vegetation structure and succession (Sharma et al. 2018; Jangra et al. 2024; Dhyani et al. 2021; Singh 2024a, b, c, d, e). Climate, soil conditions, topography, and human involvement substantially shape post-fire vegetation recovery and succession patterns (Zhong et al. 2023).

Forest fires significantly impact the structure and dynamics of vegetation, leading to distinct changes in canopy cover and tree density in affected ecosystems. Post-fire, vegetation structure undergoes a series of modifications influenced by various factors such as fire severity, ecosystem type, and regeneration mechanisms. Following a forest fire, there is a significant decrease in canopy cover due to the loss of foliage and tree crowns (Hood et al. 2018). High-intensity fires often damage the canopy, causing widespread tree mortality and opening the previously dense canopy cover. This initial reduction in canopy cover exposes the forest floor to increased light availability, promoting the growth of understory vegetation and the germination of fire-adapted plant species. Fire-induced changes in tree density are observed due to selective mortality and regeneration patterns. While some trees perish in the fire, others, particularly fire-adapted species, may survive or regenerate from seeds or root systems. This process contributes to alterations in the distribution and density of trees within the affected area. The succession processes following a fire are pivotal in dictating the pace and scope of recovery, shaping the emergence and development of fresh vegetation (Spatola et al. 2023). Over time, as the ecosystem undergoes succession, canopy cover begins to recover through the expansion of regenerating vegetation. Pioneer species, often fast-growing and adapted to disturbance, tend to dominate early successional stages, contributing to increased canopy cover. As succession progresses, the composition and structure of the vegetation community gradually evolve, with more shade-tolerant species becoming established, leading to a gradual restoration of canopy cover and changes in tree density.

Forest fires shape ecosystem dynamics by profoundly affecting vegetation structure, composition, soil characteristics, and nutrient cycling. The severity and extent of the fire, alongside factors like fire recurrence and ecosystem resilience, rule the degree of vegetation alteration. Fires often lead to immediate changes in vegetation structure, removing canopy cover, altering species dominance, and creating niches for pioneering species. Over time, this disturbance triggers a sequence of ecological responses, initiating secondary succession (Johnson and Miyanishi 2021). Early successional species with adaptive traits for rapid regeneration and colonization tend to thrive initially, gradually paving the way for re-establishing a more diverse and complex community structure.

In parallel, forest fires greatly influence soil characteristics and nutrient cycling. Fire-induced alterations in soil properties like pH, organic matter content, and nutrient availability significantly impact ecosystem functioning. Fire heat can modify soil structure, water retention capacity, and microbial activity. While high-intensity fires may increase soil hydrophobicity, hinder water infiltration, and promote erosion, moderate fires can facilitate nutrient release by breaking down organic matter, enhancing nutrient availability for post-fire vegetation recovery (Rego et al. 2021).

The severity of the fire and its duration can also lead to nutrient loss through volatilization and ash displacement, affecting the overall fertility of the soil. The interaction between fire and nutrient cycling is intricate, influencing the biogeochemical cycles within the ecosystem. Fire-induced changes in nutrient availability, particularly nitrogen and phosphorus, impact plant recolonization and growth patterns, further shaping the successional trajectories (Knelman 2015). Understanding the intricate relationships between fire, vegetation dynamics, soil properties, and nutrient cycling is vital for effective ecosystem management and post-fire restoration strategies. Further investigation is needed to understand better the enduring effects of fires on ecosystem resilience, species interactions, and the complex feedback mechanisms that influence post-fire recovery (Singh et al. 2024a, b; Yadav et al. 2019a; Singh 2021; Singh et al. 2018). This will facilitate the development of more holistic strategies for conserving and restoring landscapes impacted by fire.

## 4 Changes in Forest Composition Post-fire

The role of forest fires in shaping ecosystem dynamics is multifaceted, influencing vegetation structure, composition, succession, species dominance, and biodiversity. Fire, a natural ecological process, is pivotal in rejuvenating and transforming ecosystems. The critical ecosystem recovery and development determinants are post-fire impacts on vegetation structure and succession. Following a fire event, vegetation structure undergoes significant alterations. Immediate effects include the removal of aboveground biomass, altering canopy cover, and creating openings in the ecosystem (Brown and Smith 2000). This disturbance triggers ecological responses, paving the way for primary succession. Pioneer species, adapted to post-fire conditions, quickly colonize the burned area, initiating the process of ecosystem regeneration. After a fire, succession involves a series of vegetation transformations over time.

In the beginning, the rapid growth of herbaceous plants and shrubs took over, using accessible resources to solidify the soil. Over successive years, there is a transition towards woody species, facilitating the restoration of pre-fire community structure and ecosystem function. The impact of fire on species dominance and diversity is complex. While some species are sensitive to fire and face population decline, others exhibit resilience or require fire for regeneration. Fire-adapted species possess adaptations such as fire-resistant bark, serotinous cones, or belowground structures, allowing for rapid re-sprouting (Nowacki and Abrams 2008). The post-fire environment creates niches favoring certain species, potentially leading to species composition and diversity shifts. Species richness and diversity patterns post-fire exhibit temporal and spatial variability. Initially, species diversity might decline due to immediate habitat destruction. However, diverse habitats emerge over time, fostering the colonization of various species and potentially enhancing overall biodiversity.

Following a forest fire, vegetation experiences varying degrees of impact that affect post-fire succession and regeneration patterns. The immediate aftermath of fire frequently leads to modifications in vegetation structure due to the depletion of above-ground biomass, shifts in soil characteristics, and the emergence of varied microenvironments. The impact of fire on vegetation structure is multifold, impacting the vertical and horizontal arrangement of plant communities, altering canopy cover, and influencing species diversity. The extent of vegetation loss may vary depending on the severity of the fire, potentially resulting in substantial alterations to species composition and the emergence of pioneering species well-suited to environments prone to fire. Post-fire succession refers to the sequence of changes in vegetation over time following a fire event (Duguy and Vallejo 2008). It involves establishing and developing plant communities, from the colonization of pioneering species to the eventual recovery of mature ecosystems. Various factors, including fire severity, soil conditions, climatic factors, and the presence of seed banks or vegetative propagules, influence successional patterns. Regeneration patterns of vegetation post-fire exhibit a dynamic process where certain plant species may rely on fire for seed germination or sprouting. Some species have adaptations like serotiny (releasing seeds in response to fire-related cues) or resprouting mechanisms that facilitate rapid recovery after fire (Lamont 2022). These regeneration patterns often result in mosaic landscapes with patches of different-aged vegetation, contributing to increased biodiversity and structural complexity within the ecosystem.

Forest fires reform and change ecosystem dynamics, significantly influencing vegetation structure, composition, and functioning. While initially devastating, fires catalyze ecological processes that drive succession, regeneration, and biodiversity in various landscapes. Forest fires intensely affect vegetation structure and succession by triggering various ecosystem adaptations and transformations. Post-fire, the effects on vegetation structure are multifaceted. Fire alters soil's physical and chemical properties, impacting seed banks and germination, thereby influencing species composition and diversity (Chick et al. 2015). Succession patterns following a fire event often reveal a transition from early successional species, characterized by rapid growth and colonization, to late-successional species, which establish more slowly but contribute to long-term ecosystem stability. Understanding these post-fire successional trajectories is crucial in predicting ecosystem recovery and resilience. Many species have evolved diverse adaptations to survive and thrive in fire-prone environments (Prakash et al. 2022; Nirmal et al. 2021; Singh and Kumar 2022; Savita et al. 2017). Some plants have developed fire-resistant features such as thick bark or underground structures, ensuring their survival and allowing rapid post-fire regeneration (Vallejo et al. 2011). Other species depend on fire for seed germination or rely on post-fire conditions to outcompete other vegetation. The adjustments ecosystems make after fires are crucial in enhancing their resilience and capacity to manage repeated fire occurrences. Furthermore, fire-mediated disturbances create a mosaic landscape, fostering habitat heterogeneity that benefits many flora and fauna (Reilly et al. 2021). Certain species, particularly fire-adapted ones, thrive in post-fire environments, capitalizing on newly available resources and niches. These adaptations and fire's influence on nutrient cycling and ecological processes contribute to fire-prone ecosystems' overall health and diversity.

# 5 Functional Consequences of Forest Fires

Forest fires are pivotal in shaping ecosystem dynamics, exerting profound effects on vegetation structure, succession patterns, carbon, and nutrient cycling within forest ecosystems. Post-fire impacts on vegetation structure and succession are multi-faceted, initially triggering dramatic alterations in the composition and distribution of plant species. Fires often result in removing existing vegetation, clearing the way for new growth, and facilitating ecological succession (Miller et al. 2013). Adapted to post-fire conditions, Pioneer species flourish, gradually establishing more complex vegetation communities. These successional changes contribute to the restoration and diversification of the ecosystem, impacting habitat suitability for various organisms. Additionally, forest fires significantly influence carbon and nutrient cycling processes (Stuart Chapin III et al. 2009). Burning in fires releases carbon previously stored, thus adding to greenhouse gas emissions and impacting global carbon balances. Simultaneously, the charred organic matter from burnt vegetation becomes incorporated into the soil, influencing nutrient availability and soil fertility. While immediate nutrient loss due to fire can be substantial, the subsequent decomposition of fire-affected organic matter contributes nutrients, fostering new growth and aiding in ecosystem recovery (Pérez-Cabello et al. 2009). However, the impact of fire on carbon and nutrient cycling is complex and varies based on fire severity, frequency, ecosystem type, and local environmental conditions. High-severity fires may increase carbon emissions and nutrient losses, potentially impairing ecosystem recovery and resilience.

Conversely, low to moderate-severity fires promote a more balanced nutrient release and facilitate quicker ecosystem regeneration (Certini 2005). Understanding the intricacies of post-fire effects on vegetation, succession, carbon, and nutrient cycling is crucial for effective forest management and conservation strategies. Proper management practices, including controlled burns and monitoring post-fire recovery, can mitigate detrimental impacts while harnessing the regenerative potential of fires to maintain healthy and resilient forest ecosystems (Rego et al. 2021). Continuously research is crucial for thoroughly understanding the connections between forest fires and ecosystem dynamics. This understanding is vital for making well-informed conservation and land management decisions, particularly as fire patterns and environmental conditions evolve.

Forest fires are natural ecological processes that trigger intricate changes in the landscape, altering vegetation composition, structure, and ecosystem functioning. The influence of forest fires on the water cycle and hydrology is substantial. Fire alters soil properties, affecting infiltration rates and water retention capacities (Úbeda and Outeiro 2009). The loss of vegetative cover increases surface runoff and erosion, potentially leading to increased sedimentation in water bodies (Zuazo and Pleguezuelo 2009). Post-fire changes in vegetation structure can influence evapotranspiration rates, impacting regional water availability. The alteration in vegetation composition, with varying root systems of regenerating plants, can modify soil moisture retention and nutrient cycling, consequently affecting hydrological processes.

This complex interplay between forest fires and ecosystem dynamics is a big challenge for land managers and authorities regarding conservation strategies. Mitigation efforts, such as prescribed burns and reforestation practices, aim to harness the positive aspects of fire while minimizing its detrimental effects on ecosystems and hydrological systems. Embracing a holistic approach that integrates ecological, hydrological, and land management perspectives is imperative to strike a balance between harnessing the environmental benefits of fire and mitigating its potential hazards on ecosystems and water resources.

Forest fires wield a crucial yet intricate influence on ecosystem dynamics, affecting vegetation structure, composition, and functioning. The impacts on vegetation structure and succession following a fire encompass immediate shifts in plant community composition, patterns of regeneration, and enduring alterations in ecosystem trajectories. Following a fire, vegetation structure undergoes a transformation marked by a species dominance and abundance shift (Hart and Chen 2008). Pioneer species often take precedence in the initial post-fire environment, their rapid growth facilitating ecosystem recovery. Over time, a dynamic succession process unfolds, gradually restoring pre-fire community composition but with potential alterations in species diversity and distribution. This succession not only influences vegetation structure but also impacts ecosystem functioning. Changes in plant communities following a fire could impact processes like nutrient cycling, soil characteristics, and water retention, thereby influencing essential ecosystem functions such as carbon storage, soil integrity, and water management (Tepley et al. 2018). Fire-induced changes in vegetation structure can also impact wildlife habitat, creating challenges and opportunities for various species. While some species face habitat loss or temporary displacement, others thrive in the post-fire landscape, capitalizing on newly available resources and habitats.

## 6 Human Responses and Management Strategies

Anthropogenic interventions and climate change have led to the severity and frequency of wildfires, resulting in significant ecological and economic impacts (Fried et al. 2004). Post-fire restoration and rehabilitation efforts are essential for mitigating these consequences and restoring ecosystems to functional states. After a fire, the impact on vegetation structure and succession is manifold. The fire's intensity determines the severity of vegetation damage, altering community composition, reducing biodiversity, and disrupting ecological processes. Its most immediate effects often burn the susceptible vegetation and the destruction of wildlife habitat and soil degradation. Adapted to post-fire conditions, Pioneer species are the first to colonize the burned area (Souza-Alonso et al. 2022). These species facilitate soil stabilization and nutrient cycling and create favorable conditions for subsequent vegetation. Over time, a diverse array of plant species is established, gradually restoring the ecosystem's structure and function. Wildlife habitat faces immediate challenges following a forest fire. The loss of vegetation cover and food sources directly impacts wildlife populations, often leading to displacement or decline. However, post-fire landscapes also offer new habitats for certain species, attracting wildlife that thrives in early successional stages. Balancing the needs of different species during recovery becomes critical to maintaining biodiversity and ensuring ecosystem resilience. Forest fires also impact ecosystem services, affecting water quality, carbon sequestration, and soil fertility. The alteration of these services can have far-reaching consequences for human well-being (Lal 2013). Post-fire rehabilitation strategies mitigate these effects by implementing erosion control measures, reforestation efforts, and promoting natural regeneration. Engaging local communities, employing innovative technologies, and leveraging scientific knowledge are integral to successful post-fire restoration initiatives.

The intricate relationship between forest conservation and fire risk mitigation is critical in contemporary forest management strategies. Forest fires, while natural components of many ecosystems, can pose significant challenges to conservation efforts and ecosystem stability. This chapter explores the delicate balance required in managing forests, weighing the conservation imperatives against mitigating the risks posed by fires. Firstly, the role of forest fires in shaping ecosystem dynamics is elucidated, acknowledging their significance in maintaining ecosystem health and promoting biodiversity. The intensification of climate change has amplified fire frequency and severity, resulting in heightened challenges for conservationists and land managers. Understanding the impact of fires on vegetation structure and succession is imperative. Fires can disrupt the existing vegetation patterns, influencing species composition and altering successional pathways, necessitating a comprehensive understanding of post-fire impacts on ecosystems.

Moreover, fires can jeopardize essential ecosystem services like water regulation, soil fertility, and carbon sequestration, requiring a nuanced approach to fire management strategies to preserve these services. Effective post-fire restoration and rehabilitation efforts are pivotal in mitigating the detrimental impacts of fires. Implementing reforestation, habitat restoration, and erosion control measures are crucial steps toward ecosystem recovery and resilience-building (del Rio 2012). However, challenges exist in balancing the need for intervention with allowing natural regeneration processes to occur. Finally, balancing conservation initiatives and fire risk mitigation is paramount. Strategies like prescribed burning, land-use planning, and community engagement can help reduce fire risks while preserving essential ecological values (Mockrin et al. 2020). Nevertheless, finding this equilibrium requires interdisciplinary collaboration, incorporating scientific knowledge, indigenous practices, and community perspectives to develop comprehensive forest management policies.

## 7 Future Perspective and Challenges

Forest fires have not only been a fundamental element of ecosystems, sculpting terrains and impacting biodiversity for countless millennia, but also challenge ecosystems, economies, and communities worldwide. As climate change intensifies, the frequency, intensity, and duration of forest fires are expected to increase, posing a considerable threat to biodiversity, air quality, human health, and infrastructure. As we look to the future, understanding the effects of forest fires on ecosystems' structure, composition, and functioning becomes increasingly crucial for conservation and management efforts. Here, we have discussed the future perspectives and challenges of forest fires, highlighting critical areas of concern and potential avenues for mitigation and adaptation.

**Current Trends and Projections**: Forest fires have risen globally, with recent years witnessing unprecedented wildfires in various regions, from the Amazon rainforest to the Australian bushlands. In India, there has been an uptrend in forest fire events since 2001, as shown in Fig. 1. Climate change exacerbates this trend by altering weather patterns, increasing temperatures, and prolonging drought periods, creating conditions conducive to fire ignition and spread. According to projections, these trends are expected to continue, with some regions experiencing more frequent and severe fire seasons.

**Ecological Impacts**: Forest fires have profound ecological consequences, altering ecosystems, threatening biodiversity, and disrupting natural processes. Forest fires can potentially destroy habitats, cause species loss, cause soil erosion, and cause alterations in vegetation composition. Moreover, releasing carbon dioxide and other greenhouse gases during fires exacerbates climate change, creating a feedback loop that further intensifies fire risks.

**Socio-Economic Implications**: The socio-economic impacts of forest fires are extensive, affecting industries such as agriculture, forestry, tourism, and infrastructure. Direct costs include firefighting efforts, property damage, and loss of livelihoods. Indirect costs encompass health issues related to smoke inhalation, reduced water quality due to ash runoff, and long-term environmental degradation. Furthermore, the displacement of communities and disruption of essential services compound the socio-economic toll of forest fires.

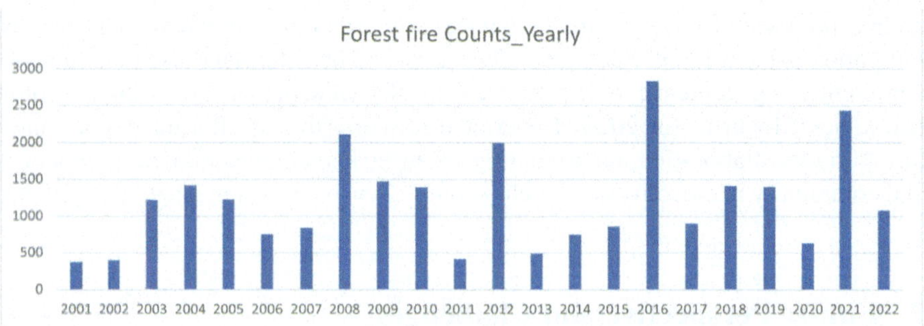


**Fig. 1** Forest fire count has been trending in the foothill area of the Indian Himalayas over the past few years

**Challenges**: Several challenges hinder effective forest fire management and mitigation efforts. These include inadequate funding for prevention and suppression, limited monitoring and early warning systems resources, and socio-political barriers to implementing sustainable land management practices. Additionally, the complexity of fire behavior and the vastness of forested landscapes pose logistical challenges for fire management agencies.

**Future Perspectives and Solutions**: Addressing the future challenges of forest fires requires a multi-faceted approach integrating science, policy, and community engagement. Key strategies include:

- *Investment in Prevention and Preparedness*: Allocating resources towards proactive measures such as prescribed burns, fuel reduction, and community education can mitigate fire risks and enhance resilience.
- *Enhanced Monitoring and Early Warning Systems*: Leveraging advancements in technology, including satellite imagery, drones, and sensor networks, can improve real-time monitoring of fire activity and facilitate timely response efforts.
- *Integrated Land Management*: Adopting holistic approaches to land management that prioritize ecosystem health, biodiversity conservation, and fire resilience can reduce the vulnerability of landscapes to catastrophic fires.
- *Community Engagement and Capacity Building*: Empowering local communities through training, outreach, and participatory decision-making fosters ownership and collaboration in fire management efforts.
- *Policy Reform and Collaboration*: Strengthening fire prevention, land use planning, and disaster management policies while fostering collaboration among stakeholders at various levels is crucial for effective fire governance.

Forest fires represent a complex and evolving challenge with far-reaching implications for ecosystems, economies, and societies. Addressing these challenges requires concerted local, national, and global efforts guided by a shared commitment to sustainability, resilience, and environmental stewardship. Adopting proactive strategies, investing in innovation, and fostering collaboration can mitigate the impacts of forest fires and build a more resilient future for future generations.

## 8 Case Studies and Research Advances

Forest fires play a crucial role in molding ecosystem dynamics, influencing the arrangement and makeup of vegetation and the progression of ecological succession. The impact of fire on vegetation is multifaceted, touching various aspects of the ecosystem, such as species diversity, forest architecture, and nutrient cycling. Following a fire, the immediate effects on vegetation structure are often profound. The intense heat generated by the blaze can lead to the loss of plant life and reshape the physical layout of the landscape. In some instances, fires clear dense undergrowth and dead vegetation, fostering the growth of species adapted to fire and revitalizing ecosystems. Nonetheless, severe fires can devastate critical habitats and trigger soil erosion, jeopardizing the stability of the landscape. This discourse delves into some of the most devastating forest fire events witnessed in the twenty-first century.

Erni et al. (2024) discuss the 2023 Canadian wildfires, introducing a standardized method to assess wildfire risk nationwide, focusing on buildings and human populations. The study emphasizes understanding the interaction between fire hazards and physical vulnerability. High and very high-risk areas cover 32.3% and 6.3% of the land, respectively, with 5.8% of buildings at high risk and 0.6% at very high risk. Approximately 283,200 individuals reside in high-risk areas, with 30,500 in very high-risk zones. Indigenous communities on reserves are particularly vulnerable, with 18.9% situated in higher-risk regions. This research provides crucial data for national wildfire risk management policies. Canosa et al. (2023) examine the 2021 Russian wildfires, highlighting how climate change has intensified wildfire activity in the Russian Arctic, prompting regional authorities to implement adaptation measures. The study reviews 12 adaptation measures implemented between 2008 and 2020, advocating for increased research and better communication to expedite adaptation efforts. Safford et al. (2022) analyze the unprecedented 2020 California wildfires, identifying factors driving high-severity burning and their implications. The study reveals 1.74 million hectares burned, economic losses of over $19 billion, and 33 direct fatalities. Despite collaborative efforts, addressing wildfire severity trends and their impacts requires a shift in management strategies, hindered by climate change and other challenges.

Filkov et al. (2020) detail the extreme 2019–2020 Australian bushfire season, known as the Black Summer fires, which devastated millions of hectares and residences and claimed numerous lives. The fires highlight long-term impact trends and global concerns over smoke emissions, with recovery expected to be prolonged and impactful on affected regions. Feng et al. (2021) scrutinize the 2019 Amazon rainforest wildfires, emphasizing the impact of deforestation and fires on biodiversity. Relaxed enforcement in 2019 reversed progress, endangering threatened species and highlighting the role of forest policies in biodiversity preservation. Gupta et al. (2018) document the 2016 Uttarakhand forest fires, attributing escalating frequency to human activities and climate change. Satellite imagery aids in real-time assessment and mitigation planning to minimize ecological damage. Devisscher et al. (2016) investigate the 2010 Bolivia forest fires, highlighting the consequences of recurrent wildfires on biomass loss and species composition shifts. This suggests a transition towards a more fire-adapted state, necessitating forest management adjustments due to drier climatic conditions and frontier development. Fire shapes ecosystems worldwide, influencing various ecological processes and dynamics. Contemporary research in fire ecology encompasses diverse areas, reflecting a growing understanding of its multifaceted impacts. We aim to encapsulate the current research trends in understanding fire ecology, highlighting vital focal points and emerging perspectives.

- Fire Regimes and Ecosystem Resilience: Recent studies emphasize the significance of fire regimes in determining ecosystem resilience. Fire frequency, severity, and spatial patterns reveal their effects on biodiversity, vegetation structure, and landscape heterogeneity. Understanding these patterns aids in predicting ecosystem responses to fires and their capacity for recovery (Johnstone et al. 2016).

- Climate Change Interactions: With escalating concerns regarding climate change, research endeavors increasingly explore the interplay between fire dynamics and climatic shifts. Studies investigate how changing temperatures, precipitation patterns, and fuel conditions affect fire behavior, frequency, and intensity. This nuanced understanding is critical in projecting future fire occurrences and their ecological consequences (Flannigan et al. 2009).
- Community Dynamics and Succession: Post-fire ecological succession and community dynamics are focal points for contemporary research. Investigations explore the intricate relationships between fire severity, species composition, and ecosystem recovery trajectories. They understand that the mechanisms governing post-fire regeneration aid in predicting species assemblages and community structures in fire-prone landscapes.
- Wildlife Responses and Habitat Restoration: Efforts to comprehend wildlife responses to fire disturbance and subsequent habitat restoration are gaining momentum. The research assesses the impacts of fire on wildlife habitat, species distributions, and population dynamics. Additionally, studies explore strategies for rehabilitating fire-affected landscapes to enhance habitat quality and connectivity for diverse fauna (Converse et al. 2006).
- Human Dimensions and Fire Management: Understanding human dimensions in fire ecology has garnered attention, recognizing the influence of human activities on fire regimes. The research evaluates socio-ecological aspects, including fire management strategies, community perceptions, and the role of indigenous knowledge in mitigating fire risks and promoting ecosystem resilience.

## 9 Conclusion

Forest fires profoundly affect ecosystem dynamics, significantly altering their structure, composition, and functioning. While fires are a natural component of many ecosystems and can even be necessary for renewal and regeneration, the frequency and intensity of wildfires have increased due to human activities and climate change, which lead to biodiversity loss, changes in species composition, and disruptions in ecological processes. It draws attention to various case studies, including the devastating 2023 Canadian wildfires and recurrent fires in different regions such as Russia, California, Australia, the Amazon rainforest, India, and Bolivia, which collectively demonstrate the significant threats wildfires pose to ecosystems, infrastructure, and human lives. These studies highlight the factors driving wildfires, including climate changes, land use patterns, vegetation dynamics, and socio-economic conditions. However, they can also create opportunities for ecological succession, promoting the growth of fire-adapted species and enhancing habitat diversity. Comprehending the intricate interplay of fire patterns and ecosystem processes is paramount in successful forest stewardship and preservation. It involves deploying approaches that balance fire suppression, controlled burns, and habitat rehabilitation. By adopting integrated approaches that consider ecological and societal needs, we can mitigate the negative impacts of wildfires and foster resilient ecosystems capable of withstanding future disturbances. Moreover, considering fire ecology in conservation planning helps mitigate risks associated with climate change-induced wildfires, thereby preserving biodiversity and maintaining ecosystem health. This holistic approach to conservation decision-making supports resilient and healthy ecosystems by comprehensively understanding and integrating natural processes.

## References


Adámek M, Jankovská Z, Hadincová V, Kula E, Wild J (2018) Drivers of forest fire occurrence in the cultural landscape of Central Europe. Landscape Ecol 33:2031–2045

Apurva K, Singh H, Kumar H (2017) Elucidating morphological and growth changes in arjun tree (*Terminalia arjuna*) grown under elevated temperature condition. Trends Biosci 10(1):332–336

Barwa S, Singh H (2024) Climate-induced changes in forest ecosystems of Indian Himalayan Region: a meta-analysis. In: Singh H (ed) Urban forests, climate change and environmental pollution. Springer, Cham. https://doi.org/10.1007/978-3-031-67837-0_39

Brockerhoff EG, Barbaro L, Castagneyrol B, Forrester DI, Gardiner B, González-Olabarria JR, Lyver PB, Meurisse N, Oxbrough A, Taki H, Thompson ID, van der Plas F, Jactel H (2017) Forest biodiversity, ecosystem functioning, and the provision of ecosystem services. Biodivers Conserv 26:3005–3035

Brown JK, Smith JK (2000) Wildland fire in ecosystems: effects of fire on flora. Gen. Tech. Rep. RMRS-GTR-42-vol 2. Ogden, UT: US Department of Agriculture, Forest Service, Rocky Mountain Research Station, p 257, 42

Canosa IV, Biesbroek R, Ford J, McCarty JL, Orttung RW, Paavola J, Burnasheva D (2023) Wildfire adaptation in the Russian Arctic: a systematic policy review. Clim Risk Manag 39:100481

Certini G (2005) Effects of fire on properties of forest soils: a review. Oecologia 143:1–10

Chick MP, Cohn JS, Nitschke CR, York A (2015) Lack of soil seed bank change with time since fire: relevance to seed supply after prescribed burns. Int J Wildland Fire 25(8):849–860

Converse SJ, Block WM, White GC (2006) Small mammal population and habitat responses to forest thinning and prescribed fire. For Ecol Manage 228(1–3):263–273

Devisscher T, Malhi Y, Landívar VDR, Oliveras I (2016) Understanding ecological transitions under recurrent wildfire: a case study in the seasonally dry tropical forests of the Chiquitania, Bolivia. For Ecol Manage 360:273–286

Dhyani K, Kuniyal HB, Singh H, Sobha (2021) Growth and physiological potential of *Terminalia arjuna* under elevated $CO_2$ levels in Open top chamber condition. J Appl Nat Sci 13(3):1121–1126. https://doi.org/10.31018/jans.v13i3.2490

Dove NC, Hart SC (2017) Fire reduces fungal species richness and in situ mycorrhizal colonization: a meta-analysis. Fire Ecol 13:37–65

Duguy B, Vallejo VR (2008) Land-use and fire history effects on post-fire vegetation dynamics in eastern Spain. J Veg Sci 19(1):97–108

Eisenhauer N, Bonkowski M, Brose U, Buscot F, Durka W, Ebeling A, Fischer M, Gleixner G, Heintz-Buschart A, Hines J, Jesch A, Schmid B (2019) Biotic interactions, community assembly, and eco-evolutionary dynamics as drivers of long-term biodiversity–ecosystem functioning relationships. Res Ideas Outcomes Open Sci J 5:e47042

Erni S, Wang X, Swystun T, Taylor SW, Parisien MA, Robinne FN, Eddy B, Oliver J, Armitage B, Flannigan MD (2024) Mapping wildfire hazard, vulnerability, and risk to Canadian communities. Int J Disaster Risk Reduction 101:104221

Feng X, Merow C, Liu Z, Park DS, Roehrdanz PR, Maitner B, Newman EA, Boyle BL, Lien A, Burger JR, Pires MM, BrandoPM, Bush MB, McMichael CNH, Neves DM, Nikolopoulos EI, Saleska SR, Hannah L, Breshears DD, Evans TP, Soto JR, Ernst KC, Enquist BJ (2021) How deregulation, drought, and increasing fire impact Amazonian biodiversity. Nature 597(7877):516–521

Filkov AI, Ngo T, Matthews S, Telfer S, Penman TD (2020) Impact of Australia's catastrophic 2019/20 bushfire season on communities and environment. Retrospective analysis and current trends. J Saf Sci Resilience 1(1):44–56

Flannigan MD, Krawchuk MA, de Groot WJ, Wotton BM, Gowman LM (2009) Implications of changing climate for global wildland fire. Int J Wildland Fire 18(5):483–507

Fried JS, Torn MS, Mills E (2004) The impact of climate change on wildfire severity: a regional forecast for northern California. Clim Change 64(1–2):169–191

González-Pérez JA, González-Vila FJ, Almendros G, Knicker H (2004) The effect of fire on soil organic matter—a review. Environ Int 30(6):855–870

Gupta S, Roy A, Bhavsar D, Kala R, Singh S, Kumar AS (2018) Forest fire burnt area assessment in the biodiversity-rich regions using geospatial technology: Uttarakhand forest fire event 2016. J Indian Soc Remote Sens 46:945–955

Hart SA, Chen HY (2008) Fire, logging, and overstory affect understory abundance, diversity, and composition in boreal forests. Ecol Monogr 78(1):123–140

He T, Lamont BB, Pausas JG (2019) Fire is a key driver of Earth's biodiversity. Biol Rev 94(6):1983–2010

Hood SM, Varner JM, Van Mantgem P, Cansler CA (2018) Fire and tree death: understanding and improving modeling of fire-induced tree mortality. Environ Res Lett 13(11):113004

Hudiburg T, Mathias J, Bartowitz K, Berardi DM, Bryant K, Graham E, Kolden CA, Betts RA, Lynch L (2023) Terrestrial carbon dynamics in an era of increasing wildfire. Nat Clim Change 13(12):1306–1316

Jackson MC, Wasserman RJ, Grey J, Ricciardi A, Dick JT, Alexander ME (2017) Novel and disrupted trophic links following invasion in freshwater ecosystems. In: Advances in ecological research. vol 57. Academic Press, USA, pp 55–97

Jangra A, Kumar K, Maikhuri S, Bhandari MS, Pandey S, Singh H, Barthwal S (2024) Unveiling stress-adapted endophytic bacteria: characterizing plant growth-promoting traits and assessing cross-inoculation effects on Populusdeltoides under abiotic stress. Plant Physiol Biochem, 108610, https://doi.org/10.1016/j.plaphy.2024.108610

Johansson MU, Abebe FB, Nemomissa S, Bekele T, Hylander K (2021) Ecosystem restoration in fire-managed savanna woodlands: effects on biodiversity, local livelihoods, and fire intensity. Ambio 50:190–202

Johnson EA, Miyanishi K (2021) Disturbance and succession. In: Plant disturbance ecology. Academic Press, USA, pp 1–15

Johnstone JF, Allen CD, Franklin JF, Frelich LE, Harvey BJ, Higuera PE, Mack MC, Meentemeyer RK, Metz MR, Perry GLW, Schoennagel T, Turner MG (2016) Changing disturbance regimes, ecological memory, and forest resilience. Front Ecol Environ 14(7):369–378

Kaushik P, Singh H (2024) Climate change-induced forest ecosystem functioning and alteration in water-atmosphere-soil–plant (WASP) relationship. In: Singh H (ed) Urban forests, climate change and environmental pollution. Springer, Cham. https://doi.org/10.1007/978-3-031-67837-0_38

Keane RE, Agee JK, Fule P, Keeley JE, Key C, Kitchen SG, Miller R, Schulte LA (2008) Ecological effects of large fires on US landscapes: benefit or catastrophe? Int J Wildland Fire 17(6):696–712

Kelkar S, Kumthekar MB (2019) There is a need of sustainable planning and development against natural disasters resulting because of climate change and human intervention. Proc Sustain Infrastruc Dev Manage (SIDM)

Knelman JE (2015) Patterns and processes underlying soil microbial community succession. Doctoral dissertation, University of Colorado at Boulder

Kumar A, Kumar P, Singh H, Kumar N (2020) Adaptation and mitigation potential of roadside trees with bio-extraction of heavy metals under vehicular emissions and their impact on physiological traits during seasonal regimes. Urb for Urb Green 58:126900

Kumar A, Kumar P, Singh H, Bisht S, Kumar N (2021a) Relationship of physiological plant functional traits with soil carbon stock in temperate forest of Garhwal Himalaya. Curr Sci 120(8):1368–1373

Kumar A, Kumar P, Singh H, Kumar N (2021b) Modulation of plant functional traits under essential plant nutrients during seasonal regime in natural forests of Garhwal Himalayas. Plant Soil 465:197–212

Kumar A, Dwivedi G K, Tewari S, Paul J, Anand R, Kumar N, Kumar P, Singh H, Kaushal R (2020a) Carbon mineralization and inorganic nitrogen pools under *Terminalia chebula* retz.-based agroforestry system in Himalayan Foothills, India. For Sci 66(5):634–643

Kumar N, Singh H, Kumar A, Singh AK, Pandey PK, Kumar A (2024) Forest-fire-induced land degradation. In: Mishra G, Giri K, Singh S, Kumar M (eds) Sustainable land management in India. Springer, Singapore. https://doi.org/10.1007/978-981-97-5223-2_4

Lal R (2013) Soils and ecosystem services. In: Ecosystem services and carbon sequestration in the biosphere, pp 11–38

Lamont BB (2022) Historical links between climate and fire on species dispersion and trait evolution. Plant Ecol 223(7):711–732

Löf M, Madsen P, Metslaid M, Witzell J, Jacobs DF (2019) Restoring forests: regeneration and ecosystem function for the future. New Forest 50:139–151

Malik A, Kumar K, Kumari G, Kumar N, Singh H (2024) Application of functional traits in modelling productivity and resilience under climate change. Springer, pp 77–96. ISBN: 978-981-97-1509-1

McLauchlan KK, Higuera PE, Miesel J, Rogers BM, Schweitzer J, Shuman JK, Tepley AJ, Varner JM, Veblen JT, Adalsteinsson SA, Balch JK, Baker P, Batllori E, Bigio E, Brando P, Cattau M, Chipman ML, Coen J, Crandall R, Daniels L, Enright N, Gross WS, Harvey BJ, Hatten JA, Hermann S, Hewitt RE, Kobziar LN, Landesmann JB, Loranty MM, Maezumi SY, Mearns L, Moritz M, Myers JA, Pausas JG, Pellegrini AFA, Platt WJ, Roozeboom J, Safford H, Santos F, Scheller RM, Sherriff RL, Smith KG, Smith MD, Watts AC (2020) Fire as a fundamental ecological process: research advances and frontiers. J Ecol 108(5):2047–2069

Midgley JJ, Bond WJ (2013) Plant adaptations to fire: an evolutionary perspective. In: Fire phenomena and the earth system: an interdisciplinary guide to fire science, pp 125–134

Miller R, Chambers JC, Pyke DA, Pierson FB, Williams CJ (2013) A review of fire effects on vegetation and soils in the Great Basin Region: response and ecological site characteristics

Miyanishi K (2001) Forest fires: behavior and ecological effects. Elsevier, Netherlands

Mockrin MH, Fishler HK, Stewart SI (2020) After the fire: perceptions of land use planning to reduce wildfire risk in eight communities across the United States. Int J Disaster Risk Reduction 45:101444

Nayak S, Zlatanova S (eds) (2008) Remote sensing and GIS technologies for monitoring and prediction of disasters. Springer Science & Business Media

Nirmal BM, Singh H (2021) Effect of $CO_2$ elevation on Shisham growth at nursery stage. Indian J Hill Farm 34:272–276

Nitschke CR, Innes JL (2007) Interactions between fire, climate change and forest biodiversity. CABI Rev (2006):9

Nowacki GJ, Abrams MD (2008) The demise of fire and "mesophication" of forests in the eastern United States. Bioscience 58(2):123–138

Pérez-Cabello F, Echeverría MT, Ibarra P, de la Riva J (2009) Effects of fire on vegetation, soil and hydrogeomorphologicalbehavior in Mediterranean ecosystems. In: Earth observation of wildland fires in Mediterranean ecosystems, pp 111–128

Popkin G (2021) Forest fight. Science 374(6572):1184–1189

Prakash V, Hunney K, Singh H (2022) Effect of elevated carbon dioxide on growth and development of *Santalum album* L. seedlings inoculated with plant growth promoting microorganisms in open top chambers. Int J Environ Agric Biotechnol 7(5):149–159

Raturi A, Singh H, Kumar P, Shukla N (2022) Characterizing the post-monsoon $CO_2$, $CH_4$, $N_2O$, and $H_2O$ vapor effluxes from a tropical wetland in the Himalayan foothill. Environ Monit Assess 194:50. https://doi.org/10.1007/s10661-021-09721-8

Rego FC, Morgan P, Fernandes P, Hoffman C, Castro Rego F, Morgan P, Fernandes P, Hoffman C (2021) Fire effects on plants, soils, and animals. In: Fire science: from chemistry to landscape management, pp 259–318

Reilly MJ, Halofsky JE, Krawchuk MA, Donato DC, Hessburg PF, Johnston JD, Merschel AG, Swanson ME, Halofsky JS, Spies TA (2021) Fire ecology and management in Pacific Northwest forests. In: Fire ecology and management: past, present, and future of US forested ecosystems, pp 393–435

del Rio CR (2012) The role of ecosystems in building climate change resilience and reducing greenhouse gases. Integrating Ecol Poverty Reduction Ecol Dimensions, 331–352

Safford HD, Paulson AK, Steel ZL, Young DJ, Wayman RB (2022) The 2020 California fire season: a year like no other, a return to the past or a harbinger of the future? Glob Ecol Biogeogr 31(10):2005–2025

Savita, Sharma SK, Singh H, Singh O (2017) Forest fire: a review. Indian Forester 143(8):719–728

Sett T, Nikam BR, Singh H, Purohit S (2022) Effect of drought stress on forest evapotranspiration—a case study on Indian forests. EGU General Assembly 2022. https://doi.org/10.5194/egusphere-egu22-510

Sett T, Nikam BR, Singh H, Purohit S (2024) Hydrological variability in indian forest ecosystem: analysis of drought resilience, recovery and water use efficiency in moist and dry deciduous forests. In: Singh H (ed) Urban forests, climate change and environmental pollution. Springer, Cham. https://doi.org/10.1007/978-3-031-67837-0_37

Sharma R, Singh H, Kaushik M, Nautiyal R, Singh O (2018) Adaptive physiological response, carbon partitioning and biomass production of *Withania somnifera* (L.) Dunal grown under elevated $CO_2$ regimes. 3 Biotech 8:267. https://doi.org/10.1007/s13205-018-1292-1

Sharma R, Singh H (2021) Alteration in biochemical constituents and nutrients partitioning of *Asparagus racemosus* in response to elevated atmospheric $CO_2$ concentration. Environ Sci Pollut Res 29:6812–6821. https://doi.org/10.1007/s11356-021-16050-3

Singh H, Sharma R, Savita SMP, Kumar M, Verma A, Ansari MW, Sharma SK (2018) Adaptive physiological response of *Parthenium hysterophorus* to elevated atmospheric $CO_2$ concentration. Indian Forester 144:1–14

Singh S, Singh H, Sharma SK, Nautiyal R (2021c) Seasonal variation in biochemical responses of bamboo clones in the sub-tropical climate of Indian Himalayan foothills. Heliyon 7(4):e06859. https://doi.org/10.1016/j.heliyon.2021.e06859

Singh H, Kumar M (2022) Climate change and its impact on Indian Himalayan Forests: current status and research needs. In: Rani S, Kumar R (eds) climate change. Springer Climate. Springer, Cham, pp 223–242. https://doi.org/10.1007/978-3-030-92782-0_11

Singh H, Kumar N, Kumar M, Singh R (2020a) Modelling habitat suitability of western tragopan (*Tragopan melanocephalus*) a range-restricted vulnerable bird species of the Himalayan region, in response to climate change. Clim Risk Manage 29:100241. https://doi.org/10.1016/j.crm.2020.100241

Singh H, Kumar N, Singh R, Kumar M (2023) Assessing the climate change impact on the habitat suitability of the range-restricted bird species (*Catreus wallichii*) in the Indian Himalayan ecosystem. Environ Sci Pollut Res. https://doi.org/10.1007/s11356-023-30789-x

Singh H, Kumar N, Kumar A (2024a) Enhancing resource use efficiency in crops through plant functional traits. Springer, Singapore, pp 97–117. ISBN: 978-981-97-1509-1

Singh M, Singh H, Kumar A, Kumar M, Barthwal S, Thakur A (2024b) Soil nitrogen availability determines the $CO_2$ fertilization effect on tree species (*Neolamarckia cadamba*): growth and physiological evidence. Environ Sustain 7:53–60. https://doi.org/10.1007/s42398-023-00300-w

Singh H (2021) Process-based carbon sequestration study in relation to the energy-water-carbon flux in a forest ecosystem. In: Kumar P, Singh RK, Kumar M, Rani M, Sharma (eds) Climate impacts on sustainable natural resource management. CRC Press, Taylor & Francis Group, United States, pp 336–361
Singh H (2024a) Climate change and forests: biological perspectives. In: Singh H (ed) Forests and climate change. Springer, Singapore. https://doi.org/10.1007/978-981-97-3905-9_1
Singh H (ed) (2024b) Forests and climate change: biological perspectives on impact, adaptation, and mitigation strategies. Springer, Singapore. https://doi.org/10.1007/978-981-97-3905-9
Singh H (2024c) Climate change and environmental pollution: biological consequences on urban forests. In: Singh H (ed) Urban forests, climate change and environmental pollution. Springer, Cham, Switzerland. https://doi.org/10.1007/978-3-031-67837-0_1
Singh H (ed) (2024d) Urban forests, climate change and environmental pollution: physio-biochemical and molecular perspectives to enhance urban resilience. Springer, Cham, Switzerland. https://doi.org/10.1007/978-3-031-67837-0
Singh H (2024e) Exploring adaptive modulation in plant functional traits and its impact on the productivity of *Acacia auriculiformis* under $CO_2$-enriched environment. Ind Crops Prod 210(2024):118186. https://doi.org/10.1016/j.indcrop.2024.118186
Souza-Alonso P, Saiz G, García RA, Pauchard A, Ferreira A, Merino A (2022) Post-fire ecological restoration in Latin American forest ecosystems: insights and lessons from the last two decades. For Ecol Manage 509:120083
Spatola MF, Borghetti M, Nolè A (2023) Elucidating factors driving post-fire vegetation recovery in the Mediterranean forests using Landsat spectral metrics. Agric for Meteorol 342:109731
Stuart Chapin III F, McFarland J, David McGuire A, Euskirchen ES, Ruess RW, Kielland K (2009) The changing global carbon cycle: linking plant–soil carbon dynamics to global consequences. J Ecol 97(5):840–850
Tepley AJ, Thomann E, Veblen TT, Perry GL, Holz A, Paritsis J, Kitzberger T, Anderson-Teixeira KJ (2018) Influences of fire–vegetation feedbacks and post-fire recovery rates on forest landscape vulnerability to altered fire regimes. J Ecol 106(5):1925–1940
Úbeda X, Outeiro LR (2009) Physical and chemical effects of fire on soil. In: Fire effects on soils and restoration strategies. CRC Press, USA, pp 121–148
Underwood R, Held A (2012) Wildfire. In: Handbook of hazards and disaster risk reduction. Routledge, UK, pp 281–293
Vallejo VR, Arianoutsou M, Moreira F (2011) Fire ecology and post-fire restoration approaches in Southern European forest types. Post-fire management and restoration of southern European forests. Springer Netherlands, Dordrecht, pp 93–119
Vasudeva SP (2018) Disastrous forest fires: management and control. Indian J Publ Adm 64(2):237–253
Vázquez-Varela C, Martínez-Navarro JM, Abad-González L (2022) Traditional fire knowledge: a thematic synthesis approach. Fire 5(2):47
Veraverbeke S, Rogers BM, Goulden ML, Jandt RR, Miller CE, Wiggins EB, Randerson JT (2017) Lightning as a major driver of recent large fire years in North American boreal forests. Nat Clim Chang 7(7):529–534
Verma S, Jayakumar S (2012) Impact of forest fire on physical, chemical and biological properties of soil: a review. proceedings of the Int Acad Ecol Environ Sci 2(3):168
Yadav SK, Singh H, Nautiyal R, Ginwal HS, Ansari SA, Barthwal S (2019a) Modulation of morpho-physiological responses in *Populus deltoides* by elevated carbon dioxide and temperature. For Sci 66:105–118. https://doi.org/10.1093/forsci/fxz048
Yadav SK, Singh H, Ginwal HS, Barthwal S (2019b) Elevated $CO_2$ enhanced growth and physiological process of *Populus deltoides* Bartr. ex Marsh Indian Forester 145(1):23–27
Zhong C, Guo M, Zhou F, Li J, Yu F, Guo F, Li W (2023) Forest succession trajectories after fires in valleys and on slopes in the Greater Khingan Mountains, China. J For Res 34(3):623–640
Zuazo VHD, Pleguezuelo CRR (2009) Soil-erosion and runoff prevention by plant covers: a review. Sustain Agric, 785–811

Zymaroieva A, Kolomiychuk V, Fedoniuk T, Goncharenko I, Borsuk O, Melnychuk T, Svenning JC (2023) Post-fire recovery of vegetation in the Chornobyl radiation and ecological biosphere reserve. Int J Environ Stud, 1–21

# Forest Fire and Climate Change Effects on Shaping Flora and Fauna

**Shubham Singh, Vishavjit Kumar, Manoj Kumar, and Bhupendra Singh Adhikari**

**Abstract** Forest fires exacerbated by climate change worsen the severe threat to forest flora and fauna. With rising temperatures and disturbed precipitation patterns, forests have become drier and more susceptible to forest fires. Frequent forest fires devastate vast expanses of vegetation and destroy crucial habitats for countless species of plants and animals. The intense heat and smoke can kill or severely injure wildlife, disrupt migration patterns, and fragment habitats, creating a challenge for species in finding food and shelter. Vegetation loss due to fires affects soil physio-chemical properties, contributes to soil erosion, and alters the composition and structure of ecosystems, impacting the delicate balance of biodiversity. Over extended time, the cumulative effects of forest fires driven by climate change can lead to irreversible damage to ecosystems and threaten the survival of numerous species. The current chapter highlights the impact of forest fires and climate change on the ecosystem, the interrelationship of these two phenomena, the pressing necessity for efficient measures, and effective mitigation and adaptation strategies.

**Keywords** Forest fire · Climate change · Flora · Fauna · Forest ecosystem

## 1 Introduction

Forest fires are a significant ecological event or phenomenon that has wide-ranging impacts on natural and human environments (Stephens and Ruth 2005; Bowman et al. 2019). In recent times, incidences of forest fires have increased dramatically, posing a threat to biodiversity, wildlife habitats, carbon reserves, and climate stability

S. Singh (✉)
ICFRE-Forest Research Institute, Dehradun, India
e-mail: shubhamrsgis@gmail.com

V. Kumar · M. Kumar
Indian Council for Forestry Research and Education, Dehradun, India

B. S. Adhikari
Wildlife Institute of India, Dehradun, India

 
H. Singh and N. Kumar (eds.), *Forest Fire and Climate Change*,
https://doi.org/10.1007/978-3-031-89967-6_6

(Pechony and Shindell 2010; Kumar et al. 2024). Each year, fires consume approximately 3–5 million square kilometers of land, releasing approximately 8 billion tonnes of $CO_2$ into the atmosphere on average (De Santis and Chuvieco 2009; Chuvieco et al. 2019). Forest Fires lower the amount of carbon sequestered in the forest by approximately 10% worldwide, significantly influencing atmospheric $CO_2$ concentrations (Kumar et al. 2021a, b; Jones et al. 2022). Approximately 11% of the Indian forest comes under moderately high fire zones (FSI 2021). Forest fires considerably impact the structure and composition of the forest, forest soil, biodiversity, nutrient cycle, and related phenomena (Nitschke and Innes 2007; Barwa and Singh 2024). Forest fires can be natural or human-induced. The natural causes of fire can be lightning and volcanic eruptions. Although natural fires are rare, forest fires are caused mainly by human activities. The human-induced fires may be intentional, such as to enhance fodder growth (Chandra and Bhardwaj 2015), or these fires may be accidental. These fires can be further divided into creeping fire, ground (peat) fire, and crown fire, depending on the part of the forest it damages (Drapalyuk et al. 2019). Fires are essential for forest health as they clear the forest floor, control pests and insects, and promote native species (Keane et al. 2008). Still, with changing climatic scenarios, forest fires have increased dramatically, causing detrimental impacts on the forest (Negi and Kumar 2016; Singh 2024a).

The ecosystem experiences significant effects due to altering climatic conditions like increasing temperatures, disrupted precipitation patterns, and shifts in plant phenology (Hoffmann et al. 2019; Singh et al. 2024a, b). Climate change and forest fires have a complex interrelationship (Flannigan et al. 2006), with fluctuating weather patterns complementing the frequency of forest fires. The increased duration of the fire season has increased the incidences and severity of fires around the globe (Flannigan et al. 2006), which, in turn, contributes to the release of greenhouse gases, further exacerbating climate change (Sett et al. 2024; Kaushik and Singh 2024). Additionally, the loss of forest cover due to wildfires reduces carbon sequestration capacity and disrupts ecosystems (Brando et al. 2019).

In the current scenario, with rapidly changing climatic conditions and increasing human intervention in the forest, the disturbances to the forest have increased to an alarming level (Barbero et al. 1990). Forest fires have become a hazard in these modern times that has still not been addressed appropriately. Climate change and forest fires should be considered exclusive phenomena, and addressing the interrelationship between forest fires and climate change requires comprehensive strategies to mitigate both the drivers and impacts of these phenomena (Singh 2024b). This chapter attempts to shed insight into the effects of forest fires and climate change on the ecosystem, the relationship between these two phenomena, and potential strategies for managing fires in the face of climate change.

## 2 Impact of Forest Fires on Ecosystems

Forest fires and ecosystems have a cascading relationship (Mishra et al. 2024). Forest fires impact on the ecosystem depends on various factors. These factors include the type of forest, the fire regime, and the ecosystem services provided. In the Indian scenario, forest fires are generally anthropogenic and thus have both advantageous and detrimental impacts on the ecosystem (Prakash et al. 2022; Nirmal and Singh 2021; Singh and Kumar 2022; Savita et al. 2017). Less frequent forest fires are somewhat considered essential and beneficial for the forest ecosystem by reducing fuel load, removing exotic species and promoting native species, breaking seed dominance, controlling infesting insects, promote species growth by opening the canopy (Keane et al. 2008). Although with an increase in the frequency and severity of fires around the globe, forest fires are becoming a severe threat to the forest ecosystem composition and structure (Valderrama et al. 2018; Kumar et al. 2020a, b). The increased frequency and severity of fires due to climate change further exacerbate these effects on ecosystems (Nitschke and Innes 2007). Disturbances that occur due to forest fires shape forest ecosystems in complex ways, influencing vegetation dynamics, biodiversity patterns, nutrient cycling, and carbon dynamics (Sharma et al. 2018; Jangra et al. 2024; Dhyani et al. 2021; Singh 2024e). Some effects of fire on forest ecosystems are nutrient cycling, habitat renewal, biodiversity, succession, soil properties and carbon dynamics. Here we will briefly discuss the impact of forest fires on forest soils, biodiversity and nutrient cycling as these factors have a major influence on shaping the flora and fauna of the forest.

### *2.1 Impact of Forest Fires on Soil Properties*

Forest fires can directly and indirectly affect soil properties (Weber and Flannigan 1997). Forest fires can change the levels of vital nutrients within the soil (Roshan and Biswas 2023). For instance, they can cause a reduction in total soil carbon (C) and nitrogen (N) concentrations (Singh et al. 2024a, b; Yadav et al. 2019a; Singh 2021; Singh et al. 2018), with varying effects, depending on the ecosystem; in broadleaf forests and savanna grasslands, higher fire frequencies reduced soil C and N, while in needle leaf forests, soil C and N increased (Pellegrini et al. 2018). High-temperature conditions caused by forest fires can lead to the volatilization of certain nutrients, such as N, which can impact plant nutrient stoichiometry or composition in the affected areas (Dale et al. 2001; Pellegrini et al. 2018; Singh 2024c). Frequent forest fires decrease the storage of C and N; however, fire also reduces the possibility for decomposition, which probably contributed to the slower cycling of C and N (Pellegrini et al. 2020). Butler et al. (2018) reported that forest fire results in much lower soil and litter Carbon:Phosphorus and Nitrogen:Phosphorus ratios and significantly greater concentrations of soil mineral Phosphorus. Also, depending on the kind of vegetation, fire has a different P-rich signature in the soil–plant system.The effects of burning on various pools of carbon (C), nitrogen (N), and phosphorus (P) in soil differed from site to site, suggesting that factors specific to the fire regime and the location control these responses (Butler et al. 2017). Scharenbroch et al. (2012) found increased levels of total nitrogen, particulate organic matter (POM), total organic carbon, pH, electrical conductivity, calcium ($Ca^{2+}$), magnesium ($Mg^{2+}$), potassium ($K^{+}$), sodium ($Na^{+}$), nitrate ($NO_3^{-}$), and potential nitrogen mineralization in the burned plots according to their study (Singh et al. 2021c; Yadav et al. 2019b; Singh et al. 2023; Sharma and Singh 2021).Repeated burning also altered the chemical properties of the organic matter, resulting in a transition toward a more significant fraction of the soil organic matter (SOM) being less accessible to microbes and more recalcitrant (Pellegrini et al. 2021). Burnt soils have significantly lower soil organic matter in comparison to the unburnt sites (Allam et al. 2020).

Although physical soil properties are generally not significantly affected by forest fires (Hungerford et al. 1991; Allam et al. 2020), the impact depends on the severity of the forest fire (Wells 1979). The amount of SOM present is essential for maintaining soil structure. The amount of SOM consumed in forest fire decides the magnitude of impact on the physical properties of the soil (Úbeda and Outeiro 2009). Fire and accompanying soil heating can degrade soil structure, reducing total porosity and pore size and changing micro- and macropore distribution in the soil's surface strata (Ekinci 2006; Xue et al. 2014; Elakiya et al. 2023). Soil pH is an essential factor affecting the availability of soil nutrients. However, the impact of forest fires on forest soil is highly variable and may depend on the vegetation type, fire severity, and related factors (Úbeda and Outeiro 2009). However, in most cases, forest fires increase the soil pH (Ekinci 2006; Aref et al. 2011).

Forest fires can impact the geomorphic process of the landscape. Forest fires lead to varying soil temperatures across burned areas due to the patchy distribution of vegetation. Elevated soil temperatures enhance sediment availability for erosion by reducing thresholds and stripping away protective soil covers (Singh 2024d). The heightened sediment availability is both sporadic and temporary. When a fire in a short time is succeeded by a storm characterized by concentrated, heavy rainfall in areas with large patches of exposed sediment, the result can be a catastrophic flood. This flood event triggers hillslope erosion, channel incision, and sediment deposition, resulting in distinct erosional and depositional features scattered throughout the scorched terrain. These characteristics endure as legacies, acting as a type of landscape memory and shaping the feedback loop for subsequent cycles of fire and flooding (Moody and Martin 2009).

Forest fires can significantly alter soil infiltration rates by destroying vegetation and litter and changing the soil's composition through chemical, physical, and biological processes (Cerda iBolinches 1995; Robichaud 2000; Inbar et al. 2014). The heat and chemical changes caused by a fire can decrease the soil's ability to absorb water (DeBano et al. 1977). This phenomenon arises from the kinetic energy of raindrops and the rapid saturation of soil, facilitating the dispersion of clay particles. These particles are subsequently carried along with the infiltrating water and block the pores just below the surface, forming a "washed-in" area. Consequently, infiltration decreases, runoff intensifies, and detached soil particles are transported away (Inbar et al. 2014). The severity of the burn and the condition of surface layers post-fire are pivotal factors in determining soil exposure and the likelihood of developing water-repellent properties. Low-severity burns typically result in temporary water repellence, whereas high-severity burns can lead to more enduring alterations (Robichaud 2000).

### 2.2 Impact of Forest Fires on Forest Biodiversity

Forest fires lead to the loss of certain plant species and ecosystems and changes in the composition and structure of plant communities. Fire creates a disturbance that can shift the balance of plant species in an ecosystem (Nitschke and Innes 2007). This can lead to changes in the overall biodiversity of an area, with some species benefiting and others being threatened or lost (Hilwan 2020). Some key impacts of forest fire on biodiversity include changes in plant composition and structure, increased species evenness, loss of biodiversity, abundance of invasive species, loss of keystone species, promotion of fire-adapted species, and habitat regeneration and succession.

In Tropical forests that are subjected to fires during each dry season, such as tropical pine forests, monsoon forests, and Savannah woodlands, tree species demonstrate adaptive features like thick bark, the capacity to recover from fire damage, the ability to regenerate through resprouting, and specialized seed characteristics. These recurring fires hold considerable ecological significance for forest ecosystems. They notably encourage the dominance of fire-adapted species, often supplanting those that typically thrive without fire disturbances (Nasi et al. 2002). A study by Slik et al. (2008) to ascertain the regeneration trajectories of forests subject to single and repeated burns over seven years in an eastern Bornean rainforest reveals significant alterations in tree diversity and composition, with implications for the overall forest structure. Single fire events result in discernible shifts in species composition and a decrease in aboveground biomass. In contrast, repeated fires exacerbate these effects, leading to further reductions in biomass and changes in species dominance (Singh 2024e). Additionally, the study highlights the importance of understanding the long-term impacts of fire on tropical rainforest ecosystems, particularly in the context of ongoing climate change and increasing fire frequency.

González-Tagle et al. (2008) reported that early successional stages post-fire exhibit reduced tree density and increased dominance of pioneer species, resulting in simplified forest structure. Over time, as the forest regenerates, there is a gradual recovery in tree density and structural complexity. Cochrane and Schulze (1999) investigates the effects of recurrent fire events on tropical forests in the Eastern Amazon, focusing on forest structure, biomass, and species composition. The study reveals significant alterations in forest structure, including changes in tree density and canopy cover, as well as shifts in species composition following fire disturbances. Recurrent fires lead to a reduction in aboveground biomass, impacting carbon storage capacity and potentially exacerbating climate change. Understanding the regeneration dynamics of the forest post-fire is crucial, with varying rates of recovery observed among different species. Despite the disruptions caused by fires, tropical forests exhibit resilience mechanisms to cope with and recover from these disturbances over time. The findings emphasize the importance of considering fire dynamics in forest management and conservation strategies to ensure the long-term health and sustainability of Amazonian ecosystems. Pati et al. (2024) found that increased fire frequency changes tree density, canopy cover, and species diversity within the forest communities.

Additionally, shifts in species composition, with fire-tolerant species becoming more prevalent, are observed in areas with higher fire frequencies. Thonicke and Cramer (2006) reveal notable shifts in vegetation patterns over time, influenced by climatic variations. Changes in temperature and precipitation regimes are identified as key drivers affecting vegetation dynamics, with implications for forest composition and structure. Additionally, the study examines trends in forest fire occurrences, highlighting increasing frequencies or intensities of fires in response to climate change. These findings underscore the importance of understanding the interactions between climate dynamics, vegetation responses, and fire regimes for effective forest management and conservation strategies in Brandenburg and similar regions facing climate-related challenges. Tessler et al. (2016) reported significant reductions in vegetation cover immediately after fires, followed by gradual recovery over time. However, the study identifies persistent impacts on species richness, with certain plant communities exhibiting decreased diversity in fire-affected areas.

Vegetation gaps generated by high-intensity fires are especially vulnerable to invasion by exotic species (Parashar and Biswas 2003). Invasive species recover swiftly after a fire and may increase their cover and decrease biodiversity through allelopathic pathways (Keeley et al. 2003; Brooks and Zouhar 2008; Franklin 2010). Recurring fires reduce the cover, richness, and diversity of native vegetation, causing a transition from indigenous to introduced species (Fisher et al. 2009). In their study, Khapugin et al. (2016) of fire-affected forests at the southern boundary of the taiga zone reported a notable increase in the establishment and spread of alien plants following fire events, with disturbed habitats providing favorable conditions for colonization. The study reported six Erigeron canadensis specie*s*, *E. annuus*, *Oenotherabiennis*, *Lactucaserriola*, S*ambucusracemose*, and *Viola arvensis*, established in the fire-affected region. Freeman et al. (2007) reported significant increases in the establishment and spread of invasive plants following fire events, with disturbances creating favorable conditions for invasion. Several factors contribute to the invasion of alien species, including fire-induced changes in soil properties, disturbance-mediated dispersal mechanisms, and the competitive advantage of invasive species in post-fire environments (Freeman et al. 2007; Fisher et al. 2009; Khapugin et al. 2016).

### 2.3 Impact of Forest Fire on Nutrient Cycling

Forest nutrient cycling refers to the exchange of elements between an ecosystem's living and non-living components (Soil Science Society of America 2001). The forest nutrient cycle encompasses processes such as nutrient absorption and retention in perennial tissues, the generation and breakdown of litter, nutrient transformations facilitated by soil organisms and plants, atmospheric nutrient influx, mineral weathering, and nutrient export via leaching and gaseous transfers (Foster and Bhatti 2006). Fire frequency has a negative relationship with productivity and N cycling rates, with frequent fires depressing the N cycle due to nitrogen losses and favouring grass dominance, depressing carbon (C) and N cycles (Reich et al. 2001). Repeated fires significantly alter the biogeochemistry and functioning of ecosystems by changing the quantity and chemistry of plant inputs and affecting the decomposition of soil organic matter (Pellegrini et al. 2020).

Extensive wildfires cause a horizontal and vertical redistribution of ecosystem resources in environments where fire is the dominant force (Foster and Bhatti 2006). The combined effects of the following mechanisms lead to redistribution: (1) Water erosion of surface soils; (2) convection of ash particles in winds generated by fire; (3) oxidation and volatilization of living and decomposing plant material; and (4) percolation of solutes through and out of the soil. In their study, Pellegrini et al. (2020) suggest that repeated burning reduces overall carbon and nitrogen storage and shifts the dynamic of these cycles, potentially contributing to long-term ecosystem changes. Forest fire can result in substantial nitrogen losses from ecosystems through volatilization and leaching but paradoxically increases soil ammonium levels, which contribute to higher rates of primary production in plant communities re-establishing after fire (Grogan et al. 2000). The impacts of forest fires on nutrient cycling can persist beyond the immediate post-fire period. Table 1 shows some significant direct and indirect effects of the severity of forest fires on nutrient cycles within forests. Changes in soil properties, vegetation composition, and ecosystem structure can have long-lasting impacts on the ecosystem's nutrient availability and cycling processes.

## 3 Climate Change: Influences on Flora and Fauna

According to the UN Climate Action website, climate change refers to changes in long-term temperature and weather patterns. These changes have been anthropogenic since the 1800's primarily due to burning fossil fuels such as coal, oil, and gas. The anthropogenic climate change has affected biological systems from gene to ecosystem scales, causing changes in population genetic variation and geographic patterns of alleles and traits, changes in vegetation community structure, and measures of ecosystem functioning (Hoffmann et al. 2019). Anthropogenic climate change already influences numerous weather and climate extremes worldwide, resulting in widespread adverse effects on food and water security, human

**Table 1** Significant direct and indirect impacts of the forest fire severity on nutrient cycles within forests

| Direct effects | Indirect effects |
|---|---|
| Nutrient loss to the atmosphere during combustion | Depletion of nutrients mobilized by fire in ash or soil can occur through erosion (caused by wind or water) or leaching |
| Soil heating | Alterations in the micro-climate within the soil and forest floor |
| Ash deposition | Changes in soil mineralization rates |
| Vegetation destruction | Altered N fixation rates<br>Altered nutrient absorption and storage by vegetation, as well as modified litter production and decomposition |

Adapted from Raison et al. (2009)

health, economies, and society. This has led to significant losses and damages to both natural ecosystems and human populations (Lee et al. 2023).

Climate change has had significant impacts on flora and fauna, such as the increased spread of wildlife diseases, frequent forest fire incidences, invasion of forests by exotic species, habitat fragmentation, water scarcity and drought, habitat fragmentation, human-wildlife conflicts, shifts in vegetation composition, loss of habitat and altered migration pattern (Nwaogu et al. 2023; Úbeda et al. 2013). The rise in the global temperature has caused a change in the species distribution, with species moving towards higher latitudes or elevations to find suitable habitat and increasing competition among species in the habitat (Shoo et al. 2006; MacLean and Beissinger 2017; Platts et al. 2019). Plant phenology is heavily influenced by climate and has thus become one of the most dependable bioindicators of climate change (Gordo and Sanz 2010). An abrupt change in the timing of natural events such as flowering, migration, and breeding has occurred due to climate change disturbing the interrelation between species that rely on each other for survival, such as pollinators and flowering plants (Menzel et al. 2006; Cleland et al. 2007; Piao et al. 2019). Precipitation patterns are significantly affected by climate change (Dore 2005; Trenberth 2011), leading to changes in the availability of water resources, which in turn affects freshwater ecosystems, wetlands, and the species that depend on them for survival (Hughes 2003; Agrawal 2011; Hoffmann et al. 2019). Extensive loss of biodiversity is reasonably attributed to the recent human-induced climate change. Climate change poses a significant threat to global biodiversity, leading to declines in population sizes, species extinctions, and changes in ecosystem functioning (Rosales 2008; Skogen et al. 2018).

## 4 Climate Change and Forest Fire: Interrelationship

A recent increase in forest fires has been reported globally (Harvey 2016), and it is essential to understand the underlying principles and effects of these fires to mitigate their risks. Understanding the interplay between forest fires and climate change is crucial for effectively managing ecosystems and mitigating their negative impacts. Climate change affects the frequency, intensity, and behaviour of forest fires, while forest fires contribute to the release of greenhouse gases and further exacerbate climate change (Nitschke and Innes 2007). Forest fires are influenced by various factors, including climate, weather patterns, vegetation type, and human activities (Flannigan et al. 2006). The recent shift in the climatic pattern and weather has resulted in drier conditions, longer fire seasons, and an increased likelihood of ignition sources (Bedia et al. 2015). These factors, combined with human activities such as land-use changes and the introduction of invasive species, create a complex and interconnected system that drives the occurrence and severity of forest fires. Rising temperature due to climate change leads to increased evaporation and moisture stress on vegetation which can dry out vegetation and increase the flammability of forests, making them more susceptible to ignition and rapid spread (Halofsky et al. 2020; Pausas and Keeley 2021).

Changes in precipitation patterns caused by climate change causes more frequent and severe droughts in some areas (Littell et al. 2016), causing favourable conditions to the onset of forest fires. Extended fire seasons due to warmer temperatures (Flannigan et al. 2006) provide a longer window to the fire incidence and more opportunity for fire to spread. Increased frequency of lightning incidences attributed to climate change (Reeve and Toumi 1999) can also lead to an increase in forest fire incidences, as lighting is one of the major natural causes of forest fires across the globe. Climate change has caused an increase in plant diseases and insect outbreaks (Jones and Barbetti 2012; Pureswaran et al. 2018), which increase the forest's plant mortality rate, providing ample fuel for the fire to outbreak. As discussed earlier, climate change affects the composition and structure of the forest, promoting invasive and fire-prone species (Gebeyehu and Hirpo 2019). This might impact the fire regime of the area and cause severe forest fires due to an increase in fuel load.

Forest fires release a substantial amount of carbon dioxide, particulate matter, and greenhouse gases into the atmosphere, contributing to the greenhouse effect and exacerbating climate change (Miranda et al. 1994; Singh 2022). As climate change progresses, it can increase the frequency and intensity of forest fires, leading to even more carbon emissions, thus creating a feedback loop (Raturi et al. 2022; Sett et al. 2022; Singh et al. 2020a; Apurva et al. 2017). Forest fires can influence atmospheric circulation patterns, leading to changes in weather and climate on regional and even global scales (Kirsanov et al. 2020). For example, large wildfires can produce smoke plumes that inject aerosols and pollutants into the atmosphere, affecting cloud formation, precipitation patterns, and temperature distribution. Forest fires can have significant ecological impacts, including habitat loss, species composition changes, and ecosystem process disruptions (Nitschke and Innes 2007). These impacts can affect the resilience and adaptive capacity of forest ecosystems in the face of climate change, potentially leading to long-term shifts in biodiversity and ecosystem services.

## 5 Protecting Biodiversity Amidst Climate and Fire Disturbances

Protecting biodiversity amidst climate and fire disturbances requires an integrated approach that combines conservation efforts with effective fire management strategies. As discussed earlier, climate and forest fires have an interlinked relationship. The management strategies for protecting and restoring biodiversity in the face of these challenges should include an integrated and holistic approach. Some key strategies are:

(a) **Ecosystem Restoration and Resilience Building**

- Ecosystem restoration projects can enhance biodiversity and ecosystem resilience. For example, restoring degraded wetlands can improve habitat quality for various species (Craft et al. 2009).
- Restoration efforts should focus on promoting native plant communities that are more resilient to climate change and fire disturbances (Hobbs et al. 2009).

(b) **Fire Management and Prevention**

- Integrated fire management strategies, including prescribed burning and suppression, are essential for reducing the risk of catastrophic wildfires and maintaining ecosystem integrity (Fernandes et al. 2013).
- Prescribed burning can promote biodiversity by reducing fuel loads, controlling invasive species, and promoting the regeneration of fire-adapted plant species (DiTomaso et al. 2006).

(c) **Climate Change Adaptation Strategies**

- Climate change adaptation strategies should prioritize protecting climate refugia and facilitating species' range shifts to track suitable habitats (Morelli et al. 2016; Sample et al. 2022).
- Assisted migration, when done cautiously and with consideration for potential risks, can help species adapt to changing climatic conditions (Ricciardi and Simberloff 2009).

(d) **Habitat Connectivity and Corridor Creation**

- Maintaining and enhancing habitat connectivity through the creation of ecological corridors can facilitate species movement and promote genetic exchange (Beier et al. 2011).
- Strategic conservation planning should prioritize areas that serve as critical connectivity hubs for species dispersal (Heller and Zavaleta 2009).

(e) **Community Engagement and Stakeholder Collaboration**

- Collaborative conservation efforts involving local communities, Indigenous peoples, and stakeholders are essential for effective biodiversity conservation (Bennett et al. 2017).
- Community-based fire management programs can build local fire prevention, monitoring, and response capacity while incorporating traditional ecological knowledge (Ballard et al. 2008).

(f) **Policy and Planning Integration**

- Integrating biodiversity conservation and fire management considerations into land-use planning and policy development is crucial for achieving conservation goals (Makiola et al. 2020).
- Adaptive governance approaches that allow for flexible decision-making and learning from feedback are essential for navigating complex socio-ecological systems (Olsson et al. 2004).

Implementing the strategies mentioned above grounded in scientific evidence and promoting collaboration among stakeholders may enhance the safeguarding of biodiversity in the face of climate change and fire disturbances. This collective effort aims to bolster the long-term resilience of ecosystems and populations of diverse species, ensuring their ability to adapt and thrive in dynamic environments.

## 6 Conclusion

The interplay between forest fires and climate change has profound and multifaceted effects on shaping flora and fauna across various ecosystems. High-intensity fires, exacerbated by climate change, lead to vegetation gaps, which become hotbeds for invasive species, weakening biodiversity. The alteration of nutrient cycling due to forest fires, particularly nitrogen cycles, affects soil productivity and can cause shifts in ecosystem structure and function, with implications for long-term ecological balance. Climate change influences extreme weather events that increase the frequency and intensity of forest fires, further stressing wildlife, disrupting habitats, and increasing competition for resources among species. Indirectly, climate change impacts water resources, phenology, and species distribution, signaling challenges for ecosystems worldwide. These intertwined dynamics between forest fires, climate change, and biological systems underscore the urgency for comprehensive strategies to mitigate their effects and preserve global biodiversity. Furthermore, adaptation tactics that can strengthen these systems' resistance to upcoming changes in climate and fire patterns include the development of native plant species that are adapted to fire and the preservation of structural, community, and species diversity. The adaptation and mitigation tactics strongly emphasize protecting species and habitats that are resistant to fire and climate change, which may help preserve biodiversity in the face of uncertain climatic change. Forest fires and climate change have direct and indirect consequences that play a major role in forming the present and future biological communities. Therefore, proactive and knowledgeable management strategies are necessary to lessen these effects and protect ecological integrity.

## References

Agrawal A (2011) Effect of global warming on climate change, flora and fauna. J Ecophysiol Occup Health, 161–174

Allam A, Borsali AH, Kefifa A, Zouidi M, Gros R (2020) Effect of fires on certain properties of forest soils in Western Algeria. Acta Technol Agric 23(3):111–117

Apurva K, Singh H, Kumar H (2017) Elucidating morphological and growth changes in Arjun Tree (*Terminalia arjuna*) grown under elevated temperature condition. Trends Biosci 10(1):332–336

Aref IM, El Atta HA, Mohamed AL, Ghamde AR (2011) Effect of forest fires on tree diversity and some soil properties. Int J Agric Biol 13(5)

Ballard HL, Fernandez-Gimenez ME, Sturtevant VE (2008) Integration of local ecological knowledge and conventional science: a study of seven community-based forestry organizations in the USA. Ecol Soc 13(2)

Barbero M, Bonin G, Loisel R, Quézel P (1990) Changes and disturbances of forest ecosystems caused by human activities in the western part of the Mediterranean basin. Vegetatio 87:151–173

Barwa S, Singh H (2024) Climate-induced changes in forest ecosystems of Indian Himalayan Region: a meta-analysis. In: Singh H (eds) Urban forests, climate change and environmental pollution. Springer, Cham. https://doi.org/10.1007/978-3-031-67837-0_39

Bedia J, Herrera S, Gutiérrez JM, Benali A, Brands S, Mota B, Moreno JM (2015) Global patterns in the sensitivity of burned area to fire-weather: implications for climate change. Agric for Meteorol 214:369–379

Beier P, Spencer W, Baldwin RF, McRAE BH (2011) Toward best practices for developing regional connectivity maps. Conserv Biol 25(5):879–892

Bennett NJ, Roth R, Klain SC, Chan K, Christie P, Clark DA, Cullman G, Curran D, Durbin TJ, Epstein G (2017) Conservation social science: understanding and integrating human dimensions to improve conservation. Biol Conserv 205:93–108

Bowman DM, Moreira-Muñoz A, Kolden CA, Chávez RO, Muñoz AA, Salinas F, González-Reyes Á, Rocco R, De la Barrera F, Williamson GJ (2019) Human–environmental drivers and impacts of the globally extreme 2017 Chilean fires. Ambio 48:350–362

Brando PM, Paolucci L, Ummenhofer CC, Ordway EM, Hartmann H, Cattau ME, Rattis L, Medjibe V, Coe MT, Balch J (2019) Droughts, wildfires, and forest carbon cycling: a pantropical synthesis. Annu Rev Earth Planet Sci 47:555–581

Brooks ML, Zouhar K (2008) Plant invasions and fire regimes. In: Wildland fire in ecosystems, fire and nonnative invasive plants Ogden, UT, USA: US Department of Agriculture, Forest Service, RMRS-GTR-42-volume 6, pp 33–45

Butler OM, Lewis T, Chen C (2017) Fire alters soil labile stoichiometry and litter nutrients in Australian eucalypt forests. Int J Wildland Fire 26:783–788

Butler OM, Elser J, Lewis T, Mackey B, Chen C (2018) The phosphorus-rich signature of fire in the soil-plant system: a global meta-analysis. Ecol Lett 21(3):335–344

Cerda iBolinches A (1995) Factores y variacionesespacio-temporales de la infiltración en los ecosistemas mediterráneos

Chandra K, Bhardwaj AK (2015) Incidence of forest fire in India and its effect on terrestrial ecosystem dynamics, nutrient and microbial status of soil. Int J Agric For 5(2):69–78

Chuvieco E, Mouillot F, Van der Werf GR, San Miguel J, Tanase M, Koutsias N, García M, Yebra M, Padilla M, Gitas I (2019) Historical background and current developments for mapping burned area from satellite Earth observation. Remote Sens Environ 225:45–64

Cleland EE, Chuine I, Menzel A, Mooney HA, Schwartz MD (2007) Shifting plant phenology in response to global change. Trends Ecol Evol 22(7):357–365

Cochrane MA, Schulze MD (1999) Fire as a recurrent event in tropical forests of the eastern amazon: effects on forest structure, biomass, and species composition 1. Biotropica 31(1):2–16

Craft C, Clough J, Ehman J, Joye S, Park R, Pennings S, Guo H, Machmuller M (2009) Forecasting the effects of accelerated sea-level rise on tidal marsh ecosystem services. Front Ecol Environ 7(2):73–78

Dale VH, Joyce LA, McNulty S, Neilson RP, Ayres MP, Flannigan MD, Hanson PJ, Irland LC, Lugo AE, Peterson CJ (2001) Climate change and forest disturbances: climate change can affect forests by altering the frequency, intensity, duration, and timing of fire, drought, introduced species, insect and pathogen outbreaks, hurricanes, windstorms, ice storms, or landslides. Bioscience 51(9):723–734

De Santis A, Chuvieco E (2009) GeoCBI: a modified version of the composite burn index for the initial assessment of the short-term burn severity from remotely sensed data. Remote Sens Environ 113(3):554–562

DeBano L, Dunn P, Conrad C (1977) Fire's effect on physical and chemical soilsproperties of chaparral soils. Department of Agriculture, Forest Service, p 65

Dhyani K, Kuniyal HB, Singh H, Sobha (2021) Growth and physiological potential of *Terminalia arjuna* under elevated $CO_2$ levels in open top chamber condition. J Appl Nat Sci 13(3):1121–1126. https://doi.org/10.31018/jans.v13i3.2490

DiTomaso JM, Brooks ML, Allen EB, Minnich R, Rice PM, Kyser GB (2006) Control of invasive weeds with prescribed burning. Weed Technol 20(2):535–548

Dore MH (2005) Climate change and changes in global precipitation patterns: what do we know? Environ Int 31(8):1167–1181

Drapalyuk M, Stupnikov D, Druchinin D, Pozdnyakov E (2019) Forest fires: methods and means for their suppression. IOP Publishing, England, p 012061

Ekinci H (2006) Effect of forest fire on some physical, chemical and biological properties of soil in Çanakkale, Turkey. Int J Agric Biol 8(1):102–106

Elakiya N, Keerthana G, Safiya S (2023) Effects of forest fire on soil properties. Int J Plant Soil Sci 35(20):8–17

Fernandes PM, Davies GM, Ascoli D, Fernández C, Moreira F, Rigolot E, Stoof CR, Vega JA, Molina D (2013) Prescribed burning in southern Europe: developing fire management in a dynamic landscape. Front Ecol Environ 11(s1):e4–e14

Fisher JL, Loneragan WA, Dixon K, Delaney J, Veneklaas EJ (2009) Altered vegetation structure and composition linked to fire frequency and plant invasion in a biodiverse woodland. Biol Cons 142(10):2270–2281

Flannigan MD, Amiro BD, Logan KA, Stocks BJ, Wotton BM (2006) Forest fires and climate change in the 21st century. Mitig Adapt Strat Glob Change 11:847–859

Foster NW, Bhatti JS (2006) Forest ecosystems: nutrient cycling. Encycl Soil Sci, 718721

Franklin J (2010) Vegetation dynamics and exotic plant invasion following high severity crown fire in a southern California conifer forest. Plant Ecol 207(2):281–295

Freeman JP, Stohlgren TJ, Hunter ME, Omi PN, Martinson EJ, Chong GW, Brown CS (2007) Rapid assessment of postfire plant invasions in coniferous forests of the western United States. Ecol Appl 17(6):1656–1665

Gebeyehu MN, Hirpo FH (2019) Review on effect of climate change on forest ecosystem. Int J Environ Sci Nat Res 17(4):126–129

Soil Science Society of America (2001) Glossary of soil science terms. Soil Science Society of America

González-Tagle MA, Schwendenmann L, Pérez JJ, Schulz R (2008) Forest structure and woody plant species composition along a fire chronosequence in mixed pine–oak forest in the Sierra Madre Oriental Northeast Mexico. For Ecol Manage 256(1–2):161–167

Gordo O, Sanz JJ (2010) Impact of climate change on plant phenology in Mediterranean ecosystems. Glob Change Biol 16(3):1082–1106

Grogan P, Burns T, Chapin F III (2000) Fire effects on ecosystem nitrogen cycling in a Californian bishop pine forest. Oecologia 122:537–544

Halofsky JE, Peterson DL, Harvey BJ (2020) Changing wildfire, changing forests: the effects of climate change on fire regimes and vegetation in the Pacific Northwest, USA. Fire Ecol 16(1):1–26

Harvey BJ (2016) Human-caused climate change is now a key driver of forest fire activity in the western United States. Proc Natl Acad Sci 113(42):11649–11650

Heller NE, Zavaleta ES (2009) Biodiversity management in the face of climate change: a review of 22 years of recommendations. Biol Cons 142(1):14–32

Hilwan I (2020) Method of estimating forest fire impact on vegetation. IOP Publishing, England, p 012004

Hobbs RJ, Higgs E, Harris JA (2009) Novel ecosystems: implications for conservation and restoration. Trends Ecol Evol 24(11):599–605

Hoffmann AA, Rymer PD, Byrne M, Ruthrof KX, Whinam J, McGeoch M, Bergstrom DM, Guerin GR, Sparrow B, Joseph L (2019) Impacts of recent climate change on terrestrial flora and fauna: some emerging Australian examples. Austral Ecol 44(1):3–27

Hughes L (2003) Climate change and Australia: trends, projections and impacts. Austral Ecol 28(4):423–443

Hungerford RD, Harrington MG, Frandsen WH, Ryan KC, Niehoff GJ (1991) Influence of fire on factors that affect site productivity, pp 32–50

Inbar A, Lado M, Sternberg M, Tenau H, Ben-Hur M (2014) Forest fire effects on soil chemical and physicochemical properties, infiltration, runoff, and erosion in a semiarid Mediterranean region. Geoderma 221:131–138

India State of Forest Report (2021) Forest survey of India

Jangra A, Kumar K, Maikhuri S, Bhandari MS, Pandey S, Singh H, Barthwal S (2024) Unveiling stress-adapted endophytic bacteria: characterizing plant growth-promoting traits and assessing cross-inoculation effects on Populusdeltoides under abiotic stress. Plant Physiol Biochem, 108610. https://doi.org/10.1016/j.plaphy.2024.108610

Jones R, Barbetti MJ (2012) Influence of climate change on plant disease infections and epidemics caused by viruses and bacteria. Plant Sci Rev 22:1–31

Jones MW, Abatzoglou JT, Veraverbeke S, Andela N, Lasslop G, Forkel M, Smith AJ, Burton C, Betts RA, van der Werf GR (2022) Global and regional trends and drivers of fire under climate change. Rev Geophy 60(3):e2020RG000726

Kaushik P, Singh H (2024) Climate change-induced forest ecosystem functioning and alteration in water-atmosphere-soil–plant (WASP) relationship. In: Singh H (ed) Urban forests, climate change and environmental pollution. Springer, Cham. https://doi.org/10.1007/978-3-031-67837-0_38

Keane RE, Agee JK, Fule P, Keeley JE, Key C, Kitchen SG, Miller R, Schulte LA (2008) Ecological effects of large fires on US landscapes: benefit or catastrophe? Int J Wildland Fire 17(6):696–712

Keeley JE, Lubin D, Fotheringham C (2003) Fire and grazing impacts on plant diversity and alien plant invasions in the southern Sierra Nevada. Ecol Appl 13(5):1355–1374

Khapugin AA, Vargot EV, Chugunov GG, Shugaev NI (2016) Invasion of alien plants in fire-damaged forests at Southern boundary of the taiga zone. For Syst 25(3):eSC13–eSC13

Kirsanov A, Rozinkina I, Rivin G, Zakharchenko D, Olchev A (2020) Effect of natural forest fires on regional weather conditions in Siberia. Atmosphere 11(10):1133

Kumar A, Dwivedi G K, Tewari S, Paul J, Anand R, Kumar N, Kumar P, Singh H, Kaushal R (2020a) Carbon mineralization and inorganic nitrogen pools under *Terminalia chebula* retz.-based agroforestry system in Himalayan Foothills, India. For Sci 66(5):634–643

Kumar A, Kumar P, Singh H, Kumar N (2020b) Adaptation and mitigation potential of roadside trees with bio-extraction of heavy metals under vehicular emissions and their impact on physiological traits during seasonal regimes. Urb For Urb Green 58:126900

Kumar A, Kumar P, Singh H, Bisht S, Kumar N (2021a) Relationship of physiological plant functional traits with soil carbon stock in temperate forest of Garhwal Himalaya. Curr Sci 120(8):1368–1373

Kumar A, Kumar P, Singh H, Kumar N (2021b) Modulation of plant functional traits under essential plant nutrients during seasonal regime in natural forests of Garhwal Himalayas. Plant Soil 465:197–212

Kumar N, Singh H, Kumar A, Singh AK, Pandey PK, Kumar A (2024) Forest-fire-induced land degradation. In: Mishra G, Giri K, Singh S, Kumar M (eds) Sustainable land management in India. Springer, Singapore. https://doi.org/10.1007/978-981-97-5223-2_4

Lee H, Calvin K, Dasgupta D, Krinner G, Mukherji A, Thorne P, Trisos C, Romero J, Aldunce P, Barret K (2023) IPCC 2023: Climate change 2023: synthesis report, summary for policymakers. Contribution of working groups I, II and III to the sixth assessment report of the intergovernmental panel on climate change (Core Writing Team, Lee H, Romero J (eds)). IPCC, Geneva, Switzerland

Littell JS, Peterson DL, Riley KL, Liu Y, Luce CH (2016) A review of the relationships between drought and forest fire in the United States. Glob Change Biol 22(7):2353–2369

MacLean SA, Beissinger SR (2017) Species' traits as predictors of range shifts under contemporary climate change: a review and meta-analysis. Glob Change Biol 23(10):4094–4105

Makiola A, Compson ZG, Baird DJ, Barnes MA, Boerlijst SP, Bouchez A, Brennan G, Bush A, Canard E, Cordier T (2020) Key questions for next-generation biomonitoring. Front Environ Sci 7:197

Menzel A, Sparks TH, Estrella N, Roy D (2006) Altered geographic and temporal variability in phenology in response to climate change. Glob Ecol Biogeogr 15(5):498–504

Miranda A, Coutinho M, Borrego C (1994) Forest fire emissions in Portugal: a contribution to global warming? Environ Pollut 83(1–2):121–123

Mishra M, Guria R, Baraj B, Nanda AP, Santos CAG, da Silva RM, Laksono FAT (2024) Spatial analysis and machine learning prediction of forest fire susceptibility: a comprehensive approach for effective management and mitigation. Sci Total Environ, 171713

Moody JA, Martin DA (2009) Forest fire effects on geomorphic processes. In: Fire effects on soils and restoration strategies. CRC Press, USA, pp 57–96

Morelli TL, Daly C, Dobrowski SZ, Dulen DM, Ebersole JL, Jackson ST, Lundquist JD, Millar CI, Maher SP, Monahan WB (2016) Managing climate change refugia for climate adaptation. PLoS ONE 11(8):e0159909

Nasi R, Dennis R, Meijaard E, Applegate G, Moore P (2002) Forest fire and biological diversity. UNASYLVA-FAO, 36–40

Negi M, Kumar A (2016) Assessment of increasing threat of forest fires in Uttarakhand, using remote sensing and GIS techniques. Glob J Adv Res 3(6):457–468

Nirmal, Bakshi M, Singh H (2021) Effect of $CO_2$ elevation on Shisham growth at nursery stage. Indian J Hill Farm 34:272–276

Nitschke CR, Innes JL (2007) Interactions between fire, climate change and forest biodiversity. CABI Reviews (2006):9

Nwaogu C, Diagi BE, Agidi VA, Okweche SI (2023) Climate change and other environmental factors as drivers of fauna and flora biodiversity in Africa. In: Sustainable utilization and conservation of Africa's biological resources and environment. Springer, Singapore, pp 441–467

Olsson P, Folke C, Berkes F (2004) Adaptive comanagement for building resilience in social–ecological systems. Environ Manage 34:75–90

Parashar A, Biswas S (2003) The impact of forest fire on forest biodiversity in the Indian Himalayas (Uttaranchal)

Pati PK, Kaushik P, Malasiya D, Ray T, Khan M, Khare P (2024) Impacts of forest fire frequency on structure and composition of tropical moist deciduous forest communities of Bandhavgarh Tiger Reserve, Central India. Trees For People 15:100489

Pausas JG, Keeley JE (2021) Wildfires and global change. Front Ecol Environ 19(7):387–395

Pechony O, Shindell DT (2010) Driving forces of global wildfires over the past millennium and the forthcoming century. Proc Natl Acad Sci 107(45):19167–19170

Pellegrini AF, Ahlström A, Hobbie SE, Reich PB, Nieradzik LP, Staver AC, Scharenbroch BC, Jumpponen A, Anderegg WR, Randerson JT (2018) Fire frequency drives decadal changes in soil carbon and nitrogen and ecosystem productivity. Nature 553(7687):194–198

Pellegrini A, Hobbie S, Reich P, Jumpponen A, Brookshire E, Caprio A, Coetsee C, Jackson RB (2020) Repeated fire shifts carbon and nitrogen cycling by changing plant inputs and soil decomposition across ecosystems. Bull Ecol Soc Am

Pellegrini A, Caprio A, Georgiou K, Finnegan C, Hobbie S, Hatten J, Jackson RB (2021) Low-intensity frequent fires in coniferous forests transform soil organic matter in ways that may offset ecosystem carbon losses. Glob Change Biol 27

Piao S, Liu Q, Chen A, Janssens IA, Fu Y, Dai J, Liu L, Lian X, Shen M, Zhu X (2019) Plant phenology and global climate change: Current progresses and challenges. Glob Change Biol 25(6):1922–1940

Platts PJ, Mason SC, Palmer G, Hill JK, Oliver TH, Powney GD, Fox R, Thomas CD (2019) Habitat availability explains variation in climate-driven range shifts across multiple taxonomic groups. Sci Rep 9(1):15039

Prakash V., Hunney K. and Singh H. 2022. Effect of elevated carbon dioxide on growth and development of *Santalum album* L. seedlings inoculated with plant growth promoting microorganisms in open top chambers. Int J Environ Agric Biotechnol 7(5):149–159

Pureswaran DS, Roques A, Battisti A (2018) Forest insects and climate change. Curr For Rep 4:35–50

Raison RJ, Khanna PK, Jacobsen KL (2009) Effects of fire on forest nutrient cycles Joan Romanya and Isabel Serrasolses. In: Fire effects on soils and restoration strategies. CRC Press, USA, pp 241–272

Raturi A, Singh H, Kumar P and Shukla N 2022. Characterizing the post-monsoon $CO_2$, $CH_4$, $N_2O$, and $H_2O$ vapor effluxes from a tropical wetland in the Himalayan foothill. Environ Monit Assess 194(50). https://doi.org/10.1007/s10661-021-09721-8

Reeve N, Toumi R (1999) Lightning activity as an indicator of climate change. Q J R Meteorol Soc 125(555):893–903

Reich PB, Peterson DW, Wedin DA, Wrage K (2001) Fire and vegetation effects on productivity and nitrogen cycling across a forest–grassland continuum. Ecology 82(6):1703–1719

Ricciardi A, Simberloff D (2009) Assisted colonization is not a viable conservation strategy. Trends Ecol Evol 24(5):248–253

Robichaud PR (2000) Fire effects on infiltration rates after prescribed fire in Northern Rocky Mountain forests, USA. J Hydrol 231:220–229

Rosales J (2008) Economic growth, climate change, biodiversity loss: distributive justice for the global north and south. Conserv Biol 22(6):1409–1417

Roshan A, Biswas A (2023) Fire-induced geochemical changes in soil: implication for the element cycling. Sci Total Environ 868:161714

Sample M, Thode AE, Peterson C, Gallagher MR, Flatley W, Friggens M, Evans A, Loehman R, Hedwall S, Brandt L, Janowiak M, Swanston C (2022) Adaptation strategies and approaches for managing fire in a changing climate. Climate 10(4)

Savita, Sharma SK, Singh H, Singh O (2017) Forest fire: a review. Indian For 143(8):719–728

Scharenbroch BC, Nix B, Jacobs KA, Bowles ML (2012) Two decades of low-severity prescribed fire increases soil nutrient availability in a Midwestern, USA oak (Quercus) forest. Geoderma 183:80–91

Sett T, Nikam BR, Singh H, Purohit S (2022) Effect of drought stress on forest evapotranspiration—a case study on Indian forests. EGU General Assembly 2022. https://doi.org/10.5194/egusphere-egu22-510

Sett T, Nikam BR, Singh H, Purohit S (2024) Hydrological variability in Indian forest ecosystem: analysis of drought resilience, recovery and water use efficiency in moist and dry deciduous forests. In: Singh H (ed) Urban forests, climate change and environmental pollution. Springer, Cham. https://doi.org/10.1007/978-3-031-67837-0_37

Sharma R, Singh H, Kaushik M, Nautiyal R, Singh O (2018) Adaptive physiological response, carbon partitioning and biomass production of *Withania somnifera* (L.) Dunal grown under elevated $CO_2$ regimes. 3 Biotech 8:267. https://doi.org/10.1007/s13205-018-1292-1

Sharma R, Singh H (2021) Alteration in biochemical constituents and nutrients partitioning of *Asparagus racemosus* in Response to elevated atmospheric CO2 concentration. Environ Sci Pollut Res 29:6812–6821. https://doi.org/10.1007/s11356-021-16050-3

Shoo LP, Williams SE, Hero J (2006) Detecting climate change induced range shifts: where and how should we be looking? Austral Ecol 31(1):22–29

Singh S (2022) Forest fire emissions: a contribution to global climate change. Front For Glob Change 5:925480

Singh H, Sharma R, Savita SMP, Kumar M, Verma A, Ansari MW, Sharma SK (2018) Adaptive physiological response of *Parthenium hysterophorus* to elevated atmospheric $CO_2$ concentration. Indian Forester 144:1–14

Singh M, Singh H, Kumar A, Kumar M, Barthwal S, Thakur A (2024b) Soil nitrogen availability determines the $CO_2$ fertilization effect on tree species (*Neolamarckia cadamba*): growth and physiological evidence. Environ Sustain 7:53–60. https://doi.org/10.1007/s42398-023-00300-w

Singh H, Kumar M (2022) Climate change and its impact on Indian Himalayan Forests: current status and research needs. In: Rani S, Kumar R (eds) Climate change. Springer Climate. Springer, Cham, pp 223–242. https://doi.org/10.1007/978-3-030-92782-0_11

Singh H, Kumar N, Kumar M and Singh R (2020a) Modelling habitat suitability of western tragopan (*Tragopan melanocephalus*) a range-restricted vulnerable bird species of the Himalayan region, in response to climate change. Clim Risk Manage 29:100241. https://doi.org/10.1016/j.crm.2020.100241

Singh S, Singh H, Sharma SK, Nautiyal R (2021c) Seasonal variation in biochemical responses of bamboo clones in the sub-tropical climate of Indian Himalayan foothills. Heliyon 7(4):e06859. https://doi.org/10.1016/j.heliyon.2021.e06859

Singh H, Kumar N, Singh R, Kumar M (2023) Assessing the climate change impact on the habitat suitability of the range-restricted bird species (*Catreus wallichii*) in the Indian Himalayan ecosystem. Environ Sci Pollut Res. https://doi.org/10.1007/s11356-023-30789-x

Singh H, Kumar N, Kumar A (2024) Enhancing resource use efficiency in crops through plant functional traits. Springer, Singapore, pp 97–117. ISBN: 978-981-97-1509-1

Singh H (2021) Process-based carbon sequestration study in relation to the energy-water-carbon flux in a forest ecosystem. In: Kumar P, Singh RK, Kumar M, Rani M, Sharma (eds) Climate impacts on sustainable natural resource management. CRC Press, Taylor & Francis Group, United States, pp 336–361

Singh H (2024a) Climate change and forests: biological perspectives. In: Singh H (ed) Forests and climate change. Springer, Singapore. https://doi.org/10.1007/978-981-97-3905-9_1

Singh H (ed) (2024b) Forests and climate change: biological perspectives on impact, adaptation, and mitigation strategies. Springer, Singapore. https://doi.org/10.1007/978-981-97-3905-9

Singh H (2024c) Exploring adaptive modulation in plant functional traits and its impact on the productivity of *Acacia auriculiformis* under $CO_2$-enriched environment. Ind Crops Prod 210(2024):118186. https://doi.org/10.1016/j.indcrop.2024.118186

Singh H (ed) (2024d) Urban forests, climate change and environmental pollution: physio-biochemical and molecular perspectives to enhance urban resilience. Springer, Cham, Switzerland. https://doi.org/10.1007/978-3-031-67837-0

Singh H (2024e) Climate change and environmental pollution: biological consequences on urban forests. In: Singh H (ed) Urban forests, climate change and environmental pollution. Springer, Cham, Switzerland. https://doi.org/10.1007/978-3-031-67837-0_1

Skogen K, Helland H, Kaltenborn B (2018) Concern about climate change, biodiversity loss, habitat degradation and landscape change: embedded in different packages of environmental concern? J Nat Conserv 44:12–20

Slik JF, Bernard CS, Van Beek M, Breman FC, Eichhorn KA (2008) Tree diversity, composition, forest structure and aboveground biomass dynamics after single and repeated fire in a Bornean rain forest. Oecologia 158:579–588

Stephens SL, Ruth LW (2005) Federal forest-fire policy in the United States. Ecol Appl 15(2):532–542

Tessler N, Wittenberg L, Greenbaum N (2016) Vegetation cover and species richness after recurrent forest fires in the Eastern Mediterranean ecosystem of Mount Carmel, Israel. Sci Total Environ 572:1395–1402

Thonicke K, Cramer W (2006) Long-term trends in vegetation dynamics and forest fires in Brandenburg (Germany) under a changing climate. Nat Hazards 38:283–300

Trenberth KE (2011) Changes in precipitation with climate change. Climate Res 47(1–2):123–138

Úbeda B, Di Giacomo AS, Neiff JJ, Loiselle SA, Guadalupe Poi AS, Galvez JA, Casco S, Cozar A (2013) Potential effects of climate change on the water level, flora and macro-fauna of a large neotropical wetland. PLoS ONE 8(7):e67787

Úbeda X, Outeiro LR (2009) Physical and chemical effects of fire on soil. In: Fire effects on soils and restoration strategies. CRC Press, USA, pp 121–148

Valderrama L, Contreras-Reyes JE, Carrasco R (2018) Ecological impact of forest fires and subsequent restoration in Chile. Resources 7(2):26

Weber MG, Flannigan MD (1997) Canadian boreal forest ecosystem structure and function in a changing climate: impact on fire regimes. Environ Rev 5(3–4):145–166

Wells CG (1979) Effects of fire on soil: a state-of-knowledge review

Xue L, Li Q, Chen H (2014) Effects of a wildfire on selected physical, chemical and biochemical soil properties in a *Pinus massoniana* forest in South China. Forests 5(12):2947–2966

Yadav SK, Singh H, Nautiyal R, Ginwal HS, Ansari SA, Barthwal S (2019a) Modulation of morpho-physiological responses in Populus deltoides by elevated carbon dioxide and temperature. For Sci 66:105–118. https://doi.org/10.1093/forsci/fxz048

Yadav SK, Singh H, Ginwal HS, Barthwal S (2019b) Elevated CO2 enhanced growth and physiological process of *Populus deltoides Bartr*. ex Marsh. Indian Forester 145(1):23–27

# Ecosystem Recovery and Resilience After Forest Fires

**Kashish Walia, Yourmila Kumari, Garima, and Aruna Mehta**

**Abstract** In the wake of forest fires, ecosystems engage in a dynamic recovery, revealing an intricate interplay between ecological factors and innate resilience mechanisms in natural landscapes. This chapter delves into the complexities of ecosystem recovery and resilience post-forest fires, exploring immediate and enduring consequences. It begins with analyzing post-fire effects and studying vegetation, soil, and biodiversity changes. Emphasis is on factors shaping fire severity and initial conditions for recovery. Ecosystem recovery spans stages, from rapid colonization by fire-adapted species to facilitating soil stabilization for complex vegetation. Resilience manifests as the ecosystem absorbs disturbances and reorganizes, with biotic interactions enhancing it. Climate, topography, and human interventions influence recovery. The chapter emphasizes interdisciplinary approaches integrating ecology, forestry, and indigenous knowledge for effective strategies. Amid rising natural disturbances, it aids researchers, land managers, and policymakers in fostering resilience post-wildfires, guiding sustainable conservation practices and policies.

## 1 Introduction

Forest fires are a regular occurrence in numerous ecosystems globally, playing a vital role in maintaining ecological balance and renewal, framing landscapes, and preserving biodiversity. However, in recent decades, the frequency and intensity of forest fires have notably increased, posing substantial challenges to ecosystems, wildlife, and human communities (Barwa and Singh 2024). This escalation in forest fires can be ascribed to various elements like climate change, human activities, and changes in land management practices (Mustafa 2009). Comprehending the origins and impact of forest fires within ecosystems is fundamental for implementing efficient management and conservation strategies. Fires pose multifaceted physical, biological, ecological, and environmental hazards and play a decisive role in nutrient

K. Walia (✉) · Y. Kumari · Garima · A. Mehta
Dr YS Parmar University of Horticulture and Forestry, College of Horticulture and Forestry, Thunag, Mandi, Himachal Pradesh, India
e-mail: kashishevs@outlook.com

 
H. Singh and N. Kumar (eds.), *Forest Fire and Climate Change*,
https://doi.org/10.1007/978-3-031-89967-6_7

cycling, seed dispersal, and creating diverse habitats for various plant and animal species (Jaiswal et al. 2002).

Climate change has increased temperatures and decreased precipitation in numerous regions, making more favourable environments for wildfires to occur and spread. Additionally, anthropogenic activities, such as land clearance, agricultural practices, and urban expansion, have increased the likelihood of ignition sources and fuel availability, further heightening the danger of forest fires (Nirmal et al. 2021; Singh and Kumar 2022; Savita et al. 2017). The effect of forest fires is influenced by ecosystem characteristics such as sensitivity to fire and adaptation as well as fire regime factors like intensity and duration (Bradstock et al. 2009). Understanding the interplay among these factors is vital for formulating efficient tactics to mitigate the impacts of forest fires on ecosystems. Notably, in India, there has been a substantial increase in reported forest fire incidents over the years, with regions like Central India, Maharashtra, Chhattisgarh, and Orissa being particularly vulnerable (FSI 2012). The impact of fires extends beyond immediate flora and fauna, profoundly reshaping ecosystems and soil properties. Forest fires impact ecosystems differently, from immediate vegetation destruction to enduring soil properties and biodiversity alterations (Prakash et al. 2022). Wildfires are recognized as significant disturbances in terrestrial ecosystems, impacting forests and woodlands by affecting major ecosystem processes such as biochemical and hydrological cycles. While forest fires traditionally play an integral role in ecological succession by acting as environmental filters and shaping ecosystem communities, frequent high-intensity fires in recent decades can result in permanent changes to ecosystems and their components (Singh et al. 2021; Yadav et al. 2019b; Singh et al. 2023; Sharma and Singh 2021).

The stringency of these impacts depends on various factors, including the intensity and duration of the fire, the type of vegetation present, and the ecosystem's resilience to fire (Wang et al. 2012). Understanding forest fires' social and ecological impacts is essential for cultivating efficient management and adaptation strategies to mitigate their effects. Effective management and conservation of forests require a holistic approach that considers the complex interactions between forest fires, ecosystems, and human communities. This approach involves implementing strategies to reduce forest fires through improved land management practices, early detection and suppression of wildfires, and restoration of fire-adapted ecosystems (Kumar et al. 2024). Additionally, it is essential to integrate traditional ecological knowledge and modern science to create sustainable strategies that enhance the resilience and adaptability of ecosystems to upcoming fire occurrences (Hiremath and Sundaram 2005). Integrating traditional ecological knowledge, scientific research, and community engagement can create a more sustainable and resilient future for forests and the ecosystems they support. Thus, this chapter critically examines the impacts of forest fires on forest ecosystems, ecosystem recovery, and resilience after forest fires.

## 2 Effect of Forest Fire on Ecosystem

Fire has been a fundamental component of forest ecosystems, significantly shaping the composition and functions of plant and animal life (Smith 1995). Opinions on the impacts of forest fires vary widely, categorizing fire incidents as detrimental and advantageous, depending on their location and characteristics. Many ecosystems benefit from periodic fires, which clear away dead organic matter, release nutrients, modify soil properties and create diverse landscapes influencing fire behaviour and ecological processes. These benefits support the survival and reproduction of certain plant and animal populations (Brown 2022). Additionally, fire plays a constructive role in preserving the health of fire-dependent ecosystems. Fire can also swiftly consume biomass, leading to adverse outcomes like soil erosion and air pollution. Acknowledging the dual nature of fire, it should be integrated as a crucial component in ecosystem management. However, the impacts of fire can vary depending on factors such as the fire regime, vegetation, climate, and spatial and temporal scales (Fowler 2003).

### *2.1 Effect on Vegetation and Habitats*

Forest fires represent a significant ecological process with numerous impacts on forests, particularly in India, where they present a prevalent challenge. Research indicates that approximately 90% of vegetation fires in the country stem from human activities. These fires are often initiated through deliberate acts of arson or unintentional ignition sources, such as those used by indigenous communities for gathering non-timber forest products (Saha and Howe 2003). Annually, these fires affect an estimated 37,300 km$^2$ of forest area (Srivastava and Garg 2013). The susceptibility of Indian forests to fire is influenced by factors such as vegetation type, climate conditions, and the timing of fire occurrences. For instance, coniferous forests in the Himalayan region, consisting of fir, spruce and cedar, are particularly vulnerable to fire. Different areas of the country experience distinct fire seasons, typically spanning from January to June. Among India's 16 types of forests, tropical deciduous forests face intense pressure from recurrent fires. The highest number of forest fires occurs in tropical dry deciduous forests, tropical moist deciduous forests, and tropical semi-evergreen forests. Factors such as high resin content in sub-tropical pine regions and dry conditions in tropical areas contribute to fire spread in India (FSI 2012). Forest fires change vegetation cover, structure, composition, density, and productivity, resulting in deforestation, population decline, edge effects, and the introduction of exotic species (Gomez-Rey et al. 2013).

The extent of damage and tree response to fire depends on factors such as fire intensity, severity, soil heating, burn season, residence time, and time since the last fire. Physical and climatic factors and biological factors like plant morphology and physiology also influence post-fire effects on plant communities (Sharma et al. 2018;

Jangra et al. 2024; Dhyani et al. 2021; Singh 2024a). While trees and shrubs are more impacted than herbaceous species, high-severity fires significantly decrease cover for all plant types. Forest regeneration is hindered by fire through the destruction of stem tissues, soil warming, and impact on growth and development attributes such as flowering, seed dispersal, germination, and seedling establishment (De Luis et al. 2005). Fire also affects the abundance and variety of tree seedling species and may cause the killing of both root-crown re-sprouters and root-sprouters. The decline in species richness after a forest fire may be attributed to eliminating early species outcompeted by rapidly growing fire-hardy species (Jhariya and Oraon 2012). Fires can detrimentally affect the diversity of native plants, potentially leading to localized extinctions (Kittur et al. 2014).

## 2.2 *Effect on Soil*

Soil is a vital natural resource pivotal in various socio-ecological systems by supporting essential functions like nutrient cycling, mineral storage, carbon sequestration, and plant growth (Osman 2013). Any degradation in the biological, chemical, and physical attributes of forest soils can significantly diminish their ability to perform these functions with effects that may be temporary or permanent. Deforestation, fires, erosion, and soil contamination are vital drivers of soil degradation within forest ecosystems. The impact of forest fires on soil properties can vary in duration, whether short-term, long-term, or permanent, depending on the specific property affected, fire severity, frequency, and post-fire climatic conditions. Furthermore, forest fires contribute significantly to soil degradation and nutrient losses through processes like volatilization and erosion (Gomez-Rey et al. 2013). These fires also degrade the biological and physicochemical quality of soils and deplete the nutrient pool through mechanisms such as volatilization, oxidation, ash transfer and erosion (Pellegrini et al. 2018). Forest fires have multifaceted effects on soil properties, both physical and chemical. Soil texture is altered due to the destruction of clay lattice structures, leading to increased sand and silt particles post-fire.

Additionally, burned soil experiences reduced moisture content and increased water repellency, exacerbating soil erosion. Forest fires also influence soil bulk density, with conflicting results reported in the literature. While some studies show a slight increase in bulk density due to soil aggregation deterioration and organic matter loss, others suggest lower bulk densities post-fire. Soil aggregate stability is affected by fire severity and soil type, with some studies indicating increased stability in burned soils due to the development of organo-mineral complexes. Chemically, wildfires result in a decline in soil organic matter content, changes in soil pH, and alterations in nutrient availability, primarily attributed to ash deposition and organic matter combustion. Nitrogen loss occurs during fires through volatilization, but biological nitrogen fixation may increase post-fire. The impact of fires on soil biological properties includes declines in microbial biomass carbon and enzymatic activities, although some studies report increased microbial biomass in burned soils

(Singh et al. 2024a, 2018; Yadav et al. 2019a; Singh 2021). These changes in soil properties significantly affect ecosystem functioning and post-fire recovery (Hinojosa et al. 2021).

## 2.3 *Effect on Wildlife*

The direct impacts of fires on animals include the destruction of those residing in the affected area and the alteration of ecological characteristics of the habitat, rendering it unfit for previously present species while potentially enabling the establishment of new species with different ecological requirements. The extent of direct destruction varies among species depending on their position in the zootaxy, with larger and quicker-moving mammals and birds generally having lower mortality rates as they can escape to safer areas. Conversely, smaller animals like snails, spiders, frogs, and lizards often experience higher mortality rates due to their limited ability to evade fires. Effective soil heat insulation enables the survival of certain organisms like ant hills and larval insects despite the intensity of forest fires. Indirect effects of fires, such as the elimination or alteration of animal habitats, result in more conspicuous and enduring changes in forest fauna, as many animal species have specific habitat requirements beyond initial assessments of the environment. The ability of fauna species to withstand fires is primarily determined by their capacity to escape, which depends on the speed of the fire at which it spreads. In rapid fires, even more extensive and faster animals may be unable to flee. Overall, nests and shelters are often destroyed in burnt areas, prompting local fauna to migrate. Consequently, the disruption of vegetation and the introduction of new fauna upset the balance of the microenvironment, leading to significant alterations in the local natural habitat (Ulery and Graham 1993).

## 2.4 *Effect on Atmosphere*

Forest fires release vast amounts of organic micro particles into the atmosphere through incomplete combustion, leading to air pollution. These fires also contribute substantial quantities of carbon dioxide ($CO_2$), exacerbating the greenhouse effect and accelerating global warming (Raturi et al. 2022; Sett et al. 2022; Singh et al. 2020; Apurva et al. 2017). The reduction in woodland territories and tree numbers, decreased photosynthetic activity, and oxygen production further worsen the situation. Incomplete burning also releases carbon monoxide (CO) into the atmosphere, posing severe health risks to animals and firefighters. Furthermore, the combustion of vegetation generates additional toxic byproducts such as volatile oils, benzene, and its derivatives, which are emitted during the burning and decomposition of wooden materials. These pollutants, including organic micro particles, pose hazards near the fire site and can be transported over long distances and higher altitudes by air movements generated by the fire, leading to pollution in remote areas (Jordán et al. 2011).

### 2.5 *Effect on Water*

Forest fires can lead to both direct and indirect pollution of water bodies. Harmful chemicals generated by fires enter the atmosphere and are subsequently added to water bodies through precipitation or sedimentation. Ash, debris, and small plant fragments from the fire pollute nearby surface waters. The wind carries pollutants over longer distances, polluting distant water bodies. Firefighting contributes to water pollution, as water used for firefighting may wash other extinguishing chemicals into nearby surface waters, contaminating rivers and lakes. Some substances, initially neutral in soil but activated in water through chemical reactions, can cause dangerous pollution over larger areas, potentially threatening or eliminating water ecosystems. The sudden spread of large volumes of water used in firefighting, particularly from aerial methods, disrupts the water balance of the affected area, leading to disturbances in the biological balance of natural surface and underground waters. This disruption results in the depletion of animal and plant populations due to rapid changes in living conditions. Extreme cases may lead to the loss of life in the affected area (Heydari et al. 2017).

## 3 Ecosystem Restoration After Fire

Ecosystems, resilient and adaptive by nature, undergo intricate processes of recovery and resilience in the aftermath of a fire. The ecological response to fire is a dynamic interplay of various factors that shape the recovery trajectory. Following a blaze, ecosystems engage in remarkable renewal processes, often called ecological succession. This intricate process involves gradually reestablishing flora and fauna and restoring the ecosystem structure and function. Additionally, pioneer species and succession dynamics are essential in this rejuvenation. Hence, understanding ecosystem recovery processes after a fire is crucial for ecological insights and informed conservation and management strategies. This introduction sets the stage for a deeper exploration of the multifaceted aspects involved in the post-fire recovery of ecosystems.

### 3.1 *Forest Land Restoration Approaches and Strategies*

Post-fire Forest Land Restoration (FLR) is a complicated process that includes planning, design, implementation, monitoring, and evaluation phases, with success depending on the chosen restoration approach. Effective restoration is crucial for reinstating ecological functions and providing economic incentives to local communities (De Jong et al. 2021). Various strategies, including active, passive, and mixed approaches, are employed based on damage severity, available resources, and restoration objectives. In less severely degraded lands, natural regeneration proves cost-effective for improving biodiversity, while active restoration may be necessary for faster recovery. Social demands and advancements in fire and restoration ecology have led to evolving approaches integrating goals like climate change adaptation, combating desertification, and biodiversity conservation (Pineda et al. 2020; Singh 2024b). Achieving long-term restoration goals requires scientific insights, technological advancements, appropriate legislation, and funding (Vallejo and Alloza 2012). Involving local communities, fostering cooperation, and exploring investment opportunities are pivotal for successful forest restoration (Nyongesa and Vacik 2019). Traditional knowledge and modern techniques are vital in ecological restoration and biodiversity conservation (Poletti et al. 2018). Considering environmental capacities, safety, and climate change expectations is essential in setting restoration goals (Chen et al. 2014; Singh 2024c). A six-step plan for sustainable FLR includes enhancing environmental knowledge, adaptive management, modeling, social improvements, participatory monitoring, and effective communication (Noulekoun et al. 2021). Understanding spatio-temporal information on forest fires aids risk reduction and informs integrated fire management policies (Chu and Guo 2014). Remote Sensing is a valuable tool for preventing and monitoring forest fires, providing insights into post-fire ecosystem dynamics (Chen and Chen 2022). It helps estimate burnt areas, assess natural recovery ability, plan restoration interventions, and monitor outcomes (Leigh et al. 2006). Landsat time series analysis and long-term data in various regions highlight the impact of topography and climatic factors on forest vegetation restoration after fires (Naficy et al. 2016). Comprehensive post-fire forest monitoring and evaluation are essential for adaptive management and redirection of restoration actions (Moreira et al. 2012; Singh 2024d).

### 3.1.1 Natural Regeneration Allows Ecological Restoration

Restoration denotes the intentional process of aiding the recovery of an ecosystem that has suffered degradation, damage, or destruction. Post-fire forest restoration focuses explicitly on restoring the ecosystem to its historical state, aiming to enhance ecological integrity and resilience (Brancalion and Chazdon 2017). This undertaking is essential for biodiversity conservation, food security, livelihood needs, and mitigating economic losses and wildfire damage (Scheper et al. 2021). To effectively reduce and respond to these consequences, a dual strategy is necessary:

- Preventing or Minimizing Degradation
- Restoring Degraded Ecological Environment (Stanturf and Mansourian 2020).

The core principles of restoration should encompass a holistic understanding and integration of ecological, economic, and social factors related to impacted forest ecosystems (De Jong et al. 2021). Considering various aspects such as recovery rate, degradation levels, land use patterns, topographic characteristics, restoration goals, potential constraints, and available resources is crucial in selecting the appropriate restoration approach. Successful forest restoration necessitates collaboration among stakeholders, including local communities, government authorities, and scientists (Derak et al. 2018). Typical approaches to forest restoration are utilized for rehabilitating extensive deforested and degraded areas. Experiences suggest that passive restoration may be as effective as active restoration in fire-affected landscapes, particularly when soils are minimally impacted and have low erosion risks (Crouzeilles et al. 2017). Passive restoration involves natural regeneration without additional interventions, while active restoration employs management techniques such as seeding or planting seedlings (Barbati et al. 2010). While passive methods are cost-effective for restoring large areas, they may entail future expenses. A mixed approach integrating passive and active restoration through assisted natural regeneration provides a balanced strategy (Noulekoun et al. 2021). Maintaining fire refugia is crucial for enabling natural ecosystem recovery as they provide havens for biodiversity and promote the regeneration of vegetation post-fire. These refugia, whether in individual trees or small forest patches, are typically less influenced by the severity or frequency of fires due to factors such as topography, isolation, or vegetation composition (Meddens et al. 2018). Identifying and preserving fire refugia is particularly important for facilitating the colonization of fire-sensitive trees and promoting biodiversity (Landesmann and Morales 2018). Albornoz et al. (2013) found that in fragmented post-fire successional areas, the size of remnant vegetation patches positively correlated with plant richness and woody species abundance, enhancing tree regeneration. Fire refugia also play a crucial role in maintaining genetic diversity by serving as reservoirs for species variability (Torres and Renison 2017). Even minor patches or individual trees within fire refugia can help in biodiversity conservation and ecosystem resilience (Gonzalez and Veblen 2007).

Furthermore, fire refugia support mycorrhizal networks, aiding nutrient uptake and the establishment of non-pioneer plant species. However, the relevance of fire refugia varies depending on ecosystem scale, heterogeneity, and contrasting land uses across regions. Topography also influences ecosystem restoration with pronounced slopes and accessibility to forest areas favoring natural regeneration, while urbanization and infrastructure inhibit it (Dos Santos et al. 2019). Additionally, elevation plays a significant role in ecosystem recovery, as temperature gradients affect species traits and growth rates (Lippok et al. 2013; Malik et al. 2024; Singh et al. 2024a, b). Also, ecological Forest Restoration, which involves carefully thinning out trees and using controlled burns, can significantly lower the chances of massive fires in forests adapted to fire, like those with conifer trees (Singh 2024e). Plus, it's good for fighting the impacts of drought and climate change. It also brings significant advantages for air and water quality, storing carbon, and providing homes for wildlife (Kumar et al. 2021a, b) (Fig. 1).

However, it's crucial to grasp how fire refuges, terrain features, and elevation changes interact in fire-prone areas. This understanding is key for successfully restoring ecosystems in these environments. Despite the importance of post-fire forest

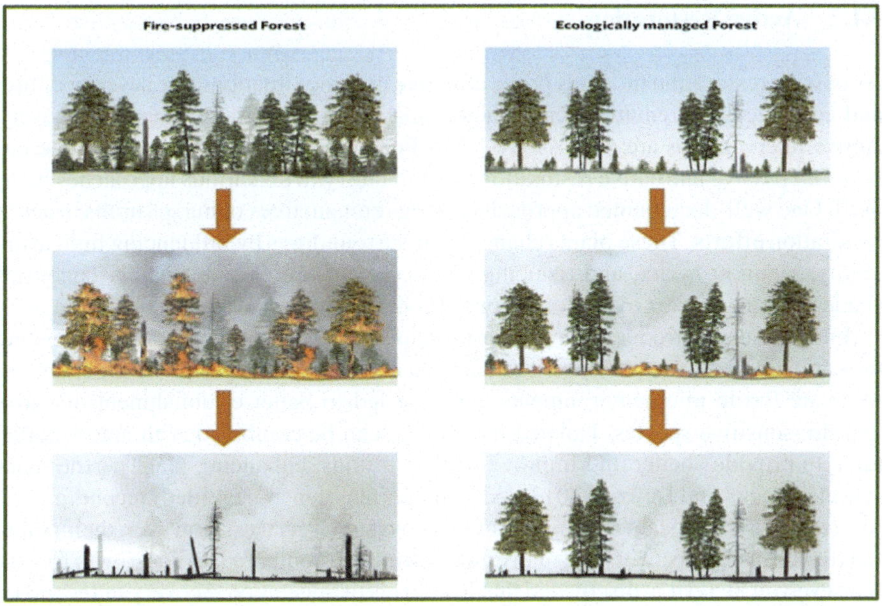


**Fig. 1** Restorative forest management for fire resilience

restoration, global efforts are limited, with less attention given to post-fire management, fire prevention, and suppression (Moreira da Silva et al. 2017). However, after a forest fire, nature undergoes a notable renewal and transformation known as ecological succession (Fig. 2). This dynamic phenomenon unfolds in stages, shaping the recovery of ecosystems.

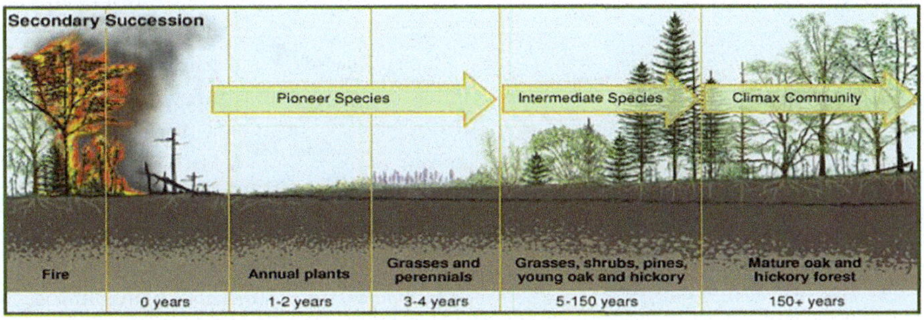


**Fig. 2** Stages of ecological succession in a terrestrial ecosystem

### 3.1.2 Active Restoration

As passive restoration methods prove effective in some situations, the severity of fires and ecological and environmental factors may hinder natural ecosystem recovery. Active interventions are necessary (USDA Forest Service 2012). However, the line between passive and active restoration is often blurry, with various approaches being used. One well-documented approach is using remnant trees or nurse shrubs in active restoration efforts. These plants change their surroundings by influencing light availability, retaining water, and reducing competition from grasses, mainly benefiting shade-tolerant species over pioneer ones (Galindo et al. 2017).

Furthermore, introducing nurse shrubs can help reduce fire incidents in restoration areas (Fig. 2). Improving site conditions, such as adding organic amendments to create fertile microenvironments, has also led to better establishment of native and fire-sensitive species. Isolated individuals can be combined with abiotic structures to provide shelter and improve soil conditions, enhancing plant performance (Urretavizcaya and Defosse 2013). Although often seen as degraded, second-growth forests play a significant role in providing ecosystem services. However, their potential for ecosystem recovery is usually underestimated compared to old-growth forests. Sansevero et al. (2017) discovered that low levels of vegetation recovery in secondary forests in the Brazilian Atlantic Forest suggested a pattern of stalled succession driven by the dominance of fire-resistant species. Strategies like enrichment plantings, nucleation techniques, and assisted natural regeneration are recommended to speed up forest restoration. Proposals focusing on ecologically-oriented forest transformation toward ecosystems resembling natural structural and compositional variability offer promising paths for increasing forest resilience. However, the success of such initiatives largely hinges on the dominance of species and ecological attributes selected to evaluate ecosystem recovery (Gonzalez and Veblen 2015) (Fig. 3).

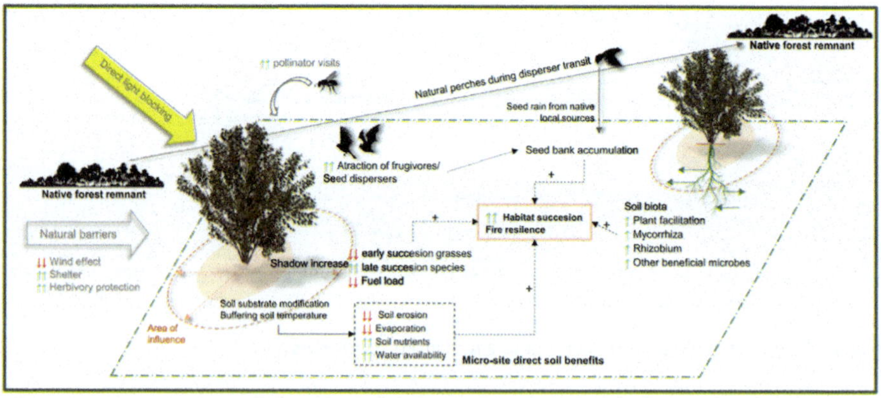


**Fig. 3** Aboveground and Belowground benefits of using nurse shrubs in ecological restoration

### 3.1.3 Attracting Seed Dispersers

Maintaining fire refugia or standing burned woody vegetation supports passive restoration by providing structural elements that attract frugivorous birds. These birds aid in seed dispersal during the early stages of post-fire regeneration (Cavallero et al. 2013). The presence of trees attracts flying visitors, increasing the dispersal of seeds by birds. These flying vertebrates play a crucial role in dispersing seeds of various fleshy-fruited broadleaf species from undisturbed cloud forest patches, thereby promoting cloud forest regeneration. This initial restoration strategy can be further enhanced with active interventions, such as planting successional tree species (Rost et al. 2015).

## *3.2 Post-fire Forest Restoration Challenges*

The significant obstacle to post-fire forest restoration is the lack of sufficient funding, identified as a critical limitation (Noss et al. 2006). Developing countries, in particular, face challenges due to insufficient funding, which hinders the effective implementation of restoration actions (Venier et al. 2018). Integrated spatial planning through national restoration programs and adaptive management is proposed to effectively manage limited resources and prioritize actions based on the current forest burning and fuel processing situation. Increasing the budget for forest fire management in respective regions is also recommended (Kganyago and Shikwambana 2019). Inconsistent and incomplete documentation of wild land fire incidents inhibits the evaluation of trends within and between countries, posing a hurdle to post-fire forest restoration (Meyer et al. 2021). Administrative obstacles, including a lack of coordination among stakeholders, management practices such as responsibility fragmentation, and the absence of planning for formulating restoration policies, contribute to limitations in effective forest restoration after fires (Arnberger 2012). Social and cognitive constraints such as landowner unwillingness, lack of awareness, and conflicts over natural resources impede optimal forest restoration efforts. Cross-border cooperation is essential to reduce the scope and number of high-risk wildfires (Sakellariou et al. 2021). Landscape heterogeneity, encompassing physical, ecological, and social characteristics, challenges Forest Land Restoration (FLR) efforts. Social constraints, including landowner reluctance and lack of awareness, contribute to barriers in FLR planning (Wotton et al. 2003). In a broader context, there are challenges in integrating various benefits, managing competing interests, and addressing different scales and priorities worsened by rapid population growth, extreme climate conditions, frequent wildfires, and shifting land use patterns. The lack of awareness about the importance of agroforestry practices and inadequate incentives hinder their inclusion in Forest Land Restoration (FLR) planning (Iniguez et al. 2022). Misguided fire management practices, such as tree felling and livestock grazing in affected forests, impede natural regeneration and prolong recovery (Drever et al. 2006). Policy and legal constraints pose challenges emphasizing the need for

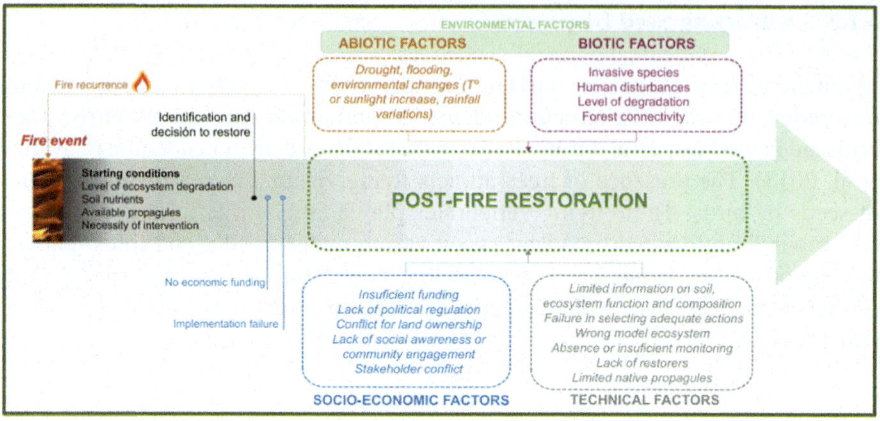


**Fig. 4** Diagram illustrating key factors for evaluating ecological restoration in fire-affected forest ecosystems

negotiation, reconciliation, and addressing power imbalances for successful forest restoration (Nerfa et al. 2021) (Fig. 4).

## 3.3 *Practical, Policy and Future Research Implications*

Post-fire forest restoration treatments have improved forest health resilience to drought and reduced future fire risks. Assessing success in Forest Landscape Restoration involves evaluating social impacts and environmental indicators across scales and over time (Stanturf and Mansourian 2020). Spatial patterns and long-term trends inform current and future restoration strategies, guiding effective remediation efforts (Hohl et al. 2020). Integrating national-scale spatial planning is crucial for managing limited resources and setting balanced restoration priorities that align with diverse goals (Oh et al. 2020). This requires a monitoring and control system to assess outcomes at each restoration stage, facilitating adaptive management for long-term resilience. Considering the complexity of ecosystems, site conditions, and diverse social and political systems is essential (Lamb et al. 2012). Forest restoration goals may extend beyond ecological aspects to encompass aesthetic amenities, social impacts, and sustainable economic development (Kauffman 2004). However, the lack of a systematic compilation of post-fire Forest Landscape Restoration (FLR) projects globally poses challenges in addressing drivers of forest loss and degradation, particularly related to agricultural progress, infrastructure development, mining, urbanization, and tenure security (Stanturf and Mansourian 2020). Inclusive legislation adaptable to new scientific insights is crucial (Iniguez et al. 2022).

Optimal post-fire management decisions should consider environmental, social, legal, and economic aspects (Venier et al. 2018). Considering multiple response variables, a holistic approach is essential for quantifying ecosystem recovery. Land managers and policymakers often lack guiding principles despite global forest restoration targets. Based on reviewed studies and best practices, a comprehensive framework with six critical criteria and 29 indicators has been compiled to guide post-fire forest restoration efforts (Morrison and Lindell 2011) (Table 1).

**Table 1** Criteria and indicators for inclusion in the comprehensive framework for post-forest fire restoration

| Sr. No. | Criteria | Indicators | References |
|---|---|---|---|
| 1 | Economic | Economic benefits (productions: NTFP + TFP)<br>True yield (outputs–inputs)<br>Restoration costs and budget<br>Funding and investment<br>Labour force Ecotourism | Barbati et al. (2010) |
| 2 | Environment | Landscape structure and condition Soil: erosion and fertility<br>Hydrology: water sources and volume<br>Environmental services: carbon sequestration<br>Disturbance types, frequency, and history conservation | Spies et al. (2017) |
| 3 | Social | Community services and social engagement<br>Livelihood opportunities<br>Life quality stakeholders | Association of Forests Development and Conservation (2008) |
| 4 | Infrastructure and Educational | Infrastructure networks and services<br>Land use<br>Technical ability and support<br>Research and educational purposes (training and programs) | Brown et al. (2004) |
| 5 | Culture and aesthetic | Culture and entertainment values<br>Spiritual values<br>Local knowledge and awareness | Uprety et al. (2012) |
| 6 | Management and legal | Governance process management (communications, collaborative participation, and restoration of silviculture practices)<br>Services to local communities<br>Monitoring and feedback<br>Policy process and key actors (sustainable restoration plans)<br>Empower laws and regulations (restoration aids, tenure disputes)<br>Institution's outcomes and plans | Sakellariou et al. (2021) |

*Source* Alayan et al. (2022)

### 3.4 Limitations

A fundamental limitation encountered in post-fire forest restoration exhibited heterogeneity with variations and occasional contradictions rather than consistent similarities. A balanced approach must address this limitation, incorporating fundamental and core criteria for restoring forests after fires. The scarcity of integrated scientific studies on forest restoration poses a potential obstacle for future systematic reviews in this field, highlighting the need to broaden the scope of interest in restoring forests after fires.

## 4 Factors Influencing Ecosystem Restoration

Forest fires are intrinsic to the functioning of many ecosystems, serving as a critical force in shaping landscapes and preserving biodiversity. They possess the dual capacity to create and destroy long-term environmental sustainability (Di Sacco et al. 2021). However, the global incidence of wildfires has escalated, leading to significant loss of forest ecosystem services and widespread degradation. The escalating frequency of wildfires, fueled by a combination of natural forces and human interventions, has sparked increasing apprehension regarding their ecological repercussions and the resilience of ecosystems to rebound. Accelerated wildfire activity and altered post-fire climate can influence forest resilience (Guz et al. 2022). The recovery of ecosystems from forest fires is a multifaceted process influenced by numerous factors.

These factors range from the fire's inherent characteristics to the affected ecosystems' intricate dynamics. Understanding these complexities is imperative for formulating effective post-fire management and restoration strategies. The key factors influencing ecosystem recovery from forest fires have been explored. Therefore, improving our understanding of how ecosystems recover from post-forest fires is essential to maintaining the ecological integrity of landscapes vulnerable to fires.

### 4.1 Factors Influencing Ecosystem Recovery

#### 4.1.1 Severity and Frequency of Fires: Wildfire Disturbances Can Significantly Affect Ecosystem Functioning and Resilience

##### 4.1.1.1 Severity of Fire

High-intensity fires can have significant ecological impacts, affecting tree mortality, soil heating, and organic matter consumption. This can result in vegetation loss, soil nutrient depletion, and changes to habitat structure. Such fires can prolong recovery, requiring extensive management interventions to restore ecosystems. Consequently, this may lead to shifts towards fire-adapted or invasive species, impeding the recovery of native plant communities (Lindenmayer et al. 2014). Furthermore, fire severity influences post-fire succession and regeneration processes, impeding natural regeneration by damaging seed banks, impeding seedling establishment, and altering plant-soil interactions (Turner et al. 1997; Kumar et al. 2020a, b). Ecosystem resilience to fire disturbances hinges on its adaptive capacity and ability to rebound from disturbances, with high-severity fires potentially surpassing resilience thresholds, triggering lasting impacts on ecosystem recovery and functioning (Pausas and Keeley 2014). Additionally, the extremity of a forest fire is an important factor in influencing forest recovery, modulated by factors such as topography, vegetation characteristics, and fuel moisture levels (Trang et al. 2023). The areas with lower burn severity, longer time since the last fire, and proximity to unburned forest edges tend to exhibit higher biodiversity and basal area, suggesting a pivotal role of burn severity in shaping post-fire forest dynamics and biodiversity patterns.

#### 4.1.1.2 Frequency of Fire

Fire frequency significantly influences ecosystem recovery dynamics post-fire. Ecosystems adapted to frequent fires often possess traits enabling rapid recovery, such as fire-resistant seeds or resprouting abilities, promoting resilience (Pausas and Keeley 2014). However, frequent fires can deplete the seed bank, hindering recovery, especially in ecosystems with limited seed dispersal. Additionally, fire frequency affects soil properties, potentially altering nutrient availability and hindering vegetation recovery (Certini 2005). Changes in species composition due to fire frequency can reduce biodiversity and impact ecosystem resilience (Bond et al. 2005). Furthermore, interactions with other disturbances, like invasive species or land use changes, can hinder recovery and alter trajectories (Hobbs et al. 2009).

### 4.1.2 Soil Characteristics and Topography

#### 4.1.2.1 Soil Characteristics

**Soil Texture and Composition**: Water retention, nutrient availability, and post-fire seed germination are all influenced by the texture of the soil. Sandy soils may experience faster water infiltration but have lower nutrient retention, while clay soils may retain water but become waterlogged, affecting plant recovery.

**Soil Organic Matter**: Fire can impact soil quality, altering organic matter content and potentially harming microbial activity and nutrient cycling. Severe fires can result in the loss of organic matter, which diminishes soil fertility and slows vegetation regrowth (Cerdà and Doerr 2008).

**Soil Moisture**: Soil moisture levels post-fire influence vegetation recovery rates. Dry soils may impede seed germination and plant establishment, while excessively wet soils can lead to erosion and nutrient leaching, affecting ecosystem recovery (Robichaud et al. 2013).

#### 4.1.2.2 Topography

**Slope Gradient**: Slope gradient affects water runoff, soil erosion, and sediment transport post-fire. Steep slopes may experience higher erosion rates and reduced soil stability, hindering vegetation recovery, while gentle slopes may promote soil retention and nutrient availability, facilitating recovery (Bodi et al. 2011).

**Aspect**: Aspect influences microclimate conditions, including temperature and moisture levels, affecting vegetation recovery rates. South-facing slopes may experience higher temperatures and drier conditions, potentially slowing recovery, while north-facing slopes may have more extraordinary, moister conditions conducive to faster recovery (Foster et al. 2015).

### 4.1.3 Climate and Weather Patterns

Climate and weather patterns are critical determinants of ecosystem recovery following a forest fire, exerting direct and indirect influences on vegetation regeneration, soil development, and overall ecosystem dynamics. Climate and weather patterns impact ecosystem recovery in various ways, including:

**Precipitation Patterns:** Precipitation timing, intensity, and duration post-fire influence soil moisture levels, seed germination, and plant establishment. Adequate rainfall can promote vegetation recovery by providing essential moisture for seedling growth, while prolonged drought conditions can hinder recovery by limiting water availability. Extreme precipitation events, such as heavy rainstorms or flash floods, can trigger soil erosion, sediment transport, and nutrient leaching, impacting soil stability and nutrient cycling post-fire. This can hinder the regrowth of vegetation and worsen ecological degradation.

**Temperature Regimes**: Temperature changes after a fire can notably impact plant growth rates, soil microbial activity, and seedling establishment. Higher temperatures may promote faster ecosystem recovery by accelerating plant growth and nutrient cycling. However, extreme heat events can cause soil moisture loss and physiological stress on recovering vegetation, hindering their growth (Pellegrini et al. 2015). Additionally, frost events can damage newly established vegetation and delay recovery by inhibiting seed germination and growth. The sensitivity to frost varies among plant species, and the timing and severity of such events can alter post-fire vegetation dynamics.

**Climate Variability**: Long-term climate variability can influence ecosystem recovery trajectories following a fire, including temperature trends, precipitation patterns, and drought frequency. Persistent drought conditions or shifts in temperature regimes may alter vegetation composition, soil properties, and hydrological processes, affecting the resilience of fire-affected ecosystems (Littell et al. 2016).

### 4.1.4 Interactions with Invasive Species and Human Activities

Interactions with invasive species and human activities can substantially impact the recovery of an ecosystem following a forest fire. Invasive species typically exhibit aggressive growth traits and lack natural predators or competitors in the affected ecosystem, enabling them to colonize burned areas rapidly. This colonization often results in out-competing native vegetation and can alter ecosystem dynamics. The presence of invasive species can hamper the recovery of native plant communities by monopolizing available resources such as sunlight, water, and nutrients, thereby impeding the establishment and growth of native species crucial for the ecosystem's resilience (Brooks et al. 2019). Moreover, human activities such as post-fire logging, road construction, and recreational activities can exacerbate ecosystem degradation and delay recovery processes. Post-fire logging operations can disturb the soil, disrupt seedbeds, and increase erosion rates, further degrading the already vulnerable post-fire landscape. Road construction for fire suppression activities or access to burned areas can fragment habitats, disrupt wildlife corridors, and introduce invasive species, compromising ecosystem recovery. These interactions highlight the complex interplay between disturbance agents and ecosystem recovery processes post-fire, emphasizing the need for integrated management approaches that consider both natural and human-induced factors. Understanding the impacts of invasive species and human activities on post-fire recovery is crucial for developing effective strategies to restore ecosystem functionality and promote long-term resilience.

## 5 Social Aspects of Restoration

Forests play a crucial role in providing diverse economic and social benefits to humanity by making significant contributions through employment, processing, and trade of forest products, energy, and investments in the forest sector. Forests serve as hosts and protectors of sites and landscapes with profound cultural, spiritual, and recreational significance. Upholding and enriching these roles is fundamental for sustainable forest management. Economic benefits are typically quantified in monetary units and comprise earnings generated by employment within the sector, the worth of goods and services produced from forests, and the sector's contribution to the national economy, energy provisions, and global trade. Measuring the social roles of forests can often be challenging and may exhibit significant variability among nations, subject to their developmental stage and cultural customs. Understanding fire's ecological and technical aspects is crucial, but the social dimensions of fire restoration are often overshadowed as forests hold significant social importance. Globally.7% of forest areas are estimated to have social services as their primary function. To ensure effective restoration initiatives, it's crucial to integrate socioeconomic considerations, catering to stakeholder needs and accommodating diverse expertise and perspectives (Bloomfield et al. 2019). Achieving this requires a thorough understanding of the socio-ecological context of forest fires, which informs biological conservation and sustainable forestry operations and the design of locally tailored management practices. Essential elements include cooperation and community engagement, as active involvement in decision-making processes and restoration activities fosters respect and appreciation for ecosystems. Effective communication is vital, involving the crafting of compelling narratives that resonate deeply with society, ultimately fostering a collective vision grounded in local values (Blignaut and Aronson 2020).

# 6 Future Perspectives

## *6.1 Promoting Collaborative Actions*

Fostering collaborative efforts such as sharing expertise, co-financing, and co-development among regions is vital to improving the efficacy of policy and restoration practices. Overcoming political boundaries and facilitating the sharing of geospatial data are essential steps in this endeavor, enabling a deeper understanding of patterns and processes in ecological, hydrological, and socio-cultural systems (Villarreal et al. 2019). Remote data acquisition emerges as a rapid, powerful, and reliable tool in this context, enabling the estimation of fire potential through analysis of fuel characteristics, assessing fire occurrence and severity, and modelling forest dynamics and vegetation recovery (Cantarello et al. 2011).

## *6.2 Improving Social Responsibility and Private Participation in ER*

Boosting social responsibility and promoting private engagement in ecological restoration (ER) is essential, especially in areas where corporations heavily extract natural resources. Elevating standards for corporate social responsibility can be achieved through strategies like public–private partnerships, payment for ecosystem services, or compensation mechanisms for ER. Companies must understand that participating in environmental restoration benefits the global community and enhances their corporate image and reputation (Souza-Alonso et al. 2022).

## 6.3 Strengthen Monitoring and the Effectiveness of Restoration

A major constraint in ecological restoration (ER) is insufficient funding, whether from public or private sources. Beyond the initial financial investment, this limitation extends to monitoring restoration progress. Restoration efforts often involve planting propagules without adequate consideration for the long-term survival of these individuals. This results in an increases rate of mortality and diminishes the capability of the funds allocated. Therefore, there is a pressing need for more effective and precise strategies that optimize resource utilization. Implementing ER projects grounded in cost-effective analyses and prioritizing natural restoration methods are imperative. When funds are scarce, efficient fund management should be prioritized to maximize impact (Souza-Alonso et al. 2022).

## 6.4 Acknowledging and Advocating the True Significance of Ecological Restoration

Traditionally, ecological restoration has been viewed as merely an aesthetic endeavor without tangible economic benefits, often perceived as conflicting with job creation. However, this perception is far from reality. Restoration is emerging as a promising avenue for sustainable economic development, providing opportunities for local stakeholders and fostering positive ecological and economic feedback loops. While activities related to ecological restoration differ from those in traditional economic sectors, they are increasingly recognized within the framework of the green economy. Studies indicate that ecological restoration creates more jobs per million invested than conventional activities and can be measured in economic output and employment rates (Bezdek et al. 2008).

## 6.5 Education and Training as a Foundation for Restoration

Successfully implementing ecological restoration (ER) depends on skilled professionals and scientists who design and monitor practical restoration activities. There is an urgent demand for trained individuals to fulfill this role, as environmental restoration needs intersect with societal demands for employment. This convergence creates an opportunity for collaboration between social work and scientific research to address both ecological and socio-economic objectives. Educational initiatives can serve as a foundation for such cooperation, bringing together multidisciplinary academic teams and stakeholders to develop strategies tailored to local socio-economic contexts. Educational projects that involve students and stakeholders in restoration efforts, utilizing methodologies like Service-Learning (S-L), offer a promising avenue to pursue. S-L projects contribute to restoration goals and provide participants direct benefits, such as learning opportunities. Given its demonstrated potential, restoration activities integrated with S-L projects should be considered viable options among the array of ER approaches available for adoption in the future (Cortina-Segarra et al. 2021).

## 7 Conclusion

While ecological restoration (ER) is a vital component of post-fire recovery and presents opportunities to reshape society around social justice, equity, and environmental stewardship principles, a significant challenge lies in understanding the human factors driving wildfires and their intricate relationships. Amid the overarching influence of climate change, various human activities such as extensive cattle ranching, forestry plantations, logging, inadequate fire management, market pressures, land speculation, and weak governance interact to influence the occurrence and severity of wildfires. Land-use changes often serve as the primary driver of these fires; this trend will likely persist without recognition, incentives or disincentives for landowners. We call upon practitioners, researchers, and ER advocates to unite to identify the major obstacles to post-fire restoration and align them with international initiatives. Significant uncertainties persist regarding forest restoration after wildfires, including post-fire soil severity assessment, intervention selection, funding shortages, regulatory inadequacies, limited social awareness, the balance between traditional and modern knowledge, and the provision of land management tools that promote sustainable land use and human welfare. Additionally, the impacts of climate change on environmental conditions and the expansion of Important Areas for Plant Conservation (IAPs) pose significant challenges. Recognizing fire as a fundamental element of many ecosystems and distinguishing between fire-sensitive and fire-prone systems is crucial for making informed decisions about restoration efforts. These challenges and their interactions consistently hinder ER progress. Nonetheless, our understanding of wildfire causes and effects has improved, accompanied by a growing recognition of the importance of forest conservation and recovery. Successful ER implementation requires the development of specific, interdisciplinary approaches that consider various dimensions, such as technical, strategic, social, economic, and political factors. Encouragingly, there is increasing acceptance of prescribed fire for managing fire-prone ecosystems, concerns about expanding IAPs, and establishing national and international networks to collaborate on ER initiatives.

## References


Alayan R, Rotich B, Lakner Z (2022) A comprehensive framework for forest restoration after forest fires in theory and practice: a systematic review. Forests 13:1354. https://doi.org/10.3390/f13091354

Albornoz AA, Resasco J, Morales JM, Naya DE (2013) Effects of forest fragmentation on the plant richness of remnant patches: a case study from the Upper Parana Atlantic Forest. Biodivers Conserv 22(6–7):1381–1397

Apurva K, Singh H, Kumar H (2017) Elucidating morphological and growth changes in Arjun tree (*Terminalia arjuna*) grown under elevated temperature condition. Trends Biosci 10(1):332–336

Association of Forests Development and Conservation (2008) Lebanon's National Forest Fire Management Strategy. Amaret Chalhoub, Lebanon

Barbati A, Arianoutsou M, Corona P de las, Heras J, Fernandes P, Moreira F, Papageorgiou K, Vallejo R, Xanthopoulos G (2010) Post-fire forest management in Southern Europe: a COST action for gathering and disseminating scientific knowledge. i Forest 3:5–7

Barwa S, Singh H (2024) Climate-Induced Changes in Forest Ecosystems of Indian Himalayan Region: A Meta-analysis. In: Singh, H. (eds) Urban Forests, Climate Change and Environmental Pollution. Springer, Cham. https://doi.org/10.1007/978-3-031-67837-0_39

Bezdek RH, Wendling RM, DiPerna P (2008) Environmental protection, the economy, and jobs: national and regional analyses. J Environ Manage 86(1):63–79

Blignaut J, Aronson J (2020) Developing a restoration narrative: a pathway towards system-wide healing and a restorative culture. Ecol Econ 168:106483. https://doi.org/10.1016/j.ecolecon.2019.106483

Bloomfield G, Meli P, Brancalion PH, Terris E, Guariguata MR, Garen E (2019) Strategic insights for capacity development on forest landscape restoration: implications for addressing global commitments. Trop Conserv Sci 12:1–11

Bodi MB, Mataix-Solera J, Doerr SH, Cerdà A (2011) The wettability of ash from burned vegetation and its relationship to Mediterranean plant species type, burn severity and total organic carbon content. Geoderma 160(3–4):599–607

Bond WJ, Woodward FI, Midgley GF (2005) The global distribution of ecosystems in a world without fire. New Phytol 165(2):525–538

Bradstock RA, Cohn JS, Gil AM, Bedward M, Lucas C (2009) Prediction of the probability of large fires in the Sydney region of south-eastern Australia using fire weather. Int J Wildland Fire 18:932–943

Brancalion PHS, Chazdon RL (2017) Beyond hectares: four principles to guide reforestation in the context of tropical forest and landscape restoration. Restor Ecol 25:491–496

Brooks ML, D'Antonio CM, Richardson DM, Grace JB, Keeley JE, DiTomaso JM, Pyke D (2019) Effects of invasive alien plants on fire regimes. Bioscience 59(7):593–601

Brown RT, Agee JK, Franklin JF (2004) Forest restoration and Fire: principles in the context of place. Conserv Biol 18:903–912

Brown T (2022) The ecological benefits of fire. The National Geographic Society. https://education.nationalgeographic.org/resource/ecological-benefits-fire

Cantarello E, Newton AC, Hill RA, Tejedor-Garavito N, Williams-Linera G, Lopez-Barrera F, Manson RH (2011) Simulating the potential for ecological restoration of dry land forests in Mexico under different disturbance regimes. Ecol Model 222(5):1112–1128

Cavallero L, Raffaele E, Aizen MA (2013) Birds as mediators of passive restoration during early post-fire recovery. Biol Cons 158:342–350

Cerdà A, Doerr SH (2008) The effect of ash and needle cover on surface runoff and erosion in the immediate post-fire period. CATENA 74(3):256–263

Certini G (2005) Effects of fire on properties of forest soils: a review. Oecologia 143(1):1–10

Chen X, Chen W (2022) Remote-sensing monitoring of post fire vegetation dynamics in the Greater Hinggan Mountain range based on long time-series data: analysis of the effects of six topographic and climatic factors. Remote Sens 14:2958

Chen W, Moriya K, Sakai T, Koyama L, Cao C (2014) Monitoring of post-fire forest recovery under different restoration modes based on time series Landsat data. Eur J Remote Sens 47:153–168

Chu T, Guo X (2014) Remote sensing techniques in monitoring post-fire effects and patterns of forest recovery in Boreal Forest Regions: a review. Remote Sens 6:470–520

Cortina-Segarra J, García-Sanchez I, Grace M, Andres P, Baker S, Bullock C, Decleer K, Dicks LV, Fisher JL, Frouz, J Klimkowska A, Kyriazopoulos AP, Moreno-Mateos D, Rodríguez-Gonzalez PM, Sarkki S, Ventocill JL (2021) Barriers to ecological restoration in Europe: expert perspectives. Restor Ecol 29(4)

Crouzeilles R, Ferreira MS, Chazdon RL, Lindenmayer DB, Sansevero JBB, Monteiro L, Iribarrem A, Latawiec AE (2017) Ecological restoration success is higher for natural regeneration than for active restoration in tropical forests. Sci Adv 3(11). https://doi.org/10.1126/sciadv.1701345

De Jong W, Liu J, Long H (2021) The Forest Restoration Frontier Ambio 50:2224–2237

De Luis M, Raventos J, Gonzalez-Hidalgo JC (2005) Factors controlling seedling germination after fire in Mediterranean gorse shrublands. Implications for fire prescription. J Environ Manage 76:159–166

Derak M, Cortina J, Taiqui L, Aledo A (2018) A proposed framework for participatory forest restoration in Semiarid areas of North Africa. Restor Ecol 26:18–25

Dhyani K, Kuniyal HB, Singh H, Sobha (2021) Growth and physiological potential of *Terminalia arjuna* under elevated $CO_2$ levels in open top chamber condition. J Appl Nat Sci 13(3):1121–1126. https://doi.org/10.31018/jans.v13i3.2490

Di Sacco A, Hardwick KA, Blakesley D, Brancalion PHS, Breman E, Cecilio Rebola L, Antonelli A (2021) Ten golden rules for reforestation to optimize carbon sequestration, biodiversity recovery and livelihood benefits. Glob Change Biol 27:1328–1348. https://doi.org/10.1111/gcb.15498

Dos Santos JFC, Gleriani JM, Velloso SGS, de Souza GSA, Do Amaral CH, Torres FTP, Medeiros NDG, Dos Reis M (2019) Wildfires as a major challenge for natural regeneration in Atlantic Forest. Sci Total Environ 650:809–821

Drever CR, Peterson G, Messier C, Bergeron Y, Flannigan M (2006) Can forest management based on natural disturbances maintain ecological resilience? Can J for Res 36:2285–2299

Foster AD, Shuman JK, Shugart HH (2015) Landscape-scale effects of fire severity on mixed-conifer and red fir forest structure in the southern Sierra Nevada, California. For Ecol Manage 338:126–137

Fowler CT (2003) Human health impacts of forest fires in the Southern United States: a literature review. J Ecol Anthropol 7(1):39–63

FSI (2012) Vulnerability of India's forests to fire. Forest Survey of India, Ministry of Environment and Forest, Government of India

Galindo JP, Fernandez C, Ribbens E (2017) Forest restoration and ecological succession along an elevation gradient in northeastern Puerto Rico. Restor Ecol 25(2):231–242

Gomez-Rey MX, Couto-Vázquez A, García-Marco S, Gonzalez-Prieto SJ (2013) Impact of fire and post-fire management techniques on soil chemical properties. Geoderma 195:155–164

Gonzalez ME, Veblen TT (2007) Tree regeneration responses to Chusquea montana bamboo die-off in mixed Notho fagus forest of the southern Andes. For Ecol Manage 242(1):299–308

Gonzalez ME, Veblen TT (2015) Forest ecosystem recovery in western Patagonia: stand dynamics after a blow down event. For Ecol Manage 339:42–52

Guz J, Sangermano F, Kulakowski D (2022) The Influence of burn severity on post-fire spectral recovery of three fires in the southern Rocky Mountains. Geography 25. https://commons.clarku.edu/faculty_geography/25

Heydari M, Rostamy A, Najaf F, Dey DC (2017) Effect of fire severity on physical and biochemical soil properties in Zagros oak (Quercus brant ii Lind l.) forests in Iran. J For Res 28(1):95–104

Hinojosa MB, Albert-Belda E, Gomez-Munoz B, Moreno JM (2021) High fire frequency reduces soil fertility underneath woody plant canopies of Mediterranean ecosystems. Sci Total Environ 752:141877

Hiremath AJ, Sundaram B (2005) The fire-lantana cycle hypothesis in Indian forests. Conserv Soc 3:26–42

Hobbs RJ, Higgs E, Harris JA (2009) Novel ecosystems: implications for conservation and restoration. Trends Ecol Evol 24(11):599–605

Hohl M, Ahimbisibwe V, Stanturf JA, Elsasser P, Kleine M, Bolte A (2020) Forest landscape restoration-what generates failure and success? Forests 11:938

Iniguez JM, Evans AM, Dadashi S, Young JD, Meyer MD, Thode AE, Bean R (2022) Comparing geography and severity of managed wildfires in California and the Southwest USA before and after the implementation of the 2009 policy guidance. Forests 13:793

Jaiswal RK, Mukherjee S, Raju KD, Saxena R (2002) Forest fire risk zone mapping from satellite imagery and GIS. Int J Appl Earth Observ Geoinformation 4:1–10

Jangra A, Kumar K, Maikhuri S, Bhandari MS, Pandey S, Singh H, Barthwal S (2024) Unveiling stress-adapted endophytic bacteria: characterizing plant growth-promoting traits and assessing cross-inoculation effects on Populus deltoides under abiotic stress. Plant Physiol Biochem 108610. https://doi.org/10.1016/j.plaphy.2024.108610

Jhariya MK, Oraon PR (2012) Regeneration status and species diversity along the fire gradients in tropical deciduous forest of Chhattisgarh. J Plant Dev Sci 4(1):49–54

Jordán A, Zavala LM, Mataix-Solera J, Nava AL, Alanis N (2011) Effect of fire severity on water repellency and aggregate stability on Mexican volcanic soils. CATENA 84(3):136–147

Kauffman JB (2004) Death rides the forest: perceptions of fire, land use, and ecological restoration of Western forests. Conserv Biol 18:878–882

Kganyago M, Shikwambana L (2019) Assessing spatio-temporal variability of wildfires and their impact on sub-Saharan ecosystems and air quality using multisource remotely sensed data and trend analysis. Sustainability 11:6811

Kittur B, Jhariya MK, Lal C (2014) Is the forest fire can affect the regeneration and species diversity. Ecol, Environ Conserv 20(3):989–994

Kumar A, Kumar P, Singh H, Kumar N (2020) Adaptation and mitigation potential of roadside trees with bio-extraction of heavy metals under vehicular emissions and their impact on physiological traits during seasonal regimes. Urb for Urb Green 58:126900

Kumar A, Kumar P, Singh H, Bisht S, Kumar N (2021) Relationship of physiological plant functional traits with soil carbon stock in temperate forest of Garhwal Himalaya. Curr Sci 120(8):1368–1373

Kumar A, Kumar P, Singh H, Kumar N (2021) Modulation of plant functional traits under essential plant nutrients during seasonal regime in natural forests of Garhwal Himalayas. Plant Soil 465:197–212

Kumar A, Dwivedi G K, Tewari S, Paul J, Anand R, Kumar N, Kumar P, Singh H, Kaushal R (2020a) Carbon mineralization and inorganic nitrogen pools under Terminalia chebula Retz.-based agroforestry system in Himalayan Foothills, India. Forest Sci 66(5):634–643

Kumar N, Singh H, Kumar A, Singh AK, Pandey PK, Kumar A (2024) Forest-fire-induced land degradation. In: Mishra G, Giri K, Singh S, Kumar M (eds) Sustainable land management in India. Springer Nature, Singapore. https://doi.org/10.1007/978-981-97-5223-2_4

Lamb D, Stanturf J, Madsen P (2012) What is forest landscape restoration? In: Forest landscape restoration. Springer, Berlin/Heidelberg, Germany. ISBN 9789400753266

Landesmann JB, Morales JM (2018) Post-fire regeneration patterns in the South American Chaco: implications for bird conservation. Biodivers Conserv 27(5):1207–1221

Leigh BL, Zachary AH, Alistair MSS, Michael JF, Hudak AT, Morgan P, Sarah AL, Gessler PE, Benso NC (2006) Remote sensing techniques to assess active fire characteristics and post-fire effects. Int J Wildland Fire 15:319–345

Lindenmayer DB, Burton PJ, Franklin JF (2014) Salvage logging and its ecological consequences. Island Press

Lippok D, Leon J, Hartung W (2013) Forest regeneration along an altitudinal gradient in the Bolivian Andes: the effect of elevation on species density. J Mt Sci 10(5):831–842

Littell JS, Peterson DL, Riley KL, Liu Y, Luce CH, Holden ZA (2016) A review of climate change impacts on fire frequencies across the US: current state of knowledge and future challenges. For Ecol Manage 294:250–261

Malik A, Kumar K, Kumari G, Kumar N, Singh H (2024) Application of functional traits in modelling productivity and resilience under climate change. Springer, pp 77–96. ISBN: 978-981-97-1509-1

Meddens AJ, Kolden CA, Lutz JA, Smith AM, Abatzoglou JT (2018) Spatial controls of prescribed and wildfires in the south western US. Int J Wild Land Fire 27(11):716–725

Meyer MD, Long JW, Safford HD (2021) Post fire restoration framework for National Forests in California. US Department of Agriculture, Forest Service, Pacific Southwest Research Station: Albany, CA, USA

Moreira da Silva AP, Schweizer D, Rodrigues Marques H, CordeiroTeixeira AM, Nascente dos Santos TVM, Sambuichi RHR, Badari CG, Gaudare U (2017) Can current native tree seedling production and infrastructure meet an increasing forest restoration demand in Brazil? Restor Ecol 25(4):509–515

Moreira F, Arianoutsou M, Vallejo VR, Heras JDL, Corona P, Xanthopoulos G, Fernandes P, Papageorgiou K (2012) Setting the scene for post-fire management. In: Post-fire management and restoration of Southern European forests, vol. 24. Springer, Berlin/Heidelberg, Germany. ISBN 978-94-007-2207-1

Morrison EB, Lindell CA (2011) Active or passive forest restoration? Assessing restoration alternatives with Avian Foraging behavior. Restor Ecol 19:170–177

Mustafa Y (2009) Forest fires. In: Proceedings of the 1st international symposium on forest fires, 7–10 Jan 2009, Antalya, Turkey pp 129–130

Naficy CE, Keeling EG, Landres P, Hessburg PF, Veblen TT, Sala A (2016) Wilderness in the 21st century: a framework for testing assumptions about ecological intervention in wilderness using a case study of fire ecology in the Rocky Mountains. J Forest 114:384–395

Nerfa L, Wilson SJ, Reid JL, Rhemtulla JM (2021) Practitioner views on the determinants of tropical forest restoration longevity. Restor Ecol 29:e13345

Nirmal, Bakshi M, Singh H (2021) Effect of $CO_2$ elevation on Shisham growth at nursery stage. Indian J Hill Farming 34:272–276

Noss RF, Franklin JF, Baker WL, Schoennagel T, Moyle PB (2006) Managing fire-prone forests in the Western United States. Front Ecol Environ 4:481–487

Noulekoun F, Mensah S, Birhane E, Son Y, Khamzina A (2021) Forest landscape restoration under global environmental change: challenges and a future roadmap. Forests 12:276

Nyongesa KW, Vacik H (2019) Evaluating management strategies for Mount Kenya Forest Reserve and National Park to reduce fire danger and address interests of various stakeholders. Forests 10:426

Oh SU, Kim EH, Kim KM, Kim MK (2020) A study on the application of successful forest greening experience for forest and landscape restoration: a comparative study of two Koreas. Sustainability 12:8712

Osman KT (2013) Forest soils. In: Soils. Springer, Dordrecht, pp 229–251

Pausas JG, Keeley JE (2014) Evolutionary ecology of resprouting and seeding in fire-prone ecosystems. New Phytol 204(1):55–65

Pellegrini AF, Hedin LO, Staver AC (2015) Fire alters ecosystem carbon and nutrients but not plant nutrient stoichiometry or composition in tropical savanna. Ecology 96(5):1275–1285

Pellegrini AF, Ahlstrom A, Hobbie SE, Reich PB, Nieradzik LP, Staver AC, Jackson RB (2018) Fire frequency drives decadal changes in soil carbon and nitrogen and ecosystem productivity. Nature 553(7687):194–198

Pineda E, Toledo T, Andresen E (2020) Comparing the success of active and passive restoration in a tropical cloud forest landscape: a multi-Taxa Fauna approach. PLoS ONE 15:e242020

Poletti C, Dioszegi G, Nyongesa KW, Vacik H, Barbujani M, Kigomo JN (2018) Characterization of forest fires to support monitoring and management of Mount Kenya Forest. Mt Res Dev 39:R1–R12

Prakash V, Hunney K, Singh H (2022) Effect of elevated carbon dioxide on growth and development of *Santalum album* L. seedlings inoculated with plant growth promoting microorganisms in open top chambers. Int J Environ, Agric Biotechnol 7(5):149–159

Raturi A, Singh H, Kumar P, Shukla N (2022) Characterizing the post-monsoon $CO_2$, $CH_4$, $N_2O$, and $H_2O$ vapor effluxes from a tropical wetland in the Himalayan foothill. Environ Monit Assess 194(50). https://doi.org/10.1007/s10661-021-09721-8

Robichaud PR, Lewis SA, Wagenbrenner JW, Ashmun LE (2013) Post-fire mulching for runoff and erosion mitigation; part I: effectiveness at reducing hillslope erosion rates. CATENA 105:75–92

Rost J, Jardel-Pelaez EJ, Bas JM, Pons P, Loera J, Vargas-Jaramillo S, Santana E (2015) The role of frugivorous birds and bats in the colonization of cloud forest plant species in burned areas in western Mexico. Anim Biodivers Conserv 38(2):175–182

Saha S, Howe HF (2003) Species composition and fire in a dry deciduous forest. Ecology 84(12):3118–3123

Sakellariou S, Sfougaris A, Christopoulou O (2021) Review of geoinformatics-based forest fire management tools for integrated fire analysis. Pol J Environ Stud 30:5423–5434

Sansevero JBB, Souza M, Rodrigues RR (2017) Arrested succession in Brazilian Atlantic Forest secondary forests: time to change the paradigm? Restor Ecol 25(1):25–31

Savita, Sharma SK, Singh H, Singh O (2017) Forest fire: a review. Indian Forester 143(8):719–728

Scheper AC, Verweij PA, Van Kuijk M (2021) Post-fire forest restoration in the humid tropics: a synthesis of available strategies and knowledge gaps for effective restoration. Sci Total Environ 771:144647

Sett T, Nikam BR, Singh H, Purohit S (2022) Effect of drought stress on forest evapotranspiration—a case study on Indian forests. EGU General Assembly 2022. https://doi.org/10.5194/egusphere-egu22-510

Sharma R, Singh H, Kaushik M, Nautiyal R, Singh O (2018) Adaptive physiological response, carbon partitioning and biomass production of *Withania Somnifera* (L.) Dunal grown under elevated $CO_2$ regimes. 3 Biotech 8:267. https://doi.org/10.1007/s13205-018-1292-1

Sharma R, Singh H (2021) Alteration in biochemical constituents and nutrients partitioning of *Asparagus racemosus* in response to elevated atmospheric $CO_2$ concentration. Environ Sci Pollut Res 29:6812–6821. https://doi.org/10.1007/s11356-021-16050-3

Singh H, Sharma R, Savita SMP, Kumar M, Verma A, Ansari MW, Sharma SK (2018) Adaptive physiological response of *Parthenium hysterophorus* to elevated atmospheric $CO_2$ concentration. Indian Forester 144:1–14

Singh S, Singh H, Sharma SK, Nautiyal R (2021) Seasonal variation in biochemical responses of bamboo clones in the sub-tropical climate of Indian Himalayan foothills. Heliyon 7(4):e06859. https://doi.org/10.1016/j.heliyon.2021.e06859

Singh H, Kumar N, Singh R, Kumar M (2023) Assessing the climate change impact on the habitat suitability of the range-restricted bird species (*Catreus wallichii*) in the Indian Himalayan ecosystem. Environ Sci Pollut Res. https://doi.org/10.1007/s11356-023-30789-x

Singh M, Singh H, Kumar A, Kumar M, Barthwal S, Thakur A (2024) Soil nitrogen availability determines the $CO_2$ fertilization effect on tree species (*Neolamarckia cadamba*): growth and physiological evidence. Environ Sustain 7:53–60. https://doi.org/10.1007/s42398-023-00300-w

Singh H, Kumar N, Kumar M and Singh R (2020) Modelling habitat suitability of western tragopan (*Tragopan melanocephalus*) a range-restricted vulnerable bird species of the Himalayan region, in response to climate change. Clim Risk Manag 29:100241. https://doi.org/10.1016/j.crm.2020.100241

Singh H, Kumar N, Kumar A (2024a) Enhancing resource use efficiency in crops through plant functional traits. Springer, pp 97–117. ISBN: 978-981-97-1509-1

Singh H (2021) Process-based carbon sequestration study in relation to the energy-water-carbon flux in a forest ecosystem. In: Kumar P, Singh RK, Kumar M, Rani M, Sharma (eds) Climate impacts on sustainable natural resource management. CRC Press Taylor & Francis Group, United States, pp 336–361

Singh H (2024a) Climate change and forests: biological perspectives. In: Singh H (ed) Forests and climate change. Springer Nature, Singapore. https://doi.org/10.1007/978-981-97-3905-9_1

Singh H (2024b) Climate change and environmental pollution: biological consequences on urban forests. In: Singh H (ed) Urban forests, climate change and environmental pollution. Springer Nature, Cham, Switzerland. https://doi.org/10.1007/978-3-031-67837-0_1
Singh H (ed) (2024c) Forests and climate change: biological perspectives on impact, adaptation, and mitigation strategies. Springer Nature Singapore. https://doi.org/10.1007/978-981-97-3905-9
Singh H (ed) (2024d) Urban forests, climate change and environmental pollution: physio-biochemical and molecular perspectives to enhance urban resilience. Springer Nature, Cham, Switzerland. https://doi.org/10.1007/978-3-031-67837-0
Singh H (2024e) Exploring adaptive modulation in plant functional traits and its impact on the productivity of *Acacia auriculiformis* under $CO_2$-enriched environment. Ind Crops Prod 210:118186. https://doi.org/10.1016/j.indcrop.2024.118186.
Smith A (1995) Adaptation to fire. In: The contribution of fire in Dramatising The Australian Landscape. Retrieved from: http://online.anu.edu.au/Forestry/fire/ecol/as20.htm
Souza-Alonso P, Saiz G, Garcia RA, Pauchard A, Ferreira A, Merino A (2022) Post-fire ecological restoration in Latin American forest ecosystems: insights and lessons from the last two decades. For Ecol Manage 509:120083
Spies TA, White E, Ager A, Kline JD, Bolte JP, Platt EK, Olsen KA, Pabst RJ, Barros AMG, Bailey JD (2017) Using an agent-based model to examine forest management outcomes in a fire-prone landscape in Oregon, USA. Ecol Soc 22:25
Srivastava P, Garg A (2013) Emissions from forest fires in India: an assessment based on MODIS fire and global land cover products. Clim Change Environ Sustain 1(2):138–144
Stanturf JA, Mansourian S (2020) Forest Landscape restoration: state of play. R Soc Open Sci 7:201218
Torres RC, Renison D (2017) Effects of habitat fragmentation on post-fire plant regeneration in montane grasslands. J Arid Environ 144:49–57
Trang PT, Andrew ME, Enright NJ (2023) Burn severity and proximity to undisturbed forest drive post-fire recovery in the tropical montane forests of northern Vietnam. Fire Ecology 19:47. https://doi.org/10.1186/s42408-023-00205-3
Turner MG, Romme WH, Gardner RH (1997) Pre fire heterogeneity, fire severity, and early post-fire plant reestablishment in subalpine forests of Yellowstone National Park, Wyoming. Int J Wildland Fire 7(2):73–82
Ulery AL, Graham RC (1993) Forest fire effects on soil color and texture. Soil Sci Soc Am J 57:135–140
Uprety Y, Asselin H, Bergeron Y, Doyon F, Boucher JF (2012) Contribution of traditional knowledge to ecological restoration: practices and applications. Ecoscience 19:225–237
Urretavizcaya MF, Defosse G (2013) Ecological restoration in degraded areas of the South American Atlantic Forest: a comparison between passive and active restoration techniques. Restor Ecol 21(1):27–34
USDA Forest Service (2012) Restoring fire-adapted ecosystems: proceedings of the 2005 national silviculture workshop. USDA Forest Service
Vallejo VR, Alloza JA (2012) Evolution post-fire management in the Mediterranean Basin. J Ecol 58:37–41
Venier LA, Walton R, Thompson ID, Arsenault A, Titus BD (2018) A review of the intact forest landscape concept in the Canadian Boreal Forest: its history, value, and measurement. Environ Rev 377:369–377
Villarreal ML, Haire SL, Iniguez JM, Montano CC, Poitras TB (2019) Distant neighbors: recent wildfire patterns of the Madrean Sky Islands of south western United States and north western Mexico. Fire Ecol 15:1–20
Wang Q, Zhong M, Wang S (2012) Meta-analysis on the response of microbial biomass dissolved organic matter, respiration, and N mineralization in mineral soil to fire in forest ecosystems. For Ecol Manage 271:91–97
Wotton BM, Martell DL, Logan KA (2003). Climate change and people-caused forest fire occurrence in Ontario. Climatic Change 60:275–295. Singh H, Kumar M (2022). Climate change and its

impact on Indian Himalayan Forests: current status and research needs. In: Rani S, Kumar R (eds) Climate change. Springer Climate. Springer, Cham, pp 223–242. https://doi.org/10.1007/978-3-030-92782-0_11

Yadav SK, Singh H, Nautiyal R, Ginwal HS, Ansari SA, Barthwal S (2019a) Modulation of morpho-physiological responses in *Populus deltoides* by elevated carbon dioxide and temperature. For Sci 66:105–118. https://doi.org/10.1093/forsci/fxz048

Yadav SK, Singh H, Ginwal HS, Barthwal S (2019b) Elevated $CO_2$ enhanced growth and physiological process of *Populus deltoides* Bartr. ex Marsh. Indian Forester 145(1):23–27

# Forest Fire Effects on the Underground Frontline: Soil Health and Underground Flora and Fauna

**Apurva Malik, Hukum Singh, Kishan Kumar, Garima Kumari, and Narendra Kumar**

**Abstract** The influence of global environmental shifts, notably climate change and human encroachment, has significantly altered forest vitality, leading to detrimental effects on humans and the environment. Fire represents a unique case underlying the intersection of natural phenomena and human activities, with its potential to become a catastrophe and a growing global concern. Forest fires are prevalent across diverse ecosystems, with some highly sensitive to fire. These fires can be characterized by three main factors: severity, time of occurrence, and frequency. Forest fires can impact aboveground and understory ecosystems, rates of nitrogen fixation, and diversity of plant species, resulting in the rapid loss of organic matter through combustion and altering soil structure and nutrient dynamics. Underground flora and fauna have developed diverse adaptive mechanisms to withstand the impact of forest fires, like dormancy or evolving fire-resistant structures such as tubers. Immediate survival mechanisms may include retreating deeper into the soil or entering a state of dormancy, whereas long-term responses often reshape the composition and structure of underground communities. Various techniques for assessment of post-fire soil health and underground ecosystems include field sampling methods, laboratory analyses, and remote sensing methods like NDVI. Post-fire restoration practices like soil stabilization measures, replanting native vegetation, and water management strategies effectively restore ecosystem health and promote resilience following wildfires. The policies and strategies that include conducting comprehensive fire risk assessments to pinpoint wildfire-prone areas and prioritize management interventions are discussed in this chapter. The chapter also focuses on the gaps and emerging technologies that promise to advance our understanding of underground ecosystems in the context of forest fires.

**Keywords** Forest fire · Underground vegetation · Soil health · Climate change · Biodiversity

A. Malik (✉) · H. Singh · K. Kumar · G. Kumari
ICFRE-Forest Research Institute, Dehradun, India
e-mail: apurva.malik14@gmail.com

N. Kumar
College of Agriculture, Central Agriculture University (I), Kyrdemkulai, Meghalaya, India

 
H. Singh and N. Kumar (eds.), *Forest Fire and Climate Change*,
https://doi.org/10.1007/978-3-031-89967-6_8

# 1 Introduction

Natural disruptions are pivotal for maintaining the ecological equilibrium of forests and facilitating rejuvenation, nutrient circulation, and biodiversity (Barwa and Singh 2024). These disruptions can arise from non-living factors like droughts or floods, living organisms like bark beetles, or human activities like deforestation (Kumar et al. 2024). The influence of global environmental shifts, notably climate change and human encroachment, has significantly altered forest vitality, leading to detrimental consequences on humans and the environment (Oldekop et al. 2020). Approximately half of the world's forests exhibit noticeable indications of human intervention, with a troubling uptick in the degradation of untouched primary forests (Curtis et al. 2018).

Fire represents a unique case underlying the intersection of natural phenomena and anthropogenic activities, with its capacity for becoming a catastrophe and a growing global concern. It has been a part of our planet's biosphere for over the past 400 million years, shaping ecosystems. Human use of fire for landscape management dates back around 300,000 years, leading to ecological changes, biodiversity promotion, and enhanced livelihoods (Singh 2024a). Presently, approximately 400 million hectares of land are consumed by fires annually, with human actions responsible for the majority, reaching up to 99% in some areas (Kumar et al. 2020a, b; Malik et al. 2024).

Climate patterns (such as seasonal variations), vegetation characteristics (such as coniferous versus deciduous), and ignition sources (whether human-induced or natural, like lightning) interact to exert significant control over fire regimes. Fire thus acts as a catalyst for evolution (Judson 2017; Singh 2024b). Despite approximately 400 million hectares burning annually worldwide, the forests encompass fewer than 100 million hectares. Grasslands are the ecosystems that experience the highest incidence of fire (Kumar et al. 2021a).

Besides, there has been a decrease in the number of global areas burned over several decades. However, fire activity has moved to regions where anthropogenic influence has led to significant changes in environmental conditions, resulting in devastating outcomes. Between 1998 and 2017, wildfire events accounted for 3.5% of recorded disasters. Although a deficiency of data indicates a trend in increased fatalities and losses, there has been a notable surge in suppression costs due to the proliferation of assets at risk in areas prone to fires. Forest fires also trigger environmental hazards and damage, hindering essential ecosystem services such as clean drinking water and societal benefits (Singh 2024c; Sett et al. 2024).

Along with this, forest fires also impact the ecosystem present beneath the ground. It affects all the floral and faunal species present in underground species. This chapter discusses how forest fires impact soil health and underground lives.

## 2 Forest Fire Dynamics and Impacts

### *2.1 Forest Fire-Overview*

Fires are a natural phenomenon in tropical forests, and their frequency may increase with human development. The impact of fires on soil, underground flora, and fauna can significantly shape the composition and structure of forests after the fire event (Jain et al. 2008). Fire is utilized globally as a management tool for various ecosystems. Alterations in slope, aspect, elevation, and soil conditions influence energy and water distribution on sites and the potential plant and animal community. These terrain characteristics also play a crucial role in fire-induced disturbances by determining moisture levels and fuel flammability across different time scales—hourly, seasonal, and successional. Forest fires are prevalent across diverse ecosystems, with some highly sensitive to fire. However, these ecosystems can often recover if extensive wildfires are avoided (Kraus and Goldammer 2007; Kumar et al. 2021b).

Forest fire science is centered on the fundamental concept of the fire triangle, which outlines the three essential components necessary for combustion: fuel, oxidant, and a heat source. When these elements are present in the correct proportions, they initiate pyrolysis. In this process, organic matter transforms at high temperatures, releasing gases that ignite upon reaching a critical temperature, leading to flaming combustion. In vegetated ecosystems, vegetation serves as the fuel (such as grassor dead wood), atmospheric oxygen is the oxidant, and human activities or lightning is the heat source, commonly called ignition. When a specific type of fire activity predominates over a significant area with slight variation over time, it establishes a fire regime, a fundamental characteristic crucial for understanding the dynamics and functioning of most vegetated landscapes worldwide (Landres et al. 1999; Chuvieco et al. 2008; Archibald et al. 2013).

### *2.2 Factors Influencing Forest Fire Severity*

The severity of forest fires varies over time and space, influenced by both the biophysical environment and the location within the fire perimeter. The severity of the fire and its effects on forest soil depend on factors like fire intensity, fuel load, and soil moisture content. Forest fires can be characterized by three main variables: intensity, frequency, and season of occurrence. Different combinations of these variables define the fire regime to which a forest is exposed. This fire regime, along with the specific characteristics of the site and the species comprising the forest, determines the level of greenhouse gas (GHG) emissions and the impact of fire on nutrient cycles and other ecosystem properties and processes (Raturi et al. 2022; Sett et al. 2022; Singh et al. 2020; Apurva et al. 2017).

The effects of fires in forests should be viewed as a continuum in terms of impact severity, often correlated with the amount of fuel consumed and fire intensity. Fuel consumption can vary widely, from a few tons of organic matter per hectare in low-intensity controlled burns to hundreds of tons in slash and wildfires. Any long-term fire frequency or intensity increase typically reduces forest carbon stocks, including soil carbon (Raison and Khanna 2011).

Forest and vegetation management practices play a significant role in influencing fire activity. For example, following the massive 1910 “Big Burn” in the Western USA, fire suppression became the standard approach for forest management, aiming to protect forests. Introducing non-native species can also significantly affect fuel loads and fire behaviour. Eucalyptus and conifer species, in particular, have been associated with numerous severe wildfire incidents that have had devastating consequences for nearby communities, although the relationship between afforestation and fire activity is not always straightforward (Wilkinson et al. 2018; Fernandes et al. 2019; Bowman et al. 2019; Ojeda 2020).

Climate change, a multiplier of threats, is emerging as a primary driver of global fire activity (AghaKouchak et al. 2020). Weather extremes, such as droughts and heatwaves, underlie fire activity and are becoming more frequent and severe. A warmer climate has been directly linked to increased fire danger: longer fire seasons, drier and warmer conditions leading to more available fuel, and heightened lightning activity resulting in more natural ignitions (Flannigan et al. 2009, 2013; Wang et al. 2017; Kumar et al. 2022). Predicted changes in vegetation composition and health, along with enhanced spread capabilities for invasive species, are expected to drive shifts in fire regimes. Fires may even accelerate the transition of vegetation toward novel ecosystems (De Rigo et al. 2017; Stralberg et al. 2018).

## 2.3 Immediate Effects of Forest Fires on Soil Health

Forest fires can have beneficial and harmful effects on soil depending on their severity and frequency. The immediate impact of forest fires on soil health is profound and multifaceted. Fires, particularly those occurring repeatedly and over shorter intervals, can impact ground and understory vegetation, nitrogen fixation rates, and plant diversity. The variability in responses to fire is influenced by factors such as the inherent variability in pre-burn resources, fire behavior, season of burning, and pre and post-fire environmental conditions such as rainfall timing, amount, and duration (Clark 2001).

Forest fires rapidly lose organic matter through combustion, altering soil structure and nutrient dynamics (Certini 2005). High temperatures from intense fires can lead to soil heating, causing physical and chemical changes such as increased pH and mineral composition alterations (Neary et al. 1999). The heat and pressure generated during fires can compact soil, reducing pore space and impeding water infiltration and root growth (DeBano et al. 1998). These alterations may affect the fertility and health of forest soils, depending on the specific forest type and the degree of deviation of the fire regime from the conditions under which the vegetation has evolved.

Plant species have evolved various traits to adapt to fire, becoming fire-resistant, fire-dependent, or exhibiting both characteristics. Examples include thick barks that provide insulation from heat or serotinous cones that require the heat from a fire to open and disperse seeds (Sharma et al. 2018; Jangra et al. 2024; Dhyani et al. 2021; Singh 2024a, b, c, d, e). Furthermore, fires do not uniformly burn across a landscape; instead, they create a mosaic of burned and unburned patches, each with different levels of fire severity, thus generating pyrodiversity. This diversity, coupled with the rejuvenating effects of fire, fosters a variety of habitats of various ages, promoting biodiversity and enhancing ecosystem resilience to future disturbances (Martin and Sapsis 1992). There is a growing suggestion that biodiversity loss could increase fire activity (Hayward et al. 2016). Conversely, altered fire regimes resulting from direct human intervention, such as fire suppression, or indirect influences like climate change, will likely lead to species extinction.

Another immediate consequence is nutrient loss, with volatile nutrients like nitrogen and phosphorus lost through volatilization. Moreover, the intense heat can disrupt soil microbial communities, reducing microbial biomass and activity (Hart et al. 2005). Low-intensity fires can lead to the combustion of litter and organic matter, increasing the availability of nutrients for plant growth and resulting in a surge in herbaceous plant growth and nutrient storage. On the other hand, high-intensity fires can cause complete loss of soil organic matter, volatilization of essential nutrients like nitrogen, phosphorus, sulfur, and potassium, and the death of soil microbes. Intense forest fires can also form organic compounds with hydrophobic properties, resulting in water-repellent soils. These immediate effects highlight forest fires' complex and significant impact on soil health, impacting ecosystem resilience and recovery.

# 3 Underground Ecosystems

## *3.1 Importance of Underground Flora and Fauna*

The underground ecosystem, often overlooked, plays a crucial role in ecosystem functioning and biodiversity. Below the soil surface, a diverse array of flora and fauna thrive, including microorganisms, fungi, bacteria, nematodes, and roots of various plant species. These organisms contribute to nutrient cycling, soil structure formation, and the decomposition of organic matter (Bardgett and van der Putten 2014; Singh et al. 2024a, b, c). Additionally, underground communities provide essential services such as nutrient mineralization, soil aeration, and water filtration, ultimately supporting aboveground vegetation and wildlife. Furthermore, these organisms are vital in carbon sequestration and storage, influencing climate regulation (Six et al. 2006). Understanding and conserving underground biodiversity is crucial for maintaining ecosystem resilience and productivity.

## 3.2 Interactions with Soil Health

Underground flora and fauna have integral roles in maintaining soil health and productivity. Microorganisms, such as bacteria and fungi, are necessary for nutrient cycling processes, breaking organic matter, and enhancing soil fertility (Philippot et al. 2013). Moreover, soil fauna, including earthworms and arthropods, contribute to soil aeration, nutrient mixing, and soil aggregates' formation, improving soil structure and water infiltration (Boraet al. 2021). The interactions between underground organisms and soil health are complex and dynamic, influencing factors such as nutrient availability, soil structure stability, and resistance to erosion. Consequently, conserving underground biodiversity is essential for sustaining healthy soils and ensuring the long-term productivity of terrestrial ecosystems.

## 3.3 Soil Health Parameters Affected by Forest Fires

The impact of forest fires and prescribed fires on forest soil is highly complex and multifaceted. It affects numerous aspects, including soil organic matter, macro and micro-nutrients, and physical properties such as texture, color, pH, bulk density, and soil biota.

The long-term effects of forest fires on soil are also significant. These fires can lead to changes in soil physical and chemical properties, such as the loss of soil structure and organic matter, reduced porosity, and increased pH (DeBano 2000; Certini 2005; Kumar et al. 2022). These soil properties alterations affect ecosystems' water dynamics, vegetation patterns, and fauna. Furthermore, these changes in soil properties can result in indirect impacts such as increased hydrophobicity (water repellency), decreased soil infiltration, and increasing runoff, often leading to heightened erosion (DeBano 2000).

## 3.4 Soil Texture and Structure

Forest fires significantly alter soil properties, affecting soil color, texture, and water repellency. Severely burned soil exhibits noticeable changes in color, with a redder hue due to the transformation of Fe-oxides and the complete removal of organic matter (Ulery and Graham 1993; Certini 2005). Black or grey ash covers the ground in low to moderate fire intensities (Certini 2005). Therefore, the spatially heterogeneous accumulation of thermally produced iron oxides in soils may serve as a detectable indicator of the long-term patterning of forest fire intensities over a landscape (Goforth et al. 2005).

Soil texture components, mainly clay, can change at high temperatures, affecting the clay's hydration and structure (Neary et al. 2005). According to Ulery and Graham

(1993), soil layers reddened after a fire tend to have significantly less clay content than unburned soils. Sand-sized aggregates formed during burning alter particle-size distribution, resulting in coarser textures. At one site, burning led to a finer texture due to increased silt fraction, resulting from the decomposition of kaolinized sand grains (Ulery and Graham 1993).

Soil water repellency is greatly affected by fire, particularly at the surface, leading to decreased water storage capacity and reduced infiltration rates (Mataix-Solera et al. 2013). Organic materials burned at high temperatures can create vapors that condense on soil particles, causing them to become water-repellent (Letey 2001). Soil water repellency is most pronounced at the soil surface in areas that have experienced high and moderate severity burns, with its strength diminishing as burn severity decreases and depth increases (MacDonald and Huffman 2004). It prevents water from wetting or infiltrating dry soils. This phenomenon is attributed to organic compounds with hydrophobic properties on the surface of soil particles (Doerr et al. 2009).

## *3.5 Soil Nutrient Availability*

Forest fires influence nutrient cycling through various mechanisms, including nutrient loss in smoke, nutrient deposition in ash, changes in soil mineralization rates, erosion of ash and soil particles, and increased nutrient leaching.

The nutrients most likely to affect site productivity and vegetation dynamics greatly interest forest managers and ecologists. Following a fire, nutrients in biomass and detritus can be lost to the atmosphere, deposited as ash, or retained in incompletely burned vegetation or detritus (Singh et al. 2024a, b, c; Yadav et al. 2019a; Singh 2021; Singh et al. 2018). Fire may also affect soil nutrient status by directly adding nutrients and indirectly altering the soil environment. Research indicates that while soil nutrient levels decrease after a forest fire, their plant-available forms increase (Kutiel and Naveh 1987). Burned soils typically have lower nitrogen but higher calcium, with potassium, magnesium, and phosphorus stocks remaining nearly unchanged (Neff et al. 2005).

The immediate effect of fire on soil macronutrients is their loss through volatilization due to high temperatures (Certini 2005; Neary et al. 1999). Nitrogen volatilization is particularly significant during prescribed fires, with approximately half of the nitrogen in organic matter being volatilized at 500 °C (Neary et al. 1999; Caldwell et al. 2002; Knicker 2007). While there has been debate over whether fire significantly alters total soil nitrogen pools, it is suggested that burning can increase the nitrogen concentration of residual material.

Micronutrients, such as Fe, Mn, Cu, Zn, B, and Mo, have received less attention in fire ecology studies. However, their availability post-fire is crucial for understanding soil and plant recovery (García-Marco and González-Prieto 2008). Some studies suggest a reduction in micronutrient levels after forest fires, with short-term changes observed in soil micronutrient availability following prescribed fires. These changes include Mn and Zn availability increases and decreases in Fe and Co availability, while Cu availability remains unaffected.

## 3.6 Microbial Communities

Fire can have both direct and indirect effects on biological organisms. Direct effects occur when organisms are exposed directly to flames, hot gases, or high temperatures, resulting in death or injury. On the other hand, indirect effects lead to long-term environmental changes that affect the welfare of organisms, such as competition for habitat and food supply.

Soil-dwelling invertebrates and microorganisms are particularly sensitive to the impacts of fire. Invertebrates play vital roles in litter decomposition, nutrient mineralization, and soil turnover, while microorganisms are essential for nutrient cycling and energy flow in forest ecosystems. Fire can significantly alter these microbial communities, leading to biomass reductions and community structure changes (Neary et al. 1999).

Studies on soil bacteria after forest fires have shown significant changes in bacterial community structure, with some bacterial groups being stimulated by fire while others are depressed. The community structure of soil bacteria in post-fire non-climax forests several years after a fire can exhibit greater heterogeneity than that in unburned climax forests (Otsuka et al. 2008).

Similarly, mycorrhizal fungi, which are crucial for nutrient uptake and overall forest health, can be negatively affected by fire. Forest fires can significantly impact ectomycorrhizal biomass, with an eight-fold reduction in total ectomycorrhizal biomass in burnt sites due to destroying the litter organic layer (Stendell et al. 1999). However, fire does not significantly affect mycorrhizal biomass in the mineral layers. Additionally, Rashid et al. (1997) suggested that forest fires can also affect arbuscular mycorrhizal (AM) fungi, leading to changes in soil conditions and altering AM proliferation. They found that burnt sites had fewer viable AM fungal propagules than nearby control areas, although the total spore count remained similar.

While some studies suggest that forest fires may not significantly affect certain bacterial groups, further research is needed to understand the long-term impacts of fire on microbial communities and their interactions with soil conditions and vegetation succession. Similarly, the relationships between mycorrhizal activities, time following fire, and soil factors require further examination, especially in tropical ecosystems.

## 3.7 Soil Organic Matter and Carbon Sequestration

There are five principal global carbon pools: oceanic, geologic, pedologic (soil), biotic, and atmospheric. The third largest terrestrial carbon pool is represented by

Soil organic matter, which is crucial for ecosystem sustainability. The most apparent change in soil during burning is the loss of organic matter. This organic horizon provides essential functions such as erosion mitigation, habitat for soil biota, nutrient availability, soil cation exchange capacity, and ions retention (Certini 2005).

The impact of fire on SOM varies widely, ranging from destruction to partial scorching, depending on factors like fire severity and soil moisture. Prescribed fires typically result in minimal soil carbon changes, but intense fires can lead to significant losses. Increasing fire frequency can increase carbon in fine soil fractions and alter stable carbon isotope ratios. Charcoal formed after fires can accelerate humus breakdown, affecting carbon cycling (Kaushik and Singh 2024). In low-intensity fires, the least affected ones are lipids, while a significant portion of cellulose, hemicelluloses, and lignin is highly impaired.

## 4 Response of Underground Flora and Fauna to Forest Fires

### *4.1 Adaptive Mechanisms*

Underground flora and fauna have developed diverse adaptive mechanisms to withstand the impact of forest fires, which are essential components of many ecosystems (Prakash et al. 2022; Nirmal and Singh 2021; Singh and Kumar 2022; Savita et al. 2017). Dormancy is a prevalent strategy among various organisms, such as certain plant seeds and insect pupae, allowing them to remain dormant underground until favorable conditions return post-fire, ensuring survival and subsequent emergence for growth and development. Additionally, some plant species have evolved fire-resistant structures like thick bark or underground storage organs such as bulbs or tubers, protecting against intense heat and facilitating regeneration post-fire (Durigan et al. 2020). Soil-dwelling animals, including insects, earthworms, and small mammals, exhibit burrowing behavior to escape the heat and flames, finding refuge underground where temperatures are lower and moisture levels are more stable. Moreover, certain species possess resilient life cycles adapted to fire disturbances; for instance, some plants necessitate fire-induced heat for seed germination, ensuring new growth in the fire's aftermath (Singh et al. 2021c; Yadav et al. 2019b; Singh et al. 2023; Sharma and Singh 2021). These adaptive strategies collectively enable underground flora and fauna to persist in the face of fire-induced challenges, contributing to the resilience and survival of ecosystems (Rainsford et al. 2020).

## 4.2 Short-Term and Long-Term Responses

The response of underground flora and fauna to forest fires exhibits both short-term and long-term effects, each influencing the dynamics of ecosystems in distinct ways. In the short term, immediate survival mechanisms come into play as underground organisms may retreat deeper into the soil or enter a state of dormancy to evade the direct impact of the fire. The destruction wrought by fires also alters resource availability underground, affecting the supply of crucial elements like food, shelter, and moisture for soil-dwelling organisms. Furthermore, in the immediate aftermath, opportunistic species may exploit the altered post-fire environment, capitalizing on newly available resources or reduced competition to establish themselves (Rossetti et al. 2022).

Conversely, long-term responses unfold over time and often reshape the composition and structure of underground communities. Forest fires can trigger successional changes, initiating ecological succession processes that lead to shifts in species composition and community dynamics. As the ecosystem recovers, new species may colonize the area, altering the underground community's makeup. Moreover, the physical and chemical properties of the soil undergo modifications due to the fire, affecting habitat suitability for underground organisms. Changes in soil structure, nutrient levels, and moisture retention can significantly impact the distribution and abundance of soil fauna and flora over the long term (Souza et al. 2021).

Additionally, recurrent fire events may drive genetic adaptation among underground organisms, leading to the evolution of traits that confer greater resilience to fire-related stresses. These genetic adaptations enhance the chances of survival for future generations, contributing to the overall resilience of underground flora and fauna in fire-prone ecosystems. Collectively, the interplay of short-term responses and long-term adaptations shapes the complex response of underground communities to forest fires, highlighting their crucial role in ecosystem dynamics and resilience.

## 4.3 Impact on Biodiversity

Forest fires have a multifaceted influence on biodiversity, encompassing detrimental and beneficial effects on underground flora and fauna. While these fires can initially lead to the loss of habitat and a decline in biodiversity due to the destruction of underground habitats and the elimination of soil-dwelling organisms, they also present opportunities for regeneration and renewal. Severe fires can devastate underground habitats, causing a short-term reduction in biodiversity. However, post-fire environments often serve as catalysts for biodiversity promotion. Forest fires create open niches and reset successional processes, facilitating colonizing disturbed areas by a diverse array of pioneer species, including underground plants and animals adapted to such conditions. The patchy nature of fire disturbances contributes to spatial heterogeneity in underground habitats, fostering diverse microhabitats and promoting species coexistence.

Moreover, fires can alter species interactions underground, reshaping predator–prey relationships, competition dynamics, and mutualistic associations. These alterations in species interactions may have cascading effects on underground communities and ecosystem function, further influencing biodiversity patterns. Thus, while forest fires pose immediate threats to underground biodiversity, they also serve as drivers of ecosystem dynamics and resilience, emphasizing fires' complex and nuanced role in shaping underground flora and fauna communities (Geraskina et al. 2022).

## 5 Techniques for Assessing Post-fire Soil Health and Underground Ecosystems

### *5.1 Field Sampling Methods*

Field sampling methods are essential for directly observing and assessing post-fire soil health and underground ecosystems, providing valuable insights into the impacts of fire disturbances. Soil core sampling allows for the extraction of soil cores from various depths, facilitating the assessment of soil properties like texture, nutrient content, and moisture levels, thereby elucidating the fire's effects on soil characteristics and composition. Pit excavation techniques enable researchers to examine soil profiles and identify fire-induced alterations, such as compaction, ash deposition, or changes in soil horizons. Collecting soil-dwelling macroinvertebrates through macroinvertebrate sampling aids in evaluating underground fauna diversity and abundance, which serve as indicators of ecosystem health and post-fire functioning. Root sampling methods involve excavating soil to access and quantify root systems, providing crucial insights into below-ground productivity and nutrient cycling, essential for ecosystem recovery assessment. Additionally, vegetation surveys contribute to understanding underground ecosystem dynamics by assessing aboveground vegetation recovery, indirectly indicating the resilience of underground ecosystems to fire disturbances through monitoring plant species composition, diversity, and cover. These field sampling methods offer comprehensive assessments of post-fire soil health and underground ecosystems, guiding effective management and restoration strategies in fire-affected landscapes (Raiesi and Pejman 2021).

### *5.2 Laboratory Analyses*

Laboratory analyses are vital complements to field sampling methods, offering in-depth assessments of soil and underground ecosystem properties through controlled experimentation. These analyses involve processing samples collected from the field in laboratory settings, enabling precise quantification and characterization of various parameters. The laboratory techniques include soil chemistry analysis, which evaluates soil pH, nutrient content (e.g., nitrogen, phosphorus, potassium), and cation exchange capacity (CEC), providing quantitative data on soil fertility and nutrient availability in the aftermath of fires (Ebel et al. 2022). Microbial biomass and activity assays, including microbial respiration and enzyme activity measurements, offer insights into soil microbial community responses to fire disturbances and their roles in ecosystem recovery processes. DNA sequencing techniques facilitate identifying changes in soil microbial community composition and diversity post-fire, aiding in understanding microbial community dynamics and functional responses to disturbance events. Additionally, chemical analysis of soil organic matter, conducted through methods like Fourier-transform infrared spectroscopy (FTIR) or pyrolysis gas chromatography-mass spectrometry (Py-GC/MS), allows for the assessment of soil carbon dynamics and decomposition processes following fire events, providing valuable information on soil health and ecosystem recovery trajectories (Jiménez-Morillo et al. 2020). These laboratory analyses collectively enhance our understanding of post-fire soil and underground ecosystem dynamics, guiding management and restoration efforts in fire-affected environments.

## 5.3 Remote Sensing Techniques

Remote sensing techniques offer a powerful and non-invasive approach to assessing post-fire soil health and underground ecosystems across extensive areas. These methods provide detailed insights into surface and subsurface conditions following fire disturbances. The remote sensing methods include the Normalized Difference Vegetation Index (NDVI), which quantifies vegetation greenness and health by comparing near-infrared and red reflectance, facilitating the monitoring of vegetation recovery and changes in vegetation cover induced by fires over time (Adagbasa et al. 2020). Thermal infrared imaging detects temperature variations in the soil surface, enabling the assessment of post-fire soil heating and moisture conditions. Changes in soil temperature profiles provide valuable insights into soil health and ecosystem recovery processes post-fire. LiDAR technology produces high-resolution 3D terrain and vegetation structure maps, enabling precise characterization of post-fire landscape topography and vegetation regrowth patterns. Hyperspectral imaging captures spectral signatures across numerous narrow electromagnetic spectrum bands, facilitating the identification and characterization of soil properties, vegetation types, and ecosystem dynamics post-fire. These remote sensing techniques offer spatially explicit information essential for understanding the extent and severity of fire impacts on soil and underground ecosystems, guiding effective management and restoration efforts in fire-affected landscapes (Sirin and Medvedeva 2022).

# 6 Case Studies: Forest Fires and Underground Ecosystems

## *6.1 Case Study 1 [Moscow, 2010]*

The Moscow oblast forest and peat fire of 2010, covering an area of 9 hectares with varying tree composition, serves as an illustrative case highlighting the challenges in assessing soil carbon loss and emissions from underground fires. By examining pre-fire soil surfaces along stump root collars and comparing peat characteristics between burned and adjacent areas, researchers determined the average burnout depth to be 15 $\pm$ 8 cm, varying from 13 $\pm$ 5 to 20 $\pm$ 9 cm across different zones. Burnout depth correlated with relative surface height and was highest in aspen-dominated areas. Researchers established relationships between carbon stock and peat thickness using data on bulk density, ash content, and carbon content in peat layers. From these relationships and considering burnout depth, they estimated carbon losses averaging 9.8 $\pm$ 5.57 kg per square meter in burned areas, ranging from 8.61 $\pm$ 3.75 to 12.9 $\pm$ 6.18 kg per square meter across different zones. This corresponds to nearly 400 tons of $CO_2$ emissions per hectare, over 1.5 times higher than the potential $CO_2$ release from the biomass of a stand with a stem wood stock exceeding 280 cubic meters per hectare. These findings highlight the significant impact of underground fires on carbon loss in boreal ecosystems, indicating a tendency to underestimate their contribution compared to tropical regions. This underscores the importance of considering forest and peatland ecosystems in assessments of atmospheric gas composition and climate influence (Sirin et al. 2020).

## *6.2 Case Study 2 [Black Saturday Bushfires and Australia, 2009]*

In February 2009, Victoria, Australia, endured the catastrophic Black Saturday Bushfires, which scorched over 4,500 square kilometers of land, including forests, grasslands, and national parks. Ignited by extreme heat and fierce winds, these infernos left a lasting impact on underground ecosystems. The intense heat of the fires inflicted significant soil degradation, leading to compaction, loss of organic matter, and alterations in soil structure, ultimately disrupting soil fertility and nutrient cycling, thereby affecting underground flora and fauna. The fires target many species of soil-dwelling organisms, resulting in a notable loss of biodiversity within underground ecosystems, particularly affecting species with limited dispersal abilities and specialized habitat requirements. Despite the devastation, signs of recovery emerged as pioneer species of plants and insects began to colonize the burned areas, gradually restoring habitat and instigating vital ecological processes necessary for the underground ecosystem's recovery (McDonald and McCormack 2022).

# 7 Management Implications and Strategies

## 7.1 Post-Fire Restoration Practices

Effective post-fire restoration practices are crucial for restoring ecosystem health and promoting resilience following wildfires. Key strategies include implementing soil stabilization measures such as mulching, seeding, and installing erosion control barriers to prevent soil erosion and promote stability, thereby protecting underground habitats and minimizing habitat loss for soil-dwelling organisms. Additionally, habitat restoration efforts involving replanting native vegetation and reintroducing key plant species crucial for supporting underground ecosystems are essential. Restoring vegetation cover helps recreate habitat structure and enhances soil stability, facilitating the recovery of underground flora and fauna. Water management strategies include installing check dams, constructing sediment traps, restoring natural hydrological processes, mitigating soil erosion impacts, enhancing moisture retention in fire-affected landscapes, and supporting underground ecosystem recovery. Furthermore, establishing long-term monitoring programs to assess restoration effectiveness and track underground ecosystem health changes over time, coupled with adaptive management approaches, allows for adjustments to restoration strategies based on monitoring data, ensuring their continued effectiveness in promoting ecosystem resilience (Scheper et al. 2021).

## 7.2 Forest Management Policies

Forest management policies are pivotal in mitigating wildfire risks and bolstering ecosystem resilience. Key policies and strategies encompass conducting comprehensive fire risk assessments to pinpoint wildfire-prone areas and prioritize management interventions, such as fuel reduction treatments and prescribed burning, to curtail fire intensity and severity, consequently mitigating impacts on underground ecosystems. Implementing sustainable logging practices is paramount, emphasizing ecosystem health and resilience through approaches like selective logging, reduced-impact logging, and maintaining buffer zones around sensitive habitats to minimize disruptions to underground ecosystems and safeguard biodiversity. Moreover, integrating measures to enhance habitat connectivity and landscape connectivity, such as preserving wildlife corridors and green infrastructure networks, fosters species dispersal and facilitates genetic exchange among populations, thereby fortifying the resilience of underground ecosystems to fire disturbances (Young et al. 2020).

### 7.3 *Incorporating Underground Ecosystems in Fire Management Plans*

Integrating underground ecosystems into fire management plans is paramount for effectively addressing their conservation needs and enhancing ecosystem resilience. Key considerations entail conducting thorough ecological assessments to identify and map underground habitats and sensitive areas, integrating this information into fire risk assessments and management plans to prioritize conservation efforts and mitigate impacts on underground ecosystems. Implementing prescribed burning practices tailored to the ecological requirements of underground flora and fauna is essential, including maintaining unburned refugia and minimizing the frequency and intensity of burns in sensitive habitats to uphold underground biodiversity while achieving fire management objectives. Additionally, conducting public education and outreach campaigns is pivotal in raising awareness about the significance of underground ecosystems and their conservation needs. Such initiatives foster support for fire management practices, prioritizing ecosystem health and resilience while promoting stewardship among local communities to participate in conservation efforts actively (Klink et al. 2020).

## 8 Future Directions and Research Needs

### 8.1 *Emerging Technologies and Methodologies*

As we look towards the future, embracing emerging technologies and methodologies holds promise for advancing our understanding of underground ecosystems in the context of forest fires. Innovative techniques such as high-resolution remote sensing, including LiDAR and hyperspectral imaging, offer opportunities to map underground habitats and assess post-fire impacts with unprecedented detail and accuracy. Furthermore, advancements in molecular biology, such as DNA meta-barcoding and metagenomics, enable comprehensive characterization of underground microbial communities and their responses to fire disturbances. Integrating these technologies into interdisciplinary research frameworks can provide novel insights into the dynamics and resilience of underground ecosystems in fire-affected landscapes (Gale et al. 2021).

### 8.2 *Gaps*

Despite its complexity, the role of fire on forest soil is less studied than its aboveground effects, highlighting the need for further research in this area. Besides significant progress, several gaps in understanding persist, necessitating further research to address key knowledge deficiencies. One critical area is the long-term effects of forest fires on underground biodiversity and ecosystem functioning. Longitudinal studies spanning multiple fire cycles are needed to elucidate the resilience mechanisms of underground flora and fauna and their capacity to withstand recurrent disturbances. Additionally, there is a lack of comprehensive understanding regarding the interactions between aboveground and underground components of ecosystems in fire-prone environments. Investigating these interactions is crucial for developing holistic management approaches encompassing the entire ecosystem.

### *8.3 Potential Mitigation Strategies*

Identifying and implementing effective mitigation strategies is essential for minimizing the impacts of forest fires on underground ecosystems. Enhanced land management practices, such as prescribed burning regimes tailored to underground biodiversity conservation objectives, hold promise for maintaining habitat heterogeneity and promoting ecosystem resilience. Furthermore, integrating underground ecosystem considerations into landscape-scale conservation planning frameworks can help identify priority areas for protection and restoration efforts. Collaborative efforts involving stakeholders from diverse sectors, including government agencies, conservation organizations, and local communities, are necessary to implement these mitigation strategies effectively and ensure the long-term sustainability of underground ecosystems in fire-prone regions.

## 9 Conclusion

Natural disturbances are essential for maintaining the ecological balance of forests, supporting rejuvenation, nutrient cycling, and biodiversity. Climate patterns, vegetation traits, and ignition sources significantly shape fire regimes. With climate change, weather extremes like droughts and heatwaves, which fuel fire activity, are becoming more frequent and severe. The wildfires not only affect the above-ground biodiversity but also impact the underground floral and faunal species. Beneath the soil surface, a diverse range of flora and fauna, including microorganisms, fungi, bacteria, nematodes, and plant roots, thrive. The impact of fires on soil and underground life significantly influences forest composition and structure post-fire. Fires lead to rapid loss of organic matter through combustion, altering soil properties and nutrient dynamics. Some plant species have evolved fire-resistant structures, such as thick bark or underground storage organs, protecting against intense heat and aiding post-fire regeneration. Recurrent fires may drive genetic adaptation among underground organisms, enhancing resilience to fire-related stresses. Field sampling methods and laboratory techniques, including soil chemistry analysis and DNA sequencing, guide effective management and restoration efforts in fire-affected areas. Enhancing habitat and landscape connectivity through preserving wildlife corridors and green infrastructure networks facilitates species dispersal and genetic exchange, bolstering underground ecosystem resilience to fire disturbances. Collaborative efforts involving various stakeholders are crucial for implementing mitigation strategies and ensuring the long-term sustainability of underground ecosystems in fire-prone regions.

## References

Adagbasa EG, Adelabu SA, Okello TW (2020) Development of post-fire vegetation response-ability model in grassland mountainous ecosystem using GIS and remote sensing. ISPRS J Photogramm Remote Sens 164:173–183

AghaKouchak A, Chiang F, Huning LS, Love CA, Mallakpour I, Mazdiyasni O, Sadegh M (2020) Climate extremes and compound hazards in a warming world. Annu Rev Earth Planet Sci 48:519–548

Apurva K, Singh H, Kumar H (2017) Elucidating morphological and growth changes in Arjun Tree (*Terminalia arjuna*) grown under elevated temperature condition. Trends Biosci 10(1):332–336

Archibald S, Lehmann CE, Gómez-Dans JL, Bradstock RA (2013) Defining pyromes and global syndromes of fire regimes. Proc Natl Acad Sci 110(16):6442–6447

Bardgett RD, Van Der Putten WH (2014) Belowground biodiversity and ecosystem functioning. Nature 515(7528):505–511

Barwa S, Singh H (2024) Climate-induced changes in forest ecosystems of Indian Himalayan Region: a meta-analysis. In: Singh H (ed) Urban forests, climate change and environmental pollution. Springer, Cham. https://doi.org/10.1007/978-3-031-67837-0_39

Bora S, Bisht SS, Reynolds JW (2021) Global diversity of earthworms in various countries and continents: a short review. Megadrilogica 26(9)

Bowman DM, Moreira-Muñoz A, Kolden CA, Chávez RO, Muñoz AA, Salinas F, Johnston FH (2019) Human–environmental drivers and impacts of the globally extreme 2017 Chilean fires. Ambio 48:350–362

Caldwell TG, Johnson DW, Miller WW, Qualls RG (2002) Forest floor carbon and nitrogen losses due to prescription fire. Soil Sci Soc Am J 66(1):262–267

Certini G (2005) Effects of fire on properties of forest soils: a review. Oecologia 143:1–10

Chuvieco E, Giglio L, Justice C (2008) Global characterization of fire activity: toward defining fire regimes from Earth observation data. Glob Change Biol 14(7):1488–1502

Clark B (2001) Soils, water, and watersheds. Fire effects guide. National Interagency Fire Center, USA

Curtis PG, Slay CM, Harris NL, Tyukavina A, Hansen MC (2018) Classifying drivers of global forest loss. Science 361(6407):1108–1111

De Rigo D, Libertà G, Durrant TH, Vivancos TA, San-Miguel-Ayanz J (2017) Forest fire danger extremes in Europe under climate change: variability and uncertainty (Doctoral dissertation, Publications Office of the European Union)

DeBano LF (1991) The effect of fire on soil properties. In: Proceedings management and productivity of western-Montane. Forest soils, pp 151–155

DeBano LF (2000) The role of fire and soil heating on water repellency in wildland environments: a review. J Hydrol 231:195–206

DeBano LF, Neary DG, Ffolliott PF (1998) Fire effects on ecosystems. Wiley

Dhyani K, Kuniyal HB, Singh H, Sobha (2021) Growth and physiological potential of *Terminalia arjuna* under elevated $CO_2$ levels in open top chamber condition. J Appl Nat Sci 13(3):1121–1126. https://doi.org/10.31018/jans.v13i3.2490

Doerr SH, Woods SW, Martin DA, Casimiro M (2009) 'Natural background' soil water repellency in conifer forests of the north-western USA: its prediction and relationship to wildfire occurrence. J Hydrol 371(1–4):12–21

Durigan G, Pilon NA, Abreu RC, Hoffmann WA, Martins M, Fiorillo BF, Vasconcelos HL (2020) No net loss of species diversity after prescribed fires in the Brazilian savanna. Front for Glob Change 3:13

Ebel BA, Moody JA, Martin DA (2022) Post-fire temporal trends in soil-physical and-hydraulic properties and simulated runoff generation: insights from different burn severities in the 2013 Black Forest Fire, CO, USA. Sci Total Environ 802:149847

Fernandes PM, Guiomar N, Rossa CG (2019) Analysing eucalypt expansion in Portugal as a fire-regime modifier. Sci Total Environ 666:79–88

Flannigan MD, Krawchuk MA, de Groot WJ, Wotton BM, Gowman LM (2009) Implications of changing climate for global wildland fire. Int J Wildland Fire 18(5):483–507

Flannigan M, Cantin AS, De Groot WJ, Wotton M, Newbery A, Gowman LM (2013) Global wildland fire season severity in the 21st century. For Ecol Manage 294:54–61

Gale MG, Cary GJ, Van Dijk AI, Yebra M (2021) Forest fire fuel through the lens of remote sensing: review of approaches, challenges and future directions in the remote sensing of biotic determinants of fire behaviour. Remote Sens Environ 255:112282

García-Marco S, González-Prieto S (2008) Short-and medium-term effects of fire and fire-fighting chemicals on soil micronutrient availability. Sci Total Environ 407(1):297–303

Geraskina AP, Tebenkova DN, Ershov DV, Ruchinskaya EV, Sibirtseva NV, Lukina NV (2022) Wildfires as a factor of loss of biodiversity and forest ecosystem functions. For Sci 5(1):97

Goforth BR, Graham RC, Hubbert KR, Zanner CW, Minnich RA (2005) Spatial distribution and properties of ash and thermally altered soils after high-severity forest fire, southern California. Int J Wildland Fire 14(4):343–354

Hart SC, DeLuca TH, Newman GS, MacKenzie MD, Boyle SI (2005) Post-fire vegetative dynamics as drivers of microbial community structure and function in forest soils. For Ecol Manage 220(1–3):166–184

Hayward MW, Ward-Fear G, L'Hotellier F, Herman K, Kabat AP, Gibbons JP (2016) Could biodiversity loss have increased Australia's bushfire threat? Anim Conserv 19(6):490–497

Jain TB, Gould WA, Graham RT, Pilliod DS, Lentile LB, González G (2008) A soil burn severity index for understanding soil-fire relations in tropical forests. AMBIO: A J Hum Environ 37(7):563–568

Jangra A, Kumar K, Maikhuri S, Bhandari MS, Pandey S, Singh H Barthwal S (2024) Unveiling stress-adapted endophytic bacteria: characterizing plant growth-promoting traits and assessing cross-inoculation effects on Populus deltoides under abiotic stress. Plant Physiol Biochem 108610. https://doi.org/10.1016/j.plaphy.2024.108610

Jiménez-Morillo NT, Almendros G, José M, Jordán A, Zavala LM, Granged AJ, González-Pérez JA (2020) Effect of a wildfire and of post-fire restoration actions in the organic matter structure in soil fractions. Sci Total Environ 728:138715

Judson OP (2017) The energy expansions of evolution. Nat Ecol Evol 1(6):0138

Kaushik P, Singh H (2024) Climate change-induced forest ecosystem functioning and alteration in water-atmosphere-soil–plant (WASP) relationship. In: Singh H (ed) Urban forests, climate change and environmental pollution. Springer, Cham. https://doi.org/10.1007/978-3-031-67837-0_38

Klink CA, Sato MN, Cordeiro GG, Ramos MIM (2020) The role of vegetation on the dynamics of water and fire in the cerrado ecosystems: implications for management and conservation. Plants 9(12):1803

Knicker H (2007) How does fire affect the nature and stability of soil organic nitrogen and carbon? A review. Biogeochemistry 85(1):91–118

Kraus D, Goldammer JG (2007) Fire regimes and ecosystems: an overview of fire ecology in tropical ecosystems. In: Schmerbeck J, Hiremath A, Ravichandran C (eds) Forest fires in India–workshop proceedings. ATREE, Bangalore, India and Institute of Silviculture, Freiburg, Germany, pp 9–13

Kumar A, Kumar P, Singh H, Kumar N (2020b) Adaptation and mitigation potential of roadside trees with bio-extraction of heavy metals under vehicular emissions and their impact on physiological traits during seasonal regimes. Urban for Urban Greening 58:126900

Kumar A, Kumar P, Singh H, Bisht S, Kumar N (2021) Relationship of physiological plant functional traits with soil carbon stock in temperate forest of Garhwal Himalaya. Curr Sci 120(8):1368–1373

Kumar A, Kumar P, Singh H, Kumar N (2021b) Modulation of plant functional traits under essential plant nutrients during seasonal regime in natural forests of Garhwal Himalayas. Plant Soil 465:197–212

Kumar A, Singh H, Kumari G, Bisht S, Malik A, Kumar N, Singh M, Raturi A, Barthwal S, Thakur A, Kaushal R (2022) Adaptive resilience of roadside trees to vehicular emissions via leaf enzymatic, physiological, and anatomical trait modulations. Environ Pollut 313:120191

Kumar A, Dwivedi GK, Tewari S, Paul J, Anand R, Kumar N, Kumar P, Singh H, Kaushal R (2020a) Carbon mineralization and inorganic nitrogen pools under *Terminalia chebula* Retz.-based agroforestry system in Himalayan Foothills, India. For Sci 66(5):634–643

Kumar N, Singh H, Kumar A, Singh AK, Pandey PK, Kumar A (2024). Forest-fire-induced land degradation. In: Mishra G, Giri K, Singh S, Kumar M (eds) Sustainable land management in India. Springer Nature, Singapore. https://doi.org/10.1007/978-981-97-5223-2_4

Kutiel P, Naveh Z (1987) The effect of fire on nutrients in a pine forest soil. Plant Soil 104:269–274

Landres PB, Morgan P, Swanson FJ (1999) Overview of the use of natural variability concepts in managing ecological systems. Ecol Appl 9(4):1179–1188

Letey J (2001) Causes and consequences of fire-induced soil water repellency. Hydrol Process 15(15):2867–2875

MacDonald LH, Huffman EL (2004) Post-fire soil water repellency: persistence and soil moisture thresholds. Soil Sci Soc Am J 68(5):1729–1734

Malik A, Kumar K, Kumari G, Kumar N, Singh H (2024) Application of functional traits in modelling productivity and resilience under climate change. Springer, pp 77–96. ISBN: 978-981-97-1509-1

Martin RE, Sapsis DB (1992) Fires as agents of biodiversity: pyrodiversity promotes biodiversity. In: Proceedings of the conference on biodiversity of northwest California ecosystems. Cooperative Extension, University of California, Berkeley, pp 150–157

Mataix-Solera J, Arcenegui V, Tessler N, Zornoza R, Wittenberg L, Martínez C, Jordán MM (2013) Soil properties as key factors controlling water repellency in fire-affected areas: evidences from burned sites in Spain and Israel. CATENA 108:6–13

McDonald J, McCormack PC (2022) Responsibility and risk-sharing in climate adaptation: a case study of bushfire risk in Australia. Clim Law 12(2):128–161

Neary DG, Klopatek CC, DeBano LF, Ffolliott PF (1999) Fire effects on belowground sustainability: a review and synthesis. For Ecol Manage 122(1–2):51–71

Neary DG, Ryan KC, DeBano LF (2005) Wildland fire in ecosystems: effects of fire on soils and water. Gen. Tech. Rep. RMRS-GTR-42-vol 4. US Department of Agriculture, Forest Service, Rocky Mountain Research Station, Ogden, UT, 250 p., 42

Neff JC, Harden JW, Gleixner G (2005) Fire effects on soil organic matter content, composition, and nutrients in boreal interior Alaska. Can J for Res 35(9):2178–2187

Nirmal, Bakshi M, Singh H (2021) Effect of $CO_2$ elevation on Shisham growth at nursery stage. Indian J Hill Farming 34:272–276

Ojeda F (2020) Pine afforestation, herriza and wildfire: a tale of soil erosion and biodiversity loss in the Mediterranean region1. Int J Wildland Fire 29(12):1142–1146

Oldekop JA, Rasmussen LV, Agrawal A, Bebbington AJ, Meyfroidt P, Bengston DN, Wilson SJ (2020) Forest-linked livelihoods in a globalized world. Nat Plants 6(12):1400–1407

Otsuka S, Sudiana I, Komori A, Isobe K, Deguchi S, Nishiyama M, Senoo K (2008) Community structure of soil bacteria in a tropical rainforest several years after fire. Microbes Environ 23(1):49–56

Philippot L, Raaijmakers JM, Lemanceau P, Van Der Putten WH (2013) Going back to the roots: the microbial ecology of the rhizosphere. Nat Rev Microbiol 11(11):789–799

Prakash V, Hunney K, Singh H (2022) Effect of elevated carbon dioxide on growth and development of *Santalum album* L. seedlings inoculated with plant growth promoting microorganisms in open top chambers. Int J Environ, Agric Biotechnol 7(5):149–159

Raiesi F, Pejman M (2021) Assessment of post-wildfire soil quality and its recovery in semi-arid upland rangelands in Central Iran through selecting the minimum data set and quantitative soil quality index. CATENA 201:105202

Rainsford FW, Kelly LT, Leonard SW, Bennett AF (2020) Post-fire development of faunal habitat depends on plant regeneration traits. Austral Ecol 45(6):800–812

Raison RJ, Khanna PK (2011) Possible impacts of climate change on forest soil health. In: Soil health and climate change. Springer, Berlin, Heidelberg, pp 257–285

Rashid A, Ahmed T, Ayub N, Khan AG (1997) Effect of forest fire on number, viability and post-fire re-establishment of arbuscular mycorrhizae. Mycorrhiza 7:217–220

Raturi A, Singh H, Kumar P, Shukla N (2022) Characterizing the post-monsoon $CO_2$, $CH_4$, $N_2O$, and $H_2O$ vapor effluxes from a tropical wetland in the Himalayan foothill. Environ Monitor Assess 194(50). https://doi.org/10.1007/s10661-021-09721-8

Rossetti I, Cogoni D, Calderisi G, Fenu G (2022) Short-term effects and vegetation response after a Megafire in a Mediterranean area. Land 11(12):2328

Savita, Sharma SK, Singh H, Singh O (2017) Forest fire: a review. Indian Forester 143(8):719–728

Scheper AC, Verweij PA, van Kuijk M (2021) Post-fire forest restoration in the humid tropics: a synthesis of available strategies and knowledge gaps for effective restoration. Sci Total Environ 771:144647

Sett T, Nikam BR, Singh H, Purohit S (2022) Effect of drought stress on forest evapotranspiration—a case study on Indian forests. EGU General Assembly 2022. https://doi.org/10.5194/egusphere-egu22-510

Sett T, Nikam BR, Singh H, Purohit S (2024) Hydrological variability in Indian forest ecosystem: analysis of drought resilience, recovery and water use efficiency in moist and dry deciduous forests. In: Singh H (ed) Urban forests, climate change and environmental pollution. Springer, Cham. https://doi.org/10.1007/978-3-031-67837-0_37

Sharma R, Singh H, Kaushik M, Nautiyal R, Singh O (2018) Adaptive physiological response, carbon partitioning and biomass production of *Withania somnifera* (L.) Dunal grown under elevated $CO_2$ regimes. 3 Biotech 8:267. https://doi.org/10.1007/s13205-018-1292-1

Sharma R, Singh H (2021) Alteration in biochemical constituents and nutrients partitioning of *Asparagus racemosus* in response to elevated atmospheric $CO_2$ concentration. Environ Sci Pollut Res 29:6812–6821. https://doi.org/10.1007/s11356-021-16050-3

Singh H, Sharma R, Savita SMP, Kumar M, Verma A, Ansari MW, Sharma SK (2018) Adaptive physiological response of *Parthenium hysterophorus* to elevated atmospheric $CO_2$ concentration. Indian Forester 144:1–14

Singh S, Singh H, Sharma SK, Nautiyal R (2021c) Seasonal variation in biochemical responses of bamboo clones in the sub-tropical climate of Indian Himalayan foothills. Heliyon 7(4):e06859. https://doi.org/10.1016/j.heliyon.2021.e06859

Singh H, Kumar N, Singh R, Kumar M (2023) Assessing the climate change impact on the habitat suitability of the range-restricted bird species (*Catreus wallichii*) in the Indian Himalayan ecosystem. Environ Sci Pollut Res. https://doi.org/10.1007/s11356-023-30789-x

Singh H, Kumar M (2022) Climate change and its impact on Indian Himalayan forests: current status and research needs. In: Rani S, Kumar R (eds) Climate change. Springer Climate. Springer, Cham, pp 223–242. https://doi.org/10.1007/978-3-030-92782-0_11

Singh H, Kumar N, Kumar M, Singh R (2020) Modelling habitat suitability of western tragopan (*Tragopan melanocephalus*) a range-restricted vulnerable bird species of the Himalayan region, in response to climate change. Clim Risk Manag 29:100241. https://doi.org/10.1016/j.crm.2020.100241

Singh M, Singh H, Kumar A, Kumar M, Barthwal S, Thakur A (2024a) Soil nitrogen availability determines the $CO_2$ fertilization effect on tree species (*Neolamarckia cadamba*): growth and physiological evidence. Environ Sustain 7:53–60. https://doi.org/10.1007/s42398-023-00300-w

Singh H, Kumar N, Kumar A (2024b) Enhancing resource use efficiency in crops through plant functional traits. Springer, pp 97–117. ISBN: 978-981-97-1509-1

Singh M, Jyoti, Kumar N, Singh H (2024c) Plant functional traits assisted crop adaptation to abiotic and biotic stress. Springer, pp 239–255. ISBN: 978-981-97-1509-1

Singh H (2021) Process-based carbon sequestration study in relation to the energy-water-carbon flux in a forest ecosystem. In: Kumar P, Singh RK, Kumar M, Rani M, Sharma (eds) Climate impacts on sustainable natural resource management. CRC Press Taylor & Francis Group, United States, pp 336–361

Singh H (2024a) Climate change and forests: biological perspectives. In: Singh H (ed) Forests and climate change. Springer Nature, Singapore. https://doi.org/10.1007/978-981-97-3905-9_1

Singh H (2024b) Climate change and environmental pollution: biological consequences on urban forests. In: Singh H (eds) Urban forests, climate change and environmental pollution. Springer Nature, Cham, Switzerland. https://doi.org/10.1007/978-3-031-67837-0_1

Singh H (ed) (2024c) Forests and climate change: biological perspectives on impact, adaptation, and mitigation strategies. Springer Nature Singapore. https://doi.org/10.1007/978-981-97-3905-9

Singh H (ed) (2024d) Urban forests, climate change and environmental pollution: physio-biochemical and molecular perspectives to enhance urban resilience. Springer Nature, Cham, Switzerland. https://doi.org/10.1007/978-3-031-67837-0

Singh H (2024e) Exploring adaptive modulation in plant functional traits and its impact on the productivity of *Acacia auriculiformis* under $CO_2$-enriched environment. Ind Crops Prod 210(2024):118186. https://doi.org/10.1016/j.indcrop.2024.118186

Sirin AA, Makarov DA, Gummert I, Maslov AA, Gul'be YI (2020) Depth of peat burning and carbon loss during an underground forest fire. Contemp Prob Ecol 13(7):769–779

Sirin A, Medvedeva M (2022) Remote sensing mapping of peat-fire-burnt areas: identification among other wildfires. Remote Sens 14(1):194

Six J, Frey SD, Thiet RK, Batten KM (2006) Bacterial and fungal contributions to carbon sequestration in agroecosystems. Soil Sci Soc Am J 70(2):555–569

Souza E, Lima AP, Magnusson WE, Kawashita-Ribeiro R, Fadini R, Ghizoni IR Jr, Fraga R (2021) Short-and long-term effects of fire and vegetation cover on four lizard species in Amazonian savannas. Can J Zool 99(3):173–182

Stendell ER, Horton TR, Bruns TD (1999) Early effects of prescribed fire on the structure of the ectomycorrhizal fungus community in a Sierra Nevada ponderosa pine forest. Mycol Res 103(10):1353–1359

Stralberg D, Wang X, Parisien MA, Robinne FN, Sólymos P, Mahon CL, Bayne EM (2018) Wildfire-mediated vegetation change in boreal forests of Alberta, Canada. Ecosphere 9(3):e02156

Ulery AL, Graham RC (1993) Forest fire effects on soil color and texture. Soil Sci Soc Am J 57(1):135–140

Wang X, Parisien MA, Taylor SW, Candau JN, Stralberg D, Marshall GA, Flannigan MD (2017) Projected changes in daily fire spread across Canada over the next century. Environ Res Lett 12(2):025005

Wilkinson SL, Moore PA, Flannigan MD, Wotton BM, Waddington JM (2018) Did enhanced afforestation cause high severity peat burn in the Fort McMurray Horse River wildfire? Environ Res Lett 13(1):014018

Yadav SK, Singh H, Ginwal HS, Barthwal S (2019a) Elevated $CO_2$ enhanced growth and physiological process of *Populus deltoides* Bartr. ex Marsh. Indian Forester 145(1):23–27

Yadav SK, Singh H, Nautiyal R, Ginwal HS, Ansari SA, Barthwal S (2019b) Modulation of morpho-physiological responses in *Populus deltoides* by elevated carbon dioxide and temperature. For Sci 66:105–118. https://doi.org/10.1093/forsci/fxz048

Young JD, Evans AM, Iniguez JM, Thode A, Meyer MD, Hedwall SJ, Huang CH (2020) Effects of policy change on wildland fire management strategies: evidence for a paradigm shift in the western US? Int J Wildland Fire 29(10):857–877

# Forest Fires and Climate Change Impacts on Water Resources

**Megha Verma and Meenakshi Sati**

**Abstract** Climate change and forest fires are interconnected phenomena that profoundly impact water resources. Climate change has resulted in more frequent and severe forest fires, droughts, storms, and outbreaks of pests and pathogens. These changes negatively affect both the quantity and quality of water.This is potentially vulnerable to catastrophic global forest fires and climate change impacts in numerous regions. Increased evaporation rates are expected to reduce water supplies in many areas. This leads to broader alterations in climate and weather patterns. Forest fires do not just consume vegetation but also devastate the forest floor, leaving soils vulnerable to precipitation. This increases the risk of erosion, sediment transport to streams, debris flows, and the release of chemicals. These factors significantly degrade water quality, enhance mineralization, and promote mineral weathering. Fire impacts catchment behavior by influencing crucial hydrological factors that regulate infiltration and surface runoff.The mineralization of organic matter induced by fire results in the liberation of substantial quantities of metals into the ash and soil matrix. This process renders these metals highly susceptible to mobilization, posing a significant risk of soil and water contamination in downstream ecosystems. This chapter delineates the intricate interplay between forest fires and climate change, examining how their synergistic effects shape water availability, quality, and quantity.There are several Adaptation and mitigation strategies to address these interconnected challenges. Implementing effective land management strategies can help mitigate the impact of emerging fires on ecosystems. Techniques such as prescribed burns and afforestation can lower the risk of uncontrolled forest fires.

**Keywords** Forest fire · Climate change · Water resources · Water quality · Hydrological processes

M. Verma (✉)
Doon P.G. College of Agriculture Science and Technology, Dehradun, India
e-mail: megha_pth@yahoo.in

M. Sati
Bharatiya Engineering Science and Technology Innovation University, Gownivaripalli, Andhra Pradesh, India



H. Singh and N. Kumar (eds.), *Forest Fire and Climate Change*,
https://doi.org/10.1007/978-3-031-89967-6_9

# 1 Introduction

Climate change denotes the sustained modification of global weather patterns, encompassing a broad spectrum from the Polar Regions to the equatorial zones. This phenomenon poses a significant threat globally, impacting various sectors (Abbass et al. 2022). A key dimension of climate change is the global escalation of temperatures, which has far-reaching implications for ecosystems, weather patterns, and human societies (Feliciano et al. 2022; Kumar et al. (2024). This shift in climate intricately influences water resources directly and indirectly by inducing alterations in the hydrological cycle (IPCC, 2007).

For millions of years, fire has been a natural phenomenon shaping the Earth's system and various biomes (Shi et al. 2022; Singh 2024a). Forest fires represent the primary disturbance impacting forest ecosystems globally. Fire is a critical environmental filter in many ecosystems, influencing species selection and community structure. However, not all fires are beneficial; some can lead to irreversible damage and increased vulnerability (Savita et al. 2017; Singh 2024b). In the country, about 36% of forests are considered vulnerable to fires, with 4% being fire-prone and 6% highly fire-susceptible. From November 2021 to June 2022, different sensors detected approximately 53,000–346,000 forest fires in India. Annually, forest fires destroy significant amounts of carbon stored in biomass, disrupt ecosystems, and transport forest-derived products. Forest fires vary based on vegetation composition, fuel availability, fire frequency, extent, and interval (Gajendiran et al. 2024).

# 2 The Link Between Forest Fire and Climate Change

Forest fires ravage approximately 3 to 5 million square kilometers of terrestrial ecosystems yearly, unleashing a staggering 8 billion tons of carbon dioxide into the atmosphere. This phenomenon devastates biodiversity and natural habitats and significantly exacerbates the global climate crisis by contributing to the accumulation of greenhouse gases (Chuvieco et al. 2019; Singh 2024c). Fires notably impact carbon storage in vegetation and contribute to atmospheric $CO_2$ levels, influencing global climate dynamics (Yin et al. 2020; Kumar et al. 2020a, 2020b). Changes in vegetation cover after forest fires impact the levels of reflective and absorptive aerosols in the atmosphere, alter surface albedo, and influence climate change feedback mechanisms, mainly through their effects on terrestrial carbon storage (Zou et al. 2020; Singh and Kumar 2022; Singh 2021).

The escalating temperatures driven by climate change are projected to prolong the fire and growing seasons, augment evaporative demands, diminish soil moisture, and reduce summer precipitation (Singh 2024d). Consequently, these changes are anticipated to expand the regions susceptible to forest fires (Prakash et al.2022). The interplay between fire and various disturbances, including drought and insect outbreaks, is poised to accelerate ecological transformations in a warming climate significantly (Reilly et al. (2017). Consequently, these interactions could lead to profound and potentially irreversible changes in ecosystem structure and function. The increase in dead wood mass and die-off events could lead to a 30% increase in fire spread.

## 3 Relation Between Forest and Water Cycle

Forests are pivotal to the global hydrological cycle due to their high evapotranspiration rates, significantly contributing to atmospheric moisture. For instance, approximately 50–80% of the water is retained within the ecosystem's water cycle in the Amazon rainforest. The dense forest canopies efficiently recycle water through evaporation, a markedly more effective process than in areas with sparse vegetation (Aragao 2012). Deforestation disrupts the local hydrological cycle by diminishing the forest's capacity for evapotranspiration and interception (Kaushik and Singh 2024). As a result, the landscape experiences altered patterns of heat release and energy fluxes (Singh 2021). Some documents explain that forests reduce the risk of floods, balance water flow, and highlight the role of forest sponges in absorbing water (Bosch and Hewlett, 1982). However, there has been a growing awareness of the importance of ecosystem services associated with Greenwater forests in the last decade. Green water is the part of rainwater absorbed and transferred by plants or evaporates from leaves, soil, and water and returns to the atmosphere.

Forests are vital for water security, as they play a key role in sustaining groundwater resources, commonly called blue water. This blue water consists of rainwater that infiltrates and replenishes aquifers, rivers, and lakes through flooding or recharge. Most of our drinking water originates from these vital groundwater sources. However, it takes 25 years for rainwater to reach groundwater. This means we are drinking water for 25 years. Nowadays, knowing the trade-offs involved in reducing blue water, green water produced by forests is considered one of the best ways to address the many benefits that can be blue water or a mixture of these (Maes et al. 2011).

## 4 Forest Fire-Induced Changes in Hydrological Processes

### *4.1 Surface Water*

Unlike simple vegetation burning, forest fires have a profound and complex effect on the forest floor, altering soil properties in the short term and over the long run. The impact of fire on soil depends on several factors, including the soil's natural composition, the fire's intensity and duration, how often fires occur, and the weather conditions following the fire. These fire-induced transformations can disrupt nutrient cycling, alter soil structure, and subsequently influence the trajectory of ecosystem recovery. The severity and nature of these impacts are intricately linked to the interplay between these variables, underscoring the complexity of fire's role in ecological dynamics (Certini 2005).

- Although heat transfers more swiftly and penetrates more deeply through moist soils than dry soils, the latent heat of vaporization plays a crucial role in tempering soil temperatures. This latent heat acts as a moderating force, ensuring that soil temperatures do not surpass 95 °C until the complete evaporation of moisture. However, soil temperatures can surge dramatically upon the culmination of this phase change, reaching levels between 200 °C and 300 °C (Franklin et al. 1997).
- Temperatures above 300 °C destroy soil organic matter, reducing the soil's water-repelling ability, which increases surface runoff and decreases infiltration(Verma and Jayakumar 2012).
- Twice-burned or over-burned plots have very low water repellency. This modification can accelerate the erosion rate and augment sediment loads, consequently leading to heightened concentrations of toxic metals in downstream water resources(Keesstra et al. 2016).
- Leaching of metals through soil can contaminate groundwater, especially where groundwater is close to the surface. Metals like Cr, Cd, Cu, Hg, Mn, Ni, Pb, Zn, and As are particularly problematic due to their tendency to accumulate and magnify in the food chain (Nunes et al. 2017).

## 5 Sediment Flows

Forest fires can increase sedimentation and turbidity while also affecting levels of organic matter, nutrients, trace metals, and other solutes due to the burning of vegetation and changes in soil and water chemistry, which impacts aquatic system quality (Finlay et al. 2012; Swindle et al. 2021). Accumulation of sediments behind dams can reduce water storage capacity, reducing the reservoir's expected benefits and physical lifetime. Removing these sediments is very costly (Wisser et al. 2013). After forest fires, there is a marked deterioration in the soil's ability to absorb water, leading to a significant decline in its infiltration capacity.

Consequently, soil erosion during heavy rainfall in these areas is markedly greater compared to other regions. Moreover, in burned areas where the soil exhibits resistance to wetting, soil movement on the surface is further impeded by increased "dry friction.“ This condition arises because the soil, having lost its wetting properties, becomes less compacted than wetted soil with weaker cohesion, resulting in diminished resistance to erosion and increased surface soil movement (Bilgin et al. 2016).

## 6 Forest Fire Impact on Water Quality

Water, a vital resource and key carrier of nutrients is susceptible to vegetation and soil disturbances. Disturbances affect water by altering flow timing, quantity, and physical parameters like temperature, sediment content, dissolved oxygen, and biological and chemical constituents (Scott et al. 2014). Forest fires also affect water chemistry, increasing macronutrients, micronutrients, and basic and acidic ions while lowering oxygen levels and raising biological oxygen demand. These changes result from the disruption of the soil and the burning of vegetation. Moreover, fire disturbances lead to changes in streamflow characteristics.(Brooks et al. 2003).

Forest fires significantly impact physical water qualities by modifying erosion rates, sediment yield, turbidity, flooding patterns, water temperature, and soil properties. These changes arise from the disruption of ground cover and the combustion of vegetation. The intensity of erosion and sediment transport typically increases due to the loss of vegetation stabilizing the soil, leading to elevated turbidity and altered flooding regimes. Additionally, exposure to bare soil surfaces can exacerbate these effects by facilitating more rapid and extensive sediment movement (Poff et al. 2010).

The combustion process releases various compounds into the environment, which can lead to increased nutrient loads in water bodies. This, in turn, results in diminished oxygen levels and an increased biological oxygen demand, reflecting the higher levels of organic and inorganic substances introduced into the aquatic systems. Initially, increased runoff and altered flow patterns are observed due to the decreased infiltration capacity of the soil and the enhanced erosion processes. Over time, the vegetation and soil structure recovery may gradually stabilize these flow characteristics. However, the trajectory and recovery rate can vary based on the fire intensity, soil type, and vegetation regrowth (Brooks et al. 2003).

## 7 Physical Characteristics of Water

### *7.1 Ph*

The pH at a given moment reflects the chemical equilibria within a water body. Following a fire, soil pH may rise (Landsberg and Tiedemann 2000), potentially leading to increased streamflow pH values. Research findings vary, with some studies showing no significant change in water pH one year after a fire (Evans et al. 2017), while others report an increase in pH within the same timeframe (Son et al. 2015). Forest fire residues typically lead to an elevation in pH levels, predominantly due to the dissolution of ash. The ash, composed of mineral oxides and other alkaline components, reacts with water to form hydroxides and carbonates, contributing to the observed rise in pH (Harper et al. 2019).

In certain instances, pH levels may decrease due to carboxyl groups in charred residues. This phenomenon is further influenced by the wet deposition of dissolved acids, which can exacerbate the environment's acidity (Evans et al. 2021). The intensity and timing of burns influence pH levels in surface waters receiving runoff from fire-affected areas. Future studies should monitor pH within a year of a forest fire, documenting burn severity and ash characteristics, to better understand how forest fires affect surface water pH (Burton et al. 2016).

## *7.2 Water Temperatures*

Water temperature is a crucial parameter influencing the quality of streams and aquatic habitats, as it regulates the survival of various aquatic organisms sensitive to temperature changes. Burning streambank vegetation can increase water temperature, resulting in thermal pollution that enhances biological activity in streams (DeBano et al. 1998). Studies on temperature changes in water bodies following forest fires indicate that 79% reported increases, with the remaining 21% reporting no change; none showed a decrease (Paul et al. 2002). Another concern associated with elevated stream temperatures is increased fish mortality. Higher temperatures can decrease dissolved oxygen ($O_2$) concentrations, posing challenges for salmonid fishes when $O_2$ levels drop below 10 ppm (less than 10 mg/L) (Brooks et al. 2003).

## *7.3 Total Suspended Solids*

The variability in Total Suspended Solids (TSS) concentrations in post-wildfire aquatic environments is intricately governed by the interplay among wildfire attributes, surface hydrological dynamics, and land characteristics (Rust et al. 2018). The initial surge in TSS is linked to the mobilization of this debris. At the same time, the subsequent decrease is likely due to the gradual stabilization of sediment sources and the attenuation of debris availability over time (Garcia-Comendador et al. 2017). Furthermore, forest fire sediments and ash often carry contaminants such as organic pollutants and heavy metals. These substances can infiltrate water systems, leading to potential degradation of water quality. The leaching of these toxic materials poses significant risks to aquatic ecosystems and human health. (Burton et al. 2016). Forest fires can significantly alter channel morphology by affecting sediment transport capacity and supply. These changes can take years to develop and fully stabilize (Shakesby & Doerr 2006).

## 7.4 Dissolved Organic Carbon

Forest fires can raise levels of organic carbon and nitrogen in water, remarkably increasing dissolved organic carbon (DOC). This concerns drinking water quality as high DOC levels can lead to disinfection byproducts (DBPs) during water treatment and enhance metal binding, impacting water safety and treatment effectiveness (Regan et al. 2017). In general, DOC increases slightly immediately after fire, mainly due to combustion, leaching of organic matter and dead vegetation, and dissolution of dissolved organic matter in runoff (Uzun et al. 2020).The dissolved organic carbon (DOC) concentration may decrease over time due to runoff processes removing forest fire debris, and forest fires that burn the upper organic soil layer above permafrost can convert soil organic carbon into more resistant and soluble organic compounds (Davidson et al. 2019).

## 7.5 Nutrients

Nutrient levels, particularly nitrogen and phosphorus, typically rise in streams following forest fires, a pattern observed across various ecosystems, including temperate, Mediterranean, boreal forests, grasslands, and wetlands (Hohner et al. 2016). Elevated nutrient inputs can enhance algal growth, potentially leading to harmful algal blooms and negatively affecting aquatic life, recreation, and water quality. Nutrient concentrations typically peak within the first year following a forest fire, subsequently diminishing over time as the effects of the fire gradually wane (Pierson et al. 2020). Spencer et al. (2003) reported a significant rise in phosphate, nitrate, and ammonium levels ranging from 5 to 60 times higher in third-order streams as a result of the Red Bench Fire, which severely affected thousands of forested acres in Glacier National Park in September 1988. The nitrogen pulses observed in the stream water were primarily attributed to the diffusion of smoke from nearby forest fires into the aquatic environment. As smoke particles containing nitrogen compounds, such as nitrogen oxides and ammonia, entered the water, they rapidly increased nitrogen concentrations.

Simultaneously, the observed spikes in phosphorus concentrations were linked to the deposition of ash, which introduced elevated amounts of phosphorus into the ecosystem. Remarkably, the elevated nitrogen and phosphorus concentrations were short-lived, with levels returning to baseline within a few weeks. This rapid return to pre-disturbance concentrations highlights the stream ecosystem's resilience in restoring equilibrium after such disruptions (Spencer and Hauer 1991).

## 7.6 Heavy Metals

Forest fires can release metals and other chemicals, which rain carries into water bodies. This can harm aquatic life, causing toxic metal buildup in tissues and reduced mobility. For example, the bioaccumulation of metals within aquatic organisms' tissues can disrupt physiological functions, while reductions in mobility may impair their ability to forage, reproduce, and evade predators. These disturbances can ultimately compromise the integrity and health of aquatic communities. (Silva et al. 2016). Following a fire, the mobilization of arsenic and cyanide non-metal inorganic chemicals can become a critical environmental concern. Forest fires, through the intense heat and chemical processes involved, have the potential to liberate arsenic from soil minerals, converting it into a more soluble and mobile form. This transformation can significantly enhance its mobility, leading to its potential runoff into surface water bodies or leaching into groundwater systems. Such processes pose substantial risks to aquatic and terrestrial ecosystems, given the toxicity of these elements and their propensity to persist and accumulate in the environment (Johnston et al. 2019).

Forest fires release heavy metals from plant biomass and soil organic matter into ash, disrupting metal–humic acid complexes. Post-fire conditions, such as lower soil organic content and higher pH from ash, affect metal release. To track changes in pollutant levels, pH measurements should be taken within 12 months of the fire, as dissolved metal concentrations may decrease while particulate metal concentrations can rise for up to five years. This temporal divergence underscores the necessity for prolonged monitoring to capture the full scope of metal dynamics and their environmental implications (Rust et al. 2018). Mercury concentrations, which are frequently observed to rise following forest fires, experience a notable increase in particulate-bound Hg(II) due to elevated levels of Total Suspended Solids (TSS)(Jensen et al. 2017).

## 7.7 Polycyclic Aromatic Hydrocarbons (PAHs)

Polycyclic aromatic hydrocarbons (PAHs) are a class of organic compounds characterized by multiple fused aromatic rings. These compounds are typically formed through incomplete combustion processes of organic materials. Forest fires, as a significant natural source of PAHs, contribute to the atmospheric burden of these compounds by pyrolyzing biomass. During such fires, complex organic matter in the vegetation is subjected to high temperatures, leading to the formation of various PAHs, including notable examples such as phenanthrene, pyrene, and retinal. Emitting these PAHs into the atmosphere can have profound implications for environmental and human health due to their persistence and potential carcinogenic properties(Muri et al. 2003). In humans, polycyclic aromatic hydrocarbons (PAHs) are predominantly sequestered in the kidneys, liver, spleen, and adipose tissue. These compounds are typically excreted from the body within a matter of days. However, chronic exposure to PAHs can have detrimental long-term effects. This carcinogenic potential is attributed to the ability of PAHs to induce genetic mutations and disrupt cellular processes, contributing to the onset of malignancies (Bunt et al.2001).

Polycyclic aromatic hydrocarbons (PAHs) exhibit remarkable environmental persistence owing to their hydrophobic nature, resistance to biodegradation, and intrinsic chemical stability. After a fire event, the concentration of PAHs in the environment is significantly influenced by the magnitude of the burn and the specific characteristics of the affected forest ecosystem. The extent of combustion is crucial in determining the amount of PAHs released, as more extensive burns generally result in higher concentrations. Additionally, whether deciduous, coniferous, or mixed, the type of forest can affect PAH distribution due to variations in the composition and structure of the biomass and differences in soil and environmental conditions (Vergnoux et al. 2011). To accurately assess total polycyclic aromatic hydrocarbon (PAH) emissions following a fire, conducting a comprehensive measurement of particulate and dissolved PAHs in surface waters for a minimum of two years post-incident is imperative. Furthermore, meticulous documentation of the fire's characteristics is crucial to account for the inherent variability in PAH concentrations and ensure a thorough and nuanced understanding of the emission patterns (Chen et al. 2018).

Moreover, careful documentation of the fire's characteristics is essential to capture the inherent variability in PAH concentrations and to provide a comprehensive and detailed understanding of the emission patterns (Chen et al. 2018).

## 8 Effects of Fire on Aquatic Life

Streams and lakes support local water supplies, and changes in water levels can negatively impact aquatic organisms. Because the forest acts as a weak zone, waste, dirt, and ash can enter the water along with rainwater and travel for miles downstream, affecting aquatic organisms and making it into feeding and breeding grounds (Smyth 2020). Moreover, firefighting foam, when dissolved in water, can be detrimental to fish populations. The combustion process during fires also releases toxic substances that pose significant risks to aquatic life (Vartan 2019).

Fish mortality is likely to increase when the intensity of vegetation fires increases and small streams decrease (Bitterroot National Forest 2000). Fish will die due to rapid water heating even if the water temperature exceeds the death limit. Rapid exposure may cause drowsiness associated with depression. The fast water flow puts these fish at greater risk from non-native fish and other animals such as birds and mammals. In addition to temperature, post-fire mercury bioaccumulation is another issue that affects fish consumption recommendations. Factors responsible for this increase include more significant mercury and methylmercury in water after fire (Riggs et al. 2017).

Research indicates that both invertebrate diversity and stream density typically decline following a fire, with particularly pronounced effects on sensitive taxa such as Ephemeroptera, Plecoptera, and Trichoptera. This pattern is observed across various biomes, including northern, temperate, and Mediterranean regions. In small, shallow bodies of water, frog and salamander eggs and larvae are particularly vulnerable to mortality or exposure to elevated temperatures. Forest fires in summer or autumn pose a significant risk, as they can readily destroy these amphibians' eggs and larvae, which may migrate to freshwater habitats or seek refuge in leaf litter for the season. (Pilliod et al. 2003).

## 9 Effect on Drinking Water

Following forest fires, the principal contaminants in drinking water sources are chemicals and microorganisms. These contaminants can infiltrate fire-damaged water systems and compromise well water quality. Fire retardants, used to manage fires, contribute additional pollutants such as phosphates or sulfates and often contain traces of iron oxide, imparting a red hue. While typically considered safe for humans and terrestrial animals, these compounds can pose a toxicity risk to aquatic organisms.

Typically, the underground elements of a well, including the well casing, pump, and septic system, remain largely intact. However, the wiring at the top of the well, the junction box, and the PVC casing are vulnerable to damage. Extensive damage to electrical wiring, control systems, pipes, tanks, and other infrastructure components can significantly impair system functionality, compromise the safety and integrity of the water supply, and obstruct the efficient disposal of wastewater.

Such degradation may lead to severe operational inefficiencies, increased vulnerability to contamination, and potential environmental hazards. Additionally, the disruption of automated control systems can fail critical monitoring and regulation processes, exacerbating the risk of system-wide malfunctions. The ensuing challenges necessitate immediate and comprehensive remedial actions to restore optimal performance, ensure public health and safety, and maintain regulatory compliance. (Waskom et al. 2013).

## 10 Climate Change and Water Availability

The impact of climate change on water resources has become a critical area of research, with recent advancements providing deeper insights into the extent and mechanisms of these changes(Borges et al. 2020). The Fourth Assessment Report (AR4) by the Intergovernmental Panel on Climate Change (IPCC) presents critical projections regarding the impacts of climate change. The report predicts a 10–30% decline in freshwater resources across many arid and central tropical regions while forecasting a substantial rise in sea levels by 2080. This projection is primarily due to the thermal expansion of seawater and the melting of glaciers and polar ice caps. (IPCC 2007). Climate change poses a profound threat to groundwater resources through its influence on hydrological cycles and the redistribution of water resources. Alterations in rainfall patterns, intensities, and frequencies are expected to disrupt the recharge rates of aquifers, potentially leading to diminished groundwater availability in certain regions. Increased temperatures and the consequent rise in evapotranspiration rates further exacerbate the strain on groundwater systems by reducing the net water input.

Moreover, shifts in the spatial and temporal distribution of precipitation can lead to more frequent and severe droughts, intensifying the pressure on groundwater as a critical water source during dry periods (Lam et al. 2021). Increased precipitation can result in elevated river and stream flow levels, significantly heightening the risk of flooding in susceptible areas. This enhanced surface runoff can also diminish water infiltration into the soil, reducing groundwater recharge rates (Sen 2021).

Climate change is poised to impact groundwater both quantitatively and qualitatively. Quantitatively, shifts in atmospheric conditions will influence soil infiltration rates, the extent of infiltration, and the subsequent flow dynamics of groundwater. Qualitatively, alterations in hydrological processes driven by climate change threaten groundwater quality by affecting contaminant transport, recharge patterns, and the overall balance of water chemistry (Klove et al. 2014). Global warming induces adverse effects on groundwater through alterations in the timing and magnitude of recharge and fluctuations in groundwater levels (Singh et al. 2024). Melting permafrost at high latitudes can lead to changes in groundwater quality due to increased connectivity to surface waters. Increased rainfall can reduce groundwater recharge in wetlands because the heavier the rainfall event, the greater the soil infiltration capacity. However, in an arid region, the change in precipitation can increase the groundwater level because only very strong rainwater can penetrate quickly before evaporating. At the same time, alluvial rivers are often recharged by overflow from floods (Doll and Florke, 2005).

# 11 Mitigation Strategies

## *11.1 Fuel Management Strategies*

### 11.1.1 Shaded Fuel Breaks

Shade oil intervention, a practice extensively utilized in Western countries, and the emerging issue of green fire, as reported by forest workers in China, are two pivotal factors in managing and controlling forest fire incidents. The use of radiation therapy has been documented in international journals (Oliveira et al. 2016). According to Green (1977), rubber oil is often described as a general product or a heavy plant, and the heavy plant is converted into a lower oil or flame. The primary methods for fire control encompass conventional fire suppression techniques, which involve the comprehensive removal of all fuel sources and the utilization of waste fuels. Shade oil, characterized by its region-specific oil dilution, is predominantly employed in the old-growth forests of North America, particularly in the arid forested regions of the western states (Cochrane et al. 2012). The width of shaded oil fracturing zones ranges from 50 to 200 m, depending on local vegetation, topography, weather, and fire conditions. While fuel explosions can help suppress or stop fires, effective control of fires initiated by oil booms still requires significant human intervention for detection and management (Varela et al. 2013).

## 12 Green Fire Barrier Technique

This strategy involves strategically planting specialized evergreen broadleaf trees within flammable coniferous forests to establish effective green fire barriers. In low to medium-intensity burns, these trees are planted in strips, effectively halting the spread of forest fires. The width of these strips, usually ranging from 20 to 35 m, is determined by several factors, including local vegetation types, oil distribution, and specific regional conditions. The approach enhances the resilience of forests against forest fires and promotes biodiversity and ecological balance within these ecosystems. In China, implementing these green fire barriers has been particularly impactful. The country has enhanced its forest fire prevention and control measures by combining traditional fire management techniques with innovative ecological solutions (Cui et al. 2019).

Choosing the suitable tree species is essential for creating effective green fire barriers, with evergreen broad-leaved trees particularly suitable. Reports from various research groups highlight the importance of water potential retention and water release by living leaves as important factors in determining the fire resistance of trees. This was confirmed by measuring cone calorimetry (Wang et al. 2020). *Schima superba* (Bamboo) is a crucial species, extensively studied for its thermophysical, chemical, and biological properties and growth characteristics (Chen et al. 1995). Throughout most seasons, the leaves of evergreen broad-leaved trees retain approximately 50% moisture, with levels rising to 100% during dry periods. Some species exhibit moisture content of up to 70%, equivalent to 233% of their dry mass. The dense canopy of these trees creates a local microclimate that maintains high moisture levels in the dead matter within the land strip(Loren and Ogle, 2009).

## 13 Improved Forest Fire Containment

In many forest fires, the fire threat to drinking water areas increases the risk of spreading corrosive organisms. Enhancing fire protection is a strategic approach to mitigating the risk of water quality deterioration to levels beyond remediation. This strategy aims to contain the fire's size and impact within manageable limits, thereby avoiding the necessity of widespread landscape removal. Recent advancements in preparedness involve delineating land into fire management zones known as Potential Fire Operational Departments (PODs). These divisions leverage infrastructure, such as roads, rivers, and gas pipelines, to optimize management efficacy and control efforts (O'Connor et al. 2016).

The value of this approach goes beyond managers' involvement in identifying and evaluating controllable resources. The resulting POD areas can be impacted for pre-fire assessment, determining the risk of forest fire and spreading forest fire. This chapter outlines response strategies based on the needs and impacts of forest fires on water resources and other natural and human resources (Thompson et al. 2016). Despite advancements achieved through engaging leaders at the grassroots level in developing and implementing PODs and associated response strategies, challenges remain (Stratton 2020). Optimizing this vein to increase water efficiency and other benefits is also considered a saving (Gannon 2020).

**Fighting fire with fire.**

The rise in fires has led to increased efforts by various stakeholders, including state and federal land managers, fire departments, utilities, and private landowners, to focus on reducing fire risk. Lawmakers are advocating for significant changes in landscape management to lower forest fire incidence. Additionally, the scientific and regulatory communities are interested in using low-energy fuel systems to produce valuable fuels in power plants. This multifaceted approach aims to integrate advanced technological solutions with traditional fire management techniques, fostering a more resilient and sustainable response to the growing forest fire threat(USDA 2022). A comprehensive analysis of two decades' worth of fire data, encompassing 124,186 square kilometers of Californian forest, reveals a significant insight into fire risk management. The study indicates that a reduction in fire frequency correlates with a substantial decrease in future fire risk. Specifically, in coniferous forests, regions that experienced fewer recent fires have exhibited a remarkable 64.0% reduction in high fire risk, with this mitigative effect enduring for up to six years. These findings advocate for a paradigm shift in fire management strategies, suggesting a move from traditional fire suppression approaches to proactive restoration methods. By embracing the strategic use of fire for ecological management and forest fire control, the study underscores the potential for enhancing forest resilience and reducing long-term fire hazards (Wu, 2023).

## 14 Global Perspective on Forest Fire and Water Resources

Fuel reduction through thinning and controlled burning has proven effective in restoring fires to their former patterns while reducing the severity and size of fires. Raising public awareness of fire restrictions can help reduce the number of human-caused fires. Although infrastructure development is expensive, measures such as the diversion of water and groundwater, parallel pipes that use untreated water for water to reduce medical needs, and integration of smart water systems into homes can make a significant difference in the energy of water supply system(Emelko et al. 2011). At the same time, decentralized urbanization trends have led to rapid growth of areas at wildland-urban interfaces, with housing built in fireweed forests or grasslands (Radeloff et al. 2005). "Converting crops into forests" The Farmland to Forest Program (CCFP) is the world's largest financial assistance funding program (Lu et al. 2012). Uncertainties about fire damage to water resources and its effects on sustainable development, economies, and water management highlight the need to consider broader environmental and social dimensions when assessing the impact of local events (Sivapalan et al. 2014).

Fire management in Europe aims to protect forest ecosystem services (Montiel-Molina 2013). The United States seeks to protect natural resources and human settlements from fire damage (Labaree 2021). It is used for emergency and rescue measures in the Mediterranean regions. It has been placed in the public protection plan. South Africa prioritizes ecosystem management activities for effective fire management, which includes liability risk for disasters. Fuel management is becoming a priority in Australia, where the safety of life and property is paramount (Tedim et al. 2015).

## 15 Analysis of Forest Fire-Related Policies and Legislation

Forest fires and critical issues related to forest fires have received significant attention in recent years, including their origins, causes, and consequences (Pivello et al. 2021). These changes are classified as fire risk management, while government-recognized targets and activities are accepted as law. To prevent risks caused by climate change, forest managers and policymakers are competing to create effective fire prevention policies (Tymstra et al. 2020). Fire risk management can reduce long-term fire risk through various strategies, including fuel management to protect people from fire, extinguish fire, and modify fire behavior (Fernandes et al. 2021). Many laws and regulations regarding fire management exist in foreign countries (Table 1).

## 16 Future Outlook

In recent years, the proliferation of satellite data on forest fires, weather patterns, and forest resources has significantly enhanced the development of models that predict the impact of fires on water systems. Despite the advancements in satellite data, information on post-fire water conditions remains scarce. Forecasting models frequently encounter validation challenges in developing countries due to a shortage of ground-truth data. However, with advances in data and physical solutions, scientists need better training to predict the impacts of natural water solutions and understand their interactions in fire, crop/agriculture, and supply and hydrological services (Robinne et al. 2018). Key concerns include the increased burned area, fire severity, safe

**Table 1** Forest fire-related Laws, Policies, and Legislation

| Country | Law, Policies and Legislation |
|---|---|
| South Africa | • Fire Brigade Services Act (Act 99 of 1987)<br>• The National Veld and Forest Fire Act (Act 101 of 1988)<br>• Fire Protection Associations (FPAs)<br>• *GEF FynbosFire*project<br>• Working on Fire<br>• National Veldfire Risk Assessment |
| USA | • National Cohesive Wildland Fire Management Strategy<br>• Federal Land Assistance, Management, and Enhancement Act, 2009 |
| Canada | • Canadian Interagency Forest Fire Center (CIFFC)<br>• FireSmart program<br>• Let burn policies |
| Australia | • The National Bushfire Management Policy Statement<br>• Stay and Defend or Leave Early<br>• Prepare, act, survive |
| United Kingdom | • The Countryside Stewardship Scheme<br>• The Met-Office plays<br>• CROW Right of Way Act |
| Portugal | • The National Plan for Integrated Fire Management (PNIGFR)<br>• The National Action Programme (PNA) |
| Italy | • Forest management in Italy is governed by the "Forest Law,"<br>• The National Framework Law 353/2000<br>• Regional Fire Management Plans<br>• Fire Prevention Plans<br>• The Italian State Forestry Corps (CorpoForestaledelloStato or CFS) |
| France | • Vulcain<br>• The Forest Code |
| Greece | • The Forest Law 998/1979<br>• The Forest Fire Prevention and Suppression Law 2690/1999<br>• The National Action Plan for Forest Fire Prevention and Management,2012<br>• The National Forest Fire Danger Rating System |
| Cyprus | • The Forest Law and the Fire Fighting Action Plan<br>• The EU's Forest Fire Prevention and Information System (EFFIS) |

drinking water availability, additional infrastructure needs, and rising operational costs. It is essential to develop adaptation and mitigation strategies that comprehensively assess forest fire impacts and protect vital water resources in surrounding areas (Bladon et al. 2014).

Strategic planning and the careful and thoughtful implementation of sustainable forest management practices can be crucial in mitigating the harmful effects of forest fires. By proactively managing forest ecosystems, including controlled burns, thinning of overgrown vegetation, and removing dead or diseased trees, forest managers can reduce the fuel available for forest fires, making fires less intense and more manageable. Such an approach protects forest ecosystems and ensures the preservation of vital water resources. Adopting a comprehensive strategy that integrates

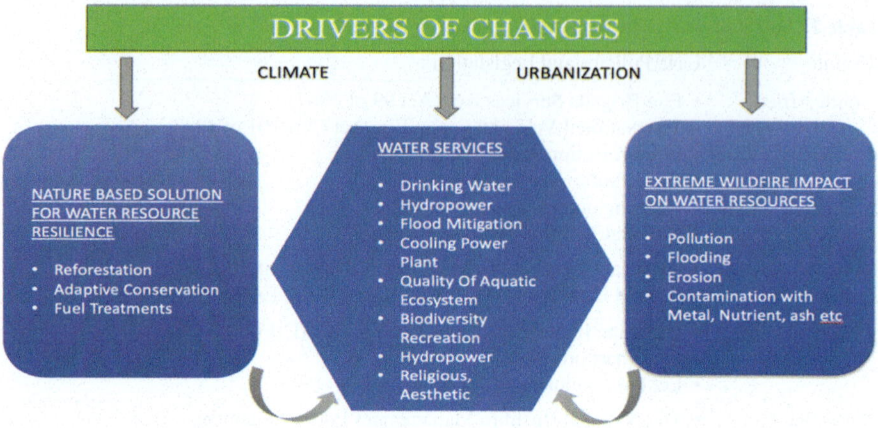


**Fig. 1** A Balanced Fire Regime for Sustainable Forest Management and Water Resource Preservation

ecological principles, land use planning, and adaptive management techniques may enhance forest resilience, minimize fire risks, and sustain the ecological balance essential for the long-term health of these vital landscapes. (Postel and Thompson 2005). Robust fire management strategies are the foundation of sustainable forests that ensure stable water supplies (Fig. 1). Shifts in fire timing, scale, drought, and insect populations are expected to alter forest structure and function. This will allow land managers to develop and implement conservation strategies focused on improving the forests' ability to manage water resources and adapt to these changes (Robinne et al. 2016).

## 17 Conclusion

Climate change has the potential to transform warfare at different spatial and physical levels, and forest fires can have social, ecological, and economic impacts. This study delivers a comprehensive quantitative dataset that advances our knowledge of post-forest fire pollutant concentration increases, elucidating the changes in water quality parameters across affected water resources. After a fire, vegetation along stream courses is reduced, and the hydrophobic soil causes large amounts of precipitation to be transferred without resistance into runoff that flows easily downstream. Fires impact the food supply, accelerate mineralization, and induce demineralization. Elevated pressure, reduced oxygen levels, and the leaching of nutrients and toxins into aquatic systems can lead to eutrophication, posing significant risks to aquatic ecosystems. Nonetheless, this study underscores the pivotal role of forest management in mitigating future forest fire conditions and decreasing forest vulnerability to fire. The interaction between climate change, fire dynamics, and water resources is expected to significantly affect water supplies across different sectors. Climate changes can alter precipitation patterns and increase forest fires' frequency and intensity, impacting water availability for domestic, agricultural, commercial, and industrial uses. Overall, shifts in these dynamics are likely to strain water resources and affect various sectors dependent on stable water supplies. Governments should strictly implement laws and regulations to address forest fire risk and assess their effectiveness. Implementing effective strategies may help protect forest ecosystems and humans from fires and minimize their impacts on water resources.

## References


Abbass K, Qasim MZ, Song H, Murshed M, Mahmood H, Younis I (2022) A review of the global climate change impacts, adaptation, and sustainable mitigation measures. Environ Sci Poll Res 29(28):42539–42559

Aragao LEOC, Shimabukuro YE (2010) Implications for REDD t. Sci 328:1275–1278

Bilgin A, Aybar M, Saglam B (2016) Effects of forest fires on water sources. CBU J Sci 12(2):173–177

Bitterroot National Forest (2000) Bitterroot Fires 2000: An assessment of postfire conditions with recovery recommendations. USDA Forest Service, Bitterroot National Forest. Unpublished report available online at http://www.fs.fed.us/r1/bitterroot/recovery/fires_2000-screen.pdf.

Borges JG, Marques S, Garcia-Gonzalo J, Rahman AU, Bushenkov V, Sottomayor M, Carvalho PO, Nordstrom EM (2017) A multiple criteria approach for negotiating ecosystem services supply targets and forest owners' programs. For Sci 63:49–61

Brooks KN, Folliott PF, Gregersen HM, DeBano LF (2003) Hydrology and the management of watersheds, 3rd edn. Iowa State University Press, Ames, Iowa, p 574

Bundt M, Krauss M, Blaser P &Wilcke (2001). Forest fertilization with wood ash: effect on the distribution and storage of polycyclic aromatic hydrocarbons (PAHs) and polychlorinated biphenyls (PCBs). J Environ Qual 30(4): 1296–1304

Burton CA, Hoefen TM, Plumlee GS, Baumberger KL, Backlin AR, Gallegos E, Fisher RN (2016) Trace elements in stormflow, ash, and burned soil following the 2009 Station Fire in Southern California. PLoS ONE 11(5):1–26

Certini G (2005) Effects of fire on properties of forest soils: a review. Oecologia 143:1–10

Chen CJ, He Z, Chen D, Huang Y, Xie J (1995) Studies on the fire-resistance of 37 species of coniferous and broadleaf trees and it's appraisal. SciSilvae Sin 31(2):135–143

Chen H, Cho AT, Li XW, Ni HG, Dahlgren RA, Zeng H, Wang JJ (2018) Wildfire burn intensity affects the quantity and speciation of polycyclic aromatic hydrocarbons in soils. ACS Earth Space Chem 2(12):1262–1270

Chuvieco E, Mouillot F, Van der Werf GR, San Miguel J, Tanase M, Koutsias N, Giglio L (2019) Historical background and current developments for mapping burned area from satellite Earth observation. Remote Sens Environ 225:45–64

Cochrane MA, Moran CJ, Wimberly MC, Baer AD, Finney MA, Beckendorf KL, Eidenshink J, Zhu Z (2012) Estimation of wildfire size and risk changes due to fuels treatments. IJWF 21(4):357–367

Cui X, Alam MA, Perry GL, Paterson AM, Wyse SV, Curran TJ (2019) Green firebreaks as a management tool for wildfires: Lessons from China. J Environ Manag 233:329–336

Davidson SJ, Elmes MC, Rogers H, van Beest C, Petrone R, Price JS, Strack M (2019) Hydrogeologic setting overrides any influence of wildfire on pore water dissolved organic carbon concentration and quality at a boreal fen. Ecohydrology 12(7):1–11

DeBano LF, Neary DG, Ffolliott PF (1998) Fire Effects on Ecosystems. John Wiley & Sons, New York

Doll P and Florke M (2005) Global-Scale Estimation of Diffuse Groundwater Recharge. Frankfurt Hydrology Paper 03, Institute of Physical Geography, Frankfurt University, Frankfurt am Main, Germany.

Emelko MB, Silins U, Bladon KD, Stone M (2011) Implications of land disturbance on drinking water treatability in a changing climate: demonstrating the need for "source water supply and protection" strategies. Water Res 45(2):461–472

Evans CD, Malcolm IA, Shilland EM, Rose NL, Turner SD, Crilly A, Monteith DT (2017) Sustained biogeochemical impacts of wildfire in a mountain lake catchment. Ecosyst 20:813–829

Evans JS, Norman AL, Reid ML (2021) Evidence of smoke from wildland fire in surface water of an unburned watershed. Water Resour Res 57(10):1–14

Feliciano D, Recha J, Ambaw G, MacSween K, Solomon D, Wollenberg E (2022) Assessment of agricultural emissions, climate change mitigation and adaptation practices in Ethiopia. Clim Policy 22(4):427–444

Fernandez-Garcia V, Beltran-Marcos D, Fernandez-Guisuraga JM, Marcos E, Calvo L (2020) Predicting potential wildfire severity across Southern Europe with global data sources. STENDL 829:154729

Franklin S B, Robertson P A&Fralish J S (1997) Small-scale fire temperature patterns in upland Quercuscommunites. J ApplEcol 613–630.

Gajendiran K, Kandasamy S & Narayanan M (2024) Influences of wildfire on the forest ecosystem and climate change: A comprehensive study. Environ Res 117537.

Gannon BM, Wei Y, Thompson MP (2020) Mitigating source water risks with improved wildfire containment. Fire 3(3):45

Garcia-Comendador J, Fortesa J, Calsamiglia A, Calvo-Cases A&Estrany J (2017) Post-fire hydrological response and suspended sediment transport of a terraced Mediterranean catchment. ESPL 42(14): 2254–2265

Green, Lisle R. 1977. Fuelbreaks and other fuel modification for wildland fire control. U.S. Dep. Agric., Agric. Handb. 499, 79, p., illus.

Harper AR, Santin C, Doerr SH, Froyd CA, Albini D, Otero XL, Perez-Fernandez B (2019) Chemical composition of wildfire ash produced in contrasting ecosystems and its toxicity to Daphnia magna. IJWF 28(10):726–737

Hohner A K, Cawley K, Oropeza J, Summers R& Rosario-Ortiz F L (2016). Drinking water treatment response following a Colorado wildfire. Water Research 105:187–198

Intergovernmental Panel on Climate Change (IPCC) (2007), Climate Change 2007: The Scientific Basis. Contribution of Working Group I to the Fourth Assessment Report of the Intergovernmental Panel on Climate Change, edited by S. Solomon et al., Cambridge Univ. Press, New York.

Jahandideh-Tehrani M, Zhang H, Helfer F, Yu Y (2019) Review of climate change impacts on predicted river streamflow in tropical rivers. Environ Monit Assess 191:1–23

Jensen AM, Scanlon TM, Riscassi AL (2017) Emerging investigator series: the effect of wildfire on streamwater mercury and organic carbon in a forested watershed in the southeastern United States. Environ. Sci: Processes Impacts 19(12):1505–1517

Johnston SG, Karimian N, Burton ED (2019) Fire promotes arsenic mobilization and rapid arsenic (III) formation in soil via thermal alteration of arsenic-bearing iron oxides. Front Earth Sci 7:139

Kaushik P, Singh H (2024) Climate change-induced forest ecosystem functioning and alteration in water-atmosphere-soil–plant (WASP) relationship. In: Singh, H. (eds) Urban Forests, Climate Change and Environmental Pollution. Springer, Cham. https://doi.org/10.1007/978-3-031-67837-0_38

Keesstra S, Wittenberg L, Maroulis J, Sambalino F, Malkinson D, Cerda A, Pereira P (2016) The influence of fire history, plant species and post-fire management on soil water repellency in a Mediterranean catchment: the Mount Carmel range, Israel. CATENA 149:857–866

Klove B, Ala-Aho P, Bertrand G, Gurdak JJ, Kupfersberger H, Kværner J, Pulido-Velazquez M (2014) Climate change impacts on groundwater and dependent ecosystems. J Hydrol 518:250–266

Kumar A, Dwivedi G K, Tewari S, Paul J, Anand R, Kumar N, Kumar P, Singh H, Kaushal R (2020a). Carbon Mineralization and Inorganic Nitrogen Pools under TerminaliachebulaRetz.-Based Agroforestry System in Himalayan Foothills, India. Forest Sci 66 (5): 634–643.

Kumar A, Kumar P, Singh H, Kumar N (2020) Adaptation and mitigation potential of roadside trees with bio-extraction of heavy metals under vehicular emissions and their impact on physiological traits during seasonal regimes. Urb for Urb Green 58:126900

Kumar N, Singh H, Kumar A, Singh AK, Pandey PK, Kumar A (2024). Forest-Fire-Induced Land Degradation. In: Mishra G, Giri K, Singh S, Kumar M. (eds) Sustainable Land Management in India. Springer Nature, Singapore. https://doi.org/10.1007/978-981-97-5223-2_4

LabareeA (2021) The US government is wasting billions on ineffective wildfire policy, Slate Magazine Slate Available at: https://slate.com/technology/2021/11/fire-industrial-complex-wildfire-policy-suppression.html.

Lam QD, Meon G, Patsch M (2021) Coupled modelling approach to assess effects of climate change on a coastal groundwater system. Groundwater for Sustain Develop 14:100633

Landsberg J D & Tiedemann A R (2000) Fire management. Drinking water from forests and grasslands: a synthesis of the scientific literature. SRS-GTR-039. U.S. G. E. Dissmeyer, Department of Agriculture Forest Service, Southern Research Station, Asheville, North Carolina 124–138

Loren St J and Dan O (2009) Green strips or vegetative fuel breaks. Tech. Note TN Plant Materials No. 16. Boise, ID; Salt Lake City, UT: U.S. Department of Agriculture, Natural Resources Conservation Service. 16 p.

Lu Y, Fu B, Feng X, Zeng Y, Liu Y, Chang R, Wu B (2012) A policy-driven large scale ecological restoration: quantifying ecosystem services changes in the Loess Plateau of China. PLoS ONE 7(2):e31782

Maes WH, Pashuysen T, Trabucco A, Veroustraete F, Muys B (2011) Does energy dissipation increase with ecosystem succession? Testing the ecosystem exergy theory combining theoretical simulations and thermal remote sensing observations. Ecol Model 222(23–24):3917–3941

Minshall GW (2003) Responses of stream benthic macroinvertebrates to fire. For EcolManag 178(1–2):155–161

Muri G, Brancelj A (2003) Seasonal Water Chemistry Variations in Three Slovenian Mountain Lakes. Acta Chim Slov 50:137–147

Nunes MR, Silva AP, Denardin JE, Giarola NFB, Vaz CMP, van Es HM, Silva AR (2017) Soil chemical management drives structural degradation of Oxisols under a no-till cropping system. Soil Res 55:819–883

O'Connor CD, Thompson MP, Rodríguez Silva F (2016) Getting ahead of the wildfire problem: Quantifying and mapping management challenges and opportunities. Geosci 6(3):35

Oliveira T M, Barros A M, Ager A A&Fernandes P M (2016) Assessing the effect of a fuel break network to reduce burnt area and wildfire risk transmission. IJWF 25(6): 619–632

Paul MJ, LeDuc SD, Lassiter MG, Moorhead LC, Noyes PD, Leibowitz SG (2022) Wildfire induces changes in receiving waters: A review with considerations for water quality management. Water Resour Res 58(9):1–28

Pierson E, Simoiu C, Overgoor J, Corbett-Davies S, Jenson D, Shoemaker A, Ramachandran V, Barghouty P, Phillips C, Shroff R, Goel S (2020) A large-scale analysis of racial disparities in police stops across the United States. Nat Hum Behav 4(7):736–745

Pilliod DS, Bury RB, Hyde EJ, Pearl CA, Corn PS (2003) Fire and amphibians in North America. ForEcolManag 178(1–2):163–181

Pivello VR, Vieira I, Christianini AV et al (2021) Understanding Brazil's catastrophic fires: Causes, consequences and policy needed to prevent future tragedies. PECON 19:233–255

Poff B, Tecle A, Neary DG, Geils B (2010) Compromise Programming in Forest Management. JANAS 42(1):44–60

Postel SL, Thompson BH Jr (2005) Watershed protection: Capturing the benefits of nature's water supply services. NRF 29(2):98–108

Prakash V., Hunney K. and Singh H. 2022. Effect of elevated carbon dioxide on growth and development of *Santalum album* L. seedlings inoculated with plant growth promoting microorganisms in Open Top Chambers. International Journal of Environment, Agriculture and Biotechnology, 7 (5): 149–159.

Pyne SJ, Andrews PL, Laven RD (1996) Introduction to Wildland Fire, 2nd edn. John Wiley and Sons, Inc., New York, NY, p 769

Regan S, Hynds P, Flynn R (2017) An overview of dissolved organic carbon in groundwater and implications for drinking water safety. Hydrogeol J 25:959–967

Reilly MJ, Dunn CJ, Meigs GW, Spies TA, Kennedy RE, Bailey JD, Briggs K (2017) Contemporary patterns of fire extent and severity in forests of the Pacific Northwest, USA (1985–2010). Ecosphere 8(3):01695

Riggs CE, Kolka RK, Nater EA, Witt EL, Wickman TR, Woodruff LG, Butcher JT (2017) Yellow perch (Percaflavescens) mercury unaffected by wildland fires in Northern Minnesota. J Environ Qual 46:623–631

Robichaud PR, Lewis SA, Brown RE, Ashmun LE (2009) Emergency post-fire rehabilitation treatment effects on burned area ecology and long-term restoration. Fire Ecolo 5:115–128

Robinne F N, Bladon K D, Miller C, Parisien M A, Mathieu J & Flannigan M D (2018) A spatial evaluation of global wildfire-water risks to human and natural systems. STENDL 1193–1206.

Robinne FN, Miller C, Parisien MA, Emelko MB, Bladon KD, Silins U, Flannigan M (2016) A global index for mapping the exposure of water resources to wildfire. Forests 7(1):22

Russell KR, Van Lear DH, GuynnJr DC (1999) Prescribed fire effects on herpetofauna: review and management implications. WildlSoc Bull 27(2):374–384

Rust W, Holman I, Corstanje R, Bloomfield J& Cuthbert M (2018) A conceptual model for climatic teleconnection signal control on groundwater variability in Europe. Earth Sci Rev 177: 164–174

Savita, Sharma SK, Singh H, Singh O (2017) Forest fire: A review. Indian Forester 143(8):719–728

Scott A C, Bowman D M, Bond W J, Pyne S J & Alexander M E (2014) Fire on earth: An Introduction. John Wiley & Sons. Alexander. Wiley Blackwell, Hoboken, New Jersey, USA. 434 pp.

Sen K, Mishra D, Debnath P, Mondal A, Kumar NM (2021) Adsorption of uranium (VI) from groundwater by silicon containing biochar supported iron oxide nanoparticle. BioresourTechnol Rep 14:100659

Shakesby RA, Doerr SH (2006) Wildfire as a Hydrological and Geomorphological Agent. Earth Sci Rev 74:269–307

Shi C, Wang S, Cai HH, Zhang HR, Long XX, Tihelka E, Spicer RA (2022) Fire-prone Rhamnaceae with South African affinities in cretaceous Myanmar amber. Nat Plants 8(2):125–135

Silva V, Abrantes N, Cost R, Keizer J J, Goncalves& Pereira J L (2016) Effects of ash-loaded post-fire runoff on the freshwater clam Corbiculafluminea. Ecol Eng 90:180–189

Singh H (2024a). Climate Change and Forests: Biological Perspectives. In: Singh, H. (eds) Forests and Climate Change. Springer Nature, Singapore. https://doi.org/10.1007/978-981-97-3905-9_1

Singh H (2024c). Climate Change and Environmental Pollution: Biological Consequences on Urban Forests. In: Singh, H. (eds) Urban Forests, Climate Change and Environmental Pollution. Springer Nature, Cham, Switzerland. https://doi.org/10.1007/978-3-031-67837-0_1

Singh H (Ed) (2024b). Forests and Climate Change: Biological Perspectives on Impact, Adaptation, and Mitigation Strategies, Springer Nature Singapore. https://doi.org/10.1007/978-981-97-3905-9

Singh H (Ed) (2024d) Urban Forests, Climate Change and Environmental Pollution: Physio-Biochemical and Molecular Perspectives to Enhance Urban Resilience, Springer Nature, Cham, Switzerland. https://doi.org/10.1007/978-3-031-67837-0

Singh H 2021. Process-based carbon sequestration study in relation to the Energy-Water-Carbon flux in a forest ecosystem. *In:* Climate Impacts on Sustainable Natural Resource Management

(Kumar P., Singh RK., Kumar M., Rani M, and Sharma (eds). CRC Press Taylor & Francis Group, United States, Pp: 336–361.
Singh H and Kumar M (2022). Climate Change and Its Impact on Indian Himalayan Forests: Current Status and Research Needs. *In*: Rani, S., Kumar, R. (eds) Climate Change. Springer Climate. Springer, Cham. Pp: 223–242; https://doi.org/10.1007/978-3-030-92782-0_11.
Singh H, Kumar N, Kumar A (2024) Enhancing Resource Use Efficiency in Crops Through Plant Functional Traits published by Springer, pp. 97–117. ISBN: 978–981–97–1509–1.
Sivapalan M, Konar M, Srinivasan V, Chhatre A, Wutich A, Scott C A& Rodríguez-Iturbe I (2014) Socio-hydrology: Use-inspired water sustainability science for the Anthropocene. Earth's Futur 2(4): 225–230
Smyth C (2020) The impacts of bushfires on coastal and marine environments. Australian Marine Conservation Society. Retrieved from https://www.marineconservation.org.au/wp-content/uploads/2020/03/Bushfire-Report_February-2 020_Final-full-for-web-1.pdf.
Son JH, Kim S, Carlson KH (2015) Effects of wildfire on river water quality and riverbed sediment phosphorus. Wat Air Soil Poll 226:1–13
Spencer CN, Hauer FR (1991) Phosphorus and nitrogen dynamics in streams during a wildfire. J-NABS 10(1):24–30
Spencer CN, Gabel KO, Hauer FR (2003) Wildfire effects on stream food webs and nutrient dynamics in Glacier National Park, USA. For Ecol Manage 178:141–153
Stratton RD (2020) The path to strategic wildland fire management planning. Wildfire 29(1):24–31
Swindle C, Shankin-Clarke P, Meyerhof M, Carlson J &Melack J (2021) Effects of Wildfires and Ash Leaching on Stream Chemistry in the Santa Ynez Mountains of Southern California. Water13 (17): 2402.
Tedim F, Leone V &Xanthopoulos G (2015) Wildfire risk management in Europe. the challenge of seeing the "forest" and not just the "trees.". Proceedings of the 13th International Wildland Fire Safety Summit & 4th Human Dimensions of Wildland Fire, Managing Fire, Understanding Ourselves: Human Dimensions in Safety and Wildland Fire 213–238.
Thompson MP, Bowden P, Brough A, Scott JH, Gilbertson-Day J, Taylor A, Haas JR (2016) Application of wildfire risk assessment results to wildfire response planning in the southern Sierra Nevada, California, USA. Forests 7(3):64
Tucca F, Luarte T, Nimptsch J, Woelfl S, Pozo K, Casas G, Galban-Malagon C (2020) Sources and diffusive air–water exchange of polycyclic aromatic hydrocarbons in an oligotrophic North-Patagonian lake. Sci Total Environ 7(38):139838
Tymstra C, Stocks BJ, Cai X, Flannigan MD (2020) Wildfire management in Canada: review, challenges and opportunities. Prog Disaster Sci 5:100045
US Department of Agriculture Forest Service, "Confronting the wildfire crisis: A strategy for protecting communities and improving resilience in America's forests" (FS-1187a, US Department of Agriculture Forest Service, 2022). www.fs.usda.gov/sites/default/files/Confronting-Wildfire-Crisis.pdf
USEPA (2000) National Water Quality Inventory: 1998 Report to Congress. EPA841-F00-006. U.S. Environmental Protection Agency, Office of Water, Washington, DC. June
Uzun H, Dahlgren RA, Olivares C, Erdem CU, Karanfil T, Chow AT (2020) Two years of post-wildfire impacts on dissolved organic matter, nitrogen, and precursors of disinfection by-products in California stream waters. Water Res 181:115891
Varela E, Giergiczny M, Riera P, Mahieu P A&Solino M (2013) Social preferences for fuel break management programs in Spain: a choice modelling application to prevention of forest fires. IJWF *23*(2): 281-289.
Vartan S (2019) What Happens to Fish After a Wildfire? Hakai Magazine. Retrieved from https://www.hakaimagazine.com/news/what-happens-to-fish-after-a-wildfire
Vergnoux A, Malleret L, Asia L, Doumenq P, Theraulaz F (2011) Impact of forest fires on PAH level and distribution in soils. Environ Res 111(2):193–198
Verma S, Jayakumar S (2012) Impact of forest fire on physical, chemical and biological properties of soil: A review. Proceed IntAcadEcol Environ Sci 2(3):168

Wang H H, Yao F Q, Tao J J& Zhu F (2020) Efficiency of sustained burning of layered leaf fuels under external radiant heat flux. FAM 44(1): 121–129

Werth D, Avissar R (2005) The local and global effects of African deforestation. Geophys Res Lett 32:12704

Wisser D, Frolking S, Hagen S, Bierkens MF (2013) Beyond peak reservoir storage? A global estimate of declining water storage capacity in large reservoirs. Water Resour Res 49(9):5732–5739

Wu X, Sverdrup E, Mastrandrea M D, Wara M W & Wager S (2023) Low-intensity fires mitigate the risk of high-intensity wildfires in California's forests. SciAdv 9(45): eadi4123.

Yin Y, Bloom AA, Worden J, Saatchi S, Yang Y, Williams M & Schimel (2020) Fire decline in dry tropical ecosystems enhances decadal land carbon sink. Nat Commun 11(1): 1900.

Zou Y, Wang Y, Qian Y, Tian H, Yang J, Alvarado E (2020) Using CESM-RES Fire to understand climate–fire–ecosystem interactions and the implications for decadal climate variability. ACP 20(2):995–1020

Zwolinski M J (2000) The role of fire in management of watershed responses. In *Proceedings of conference "Land stewardship in the 21st century. The contributions of watershed management", Tucson, AZ* (pp. 13–16)

# Climate Change and Forest Fires: Ecosystem Vulnerability and Resilience

**Hukum Singh, Narendra Kumar, Amit Kumar, Manoj Kumar, Maneesh Singh Bhandari, and Santan Barthwal**

**Abstract** Climate change and fires are critical drivers of forest ecosystem vulnerability and resilience, impacting biodiversity, carbon sequestration, and ecosystem services. This chapter explores the complex interactions between climate change and forest fires, investigating how climate change influences the frequency, intensity, and distribution of forest fires and how these fires, in turn, impact ecosystem dynamics. This study revealed that rising global temperatures and changing precipitation patterns are critical drivers of more frequent and severe forest fires. The changes in temperature and precipitation are significant to forest ecosystems, disrupting natural processes and contributing to habitat loss and species decline. Moreover, the factors, particularly vegetation type, soil moisture, and topography, play a significant role in forest fire severity. Understanding the interactions between climate change, forest fire, and society and integrating them into management practices would help develop adaptive management strategies to enhance forest resilience, mitigate climate change impacts, and conserve biodiversity. Furthermore, we discuss future research directions to address knowledge gaps and improve predictive models to effectively manage and conserve forest biodiversity amidst ongoing global climate change.

**Keywords** Climate change · Forest fire · Ecosystem vulnerability · Forest resilience · Predictive models

---

H. Singh (✉) · M. S. Bhandari · S. Barthwal
ICFRE-Forest Research Institute, P.O. New Forest, Dehradun, Uttarakhand, India
e-mail: hukumsingh97@yahoo.com

N. Kumar
College of Agriculture, Central Agriculture University (I), Kyrdemkulai, Meghalaya, India

A. Kumar
CSIR-Institute of Himalayan Bioresource Technology, Palampur, Himachal Pradesh, India

M. Kumar
Centre of Excellence on Sustainable Land Management, Indian Council of Forestry Research and Education, P.O. New Forest, Dehradun, Uttarakhand, India

 
H. Singh and N. Kumar (eds.), *Forest Fire and Climate Change*,
https://doi.org/10.1007/978-3-031-89967-6_10

## 1 Introduction

Climate change and forest fires threaten forest ecosystems globally, impacting biodiversity, carbon sequestration, and ecosystem services (Kumar et al. 2024). Changing climatic conditions such as temperature and precipitation alter forest structure and composition, affecting species distribution and ecosystem dynamics (Allen et al. 2010; Kaushik and Singh 2024; Barwa and Singh 2024). Increased temperatures and prolonged drought exacerbate forest fire frequency and severity, leading to extensive forest loss and ecosystem degradation (Bowman et al. 2021). Moritz et al. (2012) indicated that higher temperatures contribute to longer fire seasons and more intense fire events in many regions.

Forest fires destroy vegetation and influence soil properties, nutrient cycling, and water availability, compromising forest resilience (Certini 2005). Post-fire recovery processes are crucial for ecosystem recovery but are often hindered by altered climatic conditions, delaying vegetation regeneration and increasing susceptibility to invasive species (Keeley et al. 2012). Additionally, the interactions between climate change and forest fire dynamics create feedback loops that amplify ecosystem vulnerabilities (Syphard et al. 2017).

The impacts of climate change on forest ecosystems vary regionally due to diverse climatic conditions and vegetation types (Sett et al. 2024). In Mediterranean regions, for example, projected increases in temperature and decreased precipitation are expected to intensify forest fire risks, posing severe challenges to forest management and biodiversity conservation (Pausas and Fernández-Muñoz 2012). Conversely, in boreal forests, warming temperatures and changing precipitation patterns alter forest fire regimes, affecting the resilience of conifer-dominated ecosystems (Flannigan et al. 2013).

Managing forest ecosystem vulnerability requires adaptive strategies integrating climate adaptation and forest fire management practices. Restoration efforts to enhance forest resilience through prescribed burns, selective logging, and reforestation can mitigate forest fire impacts and promote ecosystem recovery (Stephens et al. 2013). Moreover, incorporating climate change projections into forest management plans enables proactive measures to enhance forest ecosystem adaptation and reduce vulnerability to future climate and forest fire scenarios (Millar et al. 2007).

Addressing climate change and forest fire issues requires improved predictive models and management strategies to cope with the evolving challenges of forest ecosystem vulnerability (Singh 2024a). Integrating advanced remote sensing technologies and climate modeling may enhance scientific understanding of spatiotemporal forest fire patterns and their ecological impacts (Cohen et al. 2016). Furthermore, interdisciplinary research is desired to monitor the socio-economic impacts of changing forest landscapes and develop holistic approaches for sustainable forest management in a changing climate (Chapin et al. 2006; Singh 2024b).

## 2 Climate Change Impacts on Forest Ecosystems

Climate change poses significant challenges to forest ecosystems worldwide, altering their structure, composition, and functioning (Singh 2024c). One of the primary impacts is the alteration of temperature and precipitation patterns, leading to shifts in species distributions and phenology (Allen et al. 2010; Singh et al. 2024, 2018; Yadav et al. 2019a; Singh 2021). Warmer temperatures have facilitated the northward migration of species, affecting forest communities adapted to specific climatic conditions (IPCC 2019; Singh 2024d). Moreover, changes in precipitation regimes, including more frequent droughts or intense rainfall events, stress forest ecosystems, impacting their ability to regenerate and sustain biodiversity (Allen et al. 2010; Williams et al. 2013; Raturi et al. 2022; Sett et al. 2022; Singh et al. 2020a; Apurva et al. 2017).

In addition to direct climate impacts, altered fire regimes have become a critical concern for forest ecosystems under climate change. Increasing temperatures and prolonged droughts enhance forest fire frequency and severity, affecting both fire-adapted and sensitive species (Moritz et al. 2012; Schoennagel et al. 2017). Forest fires disrupt forest dynamics and release substantial amounts of carbon dioxide, further exacerbating climate change (Bowman et al. 2009; Singh et al. 2021c; Yadav et al. 2019b; Singh et al. 2023; Sharma and Singh 2021). This feedback loop underscores the complexity of climate-forest interactions and the urgent need for adaptive management strategies (IPCC 2021).

Furthermore, climate change influences forest ecosystem vulnerability through interactions with biotic factors, such as pest outbreaks and diseases. Warmer temperatures accelerate pest development rates and expand their geographic ranges, causing widespread tree mortality and altering forest structure (Dale et al. 2001; Bentz et al. 2010). For example, bark beetle outbreaks in North American forests have been linked to warmer winters, leading to extensive tree mortality and shifts in species composition (Dale et al. 2001; Bentz et al. 2010). These biotic disturbances amplify the impacts of climate change on forest ecosystems, challenging their resilience and adaptive capacity (Millar et al. 2007; Sharma et al. 2018; Jangra et al. 2024; Dhyani et al. 2021; Singh 2024e).

Moreover, climate change-induced alterations in hydrological cycles affect water availability in forested landscapes, influencing ecosystem productivity and species distribution patterns (IPCC 2021). Changes in snowmelt timing and reduced summer streamflow can stress riparian habitats and aquatic species, further complicating the resilience of forest ecosystems (Barnett et al. 2008; Williams et al. 2013). These hydrological changes also interact with other stressors, such as forest fires and invasive species, amplifying the vulnerability of forest ecosystems to climate impacts (Barnett et al. 2008).

Additionally, the socio-economic impacts of climate change on forest ecosystems are profound, affecting communities dependent on forest resources for livelihoods and cultural practices (IPCC 2014). Forest productivity and composition shifts can disrupt ecosystem services such as timber production, water regulation, and recreational opportunities, posing challenges to sustainable forest management (Millar et al. 2007). Vulnerable populations, including Indigenous communities, often bear disproportionate burdens from these changes, highlighting the need for inclusive and equitable adaptation strategies (IPCC 2014; Davidson et al. 2020).

Addressing the complex challenges of climate change requires integrated approaches considering ecological, social, and economic dimensions. This involves implementing adaptive management strategies, such as promoting forest resilience through diverse species planting and reducing fire fuel loads, which enhance ecosystem resistance and adaptation capacity (Schoennagel et al. 2017; Millar et al. 2007). Furthermore, international cooperation and policy frameworks are essential for mitigating global climate impacts on forest ecosystems and ensuring long-term sustainability (IPCC 2021).

## 3 Forest Fire Dynamics and Forest Vulnerability

Forest fires play a pivotal role in shaping forest ecosystems globally. Climate change influences their frequency, intensity, and spatial extent globally. Climate variability and forest fire dynamics impact forest vulnerability, altering ecosystem structure, biodiversity, and ecosystem services (Prakash et al. 2022; Nirmal et al. 2021; Singh and Kumar 2022; Savita et al. 2017). Climate change-induced shifts in temperature and rainfall patterns affect fuel availability and moisture content and contribute to longer fire seasons and more extreme fire weather conditions (Flannigan et al. 2009; Schoennagel et al. 2017). These changes intensify the susceptibility of forests to ignition and spread, thereby increasing forest fire risk and severity (Moritz et al. 2012).

Complex feedback mechanisms within forest ecosystems influence forest fires' spatial and temporal patterns. For instance, fire-adapted species and ecosystems have evolved with natural fire regimes, often exhibiting resilience to moderate fire disturbances (Bowman et al. 2013). However, altered fire regimes driven by climate change surpass ecosystem resilience thresholds, leading to ecosystem shifts and biodiversity loss (Whitman et al. 2017). Increased fire frequency and intensity compromise forest regeneration and carbon sequestration capacities, affecting global carbon cycles and climate regulation (Bond et al. 2005; Liu et al. 2016).

Furthermore, socio-economic factors such as land use change, fire suppression policies, and human settlement patterns interact with climate-induced forest fire dynamics, influencing landscape vulnerability (Syphard et al. 2017). Human activities introduce ignition sources and alter fire regimes, exacerbating climate-driven changes' impacts on forest ecosystems (Balch et al. 2017). Therefore, effective management strategies must integrate ecological and socio-economic dimensions to mitigate forest fire risks and enhance forest ecosystem resilience (McWethy et al. 2013).

Understanding forest fire regimes' spatial and temporal variability is crucial for assessing forest vulnerability and implementing adaptive management strategies. Technological interventions such as remote sensing technologies and geographic information systems are significant in monitoring temporal and spatial patterns at various scales of fire occurrence, severity, and ecological impacts (Key et al. 2017). These tools facilitate real-time fire management decisions and contribute to long-term forest fire risk assessments under changing climatic conditions (Chuvieco et al. 2012).

The abovementioned issues require future research to improve predictive models that integrate climate projections, landscape dynamics, and socio-economic factors to enhance forest fire risk assessments (Preisler et al. 2017). Furthermore, incorporating indigenous knowledge systems and traditional fire management practices may provide valuable insights into sustainable forest fire mitigation strategies (Carter et al. 2020). Ultimately, advancing interdisciplinary research efforts is essential for mitigating the escalating impacts of climate change-induced forest fire dynamics on forest ecosystems and promoting their long-term resilience (Moreno-Mateos et al. 2015).

## 4 Synergistic Impacts of Climate Change and Forest Fire on Forest Ecosystem

The synergistic interactions between climate change and forest fire significantly exacerbate their impacts on forest ecosystems. Warmer temperatures and prolonged droughts increase the frequency of extreme fire weather conditions, expanding the geographic range of fire-prone areas and drying out vegetation, thus increasing fuel availability (Allen et al. 2015; Seidl et al. 2017; Abatzoglou and Williams 2016). These climatic shifts alter vegetation dynamics and fuel availability, favoring more frequent and intense fires (Jolly et al. 2015; Westerling 2016).

Forest fire amplifies the ecological impacts of climate change on forests by causing extensive tree mortality and altering forest structure and composition (Moritz et al. 2013). High-intensity fires also disrupt nutrient cycling processes and reduce soil fertility, impairing forest regeneration efforts (Santana et al. 2016; Cannon et al. 2018). Additionally, the loss of vegetation cover due to forest fire heightens landscapes' vulnerability to erosion and landslides, especially in steep terrain and regions experiencing heavy rainfall.

The feedback loops between climate change and forest fire further complicate forest ecosystem dynamics. Forests, when burned, release stored carbon into the atmosphere, contributing to carbon emissions that exacerbate global warming (Bowman et al. 2020). This creates a positive feedback loop where climate change increases forest fire risk, raises emissions, and amplifies climate change impacts on forests (Williams et al. 2019). These interactions underscore the urgent need for integrated approaches to mitigate climate change and forest fire impacts on forest ecosystems (Turner et al. 2022).

Understanding regional variability in climate-fire relationships is crucial. Forest types and ecosystems respond uniquely to climate-driven changes in forest fire regimes (Abatzoglou and Williams 2016). For example, boreal forests may experience increased fire frequency and severity under warmer and drier conditions, while temperate forests may undergo shifts in species composition and heightened vulnerability to insect outbreaks following forest fire (Flannigan et al. 2013; Turner et al. 2022).

Effective management strategies must account for these complex interactions to enhance forest resilience. Adaptive management approaches, such as prescribed burning and fuel management, mitigate forest fire risks and maintain ecosystem health amidst changing climate conditions (Moritz et al. 2014). Moreover, collaboration among researchers, policymakers, and local communities is essential for implementing sustainable practices that mitigate climate change impacts and promote long-term forest sustainability (Stephens et al. 2018).

## 5 Assessing Forest Ecosystem Vulnerability to Climate Change and Forest Fire

Assessing forest ecosystem vulnerability to climate change and forest fire requires a multidisciplinary approach that integrates ecological, climatological, and remote sensing techniques. This section outlines the methodologies commonly employed to understand the complex interactions and impacts on forest ecosystems.

### *5.1 Ecological Surveys and Monitoring*

Ecological surveys play a fundamental role in assessing the health and composition of forest ecosystems. Field-based methods, such as vegetation sampling, biodiversity assessments, and tree coring for age and growth analysis, provide crucial baseline data on species distribution and ecosystem structure (Franklin et al. 2016). Long-term monitoring programs, such as the Forest Inventory and Analysis (FIA) program in the United States, offer valuable insights into forest dynamics and responses to environmental changes (Bechtold and Patterson 2005).

## *5.2 Climate Data Analysis*

Climate data analysis is pivotal in understanding the environmental conditions influencing forest fire occurrence, behavior, and intensity. Researchers utilize historical climate records and statistical methods to identify trends and anomalies in temperature, precipitation, humidity, and wind patterns over time. These analyses provide insights into long-term climate variability and how it impacts fire regimes in different regions. For instance, trends indicating prolonged droughts or shifts in seasonal rainfall patterns can significantly increase the likelihood of severe forest fire by altering fuel moisture content and fire weather conditions (Flannigan et al. 2009; Abatzoglou and Williams 2016).

Various methods in climate data analysis include trend analysis, which identifies significant changes in climate variables over decades, and the use of climate indices such as the Palmer Drought Severity Index (PDSI) or the Fire Weather Index (FWI) to quantify fire risk under different climatic scenarios (IPCC 2021; Vicente-Serrano et al. 2010). Climate models, including General Circulation Models (GCMs) and Regional Climate Models (RCMs), simulate future climate conditions based on various greenhouse gas emission scenarios. These models help predict potential shifts in fire-prone areas and the frequency of extreme fire weather events, aiding in proactive fire management and adaptation planning (Flannigan et al. 2009).

## *5.3 Foret Fire Modeling and Simulation*

Forest fire modeling integrates fire behavior, fuel characteristics, and weather conditions to predict fire occurrence and spread. Techniques range from simple statistical models to complex spatially explicit simulations using Geographic Information Systems (GIS) and fire behavior models (Calkin et al. 2015). Historical fire records, fire severity assessments, and burn scar mapping provide empirical data for validating and calibrating forest fire models (Miller and Urban 2000).

Fire modeling integrates climate data with topography, vegetation types, and human activities to simulate forest fire behavior and assess fire risk. These models range from simple empirical models that estimate fire spread based on historical data to complex spatially explicit models that incorporate dynamic variables such as fuel moisture, wind speed, and terrain slope (Finney 2005; Calkin et al. 2015).

One of the primary types of fire models is the Fire Spread Model, which predicts the rate and direction of fire spread based on fuel characteristics and weather conditions at ignition. These models are crucial to predict fire behavior in real time and make informed decisions about resource allocation and evacuation orders (Finney 2005).

Coupled models, which integrate fire behavior models with climate projections and ecological data, offer a more comprehensive understanding of how future climate change may alter fire regimes and impact ecosystems. These models help identify vulnerable areas and prioritize mitigation efforts, such as fuel reduction treatments and prescribed burns, to reduce forest fire risk and enhance landscape resilience (Keane et al. 2013; Calkin et al. 2015).

## *5.4 Remote Sensing and GIS Analysis*

Remote sensing techniques, such as satellite imagery and LiDAR (Light Detection and Ranging), offer spatially explicit data on forest cover, fragmentation, and disturbance patterns (Cohen and Goward 2004). GIS-based analysis enables researchers to assess landscape-scale changes in vegetation dynamics, fire regimes, and habitat suitability (Syphard et al. 2017). Integrating remote sensing with field data improves accuracy in mapping and monitoring forest vulnerability to climate change and forest fire.

## *5.5 Integrated Assessments and Vulnerability Indices*

Integrated assessments combine ecological, climatological, and socio-economic data to develop vulnerability indices for forest ecosystems (Seidl et al. 2017). These indices quantify forests' exposure, sensitivity, and adaptive capacity to climate change and forest fire impacts. Multi-criteria decision analysis and scenario modeling help prioritize management actions and adaptation strategies based on predicted vulnerabilities (IPBES 2019).

## *5.6 Stakeholder Engagement and Adaptive Management*

Incorporating stakeholder perspectives and local knowledge is crucial for effective forest management and conservation strategies. Participatory approaches, such as workshops and collaborative planning processes, enhance the relevance and acceptance of vulnerability assessments (Reed et al. 2009). Adaptive management frameworks facilitate iterative decision-making, allowing adjustments in response to new data and changing environmental conditions (Holling 1978).

## 6 Forest Fire Management and Adaptation Strategies for Enhancing Ecosystem Resilience

Effective forest fire management involves a combination of prevention, preparedness, response, and recovery strategies to mitigate fire impacts and enhance ecosystem resilience. Prevention efforts focus on reducing fire ignition sources through public awareness campaigns, enforcing fire restrictions, and implementing landscape-scale fuel management practices such as prescribed burns and mechanical thinning (Calkin et al. 2015). These proactive measures aim to minimize the likelihood of forest fires and reduce their intensity when they occur.

Preparedness strategies involve enhancing firefighting capabilities, training personnel, and developing emergency response plans tailored to regional fire regimes and risk profiles (Moritz et al. 2014). Collaborative efforts among fire agencies, communities, and stakeholders are crucial for coordinating responses and mobilizing resources effectively during fire outbreaks (Martin et al. 2020). Early detection systems, including remote sensing technologies and community-based monitoring networks, facilitate rapid response and containment of fires, reducing their spread and associated damages (Chuvieco et al. 2012).

Adaptation strategies focus on building resilience to climate change and evolving fire regimes. These strategies include promoting ecosystem diversity, restoring fire-adapted landscapes, and integrating climate-smart land-use planning practices (Stephens et al. 2018). Adaptive management approaches involve iterative decision-making processes incorporating scientific knowledge, stakeholder input, and monitoring feedback to adjust strategies in response to changing environmental conditions (Millar et al. 2007).

Community engagement and public education are critical in forest fire management and adaptation. Educating communities about fire ecology, safe practices, and evacuation procedures enhances their capacity to coexist with fire-prone landscapes (McCaffrey et al. 2013). Community-based initiatives, such as Firewise USA® programs, empower residents to reduce forest fire risks around their homes through fire-resistant landscaping and building materials (Brenkert-Smith et al. 2016).

Investments in research and technology innovation drive advancements in fire management and adaptation. Remote sensing technologies and Geographic Information Systems enable real-time fire monitoring, behavior modeling, and post-fire damage assessment (Cohen and Goward 2004). Climate modeling and scenario planning help forecast future fire risks and inform adaptive strategies considering changing climatic conditions and socio-economic factors (Keane et al. 2017).

Another approach considers reforestation with fire-resistant species as an innovative and strategic method to mitigate the impacts of forest fires and enhance ecosystem resilience. This practice involves planting tree species with traits that make them less susceptible to fire damage, such as thick bark, high moisture content, and the ability to regenerate quickly after fire events. Incorporating fire-resistant species into reforestation efforts helps create more resilient forest ecosystems that can withstand and recover from fire disturbances more effectively.

Therefore, effective forest fire management requires a holistic approach integrating prevention, preparedness, response, and adaptation strategies. Scientific knowledge, technological innovations, and community engagement may guide stakeholders in enhancing forest fire resilience, safeguarding ecosystems, and protecting human communities in a changing climate.

## 7 Future Research Directions

Future research must focus on enhancing predictive models to effectively manage and conserve forest biodiversity in the face of escalating global climate change. One critical area for advancement is refining climate and fire behavior models to incorporate complex interactions between climate variables, vegetation dynamics, and fire regimes. Integrating advanced statistical techniques and machine learning algorithms can improve the accuracy of climate projections and enhance our ability to forecast future fire risks with greater precision (Syphard et al. 2017; Bowman et al. 2020).

Furthermore, there is a pressing need to develop spatially explicit models considering landscape heterogeneity and species-specific responses to environmental changes. These models can help identify priority areas for conservation and restoration efforts, considering habitat connectivity and forest ecosystems' resilience to multiple stressors (Stephens et al. 2018). Incorporating socio-economic factors and human dimensions into these models is also crucial for assessing the socio-ecological impacts of climate change on forest biodiversity and designing adaptive management strategies that are socially acceptable and economically feasible (IPBES 2019).

In addition to modeling advancements, future research should prioritize long-term monitoring programs for understanding ecological responses to climate change and fire disturbances over extended periods. Long-term datasets provide valuable insights into ecosystem resilience, species adaptation, and community dynamics under changing environmental conditions (Franklin et al. 2016). Emphasizing interdisciplinary collaborations between ecologists, climatologists, social scientists, and stakeholders can facilitate data sharing, integrate diverse perspectives, and foster innovative approaches to conservation planning and adaptive management (Millar et al. 2007; Reed et al. 2009).

Moreover, investing in technological innovations such as remote sensing, unmanned aerial vehicles (UAVs), and sensor networks can revolutionize monitoring efforts and enhance our capacity to detect early warning signs of ecosystem stress and biodiversity loss (Cohen and Goward 2004). These technologies enable real-time data collection, spatial mapping of habitat changes, and assessment of ecosystem health, supporting timely decision-making and intervention strategies (Chuvieco et al. 2012; Keane et al. 2017).

Addressing these research priorities and leveraging technological advancements can enhance our understanding of the complex dynamics driving forest biodiversity loss under climate change. This knowledge is essential for developing proactive conservation strategies that safeguard biodiversity, maintain ecosystem services, and promote the resilience of forest ecosystems in a rapidly changing world.

## 8 Conclusion

It is concluded that forest ecosystems are vulnerable to the synergistic impacts of climate change and forest fire, disrupting ecological processes, threatening biodiversity and ecosystem services, and loss of carbon storage capacities, ultimately affecting the resilience of forest ecosystems. The synthesis also demonstrated significant threats to livelihoods dependent on forest resources and ecosystem-based economies due to climate change and forest fires. Furthermore, understanding the interactions between climate change, forest fire, and society and integrating them into management practices guide policymakers and managers in developing adaptive management strategies to enhance forest resilience, mitigate climate change impacts, and conserve biodiversity. Stakeholders may better anticipate and respond to future challenges by prioritizing conservation efforts, promoting sustainable land-use practices, and advancements in predictive modeling and monitoring technologies. Future research should focus on closing knowledge gaps and improving predictive models to support evidence-based decision-making. Long-term monitoring programs and interdisciplinary studies are needed to track ecosystem responses, identify resilient species and habitats, and inform adaptive strategies. Integrating socio-economic considerations into the predictive models may be critical for assessing forest vulnerability's human dimensions and ensuring equitable and sustainable management practices in future climate change.

## References


Abatzoglou JT, Williams AP (2016) Impact of anthropogenic climate change on wildfire across western US forests. Proc Natl Acad Sci 113(42):11770–11775

Allen CD et al (2010) A global overview of drought and heat-induced tree mortality reveals emerging climate change risks for forests. For Ecol Manage 259(4):660–684

Apurva K, Singh H, Kumar H (2017) Elucidating morphological and growth changes in arjun tree (*Terminali aarjuna*) grown under elevated temperature condition. Trends Biosci 10(1):332–336

Balch JK et al (2017) Human-started wildfires expand the fire niche across the United States. Proc Natl Acad Sci 114(11):2946–2951

Barnett TP et al (2008) Human-induced changes in the hydrology of the western United States. Science 319(5866):1080–1083

Barwa S, Singh H (2024) Climate-induced changes in forest ecosystems of Indian Himalayan region: a meta-analysis. In: Singh H (ed) Urban forests, climate change and environmental pollution. Springer, Cham. https://doi.org/10.1007/978-3-031-67837-0_39

Bentz BJ et al (2010) Climate change and bark beetles of the western United States and Canada: direct and indirect effects. Bioscience 60(8):602–613

Bond WJ et al (2005) Fire as a global 'herbivore': the ecology and evolution of flammable ecosystems. Trends Ecol Evol 20(7):387–394
Bowman DMJS et al (2009) Fire in the Earth system. Science 324(5926):481–484
Bowman DM et al (2013) The human dimension of fire regimes on Earth. J Biogeogr 40(12):2293–2298
Bowman DM et al (2020) Human exposure and sensitivity to globally extreme wildfire events. Nat Ecol Evol 4(3):427–436
Bowman DMJS et al (2021) Vegetation fires in the Anthropocene: a pyrogeographic perspective. Prog Phys Geogr: Earth Environ 45(5):679–706
Brenkert-Smith H et al (2016) A tale of two diverse urban communities' wildfire risk interpretation, adaptation, and action. Environ Manage 57(2):475–488
Calkin DE et al (2015) Negative consequences of positive feedbacks in US wildfire management. For Ecol Manage 338:139–153
Cannon SH et al (2018) Landscape response to the 1980 eruption of Mount St. Helens: using historical aerial imagery to track ecosystem recovery. Nat Hazards Earth Syst Sci 18(1):3–16
Carter J et al (2020) Cultural burning: Indigenous fire stewardship meets climate science. Fire 3(1):2
Certini G (2005) Effects of fire on properties of forest soils: a review. Oecologia 143(1):1–10
Chapin FS et al (2006) Ecosystem stewardship: sustainability strategies for a rapidly changing planet. Trends Ecol Evol 21(7):368–377
Chuvieco E et al (2012) Remote sensing of burned areas: from theory to operational applications. Springer Science & Business Media
Cohen WB, Goward SN (2004) Landsat's role in ecological applications of remote sensing. Bioscience 54(6):535–545
Cohen WB et al (2016) An assessment of the effectiveness of a large-scale forest management experiment using a multi-scale modeling approach. Ecosystems 19(2):202–216
Dale VH et al (2001) Climate change and forest disturbances. Bioscience 51(9):723–734
Davidson K, Briggs J, Nolan E, Bush J, Håkansson I, Moloney S (2020) The making of a climate emergency response: examining the attributes of climate emergency plans. Urban Clim 33:100666
Dhyani K, Kuniyal HB, Singh H, Sobha (2021) Growth and physiological potential of *Terminalia arjuna* under elevated $CO_2$ levels in open top chamber condition. J Appl Nat Sci 13(3):1121–1126. https://doi.org/10.31018/jans.v13i3.2490
Finney MA (2005) The challenge of quantitative risk assessment for wildland fire. For Ecol Manage 211(1–2):97–108
Flannigan MD et al (2009) Forest fires and climate change in the 21st century. Mitig Adapt Strat Glob Change 14(5):489–508
Flannigan MD et al (2013) Forest fires and climate change in the 21st century. Mitig Adapt Strat Glob Change 18(7):847–859
Franklin JF et al (2016) Disturbances and the resilience of coupled human–environment systems. Ecol Soc 21(3):15
IPBES (2019) Global assessment report on biodiversity and ecosystem services of the Intergovernmental Science-Policy Platform on Biodiversity and Ecosystem Services. IPBES Secretariat
IPCC (2014) Climate change 2014: impacts, adaptation, and vulnerability. contribution of working group II to the fifth assessment report of the intergovernmental panel on climate change. Cambridge University Press
IPCC (2019) IPCC special report on climate change and land. Intergovernmental Panel on Climate Change
IPCC (2021) Climate change 2021: the physical science basis. Contribution of working group I to the sixth assessment report of the intergovernmental panel on climate change. Cambridge University Press
Jangra A, Kumar K, Maikhuri S, Bhandari MS, Pandey S, Singh H and Barthwal S (2024) Unveiling stress-adapted endophytic bacteria: characterizing plant growth-promoting traits and assessing

cross-inoculation effects on *Populus deltoides* under abiotic stress. Plant Physiol Biochem 108610. https://doi.org/10.1016/j.plaphy.2024.108610

Jolly WM et al (2015) Climate-induced variations in global wildfire danger from 1979 to 2013. Nat Commun 6:7537

Kaushik P, Singh H (2024) Climate change-induced forest ecosystem functioning and alteration in water-atmosphere-soil–plant (WASP) relationship. In: Singh H (ed) Urban forests, climate change and environmental pollution. Springer, Cham. https://doi.org/10.1007/978-3-031-67837-0_38

Keane RE et al (2013) Ecological effects of large fires on US landscapes: benefit or catastrophe? Int J Wildland Fire 22(7):893–908

Keane RE et al (2017) A review of operational, research, and future applications of the wildland fire operational risk assessment (WFRA). For Ecol Manage 404:183–198

Keeley JE et al (2012) Fire as an ecological process. In: Fire in California's ecosystems. University of California Press, pp 21–74

Keys PW, Wang-Erlandsson L, Gordon LJ, Galaz V, Ebbesson J (2017) Approaching moisture recycling governance. Glob Environ Change 45:15–23

Kumar N, Singh H, Kumar A, Singh AK, Pandey PK, Kumar A (2024). Forest-fire-induced land degradation. In: Mishra G, Giri K, Singh S, Kumar M (eds) Sustainable land management in India. Springer Nature, Singapore. https://doi.org/10.1007/978-981-97-5223-2_4

Liu, Z., et al. (2016). Climate change and wildfires in the U.S. Sci Rep 6:28543

Martin KL et al (2020) Communities and wildfire mitigation in the United States: a review and implications for research, policy, and practice. Fire 3(1):4

McCaffrey SM et al (2013) Social science research related to wildfire management: an overview of recent findings and future research needs. Int J Wildland Fire 22(1):15–24

McWethy DB et al (2013) A conceptual framework for predicting temperate ecosystem sensitivity to human impacts on fire regimes. Glob Ecol Biogeogr 22(8):900–912

Millar CI et al (2007) Climate change and forests of the future: managing in the face of uncertainty. Ecol Appl 17(8):2145–2151

Moreno-Mateos D et al (2015) Anthropogenic ecosystem disturbance and the recovery debt. Nat Commun 7:11294

Moritz MA et al (2012) Climate change and disruptions to global fire activity. Ecosphere 3(6):1–22

Moritz MA et al (2013) Learning to coexist with wildfire. Nature 515(7525):58–66

Moritz MA et al (2014) Learning to coexist with wildfire. Nature 515(7525):58–66

Nirmal, Bakshi M, Singh H (2021) Effect of $CO_2$ elevation on Shisham growth at nursery stage. Indian J Hill Farming 34:272–276

Pausas JG, Fernández-Muñoz S (2012) Fire regime changes in the Western Mediterranean Basin: from fuel-limited to drought-driven fire regime. Clim Change 110(1–2):215–226

Prakash V, Hunney K, Singh H (2022) Effect of elevated carbon dioxide on growth and development of *Santalum album* L. seedlings inoculated with plant growth promoting microorganisms in open top chambers. Int J Environ, Agric Biotechnol 7(5):149–159

Preisler HK et al (2017) Bayesian networks for wildfire risk assessment and management. Environ Model Softw 92:132–147

Raturi A, Singh H, Kumar P, Shukla N (2022) Characterizing the post-monsoon $CO_2$, $CH_4$, $N_2O$, and $H_2O$ vapor effluxes from a tropical wetland in the Himalayan foothill. Environ Monitor Assess 194(50). https://doi.org/10.1007/s10661-021-09721-8

Reed MS et al (2009) Participatory methods for process monitoring in environmental management. J Environ Manage 90(5):1656–1660

Santana VM, González-Pelayo O, Maia PA, Varela TME, Valdecantos A, Ramón Vallejo V, Jacob Keizer J (2016) Effects of fire recurrence and different salvage logging techniques on carbon storage in Pinus pinaster forests from northern Portugal. Eur J For Res 135:1107–1117

Savita, Sharma SK, Singh H, Singh O (2017) Forest fire: a review. Indian Forester 143(8):719–728

Schoennagel T et al (2017) Adapt to more wildfire in western North American forests as climate changes. Proc Natl Acad Sci 114(18):4582–4590

Sett T, Nikam BR, Singh H, Purohit S (2022) Effect of drought stress on forest evapotranspiration—a case study on Indian forests. EGU General Assembly 2022. https://doi.org/10.5194/egusphere-egu22-510

Sett T, Nikam BR, Singh H, Purohit S (2024) Hydrological variability in Indian forest ecosystem: analysis of drought resilience, recovery and water use efficiency in moist and dry deciduous forests. In: Singh H (eds) Urban forests, climate change and environmental pollution. Springer, Cham. https://doi.org/10.1007/978-3-031-67837-0_37

Seidl R, Thom D, Kautz M, Martin-Benito D, Peltoniemi M, Vacchiano G, ..., Reyer CP (2017) Forest disturbances under climate change. Nat Clim Change 7(6):395–402

Sharma R, Singh H, Kaushik M, Nautiyal R, Singh O (2018) Adaptive physiological response, carbon partitioning and biomass production of *Withania somnifera* (L.) Dunal grown under elevated $CO_2$ regimes. 3 Biotech 8:267. https://doi.org/10.1007/s13205-018-1292-1

Sharma R, Singh H (2021) Alteration in biochemical constituents and nutrients partitioning of *Asparagus racemosus* in response to elevated atmospheric $CO_2$ concentration. Environ Sci Pollut Res 29:6812–6821. https://doi.org/10.1007/s11356-021-16050-3

Singh H, Kumar M (2022) Climate change and its impact on Indian Himalayan forests: current status and research needs. In: Rani S, Kumar R (eds) Climate change. Springer climate Springer, Cham, pp 223–242

Singh H, Sharma R, Savita SMP, Kumar M, Verma A, Ansari MW, Sharma SK (2018) Adaptive physiological response of *Parthenium hysterophorus* to elevated atmospheric $CO_2$ concentration. Indian Forester 144:1–14

Singh H, Kumar N, Singh R, Kumar M (2023) Assessing the climate change impact on the habitat suitability of the range-restricted bird species (*Catreus wallichii*) in the Indian Himalayan ecosystem. Environ Sci Pollut Res. https://doi.org/10.1007/s11356-023-30789-x

Singh M, Singh H, Kumar A, Kumar M, Barthwal S, Thakur A (2024) Soil nitrogen availability determines the $CO_2$ fertilization effect on tree species (*Neolamarckia cadamba*): growth and physiological evidence. Environ Sustain 7:53–60. https://doi.org/10.1007/s42398-023-00300-w

Singh H, Kumar N, Kumar M, Singh R (2020a) Modelling habitat suitability of western tragopan (*Tragopan melanocephalus*) a range-restricted vulnerable bird species of the Himalayan region, in response to climate change. Clim Risk Manag 29:100241. https://doi.org/10.1016/j.crm.2020.100241

Singh S, Singh H, Sharma SK, Nautiyal R (2021c) Seasonal variation in biochemical responses of bamboo clones in the sub-tropical climate of Indian Himalayan foothills. Heliyon 7(4):e06859. https://doi.org/10.1016/j.heliyon.2021.e06859

Singh H (2021) Process-based carbon sequestration study in relation to the energy-water-carbon flux in a forest ecosystem. In: Kumar P, Singh RK, Kumar M, Rani M, Sharma (eds) Climate impacts on sustainable natural resource management. CRC Press Taylor & Francis Group, United States, pp 336–361

Singh H (2024a) Climate change and forests: biological perspectives. In: Singh H (ed) Forests and climate change. Springer Nature, Singapore. https://doi.org/10.1007/978-981-97-3905-9_1

Singh H (2024c) Climate change and environmental pollution: biological consequences on urban forests. In: Singh H (eds) Urban forests, climate change and environmental pollution. Springer Nature, Cham, Switzerland. https://doi.org/10.1007/978-3-031-67837-0_1

Singh H (ed) (2024b) Forests and climate change: biological perspectives on impact, adaptation, and mitigation strategies. Springer Nature Singapore. https://doi.org/10.1007/978-981-97-3905-9

Singh H (ed) (2024d) Urban forests, climate change and environmental pollution: physio-biochemical and molecular perspectives to enhance urban resilience. Springer Nature, Cham, Switzerland. https://doi.org/10.1007/978-3-031-67837-0

Singh H (2024e) Exploring adaptive modulation in plant functional traits and its impact on the productivity of Acacia auriculiformis under $CO_2$-enriched environment. Ind Crops Prod 210(2024)118186. https://doi.org/10.1016/j.indcrop.2024.118186

Stephens SL et al (2013) Managing forests and fire in changing climates. Science 342(6154):41–42

Stephens SL et al (2018) Temperate and boreal forest mega-fires: characteristics and challenges. Front Ecol Environ 16(9):511–519

Syphard AD et al (2017) Human presence diminishes the importance of climate in driving fire activity across the United States. Proc Natl Acad Sci 114(52):13750–13755

Turner MG, Braziunas KH, Hansen WD, Hoecker TJ, Rammer W, Ratajczak Z, …, Seidl R (2022) The magnitude, direction, and tempo of forest change in Greater Yellowstone in a warmer world with more fire. Ecol Monogr 92(1):e01485

Vicente-Serrano SM et al (2010) A global capacity for drought monitoring and prediction: current status and future prospects. Earth-Sci Rev 99(1–2):48–60

Westerling AL (2016) Increasing western US forest wildfire activity: sensitivity to changes in the timing of spring. Philos Trans Roy Soc B Biol Sci 371(1696):20150178

Whitman E et al (2017) Wildfire, vegetation change, and the ecology of altered futures. Anthropocene 18:59–68

Williams AP et al (2013) Forest responses to increasing aridity and warmth in the southwestern United States. Proc Natl Acad Sci 110(52):2186–2191

Williams AP, Abatzoglou JT, Gershunov A, Guzman-Morales J, Bishop DA, Balch JK, …, Lettenmaier DP (2019) Observed impacts of anthropogenic climate change on wildfire in California. Earth's Future 7(8):892–910

Yadav SK, Singh H, Nautiyal R, Ginwal HS, Ansari SA, Barthwal S (2019) Modulation of morpho-physiological responses in *Populus deltoides* by elevated carbon dioxide and temperature. For Sci 66:105–118. https://doi.org/10.1093/forsci/fxz048

Yadav SK, Singh H, Ginwal HS, Barthwal S (2019b) Elevated $CO_2$ enhanced growth and physiological process of *Populus deltoides* Bartr. ex Marsh. Indian Forester 145(1):23–27

# Advancing Forest Fire Management: Innovations in Technology and Intervention Strategies

**Saina Maharjan and Hukum Singh**

**Abstract** The escalating threats of forest fire intensified by climate change constitute a critical challenge to ecosystems, human safety, and infrastructure worldwide. Effective forest fire management demands the implementation of innovative technologies and strategic intervention. This chapter covers forest fire management scenery, emphasizing the pivotal role of technologies and interventions to control and manage forest fires. It provides insights into the latest advancements in fire detection, suppression, and mitigation strategies encompassing remote sensing, artificial intelligence, and uncrewed aerial vehicles like drones. Additionally, it explores pre-emptive measures like community initiatives and cross-border cooperation to improve preparedness and response efforts. The synthesis of recent advancements and practical insights may provide a comprehensive understanding of the evolving landscape of forest fire management. Further, this chapter may inform policymakers, practitioners, and researchers, empowering them to adopt effective strategies for sustainable forest fire mitigation measures, i.e., control and management.

**Keywords** Climate change · Forest fire · Forest fire management · Technologies and innovation

## 1 Introduction

Forest fires present a substantial and complex threat to land ecosystems (Flannigan et al. 2006), resulting in significant ecological, economic, and societal consequences (Tyukavina et al. 2022; Babu et al. 2016; Kumar et al. 2024). The changing climate patterns caused by climate change have led to an extended wildfire season in various regions (Battistoni et al. 2023; Barwa and Singh 2024). Projections presented by the

S. Maharjan (✉)
Forest Research Institute Deemed to be University, Dehradun, India
e-mail: sainamaharjan07@gmail.com

H. Singh
ICFRE-Forest Research Institute, P.O New Forest, Dehradun, Uttarakhand, India

 
H. Singh and N. Kumar (eds.), *Forest Fire and Climate Change*,
https://doi.org/10.1007/978-3-031-89967-6_11

Intergovernmental Panel on Climate Change (IPCC) indicate the potential temperature will ascend to 4.8 °C in tropical forests by the end of the century (Celis et al. 2023; Sett et al. 2024). Effective control and management of forest fires are critical to minimize the social, ecological, and financial problems they cause (Singh et al. 2024).

Typically, fire management programs encompass preventive measures to reduce human-caused fires, early detection systems to identify fires promptly, swift fire suppression, and large fire management systems to mitigate damage from uncontrolled large fires (Martell 2001; Singh 2024a). However, despite growing technological innovations like remote sensing, Unmanned Aerial and Ground Vehicles, Internet, etc., and interventions, various challenges hinder progress in forest fire management. These challenges include limited human resources, financial constraints, technological obstacles, regional concerns, and inadequate coordination of stakeholders and environmental policies (Wang et al. 2021; Kunwar 2006).

Therefore, a comprehensive approach is required to incorporate innovative technologies and strategies for effective forest fire management that consider the nature of fires and the topographic factors specific to each area.

## 2 Understanding of Forest Fires

### *2.1 2.1. Cause and Factors Contributing to Forest Fires*

Forest Fire incorporate both Natural and human-initiated factors. The natural factors contributing to forest fires include weather/climate, lightning strikes, volcanic activities, and spontaneous combustion (Flannigan et al. 2006; Bowman et al. 2009; Kaushik and Singh 2024). Climate change resulting in an increase in temperature worldwide has prolonged and intensified droughts, which have established ideal circumstances for wildfires (Jolly et al. 2015), thus accelerating the frequency and strength of occurrence, leading to severe forest fire conditions (US EPA 2016). Comprehending the meteorological factors that trigger forest fires is crucial to understanding these disastrous occurrences' causes, trends, and dynamics (Celis et al. 2023). The potential severity of a fire also relies on landform-related variables such as aspect, elevation, slope, and topographic position (Ciesielski et al. 2022). The topography influences the straight-line spread of Fire by altering wind patterns and regulating the moisture content of fuel through exposure to sunlight and humidity pooling (Rogers 2020).

In contrast to other natural disasters, forest fires are often caused by humans (Dhall et al. 2020). Only 4% of forest fires are caused by natural factors, whereas human activity is a significant cause in every other instance (Hirschberger 2016; Singh 2024b). Anthropogenic activities are the primary cause of forest fire initiation (Bahadur et al. 2017). Uncontrolled agricultural practices, forest vegetation modification, land-use change (Carta et al. 2023), negligence, and carelessness (Moore

2019) are the major human activities causing forest fires. In many regions of the world, the expansion of urbanization is reducing the edges linking urban areas and wilderness areas, which are gradually distorted (Singh 2024c). The consequences of the forest fire tragedies highlight the Wildland-Urban Interface (WUI) (NDMA 2020).

## 2.2 *2.2. Impacts of Forest Fires on Eco-Systems and Communities*

Forest fires, whether caused by natural or human factors, have profound consequences on ecosystems and communities. Fire is an infrequent event that can cause significant harm to vegetation and can extend the degradation. Understanding these impacts is crucial for developing strategies to mitigate and recover from catastrophic events.

## 2.3 *2.2.1. Ecosystem Impacts*

Forest fire is a significant cause of forest loss and soil deterioration (Mishra et al. 2023). Forest fires severely harm the composition and structure of fragile ecosystems, ultimately causing their collapse (Valderrama et al. 2018; Singh 2024d). It increases runoff, floods, and landslides (American Forest Foundation 2015), reducing the soil porosity, water retention ability, and loss of nutrients such as Nitrogen (Satendra and Kaushik 2012). Not only are factors like vegetation composition, succession, and carbon budget influenced by a forest fire (Burrell et al. 2022), but many services and products sourced from forests, i.e., water supply recreation, protection services, socio-economic status of the countries and socially vulnerable regions are also the significantly affected sectors (Kala 2023; Paudel 2021; Chas-Amil et al. 2022).

## 2.4 *2.2.2. Community Impacts.*

Globally, forest fires result in loss to society and the economy (Baranovskiy et al. 2021). The nexus between the forest and human beings is interlinked. Ecosystem services from forests are crucial to human communities. The forest supports in the form of gap filler in times of regular seasonal shortfalls (Paumgarten 2005), prevention from natural calamities, potential income generation improving livelihoods (Angelsen and Wunder 2003), social well-being, enhancing the quality of life (Doimo et al. 2020; Song et al. 2016; Nilsson et al. 2011). The communities near forests face an imminent threat from forest fires, forcing them to evacuate their population, relocate (McCaffrey et al. 2015), and destroy their infrastructure. The people's health is also adversely affected by smoke and ashes from forest fires (Nerfa et al. 2020). On the other hand, the communities near forests experience impacts on their deeply ingrained cultural and spiritual traditions (Byron and Arnold 1999; Rosa and Collado 2019).

## 2.5 *2.2.3. Long-term impacts.*

Forest fires have far-reaching implications for abiotic and biotic factors. The vegetation following the fire in the forest is more vulnerable and consequently easily burnt to the subsequent future flames (Hoffmann et al. 2012; Tepley et al. 2018). Also, the increasing fire rates in the systems result in the rapid depletion of original forest cover (McWethy et al. 2010), and the regeneration of vegetation and the return of animal species may take years or even decades. In terms of permafrost, forest fire has a long-lasting impact on the permafrost; as discussed by (Masyagina 2021), it may take 30 years for it to recover and restore to its pre-fire state. Furthermore, global forest fire emissions represent a source of $CO_2$ emissions that are irreversible on decadal to centennial time scales (Zheng et al. 2021).

It is widely recognized that forest fires have several consequences. Hence, effective prevention, mitigation, and adaptation strategies are required to minimize the damage to ecosystems and communities for future generations.

# 3 Importance of Forest Fire Management

In managing forest fires, it is crucial to have a thorough understanding of how they contribute to the functioning of ecosystems (CIFFC 2017a, b). Fire has played a fundamental role in shaping the development and management of many of the forests in the world (Satendra and Kaushik 2012). Large-scale fires can affect the climate and energy budget regionally due to their albedo alteration and vegetation destruction, which lowers evapotranspiration (Flannigan et al. 2000). Ecosystem services like carbon sequestration and biodiversity, which are essential for climate regulation and ecosystem function, are in danger due to the consequences of climate change on forest fires (Hurteau et al. 2014; Kumar et al. 2020a, 2021a). With the increased frequency of occurrence, severity, and burnt area, forest fires persist as one of the most urgent disruptions (Mohanty and Mithal 2022) and a significant issue that concerns policymakers, emergency personnel, researchers, and citizens (Miranda et al. 2009).

Each year, millions of hectares of forest cover are lost due to Fire World Wilde (Alkhatib 2014). Six million $km^2$ of forest area has already been lost worldwide, mainly due to forest fires in less than 200 years (Dimopoulou and Giannikos 2002). In 2021 alone, an area equivalent to Portugal (Europe) was damaged and lost due to the forest fire (Carta et al. 2023). The Intergovernmental Panel on Climate Change's

Fifth Assessment Report states that between 2.5 billion and 4.0 billion tons of $CO_2$ are released into the atmosphere each year due to forest fires.

The rapid forest loss due to fires may intensify the magnitude and effect of socio-ecological implications, such as a lack of fuel wood supplies, population displacement, and reduced ecosystem services (Han et al. 2020), which are essential resources for both biodiversity and human well-being (Wang et al. 2021). Comprehensive forest fire management measures are crucial in mitigating the effects of disaster and effectively addressing this issue (Dhall et al. 2020; Carta et al. 2023). In general, the programs related to fire management encounter a common challenge in different regions of the world, i.e., the need to effectively address and mitigate the increasing occurrence and impact of forest fires (Oliveira et al. 2021; Kumar et al. 2020b, 2021b).

## 4 Controlling Technologies and Interventions

Forest fires are highly intricate and lack a structured format, thus emphasizing the necessity of utilizing diverse information sources from various fields when combating such incidents (Dhall et al. 2020). Researchers persist in enhancing forest fire management through scientific and technical means, employing data analytics, advanced modelling, and simulation technologies within fire control strategies (Raftoyannis et al. 2014). These efforts improve wildfire detection, monitoring, suppression, and mitigation of ecological disasters (Kalabokidis et al. 2008).

Numerous authorities in countries like Canada, Switzerland, and South Africa, among others, have adopted a range of methods in detection and monitoring systems (Mathews et al. 2010). These systems encompass components such as patrols or monitoring towers with human observers, satellite and aerial observing, progressively more favoured methods for detecting and monitoring utilizing optical sensor cameras, and diverse sensors for detecting, used individually or in combination (Alkhatib 2014). Technological innovation is rapidly progressing through remote sensing approaches and the application of modern technologies like Internet of Things (IoT), robotics, and drones (Wang et al. 2015), particularly in the realm of detecting and monitoring forest fires (Chisholm et al. 2013).

Furthermore, diverse approaches to forest fire management are underway. National/local strategies such as Integrated Fire Management (Schmidt et al. 2018), Community-Based Forest/Fire Management (CBFiM) (Ganz et al. 2007), Indigenous Forest Management (Nikolakis et al. 2020), Joint Forest Management (JFM), among others, aim to engage various stakeholders in forest protection and conservation systematically, thereby aiding in planning and addressing forest fires. Additionally, initiatives for transboundary cooperation and disaster response networks for forest fire management have been launched globally. Moreover, there is a surge in awareness programs utilizing technologies such as the 'Fire Near Me App,' which provides bushfire information from the fire departments in Australian states and territories, as well as New Zealand (NDMA 2020). These advancements and interventions in forest fire management are further elaborated concerning global technological innovations.

## *4.1 4.1. Controlling Technologies for Forest Fire Management*

Globally, the effect of climate change will continue, causing an escalation in wildfire season, dry thunderstorms, increasing vulnerability to forest fires, and challenges associated with fire prevention. (Reneke et al. 2015). To reduce the significant losses from forest fires, using technologies to control these fires can be a crucial step to combat further disasters. Controlling technologies focuses mainly on the technological tools used in fire management. It has made significant progress in most countries, from traditional detection by humans to the use of advanced technologies like aerial firefighting and satellite-based remote sensing technologies to detect, monitor, and manage forest fires. However, researchers are exploring using advanced technologies to increase efficiency and efficacy (Battistoni et al. 2023).

The primary forest fire detection technique was manual human observation from lookout towers at the top of hills or mountains (Georgiev 2023). It was only a primitive tool such as the Osborne fire finder, which was used for locating fires, mainly by utilizing intersections between two towers (Guth et al. 2005). Further, the ground-based system was improved by introducing telephones and radios, allowing quick communication of fire sightings from lookout towers to fire departments. This prime tool for fire detection was human-based, and because human observation is unpredictable and challenging living environments, these fundamental strategies were found to be ineffective (Barmpoutis et al. 2020; Battistoni et al. 2023).

The utilization of aerial fighting for the patrolling and surveillance of forest fires was one of the significant developments using motor vehicles and aircraft. Using aircraft like planes and helicopters allowed a more comprehensive perspective and expedited survey of vast areas, particularly those with challenging ground access. This was more effective than tower lookouts alone. Nevertheless, it was limited by the fuel restrictions, meteorological conditions, and the availability of skilled/experienced pilots. Eventually, satellites were acknowledged for their ability to detect forest fires, initially intended for weather forecasting and environmental monitoring. One of the most notable advantages of using satellites is the extensive coverage, allowing monitoring disproportionately large and inaccessible ground-based methods. They offer regular monitoring, essential for early fire detection and quick action. Moreover, satellites collect valuable information about fire behavior, spread, and impact, which is helpful for research and improving future prevention strategies (Maffei and Menenti 2019).

### 4.1.1 Emerging Technologies for Forest Fire Control

**Remote Sensing**

Remote sensing supplies worldwide data, systematically monitoring the globe's surface at various spatial resolutions and spectral regions. Since the early 1970s, remote sensing has been extensively used in the forest fire sector (Chuvieco et al. 2020). Various sensor technologies, such as aerial photography and satellite imaging, are employed in the process, such as LiDAR, MODIS, etc., to identify and track wildfires across vast geographic areas. As discussed by (Dhall et al. 2020), remote sensing approaches have been grouped into four categories: (a) Ground-based systems, (b) Manned aerial vehicle-based systems, (c) Satellite-based systems, and (d) Unmanned Aerial Vehicle (UAV) based systems. The ground and manned aerial vehicles-based systems are operated by humans on the ground, which is threatening with hazardous environment and operator fatigue for pilots in aerial vehicles (Singhal et al. 2018). Many satellites, like NASA, TERRA, AQUA, and others, track frequent changes in the weather and distribution of vegetation, which is potentially valuable for predicting intense fire conditions (Alkhatib et al. 2023). Uncrewed aerial vehicle (UAV) technology, such as drones with remote sensing systems, has shown a promising response as a quick, portable, and affordable solution (Yuan et al. 2017) for all stages of fire management that comprise assessing fire hazard conditions before the Fire, identifying active fires, estimating their behaviors during the Fire, and analyzing the consequences of the fire and vegetation recovery post the Fire (Chuvieco 2009).

**Internet of Things (IoT)**

It is a paradigm in detecting fires beneficial for managing forests and preventing forest fires, providing real-time monitoring capabilities for different forest conditions, such as wind speed, humidity, and temperature (Zanella et al. 2014); (Battistoni et al. 2023). It enables direct communication and integrates sensors and devices without human involvement (Gaitan and Hojbota 2020) and wireless communication networks. In forest fire systems, the Internet of Things (IoT) facilitates forest fire systems operating 24/7, depending on deployed sensors throughout the forest to provide broad coverage (Battistoni et al. 2023). This is an effective technique for fire detection, which reduces the cost and time of human resources and also saves lives and property damage in the event of early fire detection (Benzekri et al. 2020). However, a significant challenge arises in communicating between sensors and receivers for data exchange due to the expansive forest areas far from urban networks' reach. Thus, based on the LoRA module, the network of wireless sensors is referred as WSN (Wireless Sensors Network). LoRa (LongRange) is a wireless communication technology suitable for large geographic areas that connects multiple final devices (like the placed sensor in the forest) with low power consumption (Benzekri et al., 2020) and an effective low transmission power and low data transfer rate solution for long-distance. (Gaitan and Hojbota 2020). Nevertheless, IoT with LoRa is limited to fire detection and cannot directly extinguish Fire.

Taking this void in, the new system has been created as a Cyber-Physical System (CPS) (Reneke et al. 2015). It is a system used in wildfire detection using IoT and Robot Operating System (ROS) created UA/GVs (Un-manned Aerial/Ground Vehicles equipped with fire extinguishers like water, foams/gels) for fire extinguishing (Battistoni et al. 2023) CPS is based on the heterogeneous devices and software components of IoT, robotics and drone technologies which in combination is cited as the Internet of Robotics and Drone Things (IoRT) (Singh et al. 2021).

**Artificial Intelligence (AI) and Machine-Learning**

Utilizing machine learning techniques has increased in the last several years and enabled researchers to generate precise and effective forest-fire prediction models. In data analysis and computing, there has been a significant and rapid growth of machine learning and AI, enabling applications to operate intelligently and leveraging the power of these technologies. (Alkhatib et al. 2023). Machine learning and AI can manage vast data/information and complex models, considering weather conditions, vegetation, geography, and human activity (Shi and Zhang 2023). This information can help evaluate the efficacy of firefighting tactics and guide prevention efforts. It can enhance forest fire resilience by utilizing huge data and cloud computing platforms to handle massive and complex datasets (Huot et al. 2022). Google Earth Engine (GEE) is another promising solution that offers access to available remote sensing data with various algorithms for classifying, improving images, and cloud masking, simplifying data processing and visualization tasks. (Amani et al. 2019).

## 4.2 *4.2. Interventions for Effective Forest Fire Management*

Intervention methods in forest fire management encompass the various strategies and techniques employed to intervene and actively mitigate the impact of fires. These methods can include tactics like controlled burns, firebreak construction, fire suppression methods, and strategies with enhancing policies, preparedness programs, etc.

**Fire Suppression Methods**

Research on fire suppression has prioritized creating and testing novel, safe, and effective suppressants. Water spraying is a conventional technique commonly used to control and extinguish flames (Grant et al. 2000). It works by lowering the temperature of combustible materials below their ignition point. (Bordado and Gomes 2007). Aerosols have also been used as a fire suppressant. In the combustion zone, aerosol particles containing even low volatile chemicals evaporate as vapor or destruction products at the flame front. However, it could be more practical due to its low reliability and high device cost. (Korobeinichev et al. 2012).

Superabsorbent polymers have also been used in Fire fighting since the 1960s. There has been better improvement in the development of high-class polymers in fire-fighting. The polymers are superabsorbent polyacrylate polymers with mild cross-linking that can absorb up to 100 times their weight in water. A mixed gel product, a fire retardant and suppressant of Class A (or Class 1) is created when these polymers are added to water. (Bordado and Gomes 2007). Fire-fighting polymers can be applied as a firebreak, direct suppression, or structural protection by ground engines, helicopters, or fixed-wing aircraft.

**Controlled Burns and Prescribed Fires**

Controlled burns and prescribed fires are essential tools in forest fire management. Controlled burning has been recognized as a valuable method for controlling huge forest areas in situations of limited financial resources, as it is practicable and economically beneficial for land management. Similarly, prescribed Fire is frequently employed for environmental and land management (Ning and Sun 2017). Both controlled burns and prescribed fires decrease the risk of wildfires and provide habitat for wildlife and grazing. (Fernandes et al. 2013).

This method has a significant non-ecological effect, restrictions, and numerous ecological benefits, which can be used in land management and as a restoration tool (Hmielowski et al. 2016). Although controlled burning and prescribed Fire can be excellent forest fire management tools, it is not yet entirely accepted worldwide (Francos and Úbeda 2021) due to the potential consequences for the range of ecosystems posed by wildfires. Thus, an integrated strategy that includes effective fire detection and suppression techniques, appropriate stand and fuel management strategies, and fire prevention must be implemented to mitigate the undesirable impacts (Fernandes and Botelho 2003).

**Community Preparedness and Awareness**

The rise in forest fires due to climate change partly stems from a lack of community awareness (Mohanty and Mithal 2022). Involving local communities in fire management is crucial for mitigating wildfires (Jurvelius 2004), enhancing their resilience, and improving emergency response efforts (Prior and Eriksen 2013). Effective fire management within the climate change context necessitates considering social, economic, and political factors alongside scientific knowledge (Le Goff et al. 2005). Historically, local communities and authorities collaborated to reduce forest fires, but the marginalization of local people led to knowledge gaps for future generations (Delhi 2020). Recognizing this, authorities have increasingly involved communities in forest management to improve sustainable forest use and reduce poverty, including managing forest disasters such as fires (Hajjar et al. 2016). Australia's fire management tools, which are market-based and feature objectives for carbon cycling and greenhouse gas emission reduction, are praised for their gradual integration of Indigenous fire expertise. (Fitzsimons et al. 2012; Mistry et al. 2016).

The concept of community-based fire management (CBFiM) is highlighted by (Ganz et al. 2007), emphasizing the need for community ownership and involvement in the planning process. It dwells on the importance of institutional support and partnerships at the community level for effective wildfire management. Similarly, the program launched in Brazil in 2014, Integrated Fire Management (IFM), incorporates ecological data, regional customs, and available management alternatives to determine the landscape mosaics with varied fire history to preserve biodiversity, lessen wildfires in the late dry season, safeguard vegetation that is vulnerable to fire, and lessen hostilities between locals and managers of protected areas (Delhi 2020) Various national and international forest management strategies, such as the Sustainable Forest Management recognized by the UN General Assembly in 2007, are based on not only principles to preserve and improve all types of forests' ecological, social, and economic principles that will benefit both the present and future generation but also its purposes to reduce the threats and risk of Fire and other environmental hazards. (Martynova et al. 2021; Torres-Rojo et al. 2016). Also, expediting public education and awareness campaigns on fire prevention can reduce the fire spread rate and safeguard homes, lives, and property before and after a forest fire (Delhi 2020).

# 5 Challenges and Limitations

The rising frequency and intensity of wildfires, fuelled by climate change, pose significant challenges for forest fire management. Despite advanced technologies' effectiveness being hindered by limitations, particularly in delivering timely and accurate details to ground control centers and operators in the change of drone networks (Battistoni et al. 2023), Unmanned aerial vehicles, relying on machine learning, face hurdles such as high costs and the demand for skilled human resources (Mahmudnia et al. 2022; Oliveira et al. 2021). Similarly, satellite and remote sensing systems encounter long scan times and limited flexibility, hampering early wildfire identification due to cloud cover and orbital constraints (Dhall et al. 2020; Georgiev 2023). Additionally, there is a high risk of false alarms, with MODIS data often unable to differentiate between various types of fires, including crop and waste burning and forest fires (Alkhatib 2014; Mohanty and Mithal 2022).

Efficient forest fire management necessitates collaboration among diverse sectors and stakeholders across different levels and jurisdictions (Schultz et al. 2019). However, the effectiveness of such cooperation is hindered by differing priorities among agencies and jurisdictions, as well as a limited understanding of resource utilization and suppression efficacy (Belval et al. 2020). Resource constraints are a common hurdle many countries face, limiting their firefighting capabilities and exacerbating the challenge (Tymstra et al. 2020). Moreover, reliance on highly skilled federal career firefighters on a seasonal basis risks personnel and resource shortages during off-season fires, straining the fire response system (Belval et al. 2020).

Furthermore, neglecting evolving socio-environmental dynamics, disregarding pre-emptive measures, and inadequate involvement of local stakeholders in risk decision-making can lead to overwhelming wildfire scenarios (Tymstra et al. 2020).

## 6 Case Studies and Success Stories of Forest Fire Management in the World

The best practices to govern populations at risk of wildfires have become progressively more explored by the fire management and research arenas in the past several years (McCaffrey et al. 2015). Furthermore, advancements in information and communication technologies, especially in forest fire detection systems, are already demonstrating a substantial impact (Bagheri and Hefeeda 2009).

Implementing Integrated Fire Management (IFM) techniques in Brazil in 2014 led to a 40–70% decrease in late dry season fires within protected areas (Mohanty and Mithal 2022). The IFM program successfully encouraged a departure from "zero-fire" policies, establishing protocols that can improve regulatory processes and decision-making (Schmidt et al. 2018). Community-based Fire Management (CBFiM), involving communities in decision-making and ownership of planning processes, is widely practiced globally (Moore et al. 2002). An exemplary case of CBFiM is the California Fire Safe Council, implemented across diverse contexts, from resource-dependent to suburban communities in the urban-wildland interface zone, showcasing practical efforts in uniting stakeholders to address the wildfire threat (Ganz et al. 2007). This underscores the importance of authorities prioritizing constructive partnerships at the community level, involving residents, governments, the private sector, and community institutions to effectively manage wildland fires' social, economic, and ecological risks.

Additionally, disaster response networks and transboundary collaboration are crucial strategies for forest fire management. In response to concerns over significant fires in the Amazon rainforest, seven nations in South America reached a consensus on safeguards for the Amazon River basin, including establishing satellite surveillance and a disaster response network (NDMA 2020). Similarly, the European Union (EU) has made efforts through the 'European Neighbourhood and Partnership Instrument', providing funding for international initiatives involving collaboration and project implementation among partner nations (Zaimes et al. 2016).

Likewise, significant progress has been made globally in automatic forest firefighting technologies, although much of the research is concentrated in the United States and Europe (Yuan et al. 2015). These technologies have supported forest fire detection and suppression. For instance, the regulated use of uncrewed aerial vehicles (drones) for fire service purposes was tested in 2004 in Hungary, evaluating their capability to cover forest fires (Restas 2006). Visible and infrared cameras were installed across the Netherlands in 2011 to evaluate UAVs' ability to identify and localize actual fire scenarios (Van Persie et al. 2011). Similarly, a drone named 'Sap' (in Portuguese) is one of the newest tools in battling extreme wildfires, successfully suppressing fires in their early stages (BBC 2023). However, these systems require significant investment and a skilled workforce. Adopting technologies and programs in forest fire management has significantly supported communities, promoting better practices and reducing wildfire risks.

# 7 Future Directions and Innovation

Forest wildfires present significant risks to biodiversity, human safety, and ecological systems, requiring novel fire control and prevention strategies. Sustainable and operational policies are essential for adopting integrated fire management, requiring reflexive, inclusive, and participatory practices (Fernandes and Panjaitan 2019; Ruane 2018). Moreover, understanding communities' technical capabilities and the need for quick assessment instruments are crucial for controlling fires and enabling regional comparisons (Ganz et al. 2007).

Despite increasing research into suppression tactics, pre-suppression activities, technologies, and prevention of fires, communities still need to be more adequately prepared for complicated fire behaviors (NIDM 2012; Coen et al. 2020). Strengthening disaster response capabilities involves improving real-time data exchange, establishing early warning systems, and enhancing resource-sharing mechanisms (Molina-Pico et al. 2016; Schultz et al. 2019). Furthermore, when updated and streamlined with manager feedback, decision-support models and tools could become more straightforward and easier to use in fire management (Rutherford and Schultz 2019). Additionally, continuous support from legislators and agency leadership, specialized teams, and increased financing are necessary (Schultz et al. 2019). Additionally, efforts to promote transboundary collaboration can be advanced as a beneficial and supportive approach for countries with limited experience and capacity to deal with wildfires (Zaimes et al. 2016).

# 8 Conclusion

Forest fires constitute an increasing threat in most countries, and the impacts of global warming can intensify this problem. It requires a comprehensive approach that combines effective technologies and strategic intervention. This paper explores various aspects of forest fire management, including the importance of early detection, effective communication systems, and the utilization of advanced technologies. Innovative Fire controlling technologies like satellite images, UAVs/Drones, and remote sensing play a crucial role in detecting and monitoring forest fires. These technologies enable rapid response and provide valuable data for decision-making. Intervention strategies such as prescribed burning, firebreaks, and firefighting techniques are essential for containing and extinguishing fires. A proactive approach to forest fire management is vital. This involves implementing preventive methods like fuel management, public education, and fire-adaptive land management practices. Reducing fuel loads and promoting fire-resilient ecosystems can mitigate the risk of severe wildfires. Furthermore, effective collaboration and coordination among stakeholders, including government agencies, fire departments, and local communities, are essential for successful forest fire management. Early warning systems and emergency response plans are required to improve the effectiveness of interventions significantly. However, it is necessary to acknowledge that forest fire management is an intricate and evolving area. Issues like climate change, limited resources, and more intense weather occurrences are emerging more frequently and setting ongoing difficulties. Therefore, continuous research, innovation, and adaptation of strategies are essential to address these challenges effectively.

## References


Alkhatib AAA (2014) A review on forest fire detection techniques. Int J Distrib Sens Netw 10(3):597368. Taylor and Francis Inc. https://doi.org/10.1155/2014/597368

Alkhatib R, Sahwan W, Alkhatieb A, Schütt B (2023) A brief review of machine learning algorithms in forest fires science. Appl Sci Switz Multi Digit Publ Inst (MDPI). 13, 14

Amani M, Brisco B, Afshar M, Mirmazloumi SM, Mahdavi S, Mirzadeh SMJ, Huang W, Granger J (2019) A generalized supervised classification scheme to produce provincial wetland inventory maps: an application of google earth engine for big geo data processing. Big Earth Data 3(4):378–394. https://doi.org/10.1080/20964471.2019.1690404

American Forest Foundation (2015) Western water threatened by wildfire

Angelsen A, Wunder S (2003) Exploring the forest-poverty link. CIFOR Occasional Paper, vol 40, pp 1–20

Babu KVS, Roy A, Prasad PR (2016) Forest fire risk modelling in Uttarakhand Himalaya using TERRA satellite datasets. Eur J Remote Sens. 49(1):381–395

Bahadur KK, Heyojoo BP, Karna YK, Sharma SP, Panthi S (2017) Incidences of wildfire hazard and its effects on forest cover change in Chitwan National Park Nepal. Clarion Int Multi J 6(1):1–6

Baranovskiy NV, Podorovskiy A, Malinin A (2021) Parallel implementation of the algorithm to compute forest fire impact on infrastructure facilities of JSC Russian railways. Algorithms 14(11). https://doi.org/10.3390/a14110333

Barmpoutis P, Papaioannou P, Dimitropoulos K, Grammalidis N (2020) A review on early forest fire detection systems using optical remote sensing. Sens Switz 20(22):1–26. https://doi.org/10.3390/s20226442

Barwa S, Singh H (2024) Climate-induced changes in forest ecosystems of Indian Himalayan region: a meta-analysis. In: Singh H, (eds) Urban forests, climate change and environmental pollution. Springer, Cham. https://doi.org/10.1007/978-3-031-67837-0_39

Battistoni P, Cantone AA, Martino G, Passamano V, Romano M, Sebillo M, Vitiello G (2023) A cyber-physical system for wildfire detection and firefighting. Future Internet, 15(7). https://doi.org/10.3390/fi15070237

BBC (2023) The Portuguese drone that douses wildfires from above. [Online]. Available from The Portuguese drone that douses wildfires from above—BBC Future. [Accessed 05 February, 2024]

Belval EJ, Short KC, Stonesifer CS, Calkin DE (2022) A historical perspective to inform strategic planning for 2020 end-of-year wildland fire response efforts. Fire 5(2):35

Benzekri W, El Moussati A, Moussaoui O, Berrajaa M (2020) Early forest fire detection system using wireless sensor network and deep learning. IJACSA) Int J Adv Comput Sci Appl, 11, 5. www.ijacsa.thesai.org

Bordado JCM, Gomes JFP (2007) New technologies for effective forest fire fighting. Int J Environ Stud 64(2):243–251. https://doi.org/10.1080/00207230701240578

Bowman DMJS, Balch JK, Artaxo P, Bond WJ, Carlson JM, Cochrane MA, D'Antonio CM, DeFries RS et al (2009) Fire in the earth system. Science 324:481–484

Burrell AL, Sun Q, Baxter R, Kukavskaya EA, Zhila S, Shestakova T, Barrett K (2022) Climate change, fire return intervals, and the growing risk of permanent forest loss in boreal Eurasia. Sci Total Environ 831:154885

Byron N, Arnold M (1999) What future for the people of the tropical forests? World Dev 27(5):789–805

Carta F, Zidda C, Putzu M, Loru D, Anedda M, Giusto D (2023) Advancements in forest fire prevention: a comprehensive survey. Sens Multi Digit Publ Inst (MDPI). 23, 14 https://doi.org/10.3390/s23146635

Celis N, Casallas A, Lopez-Barrera EA, Felician M, De Marchi M, Pappalardo SE (2023) Climate change, forest fires, and territorial dynamics in the amazon rainforest: an integrated analysis for mitigation strategies. ISPRS Int J Geo-Inform, 12(10). https://doi.org/10.3390/ijgi12100436

Chas-Amil ML, Nogueira-Moure E, Prestemon JP, Touza J (2022) Spatial patterns of social vulnerability in relation to wildfire risk and wildland-urban interface presence. Landsc Urban Plan 228:104577

Chisholm RA, Cui J, Shawn K, Lum Y, Chen BM (2013) UAV LiDAR for below—canopy forest surveys. J Unmanned Veh Syst 1:61–68

Chuvieco E, Aguado I, Salas J, García M, Yebra M, Oliva P (2020) Satellite remote sensing contributions to wildland fire science and management. Curr Forest Rep, 6, 2, 81–96. Springer. https://doi.org/10.1007/s40725-020-00116-5

Chuvieco E (2009) Earth observation of wildland fires in mediterranean ecosystems. In: Earth observation of wildland fires in mediterranean ecosystems. Springer Berlin Heidelberg. https://doi.org/10.1007/978-3-642-01754-4

Ciesielski M, Bałazy R, Borkowski B, Szczęsny W, Zasada M, Kaczmarowski J, Kwiatkowski M, Szczygieł R, Milanović S (2022) Contribution of anthropogenic, vegetation, and topographic features to forest fire occurrence in Poland. IForest 15(4):307–314. https://doi.org/10.3832/IFOR4052-015

CIFFC (2017a) Canada Report 2017. Canadian Interagency Forest Fire Centre, Winnipeg, MB

Canadian Interagency Forest Fire Centre—CIFFC (2017b) Canadian Wildland Fire Management Glossary. https://www.ciffc.ca/sites/default/files/2019-03/CIFFC_Canadian_Wildland_Fire_Mgmt_Glossary_2017_10_24.pdf. Accessed 2 February, 2024

Coen JL, Johnson GW, Romsos JS, Saah D (2024) A framework for conducting and communicating probabilistic wildland fire forecasts. Fire 7(7):227

Delhi N (2020) Forest fire management global best practices national disaster management authority

Dhall A, Dhasade A, Nalwade A, Mohan MR, Kulkarni V (2020) A survey on systematic approaches in managing forest fires. In: Applied geography. Elsevier Ltd. Vol 121 https://doi.org/10.1016/j.apgeog.2020.102266

Dimopoulou M, Giannikos I (2002) Towards an integrated framework for forest fire in ERS-1 SAR image intensity associated with forest fires in Alaska. IEEE Trans Geosci Remote Sens 32(1):206–210

Doimo I, Masiero M, Gatto P (2020) Forest and wellbeing: bridging medical and forest research for effective forest-based initiatives. Forests. MDPI AG. 11, 8 https://doi.org/10.3390/F11080791

EPA US (2016) Environmental Protection Agency. DRAFT Inventory of U.S. Greenhouse Gas Emissions and Sinks: 1990–2014, EPA 430-R-16-002

Fernandes PM, Botelho HS (2003) A review of prescribed burning effectiveness in fire hazard reduction. Int J Wildland Fire 12(2):117–128. https://doi.org/10.1071/WF02042

Fernandes AAR, Panjaitan RB (2019) The effect of community and company participation and implementation of good forest fire governance on the forest fire policy in Indonesia. J Sci Technol Policy Manag 10(1):102–115

Fernandes PM, Davies GM, Ascoli D, Fernández C, Moreira F, Rigolot E, Stoof CR, Vega JA, Molina D (2013) Prescribed burning in southern Europe: Developing fire management in a dynamic landscape. Front Ecol Environ. 11, 1 https://doi.org/10.1890/120298

Fitzsimons J, Russell-Smith J, James G, Vigilante T, Lipsett-Moore G, Morrison J, Looker M (2012) Insights into the biodiversity and social benchmarking components of the Northern Australian fire management and carbon abatement programmes. Ecol Manag Restor 13(1):51–57

Flannigan MD, Amiro BD, Logan KA, Stocks BJ, Wotton BM (2006) Forest fires and climate change in the 21ST century. Mitig Adapt Strat Glob Change 11(4):847–859. https://doi.org/10.1007/s11027-005-9020-7

Flannigan MD, Stocks BJ, Wotton BM (2000) Climate change and forest fires. Sci Total Environ 262

Food and Agriculture Organization of the United Nations and Project Firefight South East Asia. Yuan C, Zhang Y, Liu Z (2015) A survey on technologies for automatic forest fire monitoring, detection, and fighting using unmanned aerial vehicles and remote sensing techniques. Can J for Res 45(7):783–792

Francos M, Úbeda X (2021) Prescribed fire management. In: Current opinion in environmental science and health. Elsevier B.V. vol 21. https://doi.org/10.1016/j.coesh.2021.100250

Gaitan NC, Hojbota P (2020) Forest fire detection system using lora technology. Int J Adv Comput Sci Appl 11(5):18–21. https://doi.org/10.14569/IJACSA.2020.0110503

Ganz D, Troy A, Saah D (2007) Chapter 9 community involvement in wildfire hazard mitigation and management: community based fire management, fire safe councils and community wildfire protection plans. In: Advances in the economics of environmental resources. vol 6, pp 143–164. https://doi.org/10.1016/S1569-3740(06)06009-3

Georgiev AG (2023) An evaluation of fire detection methods: comparative analysis and performance assessment 16. In: Proceedings of University of RUSE-2023. vol 62

Goldammer JG, Fire W (2007) Forest fires – a global perspective. Global Fire Monitoring Center

Grant G, Brenton J, Drysdale D (2000) Fire suppression by water sprays. www.elsevier.com/locate/pecs

Guth PL, Craven T, Chester T, O'Leary Z, Shotwell J (2005) Fire location from a single osborne-firefinder and a dem. In ASPRS 2005 Annual conference geospatial goes global: from your neighborhood to the whole planet

Han SK, Han HS (2020) Productivity and cost of whole-tree and tree-length harvesting in fuel reduction thinning treatments using cable yarding systems. Forest Sci Technol 16(1):41–48

Hajjar R, Oldekop JA, Cronkleton P, Etue E, Newton P, Russel AJM, Agrawal A (2016) The data was not collected on community forestry. Conserv Biol 30:1357–1362

Hefeeda M, Bagheri M (2009) Forest fire modeling and early detection using wireless sensor networks. Ad Hoc Sens Wirel Netw 7(3–4):169–224

Hirschberger P (2016) Forests ablaze: causes and effects of global forest fires. WWF: Berlin, Germany

Hmielowski TL, Carter SK, Spaul H, Helmers D, Radeloff VC, Zedler P (2016) Prioritizing land management efforts at a landscape scale: a case study using prescribed fire in Wisconsin. Ecol Appl 26(4):1018–1029

Hoffmann WA, Jaconis SY, Mckinley KL, Geiger EL, Gotsch SG, Franco AC (2012) Fuels or microclimate? Understanding the drivers of fire feedbacks at savanna–forest boundaries. Austral Ecol 37(6):634–643

Huot F, Hu RL, Goyal N, Sankar T, Ihme M, Chen YF (2022) Next day wildfire spread: a machine learning dataset to predict wildfire spreading from remote-sensing data. IEEE Trans Geosci Remote Sens. https://doi.org/10.1109/TGRS.2022.3192974

Hurteau MD, Bradford JB, Fulé PZ, Taylor AH, Martin KL (2014) Climate change, fire management, and ecological services in the southwestern US. For Ecol Manag 327:280–289. https://doi.org/10.1016/j.foreco.2013.08.007

IEEE Control Systems Society, IEEE Robotics and Automation Society, & Institute of Electrical and Electronics Engineers. (n.d.). In: 2017 International Conference on Unmanned Aircraft Systems (ICUAS'17): final program and book of abstracts, June 13–16, 2017, Miami Marriott Biscayne Bay, Miami, FL 33132

Jolly WM, Cochrane MA, Freeborn PH, Holden ZA, Brown TJ, Williamson GJ et al (2015) Climate-induced variations in global wildfire danger from 1979 to 2013. Nat Commun 6:7537. https://doi.org/10.1038/ncomms8537

Jurvelius M (2004) Community-based fire management in Southern Africa. Unasylva 55:3

Kala CP (2023) Environmental and socioeconomic impacts of forest fires: a call for multilateral cooperation and management interventions. Nat Hazards Res 3(2):286–294. https://doi.org/10.1016/j.nhres.2023.04.003

Kalabokidis K, Iosifides T, Henderson M, More house B (2008) Wildfire policy and use of science in the context of a socio-ecological system on the Aegean Archipelago. Environ Sci Policy. 11: 408-421. https://doi.org/10.1016/j.envsci.2008.01.006

Kant K, Anand S, Sudeep Tanwar N, Abouhawwash M (n.d.) Advances in science, technology & innovation IEREK interdisciplinary series for sustainable development emergence of cyber physical system and IoT in smart automation and robotics computer engineering in automation. http://www.springer.com/series/15883

Kaushik P, Singh H (2024) Climate change-induced forest ecosystem functioning and alteration in water-atmosphere-soil–plant (WASP) Relationship. In: Singh H (eds) Urban forests, climate change and environmental pollution. Springer, Cham. https://doi.org/10.1007/978-3-031-67837-0_38

Korobeinichev OP, Shmakov AG, Shvartsberg VM, Chernov AA, Yakimov SA, Koutsenogii KP, Makarov VI (2012) Fire suppression by low-volatile chemically active fire suppressants using aerosol technology. Fire Saf J 51:102–109. https://doi.org/10.1016/j.firesaf.2012.04.003

Kumar A, Dwivedi GK, Tewari S, Paul J, Anand R, Kumar N, Kumar P, Singh H, Kaushal R (2020) Carbon mineralization and inorganic nitrogen pools under Terminalia Chebula Retz. based agroforestry system in Himalayan foothills India. Forest Sci 66(5):634–643

Kumar A, Kumar P, Singh H, Kumar N (2020) Adaptation and mitigation potential of roadside trees with bio-extraction of heavy metals under vehicular emissions and their impact on physiological traits during seasonal regimes. Urb for Urb Green 58:126900

Kumar A, Kumar P, Singh H, Bisht S, Kumar N (2021) Relationship of physiological plant functional traits with soil carbon stock in temperate forest of Garhwal Himalaya. Curr Sci 120(8):1368–1373

Kumar A, Kumar P, Singh H, Kumar N (2021) Modulation of plant functional traits under essential plant nutrients during seasonal regime in natural forests of Garhwal Himalayas. Plant Soil 465:197–212

Kumar N, Singh H, Kumar A, Singh AK, Pandey PK, Kumar A (2024) Forest-fire-induced land degradation. In: Mishra G, Giri K, Singh S, Kumar M (eds) Sustainable land management in India. Springer Nature, Singapore. https://doi.org/10.1007/978-981-97-5223-2_4

Kunwar RM (2006) Forest fire in the Terai Nepal, causes and community management interventions. Int Forest Fire News No. 34. ISSN1029-0864.

Le Goff H, Leduc A, Bergeron Y, Flannigan M (2005) The adaptive capacity of forest management to changing fire regimes in the boreal forest of Quebec. For Chron 81:582–592

Littell JS, McKenzie D, Peterson DL, Westerling AL (2009) Climate and wildfire area burned in western U.S. ecoprovinces 1916–2003. Ecol Appl 19(4):1003–1021

Liu JC, Mickley LJ, Sulprizio MP, Dominici F, Yue X, Ebisu K, Bell ML (2016) Particulate air pollution from wildfires in the Western US under climate change. Clim Change 138:655–666

Maffei C, Menenti M (2019) Predicting forest fires burned area and rate of spread from pre-fire multispectral satellite measurements. ISPRS J Photogramm Remote Sens 158:263–278. https://doi.org/10.1016/j.isprsjprs.2019.10.013

Mahmudnia D, Arashpour M, Bai Y, Feng H (2022) Drones and blockchain integration to manage forest fires in remote regions. Drones 6(11):331

Martell DL (2001) Forest fire management. In: Forest fires. Academic Press, pp 527–583

Martynova M, Sultanova R, Khanov D, Talipov E, Sazgutdinova R (2021) Forest management based on the principles of multifunctional forest use. J Sustain for 40(1):32–46. https://doi.org/10.1080/10549811.2020.1734025

Masyagina OV (2021) Carbon dioxide emissions and vegetation recovery in fire-affected forest ecosystems of Siberia: recent local estimations. Curr Opin Environ Sci Health. Elsevier B.V. Vol 23 https://doi.org/10.1016/j.coesh.2021.100283

Mathews S, Ellis P, Hurle JH (2010) Evaluation of three systems. Australia: Bushfire Cooperative Research Centre

McCaffrey S, Rhodes A, Stidham M (2015) Wildfire evacuation and its alternatives: perspectives from four United States' communities. Int J Wildland Fire 24(2):170–178. https://doi.org/10.1071/WF13050

McWethy DB, Whitlock C, Wilmshurst JM, McGlone MS, Fromont M, Li X, Dieffenbacher-Krall A, Hobbs WO, Fritz SC, Cook ER (2010) Rapid landscape transformation in South Island, New Zealand, following initial Polynesian settlement. Proc Natl Acad Sci USA 107:21343–21348

Miranda AI, Marchi E, Ferretti M, Millán MM (2009). Forest fires and air quality issues in southern Europe. In: Developments in environmental science Krupa SV, Bytnerowicz A, Arbaugh MJ, Riebau AR, Andersen C (eds). Wild land fires and air pollution. Elsevier, The Netherlands, vol 8, pp 209–231

Mishra B, Panthi S, Poudel S, Ghimire BR (2023) Forest fire pattern and vulnerability mapping using deep learning in Nepal. Fire Ecol, 19(1). https://doi.org/10.1186/s42408-022-00162-3

Mistry J, Bilbao BA, Berardi A (2016) Community owned solutions for fire management in tropical ecosystems: case studies from Indigenous communities of South America. In: Philosophical transactions of the royal society b: biological sciences. Royal Society of London. Vol 371, Issue 1696 https://doi.org/10.1098/rstb.2015.0174

Mohanty A, Mithal V (2022) Managing forest fires in a changing climate. Council on Energy, Environment and Water, New Delhi, India

Molina-Pico A, Cuesta-Frau D, Araujo A, Alejandre J, Rozas A (2016) Forest monitoring and wildland early fire detection by a hierarchical wireless sensor network. J Sens 2016(1):8325845

Moore PF (2019) Global wildland fire management research needs. Curr For Rep 5:210–225. https://doi.org/10.1007/s40725-019-00099-y

Moore P, Ganz D, Tan LC, Enters T, Durst PB (2002) Communities in flames: proceedings of an international conference on community involvement in fire management

NDMA (2020) Forest fire management global best practices national disaster management authority

Nerfa L, Rhemtulla JM, Zerriffi H (2020) Forest dependence is more than forest income: development of a new index of forest product collection and livelihood resources. World Dev. https://doi.org/10.1016/j.worlddev.2019.104689

Nikolakis WD, Roberts E (2020) Indigenous fire management: a conceptual model from literature. Ecol Soc 25(4)

Nilsson K, Sangster M, Konijnendijk CC (2011) Forests, trees and human health and well-being: Introduction. Springer, Netherlands, pp 1–19

Ning Z, Sun C (2017) Forest management with wildfire risk, prescribed burning and diverse carbon policies. Forest Policy Econ 75:95–102. https://doi.org/10.1016/j.forpol.2016.10.004

Oliveira AS, Soares-Filho BS, Oliveira U, Van der Hoff R, Carvalho-Ribeiro SM, Oliveira AR, Scheepers LC, Vargas BA, Rajão RG (2021) Costs and effectiveness of public and private fire management programs in the Brazilian Amazon and Cerrado. Forest Policy Econ. https://doi.org/10.1016/j.forpol.2021.102447

Paudel J (2021) Beyond the blaze: the impact of forest fires on energy poverty. Energy Econ 101:105388

Paumgarten F (2005) The role of non-timber forest products as safety-nets: a review of evidence with a focus on South Africa. GeoJournal 64(3):189–197

Prior T, Eriksen C (2013) Wildfire preparedness, community cohesion and social-ecological systems. Glob Environ Change Hum Policy Dimen 23:1575–1586

Priority Areas for Prescribed Fire Use (2016). www.landfire.gov

Raftoyannis Y, Nocentini S, Marchi E, Calama Sainz R, Garcia Guemes C, Pilas I, …, Lindner M (2014) Perceptions of forest experts on climate change and fire management in European Mediterranean forests. IForest 7:33–41

Reneke P, Grant C, Bryner NP, Jones AW, Koepke GH (2015) Res Roadmap Smart Fire Fighting. https://doi.org/10.6028/NIST.SP.1191

Restas A (2006) Wildfire management supported by uav based air reconnaissance: Experiments and results at the szendro fire department, hungary. First Int Workshop Fire Manag

Rogers BM, Balch JK, Goetz SJ, Lehmann CE, Turetsky M (2020) Focus on changing fire regimes: interactions with climate, ecosystems, and society. Environ Res Lett 15(3):030201

Rosa CD, Collado S (2019) Experiences in nature and environmental attitudes and behaviors: Setting the ground for future research. Front Psychol Front Med S.A. Vol 10, Issue APR https://doi.org/10.3389/fpsyg.2019.00763

Ruane S (2018) Using a worldview lens to examine complex policy issues: a historical review of bushfire management in the South West of Australia. Local Environ 23(8):777–795

Rutherford TK, Schultz CA (2019) Adapting wildland fire governance to climate change in Alaska. Ecol Soc, 24(1). https://doi.org/10.5751/ES-10810-240127

Satendra A, Ashutosh Kaushik ID (n.d.) Forest fire disaster management. www.nidm.gov.in

Schmidt IB, Moura LC, Ferreira MC, Eloy L, Sampaio AB, Dias PA, Berlinck CN (2018) Fire management in the Brazilian savanna: first steps and the way forward. J Appl Ecol 55(5):2094–2101

Schultz CA, Moseley C (2019) Collaborations and capacities to transform fire management. Science 366(6461):38–40

Sett T, Nikam BR, Singh H, Purohit S (2024) Hydrological variability in Indian forest ecosystem: analysis of drought resilience, recovery and water use efficiency in moist and dry deciduous forests. In: Singh H (eds) Urban forests, climate change and environmental pollution. Springer, Cham. https://doi.org/10.1007/978-3-031-67837-0_37

Shi C, Zhang F (2023) A forest fire susceptibility modeling approach based on integration machine learning algorithm. Forests, 14(7). https://doi.org/10.3390/f14071506

Singh AK, Rafeek SM, Harikrishnan PS, Wilson I (2021) Review of study on various forest fire detection techniques using iot and sensor networks. In: International conference on advances and innovations in recycling engineering. Springer Nature Singapore, Singapore, pp 29–37

Singh H, Kumar N, Kumar A (2024) Enhancing resource use efficiency in crops through plant functional traits published by Springer, pp 97–117. ISBN: 978-981-97-1509-1

Singh H (2024a). Climate change and forests: biological perspectives. In: Singh H (eds) Forests and climate change. Springer Nature, Singapore. https://doi.org/10.1007/978-981-97-3905-9_1

Singh H (2024c). Climate change and environmental pollution: biological consequences on urban forests. In: Singh H (eds) Urban forests, climate change and environmental pollution. Springer Nature, Cham, Switzerland. https://doi.org/10.1007/978-3-031-67837-0_1

Singh H (2024b) Forests and climate change: biological perspectives on impact, adaptation, and mitigation strategies, Springer Nature Singapore. https://doi.org/10.1007/978-981-97-3905-9

Singh H (2024d) Urban forests, climate change and environmental pollution: physio-biochemical and molecular perspectives to enhance urban resilience, Springer Nature, Cham, Switzerland. https://doi.org/10.1007/978-3-031-67837-0

Singhal G, Bansod B, Mathew L (2018) Unmanned aerial vehicle classification, applications and challenges: a review

Song C, Ikei H, Miyazaki Y (2016) Physiological effects of nature therapy: a review of the research in Japan. Int J Environ Res Public Health 13(8):781

Sudhakar S, Chakraborty K, Chabukdhara M, Mondal PP (2014) Forest fire assessment in northeast India. North Eastern Space Applications Center, Department of Space, Government of India, India

Tepley AJ, Veblen TT, Perry GLW, Stewart GH, Naficy CE (2016) Positive feedbacks to fire-driven deforestation following human colonization of the South Island of New Zealand. Ecosystems 19:1325–1344. https://doi.org/10.1007/s10021-016-0008-9

Tepley AJ, Thomann E, Veblen TT, Perry GLW, Holz A, Paritsis J, Kitzberger T, Anderson-Teixeira KJ (2018) Influences of fire–vegetation feedbacks and post-fire recovery rates on forest landscape vulnerability to altered fire regimes. J Ecol 106(5):1925–1940. https://doi.org/10.1111/1365-2745.12950

Torres-Rojo JM, Moreno-Sánchez R, Mendoza-Briseño MA (2016) Sustainable forest management in Mexico. In: Current forestry reports. Springer International Publishing. Vol 2, Issue 2, pp 93–105 https://doi.org/10.1007/s40725-016-0033-0
Tymstra C, Stocks BJ, Cai X, Flannigan MD (2020) Wildfire management in Canada: review, challenges and opportunities. Prog Disaster Sci 5:100045
Tyukavina A, Potapov P, Hansen MC, Pickens AH, Stehman SV, Turubanova S, Parker D, Zalles V, Lima A, Kommareddy I, Song X-P, Wang L, Harris N (2022) Global trends of forest loss due to fire from 2001 to 2019. Front Remote Sens. https://doi.org/10.3389/frsen.2022.825190
Valderrama L, Contreras-Reyes JE, Carrasco R (2018) Ecological impact of forest fires and subsequent restoration in Chile. Resources. MDPI AG. 7, 2 https://doi.org/10.3390/resources7020026
van Persie M, Oostdijk A, Fix J, van Sijl MC, Edgardh L (2012) Real-time UAV based geospatial video integrated into the fire brigades crisis management GIS system. Int Arch Photogramm Remote Sens Spat Inf Sci 38:173–175
Wang SW, Lim CH, Lee WK (2021) A review of forest fire and policy response for resilient adaptation under changing climate in the Eastern Himalayan region. For Sci Technol 17(4):180–188. https://doi.org/10.1080/21580103.2021.1979108
Wang H, Liu J, Han J (2015) RS-CPS: a distributed architecture of robotic surveillance cyber-physical system in the nature environment. In: 2015 IEEE International conference on cyber technology in automation, control, and intelligent systems (CYBER). IEEE. pp 1287–1292
Yuan C, Liu Z, Zhang Y (2017) Fire detection using infrared images for UAV-based forest fire surveillance. In: 2017 International conference on unmanned aircraft systems (ICUAS). IEEE, pp 567–572
Zaimes GN, Tufekcioglu M, Tufekcioglu A, Zibtsev S, Corobov R, Emmanouloudis DA, Uratu R, Ghulijanyan A, Borsuk A, Trombitsky I (2016) Transboundary collaborations to enhance wildfire suppression in protected areas of the Black Sea Region. J Eng Sci Technol Rev 9(2):108–114
Zanella A, Bui N, Castellani A, Vangelista L, Zorzi M (2014) Internet of things for smart cities. IEEE Internet Things J 1(1):22–32. https://doi.org/10.1109/JIOT.2014.2306328
Zheng B, Ciais P, Chevallier F, Chuvieco E, Chen Y, Yang H (2021) Increasing forest fire emissions despite the decline in global burned area. Sci Adv. 7. https://www.science.org

# Fire-Adapted Tree Species: Characteristics for Thriving in the Face of Climate-Driven Fires

**Nada Tadi, Mudang Yampi, and Tajum Doni**

**Abstract** One of the major disturbances causing phenomena in the forests is forest fire, which significantly influences the composition and structure of vegetation within a given area. Forest fires can cause severe damage to trees, leading to their mortality due to burning and exposure to intense heat. Tree species growing in places where forest fires are natural and recurring phenomena have evolved certain characteristics for their survival and resilience. These characteristics allow them to survive, regenerate, and benefit from fire disturbances. Typical features of fire-adapted tree species include characteristics like dense bark to provide protection from high temperature and damages, branches with self-pruning ability to inhibit the spreading of fire to the crown, deep root systems to bolster drought resilience, which may follow after forest fire, serotinous cones or seeds that get opened or released in response to high temperature during a forest fire, ability to resprout from dormant buds, etc. Forest fires are becoming more frequent and intense worldwide as climate patterns change. Thus, understanding the attributes of fire-adapted tree species is crucial for effective forest fire management and ecosystem restoration, enhancing the survival and resilience of forests.

**Keywords** Thick bark · Self-pruning · Deep root · Serotiny · Fire resistant foliage · Resprouting

---

N. Tadi (✉) · M. Yampi
Silviculture and Forest Management Division, ICFRE-Rain Forest Research Institute, Jorhat, Assam, India
e-mail: nadatadi05@gmail.com

T. Doni
Forest Ecology and Climate Change Division, ICFRE-Rain Forest Research Institute, Jorhat, Assam, India

 
H. Singh and N. Kumar (eds.), *Forest Fire and Climate Change*,
https://doi.org/10.1007/978-3-031-89967-6_12

# 1 Introduction

Forest fires represent a fundamental natural occurrence that has persisted globally since prehistoric times, dating back to the period approximately 443.7–416 million years ago, known as the Silurian period (Glasspool et al. 2004). Forest fires play a crucial part in ecosystem dynamics, triggering shifts in structure and function and impacting biodiversity (Savita et al. 2017; Kumar et al. 2024). The transformative effects of fires extend to both biotic and abiotic components of ecosystems, altering forests' productive and protective functions (Singh 2024a). Over prolonged periods, specific flora and fauna adapt to fire-prone environments. Fire regimes heavily influence this adaptation, encompassing long-term patterns of fire occurrence and the immediate consequences of each fire event on the surrounding ecosystem (Sett et al. 2024; Kaushik and Singh 2024). Fire regimes are shaped by three primary factors: the frequency of fires, the intensity of their burning, and the quantity of fuel consumed. Changes in fire regimes exert a profound influence on the environment, far surpassing the impact of isolated fire incidents, as emphasized by Smith (1995; Barwa and Singh 2024).

A forest fire, also known as a wildfire, denotes any fire occurring within a forest region that has not been deliberately initiated for management or protection purposes according to an approved plan. The inherent resilience of the flora and fauna, shaped by fire-adaptive traits, typically mitigates the destructive impact of naturally transpiring forest fires on the landscape (Singh 2024b). However, human intervention disturbs variables such as the fire's timing, intensity, and fuel composition, exacerbating its adverse consequences.

The plant and animal life inhabiting environments susceptible to fires have developed specific characteristics over millions of years that enable them to endure recurrent fire events within their surroundings (Singh 2024c). These fire-adapted traits have independently evolved across various plant groups, including angiosperms and gymnosperms (Verdúet al. 2007). The species adaptation to fire is intricately linked to how frequent fires occur and how intense the fires are within their habitat (Bellingham and Sparrow 2000), and is influenced by their evolutionary past (Pausas and Keeley 2009).

In addition to knowing the traits of the species, the recognition of how they interact with their environment over time and how they respond to frequency, intensity, duration of forest fire, and other aspects of fire regimes provides a more comprehensive understanding of how species adapt to fire-prone habitats and the role of their adaptive traits in ensuring their survival and persistence amidst fire disturbances (Singh 2024c, d). Variability in the severity and distribution of fires across space and time can profoundly affect the structure and composition of species in ecosystems after fire events, while influencing the potential for such disturbances to occur. Individual plants may experience multiple fire occurrences within a fire regime, each characterized by distinct fire properties and effects. This recurrent exposure has resulted in the development of various traits that help many plant species to resist and survive the fire. Such traits encompass multiple mechanisms, including the ability to sprout again from both above and below-ground tissues after the fire has occurred, the development of bark that is thick enough to withstand high temperatures, the ability to shed lower branches to inhibit the spread of fire to upper branches, and presence of serotiny and germination due to the exposure of seeds to smoke (Agee 1993; Keeley and Zedler 1998; Keeley 2012). Studies have shown that these adaptive traits often coalesce into ecologically significant strategies or syndromes (Rowe 1983; McCune 1988; Schwilk and Ackerly 2001). Furthermore, there is evidence to suggest that these fire-adapted traits are passed down within families, underscoring their significance in the persistence and adaptation of species in environments where incidences of fires are quite frequent (Verdú and Pausas 2007; Pausas et al. 2006; Singh et al. 2024; Malik et al. 2024).

Traditionally, fire has been viewed as an external disruptor in an otherwise stable ecosystem. Many studies have depicted fire as an unwanted event, even though there is growing acknowledgment of its ecological importance, and it is increasingly viewed as an internal disturbance within a generally stable system (Lyon and Stickney 1976). Thus, instead of perceiving fire solely as an external threat, its role as a natural process that contributes to the dynamics and resilience of ecosystems should be recognized. By acknowledging fire as an internal perturbation, its integral role in shaping ecological patterns and processes can be better understood.

The climate of the region influences the occurrence of forest fires. As greenhouse gas emissions increase, researchers have predicted that the change in the global climate (IPCC 2013; Singh and Kumar 2022; Singh et al. 2023) will affect forest fire activities. Studies have shown that the frequency of forest fires, intensity, and period of forest fires can be affected by climate change (Liu et al. 2010). For example, if spring and summer temperatures are higher than usual, the period of the fire season will be longer. Moreover, heightened temperatures and reduced moisture levels will render trees more vulnerable to disease and infestations by pests, thereby escalating tree mortality and the risk of fires (Westerling et al. 2006). Hence, a better understanding of the characteristics of fire-adapted tree species can help develop strategies for wildfire management and ecosystem restoration after fire, which is influenced by climate change.

## 2 Traits and Characteristics of Fire-Adapted Tree Species

Fire-adapted tree species often exhibit a combination of characteristics that work together to reduce the overall damage by fire and ensure the population's survival. These traits include adaptations in foliage, deep root system, Seed bank, bark properties, etc., that help inhibit the spread of fire from the forest floor to tree canopy, reduce fire damage, and facilitate population recovery post-fire. A fire-adapted species possesses multiple traits and characteristics. The examples of such major factors are as follows:

## 2.1 Thick Bark

Fire-adapted tree species often possess thick bark, protecting against heat and flame damage. This thick bark is an insulation layer, shielding the tree's crucial cambium layer from lethal temperatures exceeding 60 °C during surface fires. Research suggests that the thickness of bark is an important factor that enhances the ability of a tree to resist fire, with thicker bark correlating with greater fire resistance (Bold et al. 2010). *Pinus roxburghii, Shorea robusta, Syzygium cumini, Lannea coromandelica, Careyaarborea, and Bombax ceiba* are a few examples of fire-hardy tree species with thick bark. Phenols, tannins, and lignin are naturally present in the bark have low thermal conductivity (Tributsch and Fiechter 2008; Celzard et al. 2011; Huss et al. 2019; Grootemaat et al. 2015; 2017; and Kempe et al. 2018), which promote fire resistance.

Another factor to consider is the presence of sclerenchyma, a tissue type with low water affinity, which offers better thermal insulation for the cambium at the inner bark in a higher percentage. This enhances fire resistance as the heat retention capacity is reduced due to lower water content (Agee 1993).

Additionally, these phenolic compounds' bactericidal and fungicidal properties contribute to the overall resilience of fire-adapted trees by protecting them against pathogens and decay (Bold et al. 2010). In essence, the thick bark of fire-adapted trees acts as a crucial defense mechanism, ensuring the survival and longevity of the tree even in the face of fire disturbances.

## 2.2 Self-Pruning

One of the crucial characteristics of fire-adapted tree species is the self-pruning ability of the tree, which is the shedding of branches that are shaded or diseased (Schwilk and Ackerly 2001), old and dead. Fire-adapted trees usually can self-prune; while the shedding of branches increases with age. As a result of self-pruning, a gap is created between the forest floor and canopy, preventing the fire from spreading from bottom to top and reducing the amount of flammable material near the ground. Because of this, the chance of fire climbing to the canopy is decreased as the flammability of the tissues of upper branches is low, and the rate of spread of fire and heat released at the individual scale is also low (Pausas 2015, Schwilk and Ackerly 2001). Self-pruning as a fire adaptive strategy has been reported from coniferous forests throughout the western region of North America primarily, which are mainly dominated by *Pinus ponderosa* and *Pinus sylvestris* and are prone to frequent fire with low severity (Fernandes et al. 2008). Besides, self-pruning characteristics have also been reported in some broadleaved species such as *Eucalyptus marginata, E. diversicolor, E. wandoo* (Burrows et al. 1990), *E. salubris* (Gosper et al. 2013), and *Callitrisintratropica* (Trauernicht et al. 2012).

## 2.3 Deep Root Systems

The deep root system constitutes a crucial element of forest ecosystems, serving vital roles in both mechanical anchorage and physiological functions such as water and mineral absorption, which are vital for sustaining plant growth (Pregitzer et al. 2002; McCormack and Guo 2014). The deep-rooted tree species (e.g. *Acacia* spp., *Prosopis* spp., *Tectona grandis, Shorea robust,* etc.) exhibit larger vessel diameters and tracheids in the roots, leading to higher hydraulic conductivity compared to shallow roots (Christina et al. 2017). These deep roots are specialized in water absorption and transportation (Wang et al. 2022; McElrone et al. 2004) and are considered crucial for plant species to tolerate stresses like drought and fire (Canadell and Zedler 1995; Schenk and Jackson 2002). They allow fire-adapted trees to access water stored deep in the soil, even during drought conditions that may follow forest fire, thereby maintaining hydration and reducing the risk of mortality from heat stress alongside mineral supply and mechanical stability.

## 2.4 Serotiny

Botanically, **Serotiny** means ‘following’ or ‘later’. In certain tree species, seeds are encased within cones or structures, which, when exposed to fire or heat, open and release seeds, following fire incidences. In this mechanism, the heat from a fire melts the resin, which seals the seeds within the fruits or scales of cones (Beaufait 1960; Johnson and Gutsell 1993).This trait can be seen in many coniferous species, such as *Sequoia dendrongiganteum, Pinus halepensis, and P. contorta.* Serotiny ensures seeds’ dispersal under the conditions optimal for germination and establishment post-fire.

## 2.5 Fire-Resistant Foliage

Certain plant species possess foliage that is more resistant to catching fire or burning at high temperatures, which diminishes the likelihood of crown fires by preventing flames from swiftly propagating across the upper canopy of trees. The foliage of such fire-resistant plants is characterized by certain traits such as thick leaves, low oil and resin content in their leaves, and high leaf moisture levels, which do not make them 100% fire-proof but help impede the advance of fires. Plants with delicate foliage and small leaves and twigs exhibit a higher surface-to-volume ratio, making them susceptible to reaching ignition temperature more rapidly compared to those with large and thick succulent leaves that possess a smaller surface-to-volume ratio and ignite at a slower rate (Schettler 2010). Additionally, plants with such foliage are natural windbreaks, absorbing and deflecting heat while trapping embers and sparks from fires. Also, in some species like pine, fallen pine needles create a porous bed that can be easily ignited, whereas, in species like cypress, the fallen scales form a compact bed that is difficult to ignite (Neyisci 2011).

## 2.6 Resprouting Ability

Many fire-adapted trees can resprout post-fire incidence, allowing them to regenerate quickly and maintain their presence in fire-prone landscapes. Certain plant species can give rise to new shoots from meristematic buds after fire has destroyed the parts above the ground (Paula et al. 2009; He et al. 2012; Kumar et al. 2020a; Kumar et al. 2020b). Factors like bark thickness, flammability, buds' insulation from fire, and position of bud within the plant (Lawes et al. 2011; Brando et al. 2012) determine from which part new shoots will arise (Knox and Clarke 2011). Generally, in areas where severe fires are common, or tree species with thin bark are present, new shoots are produced from adventitious buds or the root collar, whereas, in areas with low-intensity fires or trees with thick bark are present, new shoots develop from axillary or terminal buds (Bond and Midgley 2001). Also, in taller trees, the upper portion, including buds, are usually safe from fires of low intensity (Keeley 2006; Vesk 2006; Brando et al. 2012). Furthermore, larger plants' ability to store more starch than smaller plants may improve fire resistance (Gurvich et al. 2005; Konstantinidis et al. 2006).

The capacity to resprout also differs among species, life stages, and disturbances of varying intensities. In woody angiosperms such as *Pterodonemarginatus, Athyanaweinmannifolia, Galipea ciliata, Psychotriapichisensis, Pterocarpus sp, Ocotea guianensis,* etc., sprouting is frequent and may have been the ancestral state (Lloret et al. 2003; Mostacedo et al. 2022). In contrast, most conifers do not resprout, with some exceptions in a few *Pinus* and *Podocarpus* species. Furthermore, the occurrence of species with the ability to resprout is lower in temperate forest sites than tropical sites, while it is higher in more humid forests (Everham and Brokaw 1996; Wells 1969). Resprouting is the important resilience capacity of the trees in frequent fire-affected areas in preserving diversity and promoting natural regeneration post-fire occurrence (Kammesheidt 1998; Pausas 2001; Peixoto et al. 2012).

## 2.7 Seed Bank

In certain plant species, seeds accumulate in soil (e.g., *P. roxburghii, Quercus lucotrichophora, S. robusta,* etc.) (Mittal et al. 2021) or in the canopy (e.g. *P. contorta, P. banksiana, Piceaabies* etc.) without losing viability and thus, seeds get protection from fire damages. Some plants, known as obligate seeders, produce fire-resistant seeds. In contrast, others, called facultative seeders, can regenerate by developing new shoots from meristematic buds and producing seeds that are resistant to fire. The population of these species can recover by germination after forest fires (Pausas et al. 2004). The fire-resistant seeds in the soil remain dormant without losing viability. There is variation in the number of stored seeds in the soil depending on the species and is affected by factors such as seed production, morphology, dormancy, and population dynamics for the fire (Thompson et al. 1993, 1998; Clemente et al. 2007). When the seeds are exposed to fire-related stimuli, like heat and smoke (Reyes and Trabaud 2009), the dormancy is broken, and germination begins. With the increase in temperature, the germination rate also increases in many species (Keeley 1991; Herranz et al. 1998; Tozer and Auld 2006), but excessively high heat can be lethal to seeds. Depending on the species of plants, there is variation in the temperature thresholds for breaking dormancy or for causing mortality of seeds (Herranz et al. 1998).

## 3 Categorization of Fire-Adapted Species

Due to exposure to fire-prone environments, fire-adapted tree species have developed specific characteristics that enhance long-term resilience and survival. They can be classified into five categories, as suggested by various studies (Lyon and Stickney 1976; Noble and Slatyer 1977; Cattelino et al. 1979; Rowe 1983; Accatino et al. 2016; Gignoux et al. 1997), based on two vital attributes: reproduction and competition. These attributes and the life cycle form dimensions of a species' "regeneration niche," as defined by Gurevitch (2006). The regeneration niche encompasses the environmental conditions for plant species to germinate and establish.

Not all tree or plant species can be neatly categorized into a single group, as they may possess multiple characteristics relevant to more than one category. This highlights the complexity and variability of plant responses to fire disturbances, underscoring the need for a nuanced understanding of their adaptive strategies and ecological roles in fire-prone environments. The categories are as follows:

### *3.1 Pioneers/Colonizers/ Invaders*

These tree species swiftly establish themselves in vacant areas following a fire, earning them the designation of pioneer species. They typically thrive in environments with abundant sunlight and are characterized by their prolific production of propagules. Pioneers, often with limited or unknown populations before the fire event, dominate recently burned landscapes. Their seeds exhibit high dispersal capabilities through various means, including wind, water, animals, or human activities. Species such as *Lantana camara* is are considered invader/invasive weeds, which take over the disturbed areas post-fire. Their rapid colonization plays a crucial role in initiating the ecological succession process following a fire, paving the way for establishing subsequent vegetation communities.

## 3.2 *Evaders/Seeders/Dispersers*

These tree species possess an adaptation strategy, storing their seeds in various locations, such as within the canopy, humus layer, or mineral soil. This storage mechanism shields the seeds from extreme heat during a fire. Subsequently, following a fire event, these seeds germinate and establish themselves in the newly burned areas. Even if the fire destroys the mature trees, new individuals can arise from the stored seeds, ensuring the continued presence of the species on the site. The seedlings of evaders benefit from the nutrient-rich environment created by the fire, as fire breaks down the organic matter, which returns to the soil as ash and other charred remains, enriching the soil with nutrients. Fire's immediate aftermath presents an optimal window for seed dispersal and germination, as there is ample space, nutrients, sunlight, and water for growth and development, while the competition is less. It is important to note that many Evaders rely on fire to create the necessary habitat conditions for their seedlings to thrive. Unlike invaders, Evaders are not new arrivals to the area; they already inhabited the site before the fire. Additionally, their population expansion is not as rapid as that of invaders, as their strategy revolves around the regeneration of individuals from existing seed banks rather than the dispersal of new seeds into unoccupied areas. This distinction underscores the nuanced roles and dynamics of different tree species in fire-affected ecosystems.

## 3.3 *Avoiders*

These tree species are susceptible to fire damage and, thus, tend to inhabit regions where fire occurrences are infrequent. Following a fire event, when colonizer species have already occupied the burnt sites, there is an accumulation of organic matter (humus) and shading from the developing vegetation, creating conditions favorable for the arrival of propagules (seeds or spores) belonging to the Avoider group. These species show rapid growth in an environment free from fire. Avoiders enhance their survival prospects through clustering behavior, which involves the spatial aggregation of trees, creating an unfavorable environment for the fire to occur in that particular area (Bacelar et al. 2014; Gignoux et al. 1997; Kennedy and Potgieter 2003; Menaut et al. 1990; Mordelet and Menaut 1995). Clustering trees results in dense vegetation and elevated humidity levels, which reduce the probability of fire igniting in the area. Also, it can create a barrier of thick vegetation and trees that impedes the spread and intensity of fire in the region while protecting from damage by flames and heat in the area.

Avoiders are typically characteristic of late successional stages and are not commonly found in recently burned areas. They possess certain physical traits such as thin bark, shallow roots, and abundant resins that contribute to their vulnerability to fire and facilitate rapid fire spread.

## 3.4 Resisters

These tree species exhibit a suite of characteristics that enable them to withstand fire-related stresses. Notably, they possess thick bark even during their juvenile stage, protecting them against high temperatures. Additionally, Resistor species may have self-pruning ability, which reduces the likelihood of fire spreading from ground level to canopy. Their deep roots are often safeguarded from fire damage, and they may shed lower branches further to prevent fire ascent. Furthermore, Resistor species usually feature moist, short needles or leaves that are resistant to ignition.

## 3.5 Endurers/Resprouters

Following a fire event, Endurer species resprout from underground organs such as rhizomes, roots, and stem bases, effectively rejuvenating their populations even when flames have consumed their aboveground biomass. The perennating organs responsible for this regeneration strategy are typically situated deep within the soil, protected by layers of humus and mineral soil, a phenomenon that Gignoux et al. (1997) call the "hide-and-resprout" strategy. The portion of the tree left after the fire, also known as a "Gulliver" (Bond and van Wilgen 1996), continues to exist as a relatively small and immature part during succeeding fire incidences. Yet, it can resprout once the area is fire-free, perpetuating the tree's life cycle. Moreover, certain Endurers have developed hard-coated seeds that depend on fire to trigger germination by breaking open their protective shells. Despite potential damage to the parent plant during a fire, the emergence of new shoots benefits from the nutrient-rich soil and reduced competition in the aftermath of the blaze. Thus, Endurer species are crucial in rejuvenating and rehabilitating ecosystems post-fire. Their resilience to recurrent fire occurrences enhances the overall robustness and stability of landscapes prone to wildfires.

Trees that have evolved to cope with fire can be sorted into distinct groups based on their flammability characteristics (Fonda 2001). This categorization broadly divides species into two main types: pyrophytes, which tend to encourage or enable high-intensity surface fires, and mesophytes, which possess traits that can dampen, extinguish, or reduce surface fires (Kane et al. 2008; Kreye et al. 2013 and Mola et al. 2014). Pyrophytic species (e.g. *Eucalyptus* sp., *Cupressus* sp., *Pinus contorta, P. ponderosa* etc.) typically exhibit features that promote the spread and intensity of fires, while mesophytic species (e.g. *Quercus incana, Mesua ferrea, Calophyllumino-phyllum, Toona ciliata, Canarium strictum, Mangifera sylvatica,* etc.) display traits that moderate fire behavior, thereby influencing the overall fire regime and dynamics within ecosystems.

Characteristics that enhance flammability can empower pyrophytic species to intensify local fire severity, simultaneously causing injury to nearby invaders lacking protective bark and wound-healing mechanisms (Bond and Midgley 1995; Kane et al.

2008; Gagnon et al. 2010). Pyrophytes are equipped with features that provide protection and help in quick recovery. For example, in such plant species, juvenile bark can accumulate swiftly (Jackson et al. 1999; Graves et al. 2014; Hammond et al. 2015), and wounds can get close rapidly (Romero et al. 2009). These traits contribute to the resilience of pyrophytic species in fire-prone environments and influence the dynamics of fire behavior and its impact on surrounding vegetation communities. Pyrophytes gain a competitive advantage in post-fire regeneration and colonization by intensifying local fire intensity and injuring competing species. The swift accumulation of protective bark and efficient wound closure mechanisms enhance the survival prospects of pyrophytic species amidst recurrent fire disturbances.

Mesophytic species possess traits that equip them to colonize areas previously prone to fire, and once they are established, they prevent the entry of pyrophytes into the area. Mesophytes hinder the spread of fire and lack the protective adaptations necessary for survival post-fire (Varner et al. 2016). Mesophytes dominate in environments where fires are suppressed, significantly changing the community's composition, structure, and inflammability. This phenomenon, known as "mesophication," describes the positive feedback loop whereby the dominance of mesophytic species reinforces their proliferation and alters ecosystem dynamics (Nowacki and Abrams 2008). Also, mesophytic species exhibit traits such as rapid growth and long leaf lifespans (Varner et al. 2016). These characteristics of mesophytes enable them to outcompete pyrophytes in fire-excluded environments, altering the ecological balance and potentially increasing the risk of catastrophic fires in the long term.

## 4 Conclusion

Forest fires have traditionally been seen as destructive events that disrupt ecological balance and biodiversity. However, perspectives are shifting, recognizing the ecological importance of forest fires. Many scholars now view fires as internal disturbances within stable systems, playing essential roles in ecosystem dynamics by transforming structure and function, impacting biodiversity, and altering forests' productive and protective functions. Flora and fauna in fire-prone environments, exposed to specific fire regimes over long periods, have developed traits that make them well-adapted to forest fires. Isolated fire incidents have a minor impact compared to changes in fire regimes, which profoundly influence the environment. Fire-adapted tree species exhibit remarkable adaptations, ensuring their survival despite periodic disturbances. Their persistence maintains biodiversity and ecosystem functionality in fire-affected environments, underscoring their ecological significance. As the global climate changes, the frequency and intensity of forest fires and the length of the fire season are increasing. Understanding fire-adaptive traits is crucial for developing effective strategies for forest fire prevention, mitigation, and post-fire ecosystem restoration. Planting and promoting fire-adapted tree species in vulnerable areas can help reduce the spread and intensity of forest fires while maintaining vegetation cover. Fire-adapted species, with fewer flammable parts and lower production of flammable materials, reduce fuel loads and create natural firebreaks, decreasing the likelihood of catastrophic fires. Additionally, replanting with fire-adapted species after fires can enhance ecosystem resilience and accelerate recovery. Promoting these species helps preserve biodiversity, as various plants and animals benefit from the tree cover. Therefore, understanding the characteristics of fire-adapted tree species is essential for effective forest fire management and enhancing ecosystem resilience in the face of climate-driven fire.

## References

Accatino F, Wiegand K, Ward D, Michele CD (2016) Trees, grass, and fire in humid savannas—the importance of life history traits and spatial processes. Ecol Model 320(135–144):0304–3800. https://doi.org/10.1016/j.ecolmodel.2015.09.014

Agee JK (1993) Fire ecology of Pacific Northwest forests. Island Press, Washington, DC

Bacelar F, Calabrese JM, Hernández-García E (2014) Exploring the tug of war between positive and negative interactions among savanna trees: competition, dispersal, and protection from fire. Ecol Complex 17:140–148

Barwa S, Singh H (2024) Climate-induced changes in forest ecosystems of Indian Himalayan region: a meta-analysis. In: Singh H (eds) Urban Forests, Climate Change and Environmental Pollution. Springer, Cham. https://doi.org/10.1007/978-3-031-67837-0_39

Beaufait WR (1960) Some effects of high temperatures on the cones and seeds of jack pine. For Sci 6:194–199

Bellingham PJ, Sparrow AD (2000) Resprouting as a life history strategy in woody plant communities. Oikos 89:409–416

Bold G, Speck T, Blömer J, Bertling J, Speck O (2010) Insulation capability of the bark of trees with different fire adaptation. J Mater Sci 45:5950–5959. https://doi.org/10.1007/s10853-010-4680-4

Bond WJ, Midgley JJ (1995) Kill thy neighbour: an individualistic argument for the evolution of flammability. Oikos 73:79–85. https://doi.org/10.2307/3545728

Bond WJ, van Wilgen BW (1996) Fire and plants. Chapman & Hall, London

Bond WJ, Midgley JJ (2001) Ecology of sprouting in woody plants: the persistence niche. Trends Ecol Evol 16:45–51

Brando PM, Nepstad DC, Balch JK, Bolker B, Christman MC, Coe M, Putz FE (2012) Fire-induced tree mortality in a neotropical forest: the roles of bark traits, tree size, wood density and fire behavior. Glob Change Biol 18(2):630–641

Burrows ND, Gardiner G, Ward B, Robinson A (1990) Regeneration of eucalyptus wandoo following fire. Aust For 53:248–258

Canadell J, Zedler PH (1995) Underground structures of woody plants in Mediterranean ecosystems of Australia, California, and Chile. In: Ecology and biogeography of Mediterranean ecosystems in Chile, California, and Australia. Springer, New York, NY, pp 177–210

Cattelino PJ, Noble IR, Slatyer RO, Kessell SR (1979) Predicting the multiple pathways of plant succession. Environ Manag 3(1):41–50

Celzard A, Fierro V, Amaral-Labat G, Pizzi A, Torero J (2011) Flammability assessment of tannin-based cellular materials. Polym Degrad Stab 96:477–482. https://doi.org/10.1016/j.polymdegradstab.2011.01.014

Christina M, Nouvellon Y, Laclau JP, Stape JL, Bouillet JP, Lambais GR, Maire G (2017) Importance of deep-water uptake in tropical eucalypt forest. Funct Ecol 31:509–519

Clemente AS, Rego FC, Correia OA (2007) Seed bank dynamics of two obligate seeders, cistus monspeliensis and rosmarinus officinalis, in relation to time since fire. Plant Ecol 190:175–188. https://doi.org/10.1007/S11258-006-9199-Z

Everham EM, Brokaw NVL (1996) Forest damage and recovery from catastrophic wind. Bot Rev 62:113–185

Fernandes PM, Vega JA, Jimenez E, Rigolot E (2008) Fire resistance of European pines. For Ecol Manag 256:246–255

Fonda RW (2001) Burning characteristics of needles from eight pine species. For Sci 47:390–396

Gagnon PR, Passmore HA, Platt WJ, Myers JA, Paine CET, Harms KE (2010) Does Pyrogenicity protect burning plants? Ecology 91:3481–3486. https://doi.org/10.1890/10-0291.1

Gignoux J, Clobert J, Menaut JC (1997) Alternative fire resistance strategies in savanna trees. Oecologia, 110(4), 576–583. http://www.jstor.org/stable/4221648

Glasspool I, Edwards D, Axe L (2004) Charcoal in the Silurian as evidence for the earliest wildfire. Geology 32:381–383. https://doi.org/10.1130/G20363.1

Gosper CR, Prober SM, Yates CJ (2013) Multi-century changes in vegetation structure and fuel availability in fire-sensitive eucalypt woodlands. For Ecol Manag 310:102–109

Graves SJ, Rifai SW, Putz FE (2014) Outer bark thickness decreases more with height on stems of fire-resistant than fire-sensitive Floridian oaks (Quercus s; Fagaceae). Am J Bot 101:2183–2188. https://doi.org/10.3732/ajb.1400412

Grootemaat S, Wright IJ, van Bodegom PM, Cornelissen JHC, Cornwell WK (2015) Burn or rot: Leaf traits explain why flammability and decomposability are decoupled across species. Funct Ecol 29:1486–1497. https://doi.org/10.1111/1365-2435.12449

Grootemaat S, Wright IJ, Van Bodegom PM, Cornelissen JH, Shaw V (2017) Bark traits, decomposition and flammability of Australian forest trees. Aust J Bot 65:327–338. https://doi.org/10.1071/BT16258

Gurevitch J (2006) The Ecology of plants, Sunderland, MA, 2nd edition, 526

Gurvich DE, Enrico L, Cingolani AM (2005) Linking plant functional traits with post-fire sprouting vigour in woody species in central Argentina. Aust Ecol 30:789–796

Hammond DH, Varner JM, Kush JS, Fan Z (2015) Contrasting sapling bark allocation of five southeastern USA hardwood tree species in a fire prone ecosystem. Ecosphere 6(7):112. https://doi.org/10.1890/ES15-00065.1

He T, Pausas JG, Belcher CM, Schwilk DW, Lamont BB (2012) Fire adapted traits of Pinus arose in the fiery Cretaceous. New Phytolog 194:751–759

Herranz JM, Ferrandis P, Martı´nez-Sa´nchez JJ (1998) Influence of heat on seed germination of seven Mediterranean Leguminosae species, Plant Ecol, 136, 95–103. https://doi.org/10.1023/A:1009702318641

Huss JC, Fratzl P, Dunlop JW, Merritt DJ, Miller BP, Eder M (2019) Protecting offspring against fire: lessons from Banksia seed pods. Front Plant Sci 10:283. https://doi.org/10.3389/fpls.2019.00283

IPCC Working Group (2013) Climate change 2013: the physical science basis; contribution of working group I to the fifth assessment report of the intergovernmental; Cambridge University Press: New York, NY, USA

Jackson JF, Adams DC, Jackson UB (1999) Allometry of constitutive defense: a model and a comparative test with tree bark and fire regime. Am Nat 153:614–632. https://doi.org/10.1086/303201

Johnson EA, Gutsell SL (1993) Heat budget and fire behaviour associated with the opening of serotinous cones in two Pinus species. J Veg Sci 4(6):745–750. https://doi.org/10.2307/3235610

Kane JM, Varner JM, Hiers JK (2008) The burning characteristics in Southeastern oaks: discriminating fire facilitators from fire impeders. For Ecol Manag 256:2039–2045. https://doi.org/10.1016/j.foreco.2008.07.039

Kammesheidt L (1998) The role of tree sprouts in the restorations of stand structure and species diversity in tropical moist forest after slash-and-burn agriculture in Eastern Paraguay. Plant Ecol 139(2):155–165

Kaushik P, Singh H (2024) Climate change-induced forest ecosystem functioning and alteration in water-atmosphere-soil–plant (WASP) relationship. In: Singh H (eds) Urban Forests, Climate Change and Environmental Pollution. Springer, Cham. https://doi.org/10.1007/978-3-031-67837-0_38

Keeley JE (1991) Seed germination and life history syndromes in the California chaparral. Bot Rev 57:81–116. https://doi.org/10.1007/BF02858766

Keeley JE (2006) Fire severity and plant age in postfire resprouting of woody plants in sage scrub and Chaparral. Madrono 53:373–379

Keeley JE (2012) Ecology and evolution of pine life histories. Ann Forest Sci 69:445–453. https://doi.org/10.1007/s13595-012-0201-8

Keeley JE, Zedler PH (1998) Evolution of life histories in Pinus. In: Richardson DM (ed) Ecology and biogeography of pinus. Cambridge University Press, Cambridge, UK, pp 219–250

Kempe A, Neinhuis C, Lautenschlager T (2018) Adansoniadigitata and Adansoniagregorii fruit shells serve as a protection against high temperatures experienced during wildfires. Bot Stud 59:1–12. https://doi.org/10.1186/s40529-018-0223-0

Kennedy AD, Potgieter ALF (2003) Fire season affects size and architecture of Colosphermum mopane in southern African savannas. Plant Ecol 167:179–192

Knox KJE, Clarke PJ (2011) Fire severity and nutrient availability do not constrain resprouting in forest shrubs. Plant Ecol 212:1967–1978

Konstantinidis P, Tsiourlis G, Xofis P (2006) Effect of fire season, aspect and pre-fire plant size on the growth of Arbutus unedo L. (strawberry tree) resprouts. For Ecol Manage 225:359–367

Kreye JK, Varner JM, Hiers JK, Mola J (2013) Toward a mechanism for eastern North American forest mesophication: the role of litter drying. Ecol Appl 23:1976–1986. https://doi.org/10.1890/13-0503.1

Kumar A, Dwivedi GK, Tewari S, Paul J, Anand R, Kumar N, Kumar P, Singh H, Kaushal R (2020) Carbon mineralization and inorganic nitrogen pools under Terminalia chebula Retz.-Based Agroforestry System in Himalayan Foothills India. Forest Sci 66(5):634–643

Kumar A, Kumar P, Singh H, Kumar N (2020) Adaptation and mitigation potential of roadside trees with bio-extraction of heavy metals under vehicular emissions and their impact on physiological traits during seasonal regimes. Urb for Urb Green 58:126900

Kumar N, Singh H, Kumar A, Singh AK, Pandey PK, Kumar A (2024) Forest-fire-induced land degradation. In: Mishra G, Giri K, Singh S, Kumar M (eds) Sustainable Land Management in India. Springer Nature, Singapore. https://doi.org/10.1007/978-981-97-5223-2_4

Lawes MJ, Richards A, Dathe J, Midgley JJ (2011) Bark thickness determines fire resistance of selected tree species from fire-prone tropical savanna in north Australia. Plant Ecol 212:2057–2069

Liu Y, Stanturf J, Goodrick S (2010) Trends in global wildfire potential in a changing climate. For Ecol Manag 259:685–697

Lloret F, Pausas,JG, Vilà M (2003) Responses of mediterranean plant species to different fire frequencies in garraf natural park (Catalonia, Spain): field observations and modelling predictions. Plant Ecol 167:223–235

Lyon LJ, Stickney PF (1976) Early vegetal succession following large Northern Rocky Mountain wildfires. Tall Timbers Fire Ecol Conf 14:355–37

Malik A, Kumar K, Kumari G, Kumar N, Singh H (2024) Application of functional traits in modelling productivity and resilience under climate change published by springer, pp 77–96. ISBN: 978-981-97-1509-1

McCormack ML, Guo D (2014) Impacts of environmental factors on fine root lifespan. Front Plant Sci 5:205

McCune B (1988) Ecological diversity in North American pines. Am J Bot 75:353–368. https://doi.org/10.2307/2443983

McElrone AJ, Pockman WT, Martínez Vilalta J, Jackson RB (2004) Variation in Xylem structure and function in stems and roots of trees to 20 m depth. New Phytol 163:507–517

Menaut JC, Gignoux J, Prado C, Clobert J (1990) Tree community dynamics in a humid savanna of the Côte-d'Ivoire:modelling the effects of fire and competition with grass and neighbours. J Biogeogr 17:471–489

Mittal A, Tewari A, Singh N, Shah S (2021) Assessment of soil seed bank on three different vegetation types in Kumaun central Himalayan forest. Indian J Ecol 48:381–386

Mola JM, Varner JM, Jules ES, Spector T (2014) Altered community flammability in Florida's Apalachicola ravines and implications for the persistence of the endangered conifer Torreya taxifolia. PLoS ONE 9(8):e103933. https://doi.org/10.1371/journal.pone.0103933

Mordelet P, Menaut JC (1995) Influence of trees on above-ground production dynamics of grasses in a humid savanna. J Veg Sci 6:223–228

Mostacedo B, Viruez A, Varon Y, Paz-Roca A, Parada V, Veliz V (2022) Tree survival and resprouting after wildfire in tropical dry and subhumid ecosystems of Chiquitania, Bolivia. Trees For People 10:100327

Neyisci T (2011) Mediterranean forest ecosystems, wildland fires, cypress and fire-resistant forests. In: Cypress and forest fires: a practical manual, pp 15–37

Noble IR, Slatyer RO (1977) Post-fire succession of plants in Mediterranean ecosystems. In: Mooney HA, Conrad CE (eds) Symposium on the environmental consequences of fire and fuel management in mediterranean ecosystems, pp 27–36. USDA Forest Service, General Technical Report WO-3.

Nowacki GJ, Abrams MD (2008) The demise of fire and "mesophication" of forests in the eastern United States. Bioscience 58:123–138. https://doi.org/10.1641/B580207

Paula S, Arianoutsou M, Kazanis D, Tavsanoglu C, Lloret F, Buhk C, Ojeda F, Luna B, Moreno JM, Rodrigo A, Espelta JM, Palacio S, Fernández-Santos B, Fernandes PM, Pausas JG, Michener WK (2009) Fire-related traits for plant species of the Mediterranean Basin. Ecology 90:1420

Pausas JG (2001) Resprouting vs seeding—a mediterranean perspective. Oikos 94(1):193–194

Pausas JG (2015) Alternative fire-driven vegetation states. J Veg Sci 26:4–6

Pausas JG, Keeley JE (2009) A burning story: the role of fire in the history of life. Bioscience 59:593–601. https://doi.org/10.1525/bio.2009.59.7.10

Pausas JG, Blade C, Valdecantos A, Seva JP, Fuentes D, Alloza JA, Vilagrosa A, Bautista S, Cortina J, Vallejo VR (2004) Pines and oaks in the restoration of Mediterranean landscapes of Spain: new perspectives for an old practice—a review. Plant Ecol 171:209–220

Pausas JG, Keeley JE, Verdú M (2006) Inferring differential evolutionary processes of plant persistence traits in Northern hemisphere Mediterranean fire-prone ecosystems. J Ecol 94:31–39. https://doi.org/10.1111/j.1365-2745.2005.01092.x

Peixoto KDS, Sanchez M, Pedroni F, Ribeiro MN, Facure KG, Gomes-Klein VL, Guilherme FAG, (2012) Tree community dynamics in a semi-deciduous seasonal forest under recurrent fires. Acta Bot Brasilica 26:697–708

Pregitzer KS, DeForest JL, Burton AJ, Allen MF, Ruess RW, Hendrick RL (2002) Fine root architecture of nine North American trees. Ecol Monogr 72:293–309

Reyes O, Trabaud L (2009) Germination behaviour of 14 Mediterranean species in relation to fire factors: smoke and heat. Plant Ecol 202:113–121. https://doi.org/10.1007/s11258-008-9532-9

Romero C, Bolker BM, Edwards CE (2009) Stem responses to damage: the evolutionary ecology of Quercus species in contrasting fire regimes. New Phytol 182:261–271. https://doi.org/10.1111/j.1469-8137.2008.02733.x

Rowe JS (1983) Concepts of fire effects on plant individuals and species, the role of fire in northern circumpolar ecosystems (eds) Wein RW, Maclean DA, New York, NY: John Wiley and Sons, pp 135–154

Savita, Sharma SK, Singh H, Singh O (2017) Forest fire: a review. Indian for 143(8):719–728

Schenk HJ, Jackson RB (2002) The global biogeography of roots. Ecolog Monogr 72(3):311–328

Schettler S (2010) Fire-resistant landscaping: a general approach and central coast perspective, fremontia. J Calif Native Plant Soc, 38:2/38:3, 46–50. https://www.cnps.org/wp-content/uploads/2018/03/Fremontia_Vol38-No2-3.pdf

Schwilk DW, Ackerly DD (2001) Flammability and serotiny as strategies: correlated evolution in pines. Oikos 94:326–336. https://doi.org/10.1034/j.1600-0706.2001.940213.x

Sett T, Nikam BR, Singh H, Purohit S (2024) Hydrological variability in Indian forest ecosystem: analysis of drought resilience, recovery and water use efficiency in moist and dry deciduous forests. In: Singh H (eds) Urban Forests, Climate Change and Environmental Pollution. Springer, Cham. https://doi.org/10.1007/978-3-031-67837-0_37

Singh H, Kumar N, Singh R, Kumar M (2023) Assessing the climate change impact on the habitat suitability of the range-restricted bird species (*Catreus wallichii*) in the Indian Himalayan ecosystem. Environ Sci Pollut Res. https://doi.org/10.1007/s11356-023-30789-x

Singh H, Kumar M (2022) Climate change and its impact on Indian Himalayan forests: current status and research needs. In: Rani S, Kumar R (eds) Climate change. Springer Climate. Springer, Cham. pp 223–242; https://doi.org/10.1007/978-3-030-92782-0_11

Singh H, Kumar N, Kumar A (2024) Enhancing resource use efficiency in crops through plant functional traits published by springer, pp 97–117. ISBN: 978-981-97-1509-1

Singh H (2024a) Climate change and forests: biological perspectives. In: Singh H (eds) Forests and climate change. Springer Nature, Singapore. https://doi.org/10.1007/978-981-97-3905-9_1

Singh H (2024c) Climate change and environmental pollution: biological consequences on urban forests. In: Singh H (eds) Urban forests, climate change and environmental pollution. Springer Nature, Cham, Switzerland. https://doi.org/10.1007/978-3-031-67837-0_1

Singh H (2024b) Forests and climate change: biological perspectives on impact, adaptation, and mitigation strategies (ed), Springer Nature Singapore. https://doi.org/10.1007/978-981-97-3905-9

Singh H (2024d) Urban forests, climate change and environmental pollution: physio-biochemical and molecular perspectives to enhance urban resilience, Springer Nature (ed), Cham, Switzerland. https://doi.org/10.1007/978-3-031-67837-0

Smith A (1995) Adaptation to fire in the contribution of fire in Dramatising The Australian Landscape, http://online.anu.edu.au/Forestry/fire/ecol/as20.htm

Thompson K, Band SR, Hodgson J (1993) Seed size and shape predict persistence in soil. Funct Ecol 7:236–241. https://doi.org/10.2307/2389893

Thompson K, Bakker JP, Bekker RM, Hodgson JG (1998) Ecological correlates of seed persistence in soil in the north-west European flora. J Ecol 86:163–169. https://doi.org/10.1046/J.1365-2745.1998.00240.X

Tozer MG, Auld TD (2006) Soil heating and germination: investigations using leaf scorch on graminoids and experimental seed burial. Int J Wildland Fire 15:509–516. https://doi.org/10.1071/WF06016

Trauernicht C, Murphy BP, Portner TE, Bowman DMJS (2012) Tree cover-fire interactions promote the persistence of a fire-sensitive conifer in a highly flammable savanna. J Ecol 100:958–968

Tributsch H, Fiechter S (2008) The material strategy of fire-resistant tree barks. WIT Trans Built Environ 97:43–52. https://doi.org/10.2495/HPSM080051

Varner J, Kane JM, Hiers J, Kreye JK, Veldman JW (2016) Suites of fire-adapted traits of oaks in the Southeastern USA: multiple strategies for persistence. Fire Ecology 12:48–64. https://doi.org/10.4996/fireecology.1202048

Verdú M, Pausas JG (2007) Fire drives phylogenetic clustering in Mediterranean Basin woody plant communities. J Ecol 95:1316–1323. https://doi.org/10.1111/j.1365-2745.2007.01300.x

Verdú M, Pausas JG, Segarra-Moragues JG, Ojeda F (2007) Burning phylogenies: fire, molecular evolutionary rates, and diversification. Evolution 61:2195–2204. https://doi.org/10.1111/j.1558-5646.2007.00187.x

Vesk PA (2006) Plant size and resprouting ability: trading tolerance and avoidance of damage? J Ecol 94:1027–1034

Wang L, Dai Y, Zhang J, Meng P, Wan X (2022) Xylem structure and hydraulic characteristics of deep roots, shallow roots and branches of walnut under seasonal drought. BMC Plant Biol 22:440. https://doi.org/10.1186/s12870-022-03815-2
Wells PV (1969) The relation between mode of reproduction and extent of speciation in woody genera of the California chaparral. Evolution 23:264–267
Westerling AL, Hidalgo HG, Cayan DR, Swetnam TW (2006) Warming and earlier spring increase western U.S. forest wildfire activity. Science 313:940–943

# Role of Local Community in the Forest Fire Prevention and Management: Insights from Uttarakhand State

**Harshit Pant and Aditi Mishra**

**Abstract** Forest fires (FFs) are complex climatic and human influences issues that disastrously impact humans, social capital, and the ecosystems of affected regions at a global scale. These fires influence the micro and macroclimate of the areas affected and, at the same time, climate change, which is expected to increase the frequency and severity of these fires, leading to post-fire recruitment failure of species and permanent ecological loss across the impacted forests. The effects of these forest fires can be detrimental or valuable, depending on their location and nature across the regions. Generally, government agencies, particularly forest departments, manage forest fire risks by managing and quantifying locality-specific biophysical plethora of variables such as land use type, land undulation, edaphic and climatic factors, and the fuel load under the forests, amongst which the fuel load plays a crucial role in determining the forest fire extent. This scenario is especially true in the Himalayan region where the Pine (*Pinus roxburghii* Sarg.) is locally known as Chir-Pine, a dominating and fire-vulnerable species. In the Himalayan state of Uttarakhand, where forest fire is a prevalent and recurring issue, these Pine forests are the dominant and most affected species, covering an area of 18% of the state forest cover. However, with the significant increase in these forest fire issues, it is suggested that the administrative and community capacities of the region should be strengthened to improve forest fire prevention and management, as public education and awareness campaigns are crucial to enhance compliance with forest fire policies at the local level and comprehending the domino impacts of the forest fire issue. This study examines the role of local communities, particularly the Van Panchayats of Uttarakhand, in forest fire prevention and management. Through case studies, the research highlights the significant contributions of these communities to forest fire management and suggests their practices be incorporated into regional decision-making processes.

**Keywords** Responsiveness · Disenchantment · Risk reduction · Awareness · Fire implications

H. Pant (✉) · A. Mishra
G.B. Pant National Institute of Himalayan Environment (NIHE), Kosi-Katarmal, Almora, Uttarakhand, India
e-mail: drharshitpant@gmail.com

 
H. Singh and N. Kumar (eds.), *Forest Fire and Climate Change*,
https://doi.org/10.1007/978-3-031-89967-6_13

# 1 Introduction

Forest fires (FFs) have been an inherent component of many terrestrial ecosystems, playing critical roles in influencing landscape configurations, species dispersal, and sustaining ecological balance (Kumar et al. 2024). Their natural occurrence facilitates essential processes such as nutrient recycling, forest regeneration, and the maintenance of species diversity. However, the prevalence of uncontrolled forest fires (hereby FFs) represents a severe threat, capable of inflicting extensive ecological damage, destroying human development, and significantly escalating atmospheric carbon levels (Bowman et al. 2020; Croker et al. 2023; Savita et al. 2017) and impacting the communities residing in the vicinity. In the context of accelerated climate change and increased human land use, the rising frequency and intensity of FFs present significant challenges for both natural ecosystems and human communities (Croker et al. 2023; Singh et al., 2023, 2020a).

In global scenarios across the terrestrial forest landscapes, fires can originate naturally (e.g., lightning, volcanic eruptions) or be caused by human activity, either intentional or accidental. These forest fires are categorized into three types: (i) surface fires (low intensity), (ii) ground fires (moderate intensity), and (iii) crown fires (high intensity). The effects of fire on vegetation can range from beneficial to harmful, depending on the fire's severity and the management practices across the regions. It is a common saying that "*Fire is a good servant but a bad master*" which can be attributed to the intensified impacts of the fire across the globe. Curbing these issues can be done by using prescribed low-intensity fires in a controlled manner in forests, so it can help reduce fuel accumulation, thereby minimizing the risk of high-intensity FFs that can harm forest ecosystems. This practice serves as an effective management tool as these prescribed and controlled burning practices have been practiced for a long time, even in the Himalayan region. According to the FAO (2007), 80% of FF are man-made, and the rest are anthropogenic. FF incidents may increase soon due to pronounced seasonality and inter-annual rainfall variability in the wake of climate change. This chapter categorically centres on the challenges posed by FF in the state of Uttarakhand, India, with its rich biodiversity and extensive Chir-Pine (*Pinus roxburghii* Sarg.) forests, which are particularly fire-prone and the case studies that are related to the human intervention in forest fire management around these forests.

# 2 The Global Dynamics of Forest Fires

Internationally, FF functions not only as a natural ecological management agent but also as agent of destruction. They contribute to ecological succession and species adaptation but, when exacerbated by climate change, lead to severe disruptions in these processes (Barwa and Singh 2024; Savita et al. 2017; Singh and Kumar 2022; Sankey et al. 2024). The complex relationship between FF and human development, particularly at the wildland-urban interface, presents significant challenges for ecological risk assessment and strategic landscape management (Singh 2021; Moritz et al. 2014; Sankey et al. 2024). The rise in surface temperatures, alterations in precipitation patterns, and the accumulation of combustible material due to land-use changes have all contributed to the increased incidence of FFs (Jolly et al. 2015; Singh 2024a, b). Globally, the recent fire outbreaks in Amazonia, the Mediterranean, Siberia, southeast Australia, the western US, and Canada have emphasized the risks and alarming situation that fires pose to humans, biodiversity, and carbon stocks of the region (Jones et al. 2022).

### *2.1 The Indian Scenario: The Prevalence of Fires in Forested Regions*

In the Indian scenario, the Forest Survey of India (FSI) reports that approximately 50% of the country's forest area is susceptible to fires, with around 6% experiencing severe fires annually. Nearly 18,000 forest fire incidents are reported yearly, affecting 1.14 million hectares and resulting in an estimated annual loss of Rs. 4,400 million. This estimate does not account for intangible losses, such as biodiversity, carbon sequestration, soil fertility, and the hydrological functions of forests. In the Indian Himalayan region, 3908 FF events occur annually with an average burnt area of 1129 $km^2$. Reports indicate that around 36% of India's forest cover is frequently affected by forest fires, impacting approximately 3.73 million hectares annually (Singh 2024c, d). Dry deciduous forests are particularly vulnerable. Central India has the most significant area affected by forest fires, accounting for about 56%, while the northwest Himalayas become especially fire-prone during the summer (FSI 2022; Sewak et al. 2021; Kala 2023). The forest inventory records of India have reported that about 54.40% of forests of the country have been subjected to sporadic fires, 7.49% to slightly frequent fires, and 2.40% to elevated prevalence stages of forest fire, whereas 35.71% of forests had nevertheless been classified as subject to significant fires in the country (Gajendiran et al. 2023).

Notably, in the Himalayan state of Uttarakhand, forest fires have become a recurrent concern, now a recurring issue in the region. According to the FSI (2024), the largest forest fire events across the Indian region have been occurring in the five major states during 2023–24 (Fig. 1) in which the highest numbers have been reported from the Uttarakhand state, while the total number of large forest fires during the fire season has ranged between 575 and 12,562 during 2019–2024 (Fig. 2).

The state's topology, climatic conditions, and the dominance of flammable Pine forests create an environment particularly susceptible to FFs. The interaction between these natural factors and human-induced alterations such as land clearing and deforestation exacerbates the risk and consequences associated with these fire events (Kodandapani et al. 2006; Toy-Opazo et al. 2024; Kumar et al. 2020a, 2021a).

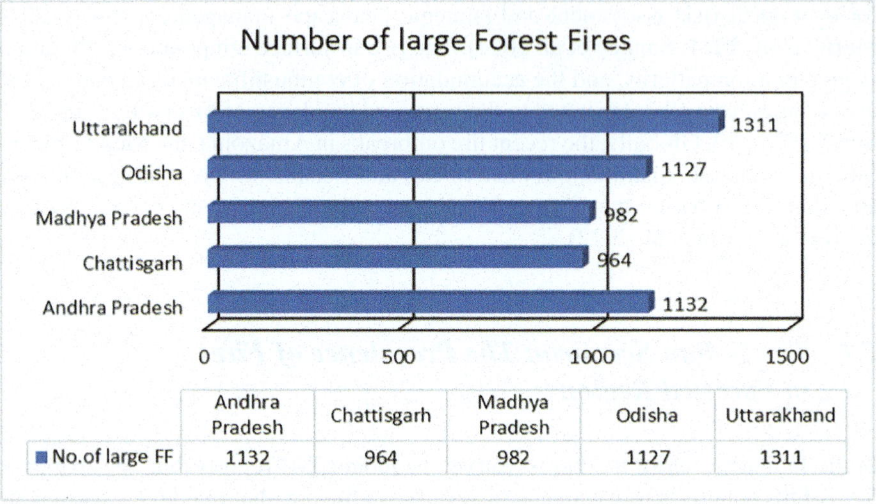


**Fig. 1** Large forest fire events during 2023–2024 (Adopted from FSI 2024)

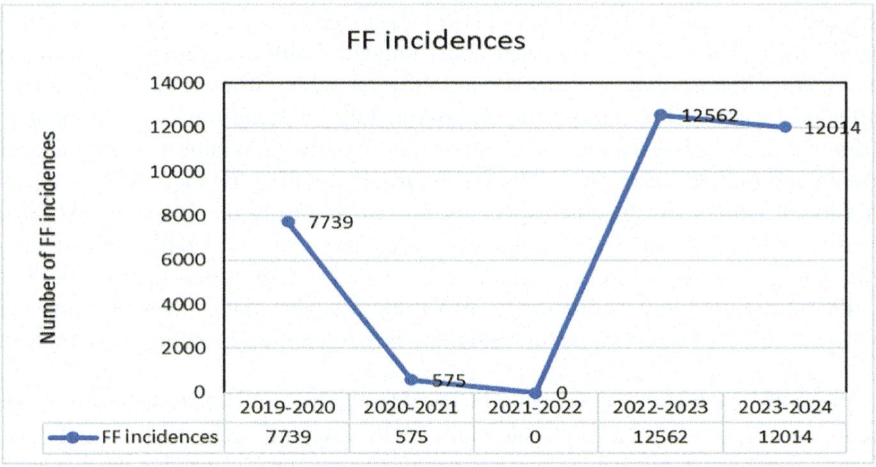


**Fig. 2** Range of large forest fire events during 2019–2024 (Adopted from FSI 2024)

## *2.2 State's Stand on Forest Fires*

In the Western Himalayan region, Chir Pine (Pinus roxburghii Sarg.) and conifer forests are especially susceptible to FFs. This is due to the accumulation of resin-rich, inflammable leaf litter on the forest floor during the summer peak leaf drop period and occasionally in the late dry winter. In the Sal (*Shorea robusta* Roth.) forests, FF leads to the invasion of *Lantana camara* L., an obnoxious weed. Lately, FF has occasionally spread to higher altitudes in the region during the hot summer months. Notably, natural fires are rare in this region, with most FFs being anthropogenic. In 1999, the Forest Survey of India recorded the largest FF in Uttarakhand's history, impacting approximately 23% of the forest area. In 1995, the National Remote Sensing Agency (NRSA) reported another significant forest fire that affected 19.3% of Uttarakhand's total forest area. As per the report of FSI (2021), the state of Uttarakhand has 0.20% area of forest cover under a highly high fire-prone zone while it has 3.12% area of forest cover under a very high fire-prone zone, which makes the state's forest resources susceptible for the fire incidence.

In 2016, between April and early June, the Forest Department recorded a major FF in Uttarakhand, with 2,069 fire incidents affecting 4,423 hectares of forest. The estimated financial loss from these fires was Rs. 4.62 million. However, several newspaper reports have estimated this to range from Rs. 40 lakhs to 50 billion. During this period, six people lost their lives, 31 were injured, and seven animals perished due to the forest fires (http://forest.uk.gov.in/contents/view). In response, the government deployed three National Disaster Response Force teams, increased the number of forest officials, and used two Indian Air Force helicopters to combat the fires in May 2016. The severity of these fires in Uttarakhand drew widespread national concern. Following the disaster, the Parliamentary Standing Committee on Science & Technology, Environment & Forests of the Government of India visited the region in June 2016. It consulted with a wide range of stakeholders and underscored that arriving at a realistic monetary loss due to FF is difficult without proper methodology and baseline data.

In 2024, the reported division-wise forest fire incidences in the Uttarakhand state were 1276 in numbers, of which 847 incidents were in the reserve forest region while 429 incidents were in the civil soyam and Van Panchayat region. These incidents have affected 1771.6 ha. Area in the region, which was 1228.58 ha. For the reserve forest region and 543.08 ha. For the civil soyam and Van Panchayat region (UKFD 2024). Though in media coverage, Since November 1,2023, a total 1368 fires in Uttarakhand have been recorded, including 789 detected between April 1 and April 30, 2024, according to the Forest Survey of India (FSI) (Satellite captures extent of Uttarakhand forest fire—India Today).

## 2.3 *The Societal Impact of Forest Fires*

The impact of FFs extends beyond ecological loss to encompass socio-economic dimensions. They bear upon the fabric of communities, influence economic activities, and often disproportionately affect the most vulnerable segments of society. Integrating indigenous knowledge and practices into fire management strategies is increasingly seen as crucial for developing sustainable and effective mitigation approaches (Souza et al. 2024). In Uttarakhand, the sociocultural relationship with the forest is complex and intimate, with communities relying on forest resources for their livelihoods, which are put at risk by recurrent fires (Rodriguez-Trejo and Fulé 2003).

The harmful effects of FFs are not limited to the immediate destruction of flora and fauna; they also have long-term consequences for soil stability, water quality, and air purity, which in turn affect agricultural productivity and health outcomes (Stephens et al. 2014; Singh et al. 2024a, b, c, d, a, b) of communities residing in the vicinity. A multisectoral approach to FF management, including the health, agricultural, and environmental sectors, is critical for a holistic response to this issue (FAO and UNEP 2020). These forests are hubs of biodiversity and natural resources and tend to provide various ecosystem services, including cultural, regulating, provisioning, and supporting services (Kala 2013; Sett et al. 2024; Kaushik and Singh 2024a, b, c, d, a, b). Multiple studies have shown that the human populace regularly and unsustainably uses these forests to provide significant resources for growing human demands, resulting in forest fragmentation, degradation, and shrinking in portion in a definite way (Kala 2015; Barwa and Singh 2024).

It has been widely discussed that FF's major causes are anthropogenic. Still, it has to be considered that these FF can be detrimental or valuable depending upon the location and circumstances of how and where they occur (Hardesty et al. 2005). The most vulnerable regions that are usually sensitive to FF are those of the forests, situated in the fringe of agriculture and pasture areas, along with road connectivity and are nearby community settlements (Reis et al. 2021), which are more vulnerable to fires, which are often accidental and sometimes intentional. These variations, varied intensity, and regular frequency of FF accelerate the rate of forest degradation in the impacted regions and result in the decline of quality and quantity of ecosystem services, which are essential for human beings. It has also been reported that globally, these FF are also displacing millions of people, negatively impacting their livelihoods and valuable national properties (Arnold 2018; Xu et al. 2020; Kala 2023; Malik et al. 2024; Singh et al. 2024a, b, c, d, a, b). Moreover, these fires also impact the regional socio-economic status, household unit characteristics, accessibility, and rates of dependence on social programs, which are the prominent factors that contribute the most to social vulnerability. FF significantly impacts socially vulnerable communities, which is highly significant for the Himalayan region (Chas-Amil et al. 2022).

## *2.4 Community Engagement in Forest Fire Management*

Uttarakhand's unique topography and variable climate and vegetation present distinct challenges in FF management. The intricate relationship between FFs and the socio-ecological system underscores the need for a nuanced understanding of fire dynamics. Strengthening local administrative capacities and fostering public awareness and education are pivotal for enhancing fire management strategies. We discuss the dichotomy of fire's ecological roles—valuable and disparaging—and the conditions under which they oscillate.

Emphasizing community engagement in forest fire management is based on acknowledging the value of local knowledge and the direct impact that fire events have on communities. Exploring the interplay between environmental factors and human elements contributes to the discourse on sustainable forest management. It underscores the importance of community involvement in mitigating the risks associated with FFs.

## 3 Case Studies

In Uttarakhand, FF frequency of FFs is significantly higher than in other Himalayan states. The case studies in the present study are based on the fire history and the level of community intervention toward FF fighting.

### *3.1 Case Study 1: Women-Driven Forest Fire Management in Pali Van Panchayat, Pauri-Garhwal, Uttarakhand*

The village Pali is located in the Kot development block of Pauri district of Uttarakhand state. The village's forest is community-managed (*Van Panchayat),* having an area of 4.9 ha, and was formed in 1999. The forest area is dominated by *Pinus roxburghii.* It has a lesser understory of *Myrica esculenta* Buch.ham, dominated by the undergrowth of *Lantana camara* L. and *Ageratina Adenophora* (Spreng.) King. The resource dependency on the forest is moderately high, while the forest area is significantly less able to sustain the day-to-day demands of fuel and fodder. The FF incidents are prominent in the region (Fig. 3).

The women-led management committee plays a crucial role in addressing FF issues, with their women's group, the Mahila Mangal Dal, being particularly active

**Fig. 3** Forest fire incidences in Pali Van panchayat (Pauri-Garhwal), Uttarakhand

**Fig. 4** Interventions of women groups towards extinguishing forest fire

in this effort. The women's team takes turns to patrol the boundaries of the forest and try to curb the unsolicited incidents of fire initiation in deliberate cases (Fig. 4).

Despite these challenges, FF management in the region is hampered by inadequate policies, limited capacity building, and a lack of mechanized tools. The rugged terrain and harsh climatic conditions, including strong winds typical of the district, further complicate fire management efforts. Although women-led groups are dedicated to raising awareness and managing forest fires while striving to sustain their natural resources, these factors make effective fire management challenging, and timely interventions for fire extinguishing take a toll on women's health and holistic well-being.

### 3.2 *Community-Driven Forest Fire Management in Syahi Devi, Sitlakhet, Almora, Uttarakhand*

Sitlakhet forest region is located in the Hawalbagh development block of Almora district at 29°34′742″ N latitude and 79°36′89″ E longitude at 1900–2200 m asl altitude. The forest is dominated by *Pinus roxburghii* and has a history of fire proneness. These fire incidents significantly influence the region's macro and micro climatic conditions and are a major concern for the community (Fig. 5).

The case study of the Syahidevi forest region in Almora, India, exemplifies how local communities can play a crucial role in forest fire management and environmental conservation. In 2003–04, Almora district faced a severe water crisis when the flow in the Kosi River drastically reduced. This situation was mirrored in the Syahi Devi Sitalakhet Reserved Forest, a vital water source since 1926 for Almora and nearby villages. A group of young locals formed the Syahi Devi Development Forum to tackle the crisis, discovering that the uncontrolled exploitation of trees for firewood and tools and recurrent forest fires depleted the water sources. In response,

**Fig. 5** Forest fire incidents in Syahidevi forest region, Sitlakhet (Almora)

they initiated the "Save Forests-Save Life" campaign, partnering with the Forest Department to educate and mobilize communities, especially women, to preserve broad-leaved trees and prevent forest fires.

The campaign's educational efforts led to the local women agreeing to stop using trees for firewood and actively participating in fire prevention. In 2005, a significant milestone was achieved when the forest was dedicated to the local goddess *Syahi Devi*, resulting in its rapid rejuvenation. Concurrently, the Forest Department constructed water conservation structures like pits and trenches and motivated towards controlled burning of the nearby forests.

Despite successful forest conservation, the tradition of burning dry grass and weeds (locally known as *Auon*) for crop preparation remained a fire hazard. This forest area also sustained damage during the severe forest fires of 2012 and 2016 and subsequently in 2024. Research indicated that the primary cause of these fires was the practice of random stubble burning (locally known as *Auon*) in March to prepare farmlands for the upcoming monsoon crop. Although women typically exercise caution when burning stubble, the fire sometimes spreads to nearby farms and reserved/panchayat/civil forests, leading to forest fire incidents. With the Forest Department's support, the community has since introduced "*Auon Day*" to complete the burning by April 1st each year, drastically reducing fire incidents, which has also been opted for by the nearby villages.

This "Sitlakhet Model" comprises several strategies, including the protection of mixed forests, public cooperation with the Forest Department, the pivotal role of Women's Welfare Groups, controlled burning of grass, quick response to fires, the provision of alternative agricultural tools, and proper compensation and support for forest fire personnel. This model has revived approximately 1100 hectares of degraded forest and became a state-wide strategy endorsed by the Uttarakhand government, showcasing the power of community engagement in environmental stewardship.

## 4 Results

A comprehensive review of literature and perception-based studies confirms that FFs are a global issue, influenced by various factors such as climatic and regional weather conditions, temperature anomalies, prolonged droughts, heat waves, and shifting weather patterns. As reported by Kala (2023), The primary causes of FFs are anthropogenic, and their impact can vary based on the location and nature of the fire, ranging from harmful to beneficial or benign. While certain forest types, such as coniferous forests in the northern hemisphere across the USA, Canada, Russia, China, and India, are highly sensitive to fire, others, like savannahs, benefit from frequent fires that help maintain their ecosystem.

As reported by Rawat et al. (2017), it has been suggested that in the context of Pine forest-based FF, reducing fuel load from the forest by needle removal for livelihood generation can be an excellent way for socio-economic interventions. The interventions of pine needle-based biomass briquette and the development of pine needle-based crafts, jewelry, and decorative items can also be sustainable livelihood options (Negi et al. 2021) for the rural regions in Pauri and Almora districts where outmigration is a major issue.

This situation necessitates a dynamic approach to fire management, particularly in Uttarakhand. A comprehensive policy guideline that encompasses various fire management interventions involving multiple stakeholders should be developed, including the forest department, revenue department, and local communities. To achieve proactive results, adequate financial resources should be allocated for fire management activities and related research. Given the rugged terrain and limited communication infrastructure in the Himalayan region, it is crucial to prioritize community mobilization and training through policies that combine modern fire-fighting techniques with community-based strategies, as highlighted by Sharma and Pant (2017). An integrated approach to forest fire management, incorporating climate-resilient adaptation pathways, offers a sustainable solution. Such pathways can drive transformative changes in current strategies and policies, impacting human institutions as well as technological and biological systems.

## 5 Discussion

The case studies and findings underscore the need for a multifaceted approach to forest fire management, especially in regions like Uttarakhand, prone to such disasters. It has also been noted that community education is vital for raising awareness and promoting fire management strategies. Local communities should be well informed about the negative impacts of forest fires and must be educated on best practices for forest fire management. Outreach programs and workshops can foster a greater understanding of the importance and outcomes of FF management. Creating forest-friendly agricultural practices is essential, especially in regions where agriculture is a primary livelihood, and the agricultural fields lie near forests. Moreover, it is crucial to establish rapid response teams from the forest department to manage FF when they occur. These teams can be trained to safely and humanely address situations involving FFs threatening human safety or agricultural interests while ensuring their safety. Based on these insights, the following recommendations are proposed to improve community involvement in forest fire prevention and management.

## *5.1 Empower Local Stewardship*

Local communities must be actively involved in forest management decisions, including participating in fire risk assessments, developing prevention strategies, and engaging in fire-fighting efforts. Engaging community members leverages their local knowledge and fosters a sense of ownership and responsibility toward forest conservation.

### 5.1.1 Bottom-Up Approach

Adopting a bottom-up approach that empowers local communities to lead initiatives can create frameworks to express their needs and contribute to policy-making, ensuring that management strategies are tailored to each forest area's specific ecological and social context.

### 5.1.2 Key Stakeholders

Recognize the local communities as key stakeholders by legally incorporating their rights to use and manage forest resources. This legal empowerment can lead to more sustainable forest management as the community becomes directly responsible for the consequences of their actions.

### 5.1.3 Community Empowerment

Providing training in sustainable agriculture, forest management, and emergency response to empower communities can help transform local inhabitants from passive observers to active participants in forest fire prevention and management.

## 5.2 *Enhance Public Awareness and Education*

Boosting public awareness and education is crucial as the primary defence, equipping communities with knowledge about the risks and complexities of forest fires. Through awareness, individuals can be transformed into vigilant guardians of their environment, able to recognize the precursors of FFs and take pre-emptive action. Education programs can build a foundation of understanding, enabling communities to engage with fire management practices not as passive subjects but as informed participants. This can also be achieved through exploring the strategies and methodologies for broadening the public's understanding of FFs and nurturing a culture of proactive environmental stewardship.

### 5.2.1 Risk Communication

Develop a consistent and clear communication strategy that educates the public about forest fire risks and the behaviours that can mitigate them, including information about the causes of FFs and the best practices to prevent them.

### 5.2.2 Awareness Campaigns

Use targeted awareness campaigns to reach different population segments to spread messages about fire safety and management utilizing various media, including radio, television, social media, and public gatherings.

### 5.2.3 Educational Programs

Implement structured educational programs within schools and local institutions to build a generation knowledgeable about forest ecology and the role of fire in forest ecosystems. These programs should focus on the risks of uncontrolled fires and the benefits of controlled burns.

### 5.2.4 Information Accessibility

It ensures that information regarding forest fire management is available in local languages and accessible to all community members, including those in remote areas. Accessibility includes language translation and the use of appropriate technologies to disseminate information and its outreach to even the most remote areas.

## *5.3 Integrate Traditional and Scientific Knowledge*

Combining local practices with scientific knowledge can lead to better FF management, as the local knowledge developed over generations can be valuable when paired with modern science. Blending this knowledge to improve the community's approach to preventing and managing FFs can pave the way for FF mitigation in the region.

### 5.3.1 Knowledge Synthesis

Creating platforms where traditional forest management practices are evaluated alongside scientific research to develop best practices as this synthesis would respect the value of both types of knowledge and seek to integrate them into a cohesive management strategy towards FF management.

### 5.3.2 Respect for Traditions

Valuing and integrating Indigenous and local communities' traditional practices and cultural norms into the broader forest management practices as many traditional practices have evolved to manage fire risk effectively and should be preserved and promoted for the coupled management of TKS and forests in the larger perspective.

### 5.3.3 Ecological Fire Management

Develop ecological fire management practices incorporating controlled burns to reduce fuel loads and scientific monitoring to assess their impact. This strategy would aim to maintain biodiversity, soil quality, and forest health and help proactively manage forests for sustainable management.

### 5.3.4 Participatory Research

Encouraging local communities to engage in scientific research on FFs, as this citizen science-based approach of involvement can foster a better understanding of fire ecology and contribute to the development of more effective management practices, blending scientific insights with local knowledge and will lead to more robust and dynamic FF management protocols.

## 5.4 *Policy Support and Sustainable Development*

Managing FFs is daunting, and policy-based prescriptions can play a significant role as a proactive approach. Policy frameworks and sustainable development initiatives provide the structural support for effective FF management as they are the scaffolding for realizing community efforts and scientific recommendations. Policies that are responsive to the needs of local ecosystems and economies can empower communities and incentivize sustainable practices. Similarly, fostering sustainable development strategies can alleviate the pressures that lead to fire-prone conditions, examining the need for robust policy support and outlining the pathways toward sustainable economic growth that can reduce the risk and impact of forest fires.

### 5.4.1 Policy Frameworks

Advocate for policy frameworks that support sustainable forest management and provide clear guidelines for community-led fire management, as these policies would include incentives for preserving forest health and penalties for practices that increase fire risk.

### 5.4.2 Economic Incentives

Establishing pathways for economic incentives such as subsidies for sustainable land-use practices, grants for community fire management initiatives, and financial support for developing and selling non-timber forest products through their community-managed forests.

### 5.4.3 Alternative Livelihoods

It is creating programs that support the development of alternative livelihoods, which can reduce the economic pressures that lead to practices contributing to fire risk. These might include agroforestry, ecotourism, the sustainable harvesting of forest products, and the development of sustainable forest resource management.

### 5.4.4 Intersectoral Collaboration

Building on an intersectoral framework that combines the forestry, agriculture, health, and environmental sectors through a collaborative approach can ensure that FF management is integrated into broader land-use planning and development strategies.

### 5.4.5 Collaborative Efforts for Fire Management

Collaborative efforts bring together various stakeholders' diverse strengths and perspectives, from local communities to international agencies, and can act as a cohesive and comprehensive approach to fire management that can be established by uniting the efforts of different sectors. This will also contribute towards highlighting the importance of partnership, shared responsibility, and cross-boundary coordination in creating a unified front against the destructive force of forest fires, aiming for a future where the risk of wildfire is a challenge met with solidarity and shared resolve.

## 5.5 *Multi-Stakeholder Partnerships*

Building partnerships with various stakeholders, from local communities to international organizations, from a broader perspective, as these partnerships can share resources, expertise, and information to tackle the complex challenge of forest fire management.

### 5.5.1 Shared Responsibility

It encourages a culture of shared responsibility where forest fire management is not seen as the sole responsibility of any single entity, either a van panchayat or the forest department, but rather as a collective effort that requires input and action from all stakeholders involved.

### 5.5.2 Cross-Boundary Coordination

Since fires do not respect administrative boundaries, fostering coordination across different jurisdictions is highly necessary, including developing agreements for sharing resources and information during cross-boundary fire events.

### 5.5.3 Information Sharing

It is developing information-sharing systems that enable stakeholders to access real-time data on fire risks, ongoing fires, and resources available for fire management. This will support more effective planning and response strategies for the local and regional government.

## 6 Conclusions

The present study shows that existing scientific evidence plays a secondary role in legitimizing policy proposals for FF management. Instead, the allocation of responsibility, the timeframe of interventions, and the involvement of the affected community in both pre- and post-fire management are crucial and play the primary factor for FF management. Additionally, the community's perception of fire-based natural resource management significantly influences the effectiveness of these policies on the ground level.

Therefore, as we consider the future of FF management, particularly in regions like Uttarakhand, a strategic, inclusive, and proactive approach is paramount. Establishing a continuous cycle of assessment, action, and adaptation is essential. This means regular monitoring of forest health, swift implementation of fire prevention strategies, and the flexibility to adjust these strategies based on outcomes and evolving environmental conditions. Strengthening the collaboration between communities, policymakers, and scientists will ensure that actions taken are well-informed and sustainable. It is also vital to secure funding and political support for fire management initiatives, ensuring that they are not just sporadic responses to crises but are integrated into long-term regional planning, which involves the decision-making from the state, regional, national, and global levels in a proactive way.

Innovation will also play a vital role in the advancement of FF management in the region as the increasing use of Machine learning (ML) techniques for FF detection and adoption of new early detection and monitoring technologies, such as satellite imagery and AI-driven predictive modelling will improve the FF management efficiency at the governmental level. The development of these tools must be accompanied by training programs for local fire managers and community leaders, enhancing their capacity to utilize such technologies effectively and suitably. Additionally, integrating FF management into broader climate change mitigation and adaptation discussions will be critically necessary as aligning fire management goals with climate objectives will create a cohesive approach that addresses the immediate risks of FF and contributes to the overall resilience of forest ecosystems in the face of a changing climate, particularly in the Himalayan region.

**Acknowledgements** The authors are thankful to Prof. (Dr.) Sunil Nautiyal, Director (GBPNIHE) for providing research support. The authors are also grateful to the community resource persons, especially Mrs. Rameshwari Devi (Van Panchayat Sarpanch, Pali), Mr. Gajendra Pathak (Sitlakhet) and the women of Mahila mangal dals from the study region for their endless support and provision of important information.

## Abbreviations

| | |
|---|---|
| FF/FFs | Forest fire/forest fires |
| FSI | Forest Survey of India |

| | |
|---|---|
| FAO | Food and Agriculture Organization |
| UKFD | Uttarakhand Forest Department |
| ML | Machine Learning |
| AI | Artificial Intelligence |
| UNEP | United Nations Environment Program |
| NRSA | National Remote Sensing Agency |

## References

Arnold J (2018) Feeling the fires of climate change. In: Bladow K, Ladino J (eds) Affective ecocriticism: emotion, embodiment, environment. University of Nebraska, USA, pp 95–116

Barwa S, Singh H (2024) Climate-induced changes in forest ecosystems of Indian Himalayan region: a meta-analysis. In: Singh H (eds) Urban forests, climate change and environmental pollution. Springer, Cham. https://doi.org/10.1007/978-3-031-67837-0_39

Bowman D, Williamson G, Yebra M, Lizundia-Loiola J, Pettinari ML, Shah S, Chuvieco E (2020) FFs: Australia needs national monitoring agency. Nat 584(7820):188–191

Cantin AS, De Groot WJ, Wotton M, Newbery A, Gowman LM (2013) Global wildland fire season severity in the 21st century. For Ecol Manag 294:54–61

Chas-Amil M, Nogueira-Moure E, Prestemon JP, Touza J (2022) Spatial patterns of social vulnerability in relation to wildfire risk and wildland-urban interface presence. Landsc Urban Plann 228:104577

Croker AR, Woods J, Kountouris Y (2023) Changing fire regimes in East and Southern Africa's savanna-protected areas: opportunities and challenges for indigenous-led savanna burning emissions abatement schemes. Fire Ecol 19(1):63

FAO (2007) Fire management—global assessment 2006. A thematic study prepared in the framework of the global forest resources assessment 14 2005. FAO, Rome. Retrieved on December 4, 2009 from http://www.fao.org/docrep/009/a0969e/a0969e00.htm

FAO & UNEP (2020). State of the world's forests 2020: forests, biodiversity and people. Rome (Italy): Food and Agriculture Organization of the United Nations and UN Environment Programme. (Accessed 2020 Sep 10). Available from: http://www.fao.org/state-of-forests/en/ Flannigan

FSI (2021) State of forest report. Forest survey of India. Chap 5, Page 1–32. ISBN 978-81-950073-1-8

FSI (2022) Forest fire and climate change. https://www.forestresearch.gv.uk/research/climate-change-impacts/forest-fires-and-climate-change/

FSI (2024) Large forest fires monitoring programme | forest survey of India (fsiforestfire.gov.in)

Gajendiran K, Kandasamy S, Narayanan M (2023) Influences of wildfire on the forest ecosystem and climate change: a comprehensive study. Environ Res, 117537

Hardesty J, Myers R, Fulks W (2005) Fire, ecosystems, and people: a preliminary assessment of fire as a global conservation issue. Geo Wri for 22:78–87

Jolly WM, Cochrane MA, Freeborn PH, Holden ZA, Brown TJ, Williamson GJ, Bowman DM (2015) Climate-induced variations in global wildfire danger from 1979 to 2013. Nat Comm 6(1):7537

Jones MW, Abatzoglou JT, Veraverbeke S, Andela N, Lasslop G, Forkel Smith AJP, Burton C, Betts RA, van der Werf GR, Sitch S, Canadell JG, Santín C, Kolden C, Doerr SH, Le Quéré C (2022) Global and regional trends and drivers of fire under climate change. Rev Geophys 60(3):e2020RG000726

Kala CP (2015) Forest structure and anthropogenic pressures in the pachmarhi biosphere reserve of India. J for Res 26(4):867–874

Kala CP (2023) Environmental and socioeconomic impacts of forest fires: a call for multilateral cooperation and management interventions. Nat Haz Res 3:286–294

Kala CP (2013) Climate change and challenges of biodiversity conservation. In: Kala CP, Silori CS (eds), Biodiversity, communities and climate change. The Energy and Resources Institute, New Delhi, pp 259–269

Kodandapani N (2006) Fire regimes and their ecological effects in seasonally dry tropical ecosystems in the Western Ghats. Michigan State University, India

Kumar A, Kumar P, Singh H, Kumar N (2021) Modulation of plant functional traits under essential plant nutrients during seasonal regime in natural forests of Garhwal Himalayas. Plant Soil 465:197–212

Kumar A, Dwivedi GK, Tewari S, Paul J, Anand R, Kumar N, Kumar P, Singh H, Kaushal R (2020) Carbon mineralization and inorganic nitrogen pools under Terminalia Chebula Retz.-based agroforestry system in Himalayan foothills India. Forest Sci 66(5):634–643

Kumar N, Singh H, Kumar A, Singh AK, Pandey PK, Kumar A (2024). Forest-fire-induced land degradation. In: Mishra G, Giri K, Singh S, Kumar M. (eds) Sustainable land management in India. Springer Nature, Singapore. https://doi.org/10.1007/978-981-97-5223-2_4

Malik A, Kumar K, Kumari G, Kumar N, Singh H (2024) Application of functional traits in modelling productivity and resilience under climate change published by springer, pp 77–96. ISBN: 978-981-97-1509-1

Moritz MA, Batllori E, Bradstock RA, Gill AM, Handmer J, Hessburg PF, Syphard AD (2014) Learning to coexist with wildfire. Nat 515(7525):58–66

Negi GCS, Chauhan DS, Bisht DS, Pant H, Arya SC, Deorari M, Rawat DS, Sundriyal RC (2021) Use of pine needles for making environment-friendly products and avoidance of forest fire. Pant GB National Institute of Himalayan Environment, Kosi-Katarmal, Almora, 263643, pp 1–40. ISBN 978-93-5593-715-5

Rawat N, Purohit S, Negi GS, Pant D (2017) Perception of local people towards blazing mountain due to forest fire in changing climatic scenario of Uttarakhand Himalaya: a need for scientific intervention for sensitization and conservation and livelihood sustainability of region. pp 28–32

Reis M, Graça PMLA, Yanai AM, Ramos CJP, Fearnside PM (2021) Forest fires and deforestation in the central Amazon: effects of landscape and climate on spatial and temporal dynamics. J Environ Manag 288:112310

Rodríguez-Trejo DA, Fulé PZ (2003) Fire ecology of Mexican pines and a fire management proposal. Int J Wildland Fire 12(1):23–37

Sankey T, Tango L, Tatum J, Sankey JB (2024) Forest fire, thinning, and flood in wildland-urban interface: UAV and lidar-based estimate of natural disaster impacts. Landscape Ecol 39(3):58

Savita, Sharma SK, Singh H, Singh O (2017) Forest fire: a review. Indian for 143(8):719–728

Sett T, Nikam BR, Singh H, Purohit S (2024) Hydrological variability in Indian forest ecosystem: analysis of drought resilience, recovery and water use efficiency in moist and dry deciduous forests. In: Singh H (eds) Urban Forests, Climate Change and Environmental Pollution. Springer, Cham. https://doi.org/10.1007/978-3-031-67837-0_37

Sewak R, Vashisth M, Gupta L (2021) Forest fires in India: a review. J Univ Shanghai Sci Technol 23(7):247–259

Sharma S, Pant H (2017) Vulnerability of Indian Central Himalayan forests to fire in a warming climate and a participatory preparedness approach based on modern tools. Curr Sci 2100–2105

Singh H, Kumar N, Singh R, Kumar M (2023) Assessing the climate change impact on the habitat suitability of the range-restricted bird species (*Catreus wallichii*) in the Indian Himalayan ecosystem. Environ Sci Pollut Res. https://doi.org/10.1007/s11356-023-30789-x

Singh H, Kumar M (2022). Climate change and its impact on Indian Himalayan forests: current status and research needs. In: Rani S, Kumar R (eds) climate change. Springer Climate. Springer, Cham. pp 223–242, https://doi.org/10.1007/978-3-030-92782-0_11

Singh H, Kumar N, Kumar M, Singh R (2020) Modelling habitat suitability of western tragopan (*Tragopan melanocephalus*) a range-restricted vulnerable bird species of the Himalayan region,

in response to climate change. Climate Risk Manag 29:100241. https://doi.org/10.1016/j.crm.2020.100241
Singh H, Kumar N, Kumar A (2024) Enhancing resource use efficiency in crops through plant functional traits published by springer, pp 97–117. ISBN: 978-981-97-1509-1
Singh M, Jyoti, Kumar N, Singh H (2024) Plant functional traits assisted crop adaptation to abiotic and biotic stress. Springer, pp 239–255. ISBN: 978-981-97-1509-1
Singh H (2021) Process-based carbon sequestration study in relation to the energy-water-carbon flux in a forest ecosystem. In: Climate impacts on sustainable natural resource management Kumar P, Singh RK, Kumar M, Rani M, Sharma (eds). CRC Press Taylor & Francis Group, United States, pp 336–361
Singh H (2024a) Climate change and forests: biological perspectives. In: Singh H (eds) Forests and climate change. Springer Nature, Singapore. https://doi.org/10.1007/978-981-97-3905-9_1
Singh H (2024c) Climate change and environmental pollution: biological consequences on urban forests. In: Singh H (eds) urban forests, climate change and environmental pollution. Springer Nature, Cham, Switzerland. https://doi.org/10.1007/978-3-031-67837-0_1
Singh H (2024b) Forests and climate change: biological perspectives on impact, adaptation, and mitigation strategies (ed), Springer Nature Singapore. https://doi.org/10.1007/978-981-97-3905-9
Singh H (2024d) Urban forests, climate change and environmental pollution: physio-biochemical and molecular perspectives to enhance urban resilience (ed), Springer Nature, Cham, Switzerland. https://doi.org/10.1007/978-3-031-67837-0
Souza-Alonso P, Omil B, Sotelino A, García-Romero D, Otero-Urtaza E, Lorenzo Moledo M, Merino A (2024) Service-learning to improve training, knowledge transfer, and awareness in forest fire management. Fire Ecol 20(1):1–19
Stephens SL, Burrows N, Buyantuyev A, Gray RW, Keane RE, Kubian R, Van Wagtendonk JW (2014) Temperate and boreal forest mega-fires: characteristics and challenges. Front Ecol Environ 12(2):115–122
Toy-Opazo O, Fuentes-Ramirez A, Palma-Soto V, Garcia RA, Moloney KA, Demarco R, Fuentes-Castillo A (2024) Flammability features of native and non-native woody species from the southernmost ecosystems: a review. Fire Ecol 20(1):21
UKFD (2024) Uttarakhand forest department Uttarakhand Fire Report

# Cross-Border Collaborative Forest Fire Management: An Integrated Approach for Forest Fire Management in the Era of Climate Change

**Tannu Ruhil, Hukum Singh, Manisha Thapliyal, and Santan Barthwal**

**Abstract** Natural calamities transcend political borders, necessitating international collaboration for effective mitigation. Climate change has extended fire seasons, putting societies at increasing risk from wildland fires. This has prompted countries and international organizations to collaborate on sustainable fire management strategies. Such efforts require a structured and coordinated global approach, with resource sharing to support national fire suppression endeavors. Forest fires are influenced by climate, topography, vegetation, land-use changes, and human activities. The complex geopolitics of international border areas make managing these fires more challenging than those within national boundaries. This highlights the urgent need for enhanced cooperation among neighboring countries to effectively tackle forest fires and protect wildlife. This chapter explores international cooperation in fire management, identifying associated challenges and proposing solutions. Setting aside political tensions is essential for efficient forest fire management. Well-designed forest fire policies can reduce the impact of these fires, but the lack of globally accepted protocols exacerbates the situation. Although several organizations focus on wildland fires, their number is limited, and numerous studies predict an increase in such fires. Therefore, urgent global measures are required to address this growing threat.

**Keywords** Forest · Fire management · Climate change · International cooperation

---

T. Ruhil (✉) · H. Singh · S. Barthwal
Plant Physiology Discipline, Genetics and Tree Improvement Division, Forest Research Institute, Dehradun, India
e-mail: tannu15m@gmail.com

M. Thapliyal
Forest Tree Seed Laboratory, Silviculture and Forest Management Division, Forest Research Institute, Dehradun, India



H. Singh and N. Kumar (eds.), *Forest Fire and Climate Change*,
https://doi.org/10.1007/978-3-031-89967-6_14

# 1 Introduction

Fire plays a crucial role in forest ecosystems, where controlled forest fires can benefit the regeneration and succession of tree species (Chandra and Bhardwaj 2015). Still, large-scale uncontrolled fires cause enormous harm to wildlife, their habitats, forest environment, and structure, reducing biodiversity and putting human life and property at risk (Blanchi et al. 2014; Singh 2024d). The recent rise in severe climatic conditions and population growth has increased forest fire cases worldwide, threatening forest ecosystems (Boer et al. 2020; Kumar et al. 2024; Singh 2024c). This made studying several driving factors of forest fires significant and the requirement to set up productive prediction models to aid in forest fire management (Quan et al. 2022). The increased global fire activity in the last decades is associated with increased extreme events and lengthened fire season (Jolly et al. 2015). The heightened risk of fires over large areas has led to unprecedented fire seasons in various regions, such as the United States and Canada (Higuera and Abatzoglou 2020).

Additionally, impactful fires are now co-occurring across diverse geographic locations (Abatzoglou et al. 2021). Climate change contributes to this and is anticipated to increase the likelihood of forest fires, as illustrated in Fig. 1. The changing patterns of fire risk have implications for international cooperation regarding the allocation of resources for fire suppression efforts (Singh 2024b). The increased danger leads to greater requirements for equipment, firefighting crews, aircraft, and good fire management between neighboring countries. Increased incidences of forest fires are making forests net carbon emitters, especially in regions like Indonesia, the United States, and Australia (Harris et al. 2021).

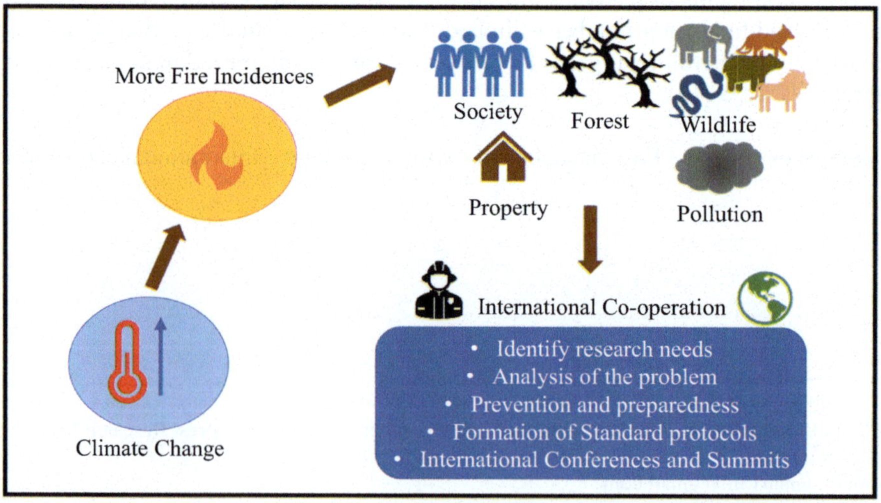


**Fig. 1** Climate change is increasing fire incidences, negatively impacting the environment and human lives

Bloem et al. (2020) proposed two strategies to address this issue: proactive prevention and reactive response. The proactive measures involve adaptation and mitigation efforts like selective thinning, controlled burning, and prime forest management to minimize the impact of fires and prospective emissions (Houtman et al. 2013). The fire response activities include fire management and suppression. The researchers have indicated that various factors facilitate the start and expansion of forest fires (Savita et al. 2017; Singh and Kumar 2022; Singh 2021). Varied fire management systems in transboundary regions led to different fire paths. A consistent data source and statistical approach are essential for analyzing variations in forest fire triggers across border regions in countries. This leads to assessing different management strategies and how these triggers contribute to fire occurrences within varied political and socio-economic settings.

## 2 Current Fire Management System

Variations in topography influence the nearby climate and the types and arrangement of flammable materials, leading to diverse fire conditions (Lee et al. 2018). Meanwhile, combustible vegetation is essential for forest fires to ignite and spread (Ye et al. 2017). Various meteorological indicators are frequently utilized to evaluate forest fire risk globally. Notable examples include Australia's Forest Fire Danger Index (FFDI) and Canada's Forest Weather Index (FWI), both of which are extensively used around the world for this assessment (Sanabria et al. 2013). Human actions can lead to the initiation of fires, and they are mainly responsible for forest fires worldwide. However, anthropogenic construction of roads hinders fire spread and serves as conduits for firefighting (Renard et al. 2012; Singh et al. 2024; Malik et al. 2024).

Charnley et al. (2017) examined factual data on forest and forest fire regulation carried out by different organizations (state, federal, and private corporations) in southcentral Oregon. Their study investigated how variations in response and management approaches impact the resilience of forest ecosystems vulnerable to severe forest fires. They assessed the resilience of the ecosystem by employing a range of metrics related to forest structure and the potential for high-severity forest fires within an agent-based model (Spies et al. 2017). Quantitative data concerning fire and forest management were extracted from agency databases.

CHANS, which stands for Coupled Human and Natural Systems, aids in enhancing ecosystem management and supporting humans' adaptation to social and environmental changes (Leslie et al. 2015). This method is rarely used to study fire-prone ecosystems. Applying this approach to forest fire management in fire-prone forests could aid in developing adaptation strategies and policy adjustments for private and public lands, promoting a balanced coexistence between humans and forest fires (Moritz et al. 2014). Further research also helps identify important factors limiting stakeholders' ability to handle disruptions and environmental changes effectively, highlighting critical points in the system where targeted interventions can be made.

Numerous researchers suggest that dry forest ecosystems exhibit feedback loops that are positive yet maladaptive. The feedback mechanisms, such as forest fire risk reduction via insurance policies, fire suppression efforts, and afterfire government-aided programs, may worsen the issue (Calkin et al. 2015). They do so by increasing forest flammable situations, raising the risk of severe forest fires, and improving safety for residents living in the Wildland-Urban Interface (WUI). Facilitating population expansion in the WUI amplifies the risk of forest fire exposure, the expenses associated with fire suppression for safeguarding residences and properties, and potential forest fire damages. Nevertheless, the diversity in ecosystem characteristics offers the prospect that various interactions within CHANS and multiple feedback mechanisms are in effect, but not all are disadvantageous. Some researchers propose conducting comparative assessments of CHANS in different geographical settings to gather insights to improve ecosystem management. Some suggested conducting experiments related to management via employing several approaches for addressing challenges concerned with natural resource management, aiming to determine the most effective way to enhance resilience (Cumming et al. 2013).

Integrating detection and initial attack system planning is necessary to establish a suitable equilibrium and effectively meet the overarching objectives of fire management (Martell 2015). Fire management agencies typically utilize a combination of detection methods, which can involve manned lookout towers, digital cameras, sensors, aerial patrols, satellite imagery, and reports from the public who spot fires while traveling through forested regions. There is a need to decide the place of towers, their activation timings, and the timing and place to direct aerial detection patrols by fire managers. A tremendous surge has been observed over the last few years in publications concerning image processing systems and remote sensing devised for fire detection (Mahdipour and Dadkhah 2014). The available literature dealing with fire detection decision support systems is limited. Doubts persist about the effectiveness of satellite technology present with fire managers for detection (Martell 2015). However, it is logical to expect that the utilization of drones for mapping extensive fires within the restricted airspace enforced by fire management agencies over significant active fires will grow. Eventually, drones will likely replace piloted fixed-wing aircraft for aerial detection patrols (Martell 2015).

It is vital to develop thorough methods for evaluating detection systems' effectiveness, which aerial patrols, lookout towers, and camera systems can do. Additionally, the challenge of public reporting of fires across various jurisdictions remains a significant issue that detection researchers have not sufficiently addressed. Improving understanding of when and where to prioritize aerial patrols, lookout towers, and public reports is essential for effective fire management. Strategic management encompasses making decisions about the allocation and acquisition of aircraft and the recruitment and deployment of fire crews. One early example of Operations Research/Management Science (OR/MS) being used for strategic air tanker management is exemplified in a simulation model developed by Martell et al. (1984), later refined by McAlpine and Hirsch (1999), and still utilized by the Ontario Ministry of Natural Resources and Forests (OMNRF). There are several simulation models available for strategic planning purposes.

## 3 Cross-Border Cooperation and Establishment

Recent research shows that forest fires are a frequent natural disturbance in the Northeast Asia forest region (Ostroukhov et al. 2020). According to the Forest Survey of India report, approximately 55% of the nation's forest areas are at risk of forest fires, with 6% facing a high risk of severe fire damage. Most of these fires are attributed to human activities (Attri et al. 2020). In India, at the village level, Joint Forest Management Committees have been created to engage local communities in protecting and conserving forests, including preventing fire damage.

The international wildland fire community has developed preliminary concepts, guidelines, and templates for fire management (Alexandris 2011). Comprehensive operational standards are essential to facilitate the sharing firefighting resources, including management personnel, aviation, and equipment. Recent advancements within various multilateral organizations, like the Association of Southeast Asian Nations (ASEAN), the UN Economic Commission for Europe (UNECE), the Asia–Pacific Economic Cooperation (APEC), the Council of Europe (EUR-OPA), the European Union (EU), and the Southern African Development Community (SADC), demonstrate an increasing trend among countries to enhance regional and cross-border collaboration in fire management capacities.

In the 1990s, the Global Wildland Fire Network (GWFN) was established as a voluntary effort led by the Global Fire Monitoring Center (GFMC) and the UNECE/FAO Team of Specialists on Forest Fire, as noted by Kaiser et al. (2012). Supported by the United Nations International Strategy for Disaster Reduction (UNISDR), the GWFN, with the facilitation of the GFMC, aims to improve international collaboration in managing wildland fires. This includes improving forest fire preparedness, prevention, and suppression and developing standardized response procedures for forest fires.

Following prolonged episodes of widespread fires and smoke, member states of the ASEAN commenced discussions for an agreement to address regional air pollution stemming from forest and land-use fires. The Transboundary Haze Pollution Agreement was ratified in June 2002. It became effective on November 25, 2003, and was the first ever regional arrangement with nine contiguous states with a common motive to handle transboundary haze pollution from forest and land fires.

The devastating forest fire in the Greater Khingan Mountains in China in 1987 led to significant damage to natural resources and threatened human safety (Huang et al. 2020). After this incident, China has improved forest fire control in several ways, including preventing border fires. Nevertheless, the rise in forest fire occurrences globally and the complexity within the cross-border areas make exploring driving mechanisms and establishing cooperation in the essential areas (Wotton et al. 2010).

The integration of Indian Remote Sensing Satellite (IRS) systems like IRS-P6 and IRS-1D, along with data from TERRA/AQUA Moderate Resolution Imaging Spectroradiometer (MODIS), National Oceanic and Atmospheric Administration—Advanced Very High-Resolution Radiometer (NOAA-AVHRR), Defense Meteorological Satellite Program-Operational Line Scan System (DMSP-OLS), and Environment Satellite (ENVISAT), plays a crucial role in detecting forest fires, monitoring their progression in real-time, promptly assessing damage, and planning mitigation efforts. In India, to enhance disaster management, the Decision Support Center (DSC) has been established at the National Remote Sensing Centre (NRSC) under the Disaster Management Support Programme of the Department of Space (DOS). Recognizing the significance of forest fire management, the Indian Forest Fire Response and Assessment System (INFFRAS) has been implemented as part of NRSC's DSC initiatives. INFFRAS integrates data from satellite sensors and ground sources using a spatially and temporally explicit GIS analysis framework. It provides fire alerts, including daily daytime fire locations from TERRA/AQUA MODIS and nightly fire locations derived from DMSP-OLS. The advancement of fires can be tracked using daily fire location data from both day and night provided by MODIS/DMSP-OLS, along with the expansion of burnt areas obtained from temporal high-resolution datasets. Mapping episodic fire events and assessing burnt areas can be achieved using moderate and high-resolution optical datasets.

Fire Hawk is a computer-aided electronic forest fire detection system controlled by a human operator. The optical sensor system is also used that is subsequently evaluated by an operator (Yadav et al. 2018). Firewatch is a digital, terrestrial, remote surveillance system that can observe large wooded regions and analyze, evaluate, link, and store the collective data. The system's sensitivity, precision, and dependability allow for the early detection of forest fires. The radio-acoustic sounding system is used to detect surface and crown fires early. MODIS is a satellite system NASA (USA) launched for real-time detection of active forest fires.

The EU Civil Protection Mechanism was created in 2001 and subsequently strengthened in 2007, serving as a pivotal entity in preventing forest fire risks and coordinating forest firefighting efforts within the EU. A total of 32 countries are currently a member of this mechanism (Goldammer 2013).

In October 2010, the Russian Federation convened the first "International Conference on Forest Fires: Management and International Collaboration to Prevent Forest Fires in the APEC Region" to promote unity among member nations in addressing disaster risks (Vidal et al. 2011). After thoroughly analyzing forest fires regionally and globally, the conference highlighted the vital need for collaborative efforts, mutual assistance, and international cooperation to reduce forest fire risks (Goldammer 2013). This led to the creation of the 'Khabarovsk Recommendations on Forest Fire Management and International Cooperation in the APEC Region.

Forest fires are among the most frequent disasters in Southern Africa, "The Fire Continent", leading to substantial financial loss and human fatalities (Archibald et al. 2010). The rising population growth and dense urban planning led to elevated fire disasters and higher human casualties (Strydom et al. 2018; Kumar et al. 2020a, 2020b). Strydom et al. (2018) examined the temporal and spatial patterns of forest fires in the SADC by analyzing an 11-year dataset of satellite-derived active fire hotspots using a freely available geographic information system. The study found significant variability in annual fire frequency across most SADC countries. Increased anthropogenic agroforestry activities were identified as a primary factor contributing to heightened fire activity in various SADC eco-regions, particularly in woodlands, savannahs, and montane grassland areas, which are the most susceptible to fires. In response to these findings, a regional fire management program for SADC was proposed in 2010, aiming to foster cooperation in addressing fire-related challenges (Goldammer 2013). Table 1 gives insight into worldwide studies on fire impacts in border areas.

**Table 1** Cross-border studies to study fire impacts in different regions of the world

| S. No | Study area | Method | Aim | References |
|---|---|---|---|---|
| 1 | Cross-border area between North Korea, Russia, and China | Forest fire probability prediction model | To evaluate the relative influence of factors contributing to fires, thereby aiding in creating forest fire risk assessment and management strategies | Quan et al. (2022) |
| 2 | Fire-prone landscape of southcentral Oregon | Agent-based landscape model | Assessment of trends in forest structure and fire metrics based on ownership | Charnley et al. (2017) |
| 3 | Italy-France maritime cooperation area | Forest fire simulation modeling | To evaluate fine-scale forest fire hazard, spread, and community exposure | Salis et al. (2023) |
| 4 | West Kalimantan, Indonesia | Qualitative research method employing a descriptive and analytical approach | To examine the collaboration between local government, West Kalimantan BPBD, and other pertinent agencies in addressing forest and land fires | Pratiwiand Nugroho (2021) |
| 5 | The border between Turkey (Yayladağı) and Syria | Geographic Information Systems were used to identify and assess the origins of forest fires and the impacted areas | To investigate the causes of fire and their evaluation using statistical methods | Atasoy and Geçen (2014) |
| 6 | Northern Olomouc Region in the Czech Republic and Opole Region in Poland | The data were analysed through set theory and data organization techniques | Identification and investigation of important factors affecting the success of cross-border cooperation | Molak and Struk (2021) |

## 4 Intricacies and Possible Solutions

Climate change has resulted in drought conditions (Singh 2024a), leading to increased forest fire seasons (Woolford et al. 2014), events (Wotton et al. 2010), intensity, and burned area (Wotton et al. 2017). The suppression costs are expected to rise depending on climate change scenarios (Hope et al. 2016). The operational consequences of climate change on forest fire management agencies include a decrease in suppression effectiveness, as they encounter more instances where their current efforts are insufficient to control escaped forest fires due to anticipated increases in forest fire intensities (Wotton et al. 2017; Kumar et al. 2021a, 2021b). Additionally, changes in forest fuel composition, structure, and load resulting from decades of forest fire suppression further compound the difficulty in managing forest fires. This phenomenon is known as the "fire paradox", where the efforts to suppress forest fires that maintain natural stands inadvertently encourage the accumulation and continuity of fuel. In the boreal forest, fire suppression has resulted in a decrease in landscape-level diversity and mosaic patterns.

The rising susceptibility of societies towards immediate and indirect impacts of forest fires, along with the cross-border extent and repercussions of such events, are compelling nations and global organizations to articulate shared objectives in bolstering their capacity for sustainable and coordinated forest fire management. This requires more formal and systematic cooperation across the borders. Direct impacts, such as lives, firefighting costs, and asset loss, are captured in developed countries. Still, collecting indirect expenses, such as health impacts, ecosystem impacts, injuries, etc., is less common (Moore 2019). Because of the unique geopolitical factors in border regions, it becomes challenging for countries to engage in cross-border cooperation regarding fire management when a forest fire occurs.

Consequently, controlling the fire promptly and effectively becomes difficult, increasing the likelihood of the fire posing threats to local forest resources and residents (Quan et al. 2022). Reports from agencies, governments, and the media and public perception of forest fires often highlight the visible destruction and losses. The data readily accessible usually centers on damage to human property and expenses related to fire fighting, but it frequently overlooks casualties, injuries, health effects, and declines in ecosystem services (Doerr and Santín 2016).

In contemporary times, forest fire early detection methods include utilizing fire lookouts stationed in towers, ground and aerial patrols, and public hotlines. However, operator fatigue, day and year time, and geographic location constrain accurate human observation (Attri et al. 2020). While certain fire management agencies employ satellites for fire monitoring in remote regions, the technology's limitations restrict its effectiveness for detecting new fires outside these remote areas, especially when fires present minimal risk to public safety and forest resources (Martell 2015). Fire managers now face political obstacles, including issues like forest fire risks to the health of both the public and firefighters, as well as concerns regarding smoke and emissions of greenhouse gases (Goldammer and Pronto 2016).

In the past, international legal agreements for environmental protection, such as the Rio and Ramsar Conventions, have also been made. However, the ones exclusively addressing wildland fires are not available. Many studies have been limited to a single region or country, highlighting issues with comparability due to differences in analytical methods, data sources, and study scales (Quan et al. 2022). Additionally, no universally recognized established protocols offer standardized processes and procedures for nations involved in providing and receiving aid during wildland fire emergencies. These protocols guarantee interoperability, efficiency, and safety among collaborating parties (Goldammer 2013). Numerous terms are used to describe or categorize fires based on the type of landscape burned, the kind of fuel involved, the social and cultural context, and the regulatory framework (Robinne et al. 2018). The fire nature decides fire science research requirements (Moore 2019). Those extreme fire events where the assets are at higher risk need proper response research to reduce fatalities and losses. Until now, less focus has been given to addressing various fire causes and preventing damaging patterns of recurring fire and deterioration in burnt areas.

Ongoing interdisciplinary research highlights the intersection of land management with economic, public, and environmental challenges. Incorporating socio-cultural contexts and ecological needs into fire-prone landscapes is becoming more critical. There is an increasing need to integrate socio-cultural contexts and ecological requirements into landscapes susceptible to fires. Additionally, the history and comprehensive analysis of fire research remain underexplored. National-level leadership is still lacking in driving forward fire research initiatives. Most of the fire research designed and organized is predominantly conducted in nations with well-developed economies and robust, diverse research capabilities, facilitated through various entities such as government agencies, universities, the private sector, and individual researchers. There is a growing recognition of the importance of understanding both the causes of intentional and accidental fires.

Furthermore, understanding the complexities of land use, urban planning, management strategies, community resilience, and the wider societal reactions towards fire incidents. As a result, there is a growing focus on social sciences and related disciplines. This imperative is universal, applying to both developed and developing countries alike.

Good fire management practices have been documented (Goldammer and De Ronde 2004). However, integrated fire management mainly details the requirements, approaches, and needs for fire preparedness and response. A few references on fire management are available from various researchers and managers who have given insights into fire management topics (Scott et al. 2013; Leblon and Alexander 2015). However, these reflect only the national perspective and the work of developed countries. FAO has acknowledged that global forest fires are not yet effectively handled and are risking human lives, assets, ecosystems, and food security. Fires also affect the environment by creating ample amounts of greenhouse gases. Forest fires disrupt the equilibrium of natural fires that support and manage ecological processes and community and traditional fire practices.

Many challenges in fire suppression stem from policies, organizational structures, and budgetary constraints rather than from a lack of knowledge about suppression techniques and practices. Forest fire management agencies handle natural and human-caused fires, which can have beneficial and detrimental effects (Moritz et al. 2014). Agencies face decisions about whether to suppress fires or allow them to burn, navigating obstacles such as conflicting land uses, species conservation efforts (Donovan et al. 2017; Skatter et al. 2017), urban development in wildlands (Johnston and Flannigan 2017), non-market valuation (Venn and Calkin 2011), smoke management, climate change impacts, and upfront and ongoing suppression costs (Stocks and Martell 2016). Agencies must tackle these obstacles to ensure government accountability while efficiently handling their financial resources. Securing adequate funding is a significant barrier in fire management, and establishing connections across ministries presents another challenge. However, adopting a risk-based approach to forest fire response, particularly for larger fires, can save costs (Calkin et al. 2014). Implementing Fire Smart initiatives and expanding forest fire management across landscapes will require substantial time and effort to build capacity and capabilities (Tymstra et al. 2020).

## 5 Future Perspective

Historically, traditional fire suppression efforts focused primarily on prevention, detection, and basic suppression techniques. In contrast, modern fire management faces more complex decision-making challenges. Advances in transportation and telecommunications have facilitated national and international collaborative agreements, enabling fire managers to mobilize larger and more sophisticated suppression resources (Martell 2015). Magnussen and Taylor (2012) examined fire weather patterns throughout Canada, focusing on the temporal and spatial variations in fire regime variables that affect the distribution of suppression resources across the country. Tsang et al. (2013) used game theory methods and interviews with fire management agency representatives to develop a strategic model for sharing fire suppression resources. These studies underscore the critical need for developing Fire Management Decision Support Systems (FMDSS) to assist decision-making within extensive and intricate collaborative networks of fire management agencies. These networks are tasked with managing fires across diverse and widely distributed landscapes, answering to various political authorities appointed by governments.

Islam et al. (2009) developed a time-dependent queueing model to assess daily air tanker deployment plans. However, the complex optimization of daily deployment and re-deployment of air tankers needs computationally challenging stochastic dynamic programming models, which leads to the development of heuristics for tackling these deployment and re-deployment problems. In the future, the GWFN will also seek to establish an International Wildland Fire Accord, which could be voluntary or legally binding internationally. This accord would be founded on the principle that safeguarding global vegetation cover from degradation or destruction is a shared international concern, and collaborative efforts in fire management will aid in reducing disaster risks (Goldammer 2013).

Additional fundamental research is necessary to expand our comprehension of the ecological, physical, social, and economic factors influencing fire's economic, social, and ecological consequences and management. Fire management is already growing more intricate, and alterations in fuel compositions, climate conditions, societal demands, land usage trends, and financial limitations will further convolute fire management efforts. Predictions indicate that the fire situation will deteriorate in the foreseeable future, marked by prolonged fire seasons, progressively challenging fire weather conditions, and more enormous forest fires (Robinne et al. 2018). A study by Moore (2019) suggested that the burned area may rise by 100% by 2100 and the number of fires by 75% in Canada (Sankey 2018). Climate change is believed to extend the length, alter the commencement and conclusion of fire seasons, and impact the scale, intensity, and consequences of fires (Goldammer et al. 2011).

The portrayal of severe forest fires often suggests to the public and decision-makers that firefighting is the primary solution for addressing dangerous wildland fires. This perception is misleading, as the issue is complex and requires long-term sustainable solutions. Effective resolution involves incorporating fire management into the management of land and routine planning and operations (Scott et al. 2013; Leblon and Alexander 2015). The effectiveness of fire management relies on successful fire prevention, detection, and pre-suppression measures, as well as sufficient fire suppression capabilities and considering the relationships within fire ecology. Efficient and effective fire management is based on previous learning experience and needs the involvement of various concerned stakeholders. The seminar proceedings on control and prevention of forest fire focused on some key research elements, like preparing a list of research institutes and directing the attention of IUFRO and other institutes towards the requirement for research into forest fire behaviour and silviculture, fire impact on forest ecosystems, mapping and modeling for fire danger risks and predictions along with knowing the reason and people attitude for causing forest fires.

## 6 Conclusion

Several international organizations are actively working to mitigate the negative impacts of forest fires and improve global cooperation in wildland fire management. Insights from numerous global conferences and summits indicate a strong willingness among countries to establish formal agreements for wildland fire management. A systematic analysis of the fire problem is essential for effectively integrating societal factors and fire management technologies. Forest fires are primarily a landscape issue fuelled by continuous fuel supply and accumulation. Altered landscapes have increased societal risk, making it crucial to assess the factors leading to fire disasters accurately. Various studies predict the situation will worsen in the coming decades, with a warming climate intensifying forest fire impacts. To address this challenge, departments, organizations, and governments must collaborate globally to develop suitable guidelines and strategies. Effective protocols and policy formation require data sharing and collaborative approaches. Addressing research gaps and ensuring adequate funding are also essential. The increasing frequency and complexity of forest fire events necessitate enhanced international collaboration and improved training and education in decision support tools. Strengthening forest fire management capabilities will be vital in mitigating the future impact of forest fires.

## References

Abatzoglou JT, Juang CS, Williams AP, Kolden CA, Westerling AL (2021) Increasing synchronous fire danger in forests of the western United States. Geophys Res Lett 48(2):e2020GL091377

Alexandris N (2011) Urned area mapping via non-centered PCA using public domain data and free open source software (Doctoral dissertation, Albert-Ludwigs-Universität Freiburg)

Archibald S, Nickless A, Govender N, Scholes RJ, Lehsten V (2010) Climate and the inter-annual variability of fire in southern Africa: a meta-analysis using long-term field data and satellite-derived burnt area data. Glob Ecol Biogeogr 19(6):794–809

Atasoy A, Geçen R (2014) Forest fires occurred on border area between Turkey (Hatay) and Syria. Proced Soc Behav Sci 120:680–5

Attri V, Dhiman R, Sarvade S (2020) A review on status, implications and recent trends of forest fire management. Arch Agric Environ Sci 5(4):592–602

Blanchi R, Leonard J, Haynes K, Opie K, James M, de Oliveira FD (2014) Environmental circumstances surrounding bushfire fatalities in Australia 1901–2011. Environ Sci Policy 37:192–203

Bloem S, Cullen AC, Mearns LO, Abatzoglou JT (2020) The role of international resource sharing arrangements in managing fire in the face of climate change. Fire 5(4):88

Boer MM, Resco de Dios V, Bradstock RA (2020) Unprecedented burn area of Australian mega forest fires. Nat Clim Chang 10(3):171–2

Calkin DE, Cohen JD, Finney MA, Thompson MP (2014) How risk management can prevent future forest fire disasters in the wildland-urban interface. Proc Natl Acad Sci 111(2):746–51

Calkin DE, Thompson MP, Finney MA (2015) Negative consequences of positive feedbacks in US forest fire management. Forest Ecosyst 2:1

Chandra KK, Bhardwaj AK (2015) Incidence of forest fire in India and its effect on terrestrial ecosystem dynamics, nutrient and microbial status of soil. Int J Agric for 5(2):69–78

Charnley S, Spies TA, Barros AM, White EM, Olsen KA (2017) Diversity in forest management to reduce forest fire losses: implications for resilience. Ecol Soc 22(1)

Cumming GS, Olsson P, Chapin FS, Holling CS (2013) Resilience, experimentation, and scale mismatches in social-ecological landscapes. Landscape Ecol 28:1139–50

Doerr SH, Santín C (2016) Global trends in forest fire and its impacts: perceptions versus realities in a changing world. Philos Trans R Soc B Biol Sci 371(1696):20150345

Donovan VM, Brown GS, Mallory FF (2017) The impacts of forest management strategies for woodland caribou vary across biogeographic gradients. PLoS ONE 12(2):e0170759

Goldammer JG, Stocks BJ, Fires V (2011) Migration and global environmental change. Spec Rep 10

Goldammer JG, Pronto L (2016) Global vegetation fire challenges and outlook. CFA Newslett Commonwealth for Assoc 74:4–6

Goldammer JG, De Ronde C, editors(2004) Wildland fire management handbook for sub-Sahara Africa. African Minds.

Goldammer JG (2013) 23 International protocols and agreements on cooperation in wildland fire management and forest fire disaster response: needs, current status, and the way ahead. Vegetation Fires and Global Change 313

Harris NL, Gibbs DA, Baccini A, Birdsey RA, De Bruin S, Farina M, Fatoyinbo L, Hansen MC, Herold M, Houghton RA, Potapov PV (2021) Global maps of twenty-first century forest carbon fluxes. Nat Clim Chang 11(3):234–40

Higuera PE, Abatzoglou JT (2020) Record-setting climate enabled the extraordinary 2020 fire season in the western United States. Glob Change Biol

Hope ES, McKenney DW, Pedlar JH, Stocks BJ, Gauthier S (2016) Forest fire suppression costs for Canada under a changing climate. PLoS ONE 11(8):e0157425

Houtman RM, Montgomery CA, Gagnon AR, Calkin DE, Dietterich TG, McGregor S, Crowley M (2013) Allowing a forest fire to burn: estimating the effect on future fire suppression costs. Int J Wildland Fire 22(7):871–82

Huang Z, Cao C, Chen W, Xu M, Dang Y, Singh RP, Bashir B, Xie B, Lin X (2020) Remote sensing monitoring of vegetation dynamic changes after fire in the Greater Hinggan Mountain Area: the algorithm and application for eliminating phenological impacts. Remote Sens 12(1):156

Islam KS, Martell DL, Posner MJ (2009) A time-dependent spatial queueing model for the daily deployment of Airtankers for forest fire control. INFOR Inform Syst Oper Res 47(4):319–33

Johnston LM, Flannigan MD (2017) Mapping Canadian wildland fire interface areas. Int J Wildland Fire 27(1):1–4

Jolly WM, Cochrane MA, Freeborn PH, Holden ZA, Brown TJ, Williamson GJ, Bowman DM (2015) Climate-induced variations in global forest fire danger from 1979 to 2013. Nat Commun 6(1):7537

Kaiser JW, Heil A, Andreae MO, Benedetti A, Chubarova N, Jones L, Morcrette JJ, Razinger M, Schultz MG, Suttie M, Van Der Werf GR (2012) Biomass burning emissions estimated with a global fire assimilation system based on observed fire radiative power. Biogeosciences 9(1):527–54

Kumar A, Dwivedi GK, Tewari S, Paul J, Anand R, Kumar N, Kumar P, Singh H, Kaushal R (2020) Carbon mineralization and inorganic nitrogen pools under Terminalia Chebula Retz.-Based agroforestry System in Himalayan Foothills India. Forest Sci 66(5):634–643

Kumar A, Kumar P, Singh H, Kumar N (2020) Adaptation and mitigation potential of roadside trees with bio-extraction of heavy metals under vehicular emissions and their impact on physiological traits during seasonal regimes. Urb for Urb Green 58:126900

Kumar A, Kumar P, Singh H, Bisht S, Kumar N (2021) Relationship of physiological plant functional traits with soil carbon stock in temperate forest of Garhwal Himalaya. Curr Sci 120(8):1368–1373

Kumar A, Kumar P, Singh H, Kumar N (2021) Modulation of plant functional traits under essential plant nutrients during seasonal regime in natural forests of Garhwal Himalayas. Plant Soil 465:197–212

Kumar N, Singh H, Kumar A, Singh AK, Pandey PK, Kumar A (2024) Forest-fire-induced land degradation. In: Mishra G, Giri K, Singh S, Kumar M. (eds) Sustainable Land Management in India. Springer, Singapore. https://doi.org/10.1007/978-981-97-5223-2_4

Leblon B, Alexander ME (eds) (2015) Current international perspectives on wildland fires. Nova Publishers, Mankind and the Environment

Lee HJ, Choi YE, Lee SW (2018) Complex relationships of the effects of topographic characteristics and susceptible tree cover on burn severity. Sustainability 10(2):295

Leslie HM, Basurto X, Nenadovic M, Sievanen L, Cavanaugh KC, Cota-Nieto JJ, Erisman BE, Finkbeiner E, Hinojosa-Arango G, Moreno-Báez M, Nagavarapu S (2015) Operationalizing the social-ecological systems framework to assess sustainability. Proc Natl Acad Sci 112(19):5979–84

Magnussen S, Taylor SW (2012) Inter-and intra-annual profiles of fire regimes in the managed forests of Canada and implications for resource sharing. Int J Wildland Fire 21(4):328–41

Mahdipour E, Dadkhah C (2014) Automatic fire detection based on soft computing techniques: review from 2000 to 2010. Artif Intell Rev 42:895–934

Malik A, Kumar K, Kumari G, Kumar N, Singh H (2024) Application of functional traits in modelling productivity and resilience under climate change published by Springer, pp 77–96. ISBN: 978-981-97-1509-1

Martell DL (2015) A review of recent forest and wildland fire management decision support systems research. Curr for Rep 1:128–37

Martell DL, Drysdale RJ, Doan GE, Boychuk D (1984) An evaluation of forest fire initial attack resources. Interfaces 14(5):20–32

McAlpine RS, Hirsch KG (1999) An overview of LEOPARDS: the level of protection analysis system. For Chron 75(4):615–21

Molak M, Struk M (2021) Cross-border cooperation: an alternative for remote border regions? Case study of fire protection in Olomouc Region, CZE

Moore PF (2019) Global wildland fire management research needs. Curr for Rep 5:210–25

Moritz MA, Batllori E, Bradstock RA, Gill AM, Handmer J, Hessburg PF, Leonard J, McCaffrey S, Odion DC, Schoennagel T, Syphard AD (2014) Learning to coexist with forest fire. Nature 515(7525):58–66

Ostroukhov AV, Klimina EM, Kuptsova VA (2020) Landscape mapping of hard-to-reach areas. A case study for the Bolonsky State nature reserve (Russia). 47–63

Pratiwi UN, Nugroho A (2021) Synergy of local government institutions in dealing with forest and land fires as a non-military threat at border regions in Indonesia

Quan D, Quan H, Zhu W, Lin Z, Jin R (2022) A comparative study on the drivers of forest fires in different countries in the cross-border area between China North Korea and Russia. Forests 13(11):1939

Renard Q, Pélissier R, Ramesh BR, Kodandapani N (2012) Environmental susceptibility model for predicting forest fire occurrence in the Western Ghats of India. Int J Wildland Fire 21(4):368–79

Robinne FN, Burns J, Kant P, Flannigan M, Kleine M, de Groot B, Wotton DM (2018) Global fire challenges in a warming world. IUFRO

Salis M, Del Giudice L, Alcasena-Urdiroz F, Jahdi R, Arca B, Pellizzaro G, Scarpa C, Duce P (2023) Assessing cross-boundary forest fire hazard, transmission, and exposure to communities in the Italy-France Maritime cooperation area. Front for Glob Change 6:1241378

Sanabria LA, Qin X, Li J, Cechet RP, Lucas C (2013) Spatial interpolation of McArthur's forest fire danger index across Australia: observational study. Environ Modell Softw 50:37–50

Sankey S (2018) Blueprint for wildland fire science in Canada (2019–2029)

Savita, Sharma SK, Singh H, Singh O (2017) Forest fire: a review. Indian for 143(8):719–728

Scott AC, Bowman DM, Bond WJ, Pyne SJ, Alexander ME (2013) Fire on earth: an introduction. John Wiley & Sons

Singh H, Kumar M (2022) Climate change and its impact on Indian Himalayan forests: current status and research needs. In: Rani S, Kumar R (eds) climate change. Springer Climate. Springer, Cham. pp 223–242, https://doi.org/10.1007/978-3-030-92782-0_11

Singh H, Kumar N, Kumar A (2024) Enhancing resource use efficiency in crops through plant functional traits published by Springer, pp 97–117. ISBN: 978-981-97-1509-1

Singh H (2021) Process-based carbon sequestration study in relation to the Energy-Water-Carbon flux in a forest ecosystem. In: Climate impacts on sustainable natural resource management (Kumar P., Singh RK., Kumar M., Rani M, and Sharma (eds). CRC Press Taylor & Francis Group, United States, pp 336–361

Skatter HG, Charlebois ML, Eftestøl S, Tsegaye D, Colman JE, Kansas JL, Flydal K, Balicki B (2017) Living in a burned landscape: woodland caribou (Rangifer tarandus caribou) use of postfire residual patches for calving in a high fire–low anthropogenic Boreal Shield ecozone. Can J Zool 95(12):975–84

Spies TA, White E, Ager A, Kline JD, Bolte JP, Platt EK, Olsen KA, Pabst RJ, Barros AM, Bailey JD, Charnley S (2017) Using an agent-based model to examine forest management outcomes in a fire-prone landscape in Oregon, USA. Ecol Soc 22(1)

Stocks BJ, Martell DL (2016) Forest fire management expenditures in Canada: 1970–2013. For Chron 92(3):298–306
Strydom S, Savage MJ (2018) A spatio-temporal analysis of fires in the Southern African development community region. Nat Hazards 92:1617–32
Tsang A, Larson K, McAlpine R (2013) Resource sharing for control of wildland fires. Proceed AAAI Conf Artif Intell 27(1):1355–1361
Tymstra C, Stocks BJ, Cai X, Flannigan MD (2020) Forest fire management in Canada: review, challenges and opportunities. Prog Disaster Sci 5:100045
Venn TJ, Calkin DE (2011) Accommodating non-market values in evaluation of forest fire management in the United States: challenges and opportunities. Int J Wildland Fire 20(3):327–39
Vidal R, Harbour T, Jorda L (2011) How lessons learned by high reliability organizations can improve incident management. In: Fifth international wildlife conference, Sun City, Africa
Woolford DG, Dean CB, Martell DL, Cao J, Wotton BM (2014) Lightning-caused forest fire risk in Northwestern Ontario, Canada, is increasing and associated with anomalies in fire weather. Environmetrics 25(6):406–16
Wotton BM, Nock CA, Flannigan MD (2010) Forest fire occurrence and climate change in Canada. Int J Wildland Fire 19(3):253–71
Wotton BM, Flannigan MD, Marshall GA (2017) Potential climate change impacts on fire intensity and key forest fire suppression thresholds in Canada. Environ Res Lett 12(9):095003
Yadav PK, Srivastava R, Chaurasiya N (2018) Optical fiber sensor: review and applications. Int J Creative Res Thoughts 6(1):1358–62
Ye T, Wang Y, Guo Z, Li Y (2017) Factor contribution to fire occurrence, size, and burn probability in a subtropical coniferous forest in East China. PLoS ONE 12(2):e0172110

# Advancing Forest Fire Management Practices: Policies and Strategies

**Sarita Bisht and Hukum Singh**

**Abstract** Forest fires are a global concern, endangering ecosystems, human lives, and economic stability. Forest fires have recently become more common and intense, exacerbated by climate change, land-use patterns, and human activity. As a result, effective implementation of robust policies and strategies is required to enhance management approaches for forest fire control. This chapter discussed key initiatives and approaches to strengthen forest fire management. Firstly, it explores the role of policy frameworks in shaping fire management practices, emphasizing the importance of interdisciplinary collaboration, community engagement, and adaptive governance structures. Effective policies prioritize prevention, preparedness, response, and recovery phases, integrating scientific knowledge, indigenous perspectives, and stakeholder involvement. Furthermore, the chapter emphasizes the significance of capacity building and knowledge exchange in enhancing fire management practices. Training programs, skill development initiatives, and information dissemination campaigns empower stakeholders at various levels to mitigate fire risks effectively. Collaborative research efforts and data-sharing mechanisms foster a culture of learning and innovation, facilitating continuous improvement in fire management strategies. We also highlighted that advancing forest fire management practices requires a multifaceted approach encompassing robust policies, innovative strategy, and capacity-building initiatives. Through integrating scientific expertise, technological advancements, and community participation, stakeholders can enhance resilience to forest fires, safeguarding ecosystems and livelihoods for future generations.

**Keywords** Policy frameworks · Forest fire management strategies · Fire management practices

S. Bisht (✉) · H. Singh
ICFRE-Forest Research Institute, Dehradun, Uttarakhand, India
e-mail: bishtaaru@gmail.com



H. Singh and N. Kumar (eds.), *Forest Fire and Climate Change*,
https://doi.org/10.1007/978-3-031-89967-6_15

## 1 Introduction

Forest fires are a global concern, endangering ecosystems, human lives, and economic stability (Flannigan et al. 2006; Savita et al. 2017). The demand for timber and non-timber forest products has increased, and forest fires have significantly lost them (Kumar et al. 2024). Furthermore, excessive resource exploitation has deteriorated significant forest areas (Sett et al. 2024). With climate change leading to biomass drying and additional organic buildup over time, forest fires are massive and no longer beneficial to forest health (NDMA 2020). Over recent years, forest fire incidences have surged, fueled by rising temperature-induced droughts, changes in land use, and forest mismanagement (Duane et al. 2021; Barwa and Singh 2024). The devastating consequences of these forest fire incidences extend beyond immediate destruction. This includes long-term ecological damage, loss of biodiversity, and adverse effects on air quality, environmental quality, and human health (Kalogiannidis et al. 2023; Kaushik and Singh 2024). Nowadays, large areas of damaged forest are burned annually or semi-annually.

India continues to struggle more and more to find better ways to prevent and control unwanted forest fires, as they are destructive to forest ecosystem health. India is one of many countries dealing with this issue (Jolly et al. 2015). According to the IPCC Fifth Assessment Report, exposure to large areas of forest fires is expected to result in 260,000–600,000 premature fatalities yearly. Further, it is anticipated that carbon emissions on an annual basis due to forest fires vary between 2.5 and 4.0 billion tons of carbon dioxide, releasing significant amounts of greenhouse gases (GHGs) into the atmosphere (MoEF & CC 2018; Singh 2024a). This chapter discusses enhancements in forest fire management, emphasizing major policies and techniques for reducing forest fire risk and increasing resilience in fire-prone areas (Lelouvier et al. 2021). We also discussed significant initiatives and approaches to improving forest fire management. Initially, it investigates the impact of policy frameworks in defining fire suppression techniques, highlighting the significance of interdisciplinary collaboration, community participation, and adaptable governance structures (Smith et al. 2016). Effective policies prioritize prevention, preparedness, response, and recovery phases, integrating scientific knowledge, indigenous perspectives, and stakeholder involvement.

Additionally, prescribed burning and forest restoration techniques are essential for ecosystem-based fire management, promoting landscape resilience and biodiversity conservation (Lelouvier et al. 2021; Kumar et al. 2021a, 2021b). Therefore, enhancing forest fire management practices by implementing robust policies and strategies is imperative (Singh 2024b). Developing and implementing effective strategies and techniques for managing forest fires is necessary to address these problems.

## 2 Principles of Forest Fire Management

Forest fire management policies are formulated and implemented to prevent, mitigate, and manage wildfires in forested areas. These policies are built upon several fundamental principles that guide decision-making and actions taken by government agencies, landowners, communities, and other stakeholders. Here are some principles commonly found in forest fire management:

**Ecological Resilience**

Ecological resilience is a fundamental principle in forest fire management, emphasizing the importance of maintaining the natural resilience of forests to fire incidents. Resilient ecosystems can absorb and recover from fire events while retaining their essential functions and structures. This principle underscores the ecological significance of fire as a natural process that shapes forest dynamics, promotes biodiversity, and maintains ecosystem health (Chang 1996; Singh 2024c). Allowing fire to play its natural role may help policymakers develop adaptive strategies, such as fire-resistant species and fire-adapted vegetation, which enhance their ability to withstand and recover from fire events. Furthermore, ecological resilience emphasizes preserving the flow of various ecosystem mechanisms, particularly biogeochemical cycling and ecological succession, which are essential for ecosystem recovery following fire disturbances (Whelan 1995; Singh 2024d). Forest managers may be able to enhance the capacity of forest ecosystems to adapt to changing patterns and behaviors of fires, as well as to climate change, through proactive management practices that promote ecological resilience, such as prescribed burning, fuel reduction, and habitat restoration (Vose et al. 2021; Kumar et al. 2020a, 2020b). This approach ensures forest ecosystems' long-term sustainability and vitality in the face of fire disturbances.

**Prevention**

Prioritize proactive measures to prevent wildfires and prepare for fire events. Focus on community education, land-use planning, and fuel management to minimize fire risk and improve response readiness. Prevention is the cornerstone of forest fire management (Sreelakshmi et al. 2023). Prevention efforts also include measures to address human interventions that may induce forest fires, mainly campfires, smoking, equipment use, etc. Prescribed burning is a key tool used in forest fire management to minimize fuel loads, restore ecosystem health, and mitigate the consequences of wildfires (Francos and Úbeda 2021).

**Public Safety**

Ensure the safety of human lives and property by implementing policies and strategies prioritizing public safety during forest fires (Dombeck 2004). Develop robust evacuation plans, early warning systems, and emergency response protocols to protect at-risk communities. Early warning systems are crucial in alerting residents to wildfire threats and providing timely information for evacuation decisions. These systems may include sirens, text alerts, social media notifications, and automated phone calls (Blake 2020). Leveraging advanced technologies such as satellite monitoring and weather forecasting may enhance the accuracy and efficiency of early warning systems. Furthermore, societal engagement and preparedness are essential for public safety during forest fires (Wise and McGuire 2009).

**Collaboration and Partnerships**

It is crucial to foster collaboration among government agencies, local communities, and other end-users involved in forest fire management. Engage in partnerships that leverage collective expertise, resources, and capacities to address complex fire-related challenges effectively (Spencer et al. 2015).

**Adaptive Management**

Embrace adaptive management principles that allow for flexible and responsive approaches to forest fire management. Monitor fire behavior, ecosystem dynamics, and community needs to adjust strategies and actions accordingly. Forest fire management policies acknowledge the fire-prone ecosystem's dynamic nature and emphasize the necessity for adaptive management approaches (Bacciu et al. 2022). Policies encourage monitoring and evaluating fire management activities to assess their effectiveness and adapt strategies based on lessons learned. This may involve incorporating new scientific knowledge, adjusting management practices in response to changing conditions, and engaging in continuous improvement processes.

## 3 Forest Fire Management Policy Framework

Forest fire management policies are developed and implemented to prevent, mitigate, and manage wildfires in forested areas. Developing a comprehensive policy framework for forest fire management is essential to guide governmental, organizational, and community actions (Schultz et al. 2019). Such a framework typically involves a set of principles, objectives, strategies, and implementation mechanisms aimed at preventing, detecting, responding to, and recovering from forest fires.

The objectives of the forest fire management policy should be identified. These objectives may include preventing property damage and loss of life, maintaining ecosystems and biodiversity, safeguarding vital infrastructure, encouraging sustainable forest management, and building community resilience (Schultz et al. 2019). Comprehensive risk assessments should identify areas prone to forest fires based on various indicators, mainly climate, vegetation type, topography, human activities, etc. Developing strategic fire management plans to prioritize prevention, early detection, and effective response to minimize potential damage. Various techniques, including remote sensing and geographical information systems (RS&GIS) and modeling techniques, should be adopted to assess fire risk and predict fire behavior (Scott et al. 2013).

Early detection of forest fires is crucial for effective fire management. Policies focus on developing and maintaining systems for early detection, such as fire lookout towers, aerial surveillance, satellite monitoring, and citizen reporting (Bowman et al. 2013). Further, resources for forest fire suppression equipment, including fire trucks, aircraft, and firefighting crews, should be allocated to enhance response capabilities. Investing in maintaining and expanding firebreaks, fuel breaks, and other infrastructure to create defensible space, facilitate fire containment, and support research and development efforts to improve firefighting technologies and techniques, such as fire-resistant materials and fire retardants (Ascoli et al. 2020).

Forest fire management policies emphasize the importance of conducting risk assessments to identify areas with high fire risk and prioritize resources accordingly. Policies also stress the need for comprehensive fire management plans that outline strategies for fire prevention, suppression, and recovery (Mateus and Fernandes 2014). These plans consider fire behavior, weather patterns, terrain, ecological values, and community assets.

Effective forest fire management requires integrated strategies considering ecological, societal, and economic factors. Policies encourage collaboration and coordination among various stakeholders, including government agencies, landowners, Indigenous communities, firefighters, researchers, and non-governmental organizations. Integrated management approaches may involve ecosystem-based management, community-based fire management, and partnerships with neighboring jurisdictions (Thompson and Calkin 2011). The framework establishes mechanisms for interagency cooperation, information sharing, and joint planning to optimize resources and achieve common goals.

Forest fire management policies promote using prescribed fire as a proactive measure to create defensible space around communities, reduce fuel continuity, and maintain fire-resilient landscapes (Francos and Úbeda 2021). Policies also support fuel management practices such as thinning, mechanical treatments, and grazing to reduce fire risk (Husari et al. 2006).

Public engagement and education are essential components of forest fire management. Policies emphasize the importance of raising awareness about fire risk, promoting responsible behavior, and building community resilience to forest fires (Yousefi et al. 2023). This may involve outreach campaigns, educational programs, community meetings, and partnerships with schools, media outlets, and other organizations to foster a fire safety and preparedness culture.

The framework outlines mechanisms for policy implementation, including the roles and responsibilities of key agencies and stakeholders, budgetary considerations, and performance metrics for evaluating the effectiveness of forest fire management efforts (Schultz et al. 2019). Regular reviews and updates ensure the policy framework's relevance and responsiveness to changing conditions and emerging challenges.

After a forest fire event, efforts focus on recovery and rehabilitation to restore ecosystems, rehabilitate affected landscapes, and support affected communities in rebuilding and recovery efforts. This may include post-fire assessments, erosion control measures, reforestation initiatives, and assistance programs for impacted individuals and businesses (Ryan and Hamin 2008). Forest fire management plans that adhere to these policy frameworks seek to minimize the adverse consequences of forest fires on ecosystems, communities, and economies while encouraging fire's ecological function and preserving fire-adapted landscapes.

# 4 National and International Forest Fire Management Policies and Initiatives

## 4.1 National Fire Management Policies and Initiatives

Forest fire management policies in India are designed to address the unique challenges of wildfires in diverse forest ecosystems across the country. These policies aim to prevent, detect, and effectively manage forest fires while promoting sustainable forest management practices and safeguarding biodiversity, livelihoods, and ecosystem services (Schmerbeck and Fiener 2015). In this regard, an initiative was taken by the Government of India in 1952 to establish the National Forest Policy under the umbrella of the Ministry of Agriculture and Farmers' Welfare (formerly the Ministry of Food and Agriculture) to manage and increase the country's forest cover. Furthermore, India's Constitution established guidelines for protecting forests and wildlife through the established National Forest Policy. This policy states that all states will strive to protect and improve the environment, forests, and wildlife.

Moreover, this policy encourages citizens to safeguard and enhance the natural ecosystem and environment, including forests, lakes, rivers, and wildlife, while also advocating compassion toward all living beings (NIDM 2012). The National Forest Policy (1988) recognizes the importance of preventing forest fires and emphasizes the need for integrated fire management approaches. It advocates establishing fire protection measures, including fire lines, firebreaks, and controlled burning to minimize forest fire spread. The policy also emphasizes community participation in forest fire prevention and management activities (Ghosh 1993). The FAO implemented an innovative Technical Cooperation Programme in India, focusing on human resource and capacity development for managing forest fire spread.

The Food and Agriculture Organization assisted the Technical Cooperation Programme in initiating training plans on managing and controlling forest fires in May 1995, which concluded in December 1995. This initiative aimed to assess the forest fire situation in India and provide training to frontline forest officials at the state and national levels. These trainings emphasized strategic fire planning, enabling officials to develop plans considering their respective states' social, economic, and ecological dimensions and create model forest fire management plans at the state level (NIDM 2014). Furthermore, the Ministry of Environment, Forest, and Climate Change (MoEF & CC) drafted the National Action Plan on Forest Fire (2000) to mitigate forest fires through various measures. These include informing, enabling, and empowering communities living near forests and encouraging collaboration with Forest Departments.

Furthermore, the plan focuses on decreasing forest vulnerability to fire risks, improving the firefighting capacities of forest personnel and institutions, and expediting post-fire recovery efforts. To ensure effective execution, the Ministry has established the Central Monitoring Committee (CMC) to oversee the plan's implementation. The National Action Plan on Forest Fire provides a comprehensive framework for managing and controlling forest fires in the country. It outlines strategies for fire prevention, early detection, suppression, and rehabilitation of fire-affected areas. The plan emphasizes the importance of capacity-building, research, and awareness-raising initiatives to enhance forest fire management capabilities (MoEF & CC 2018).

The National Forest Fire Prevention and Management Scheme (NFFPMS) was launched in 2003 by MoEF & CC, Government of India, to address the increasing incidence of forest fires across India. The scheme aims to strengthen the institutional capacity of state forest departments for effective forest fire prevention, detection, suppression, and rehabilitation. It provides financial assistance to states for implementing forest fire management activities, including procuring fire detection and suppression equipment, training personnel, and awareness campaigns.

The Ministry of Environment, Forest and Climate Change (MoEF & CC) has developed comprehensive guidelines to prevent and manage forest fires, encompassing standard operating procedures (SOPs) for various aspects of fire management. These guidelines recommend fire prevention measures, early detection systems, firefighting techniques, community participation, and post-fire rehabilitation. The Forest Survey of India (FSI) also operates the National Forest Fire Monitoring System (NFFMS) to monitor and assess forest fire incidents nationwide. Leveraging remote sensing and geographic information systems (RS&GIS) technologies, the NFFMS detects and develops maps for analyzing forest fire hotspots in near real-time. This system offers early warning alerts, facilitates decision-making, and aids resource allocation during wildfire events. The involvement of indigenous communities in forest fire management and prevention is actively promoted through the Joint Forest Management (JFM) program. Community forest protection committees engage in various fire prevention activities such as developing firebreaks, controlling burning, and creating awareness regarding fire safety. The JFM approach fosters cooperation between forest department officials and Indigenous communities, thereby nurturing ownership and responsibility for forest resources (MoEF & CC 2018).

## 4.2 International Fire Management Policies and Initiatives

Improving vegetation health and sustainability is crucial in the era of forest fires and climate change (Trumbore et al. 2015). Developing and implementing robust forest fire management and planning strategies through international collaboration is imperative to address these challenges effectively. Local communities should be integral to these efforts, as they often serve as the first defense against forest fires. Further,

It is essential to facilitate the exchange of experiences and expertise among stakeholders to enhance the capacity for forest fire management. Joint programs are to be established to strengthen existing systems and develop new approaches by leveraging global knowledge and experiences (Carta et al. 2023). The collaborative approach fosters learning and ensures that diverse perspectives and best practices are incorporated into forest fire management strategies. Hence, effective forest fire management requires international cooperation, active engagement of local communities, and the exchange of knowledge and experiences on a global scale.

International forest planning encompasses diverse strategies and initiatives to prevent, detect, and effectively manage wildfires across borders. These policies recognize the transboundary nature of forest fires and the need for collaborative approaches to address this global challenge effectively. A German-based organization, i.e., Global Fire Monitoring Centre (GFMC), assists and provides information to strengthen collaboration between countries. The Fire Ecology and Biomass Burning Research Group (FEBBRG) of the Max Planck Institute of Chemistry, Bio-geochemistry Department, Germany, hosts this center. Further, The German government initially created a Centre as a German contribution to the UN-sponsored International Decade of Natural Disaster Reduction (IDNDR) 1990–2000 program. The GFMC conducts comprehensive research and demonstrations on forest fires and provides consultancy services to many countries (Goldammer et al. 2003).

Furthermore, the United Nations Framework Convention on Climate Change (UNFCCC) plays a pivotal role in global efforts to address the multifaceted challenges of climate change, including forest fire management. Forest fires are acknowledged as significant contributors to carbon emissions and major drivers of climate change, underscoring the urgent need to integrate fire management into broader climate mitigation and adaptation strategies.

Initiatives such as Reducing Emissions from Deforestation and Forest Degradation (REDD + ) highlight the critical importance of sustainable forest management practices, including fire prevention and suppression, in mitigating greenhouse gas emissions from forests. The UNFCCC provides a crucial platform for international cooperation and action on forest fire management by advocating for the conservation and sustainable use of forests and enhancing the resilience of forest ecosystems to climate change impacts. Through collaborative efforts within the UNFCCC framework, countries can effectively address the complex challenges of forest fires while advancing global climate goals and promoting the long-term health and sustainability of forests worldwide.

Furthermore, the UNFCCC's financial and technical support provisions to developing countries further facilitate capacity-building efforts and technology transfer to strengthen their forest fire management capabilities. Through its inclusive and collaborative approach, the UNFCCC underscores the interconnectedness of forest fires with broader climate change challenges, highlighting the need for integrated and holistic approaches to safeguarding forests and mitigating climate risks.

The Sustainable Development Goals (SDGs) offer a comprehensive global framework for fostering sustainable development. Among these goals, SDG 15 specifically targets the protection, restoration, and sustainable management of terrestrial ecosystems, including forests. Target 15.2 of the SDGs advocates for facilitating sustainable management across all forest types, encompassing forest fire and restoration initiatives (Goldammer and Scholz 2004).

The United Nations International Strategy for Disaster Reduction (UNISDR) is dedicated to reducing disaster risk and fostering resilience to various hazards, including wildfires. The Sendai Framework for Disaster Risk Reduction 2015–2030 underscores the necessity of integrating disaster risk reduction into forest management policies and practices to mitigate the impacts of forest fires on communities and ecosystems (Petrovic et al. 2017).

In addition, the International Association of Wildland Fire (IAWF) serves as a global network uniting wildfire professionals to exchange knowledge and best practices in wildland fire management. This organization aims to enhance forest fire management policies and practices worldwide by facilitating collaboration among fire practitioners, researchers, and policymakers.

Furthermore, the Global Wildland Fire Network (GWFN), coordinated by the Food and Agriculture Organization (FAO) of the United Nations, promotes international cooperation in wildland fire management. By facilitating the exchange of information, expertise, and resources among countries, the GWFN strengthens its capacity to prevent, detect, and respond to forest fires (Goldammer and Fire 2007).

The International Cooperation on Forest Fire Management Many countries engage in bilateral and multilateral partnerships to strengthen their forest fire management capabilities. These collaborations may involve sharing technologies, expertise, and resources for early detection, suppression, and rehabilitation of fire-affected areas. Regional initiatives, such as the European Forest Fire Information System (EFFIS) and the ASEAN Specialized Meteorological Centre (ASMC) on Forest Fires, serve as vital platforms for fostering cooperation and sharing information among neighboring countries facing similar fire risks. These initiatives facilitate the exchange of critical data and expertise, enabling more effective coordination in wildfire management efforts across borders.

Moreover, international research collaborations and partnerships are crucial in driving innovation in forest fire management. Organizations like the Global Fire Monitoring Center (GFMC) and the International Fire Aviation Working Group (IFAWG) are instrumental in advancing research, training, and building fire science and management capacity. These organizations contribute to developing cutting-edge technologies and strategies for mitigating the impact of forest fires by leveraging global expertise and resources.

International forest fire management policies and frameworks emphasize the importance of cooperation, coordination, and knowledge sharing to address the multifaceted challenges of forest fires worldwide. Countries should enhance their capacity to prevent, detect, and effectively manage forest fires by fostering collaboration and mutual support. This collective effort is essential for safeguarding lives, livelihoods, and ecosystems from the destructive consequences of forest fires.

## 5 Strategies for Forest Fire Management

Forest fire management strategies comprise a spectrum of proactive and reactive measures to mitigate forest fire risks and reduce their adverse effects on ecosystems, communities, and the economy (Fig. 1).

### 5.1 Prevention

Prevention strategies focus on reducing the likelihood of forest fires by addressing key factors that contribute to fire ignition and spread. The plan, particularly management, to reduce the buildup of flammable biomass and create firebreaks may help limit the spread of wildfires (Sreelakshmi et al. 2023). Moreover, enforcing seasonal fire bans and implementing restrictions on activities such as campfires, outdoor burning, and fireworks during high fire danger helps prevent accidental ignitions. Public education is crucial in raising awareness about wildfire risks, promoting fire-safe behaviors, and providing guidance on fire prevention measures, evacuation procedures, and emergency preparedness (Carta et al. 2023). Furthermore, establishing and enforcing regulations related to building construction, landscaping, and land use planning helps minimize communities' vulnerability to forest fires and ensures the implementation of fire-resistant practices. Monitoring weather conditions, fire danger ratings, and meteorological forecasts also plays a significant role in prevention by enabling the anticipation of periods of elevated fire risk and implementing proactive measures to prevent fire ignitions (Wunder et al. 2021). When integrated into comprehensive forest fire management plans, these prevention strategies contribute to a more resilient

**Fig. 1** Forest fire management strategies cycle

and fire-adapted landscape, reducing the potential for catastrophic forest fires and protecting lives, property, and natural resources.

## 5.2 *Suppression*

Suppression strategies are pivotal in forest fire management, focusing on promptly containing and extinguishing forest fires to minimize their impacts on lives, property, and natural resources (MoEF & CC 2018). Critical components of these strategies encompass early detection, whereby advanced fire detection technologies such as lookout towers, aerial patrols, satellite imagery, and automated surveillance systems are utilized to identify wildfires in their emerging stages, enabling rapid response efforts (Gebert and Black 2012). Mobilizing firefighting resources is crucial, involving the deployment of firefighting crews, equipment, aircraft, and other assets to the wildfire location to suppress fire activity, establish control lines, and safeguard vulnerable areas. Implementing an incident command system is vital in coordinating firefighting operations, efficiently allocating resources, and facilitating communication among responding agencies, personnel, and stakeholders (Stephens and Ruth 2005). Tactical operations are diverse, employing strategies such as direct attacks using hand crews and bulldozers, indirect attacks involving back burning and aerial water drops, and structure protection to contain and extinguish forest fires while minimizing risks to responders and communities. Moreover, fostering a unified response is essential, necessitating collaboration with neighboring jurisdictions, agencies, and organizations to provide mutual aid and support during large-scale wildfires that demand coordinated suppression efforts across multiple regions (Backer et al. 2004). Integrating these critical components into suppression strategies would help stakeholders effectively combat wildfires, mitigate their impacts, and safeguard ecosystems and communities from the devastating consequences of uncontrolled fire incidents.

### 5.2.1 Detection

Wildfires occur annually worldwide, leading to substantial loss of life and property damage. Satellite-based remote sensing plays a vital role in detecting forest fires in India. However, there is scope for enhancing the effectiveness of fire alert systems developed by the Forest Survey of India (FSI) and state authorities (MoEF & CC 2018). Despite notable advancements in science and technology, accurately predicting wildfires remains challenging. Many countries utilize various wildfire monitoring systems, often relying on post-event analysis of photographs or videos to identify wildfire occurrences. Unfortunately, confirmation through these methods can involve significant delays (Sreelakshmi et al. 2023). In such circumstances, reliance on a wildfire early warning system based on wireless sensors becomes crucial. Numerous studies are underway in this field, with some proposing using uncrewed aerial vehicles (UAVs) to conduct regular surveillance flights along predefined routes in forested areas, adhering to strict protocols. These UAVs would be equipped with sensors affixed to trees, gathering environmental data to monitor and forecast wildfires. The UAVs would issue early warnings before wildfire ignition if a potential threat arises. Bluetooth Low Energy (BLE) technology facilitates seamless data exchange between the UAV and the central network. The collected data, encompassing temperature and humidity readings, offers a real-time depiction of forest conditions, augmenting community readiness and anticipation of wildfires. This innovative system promises to significantly enhance wildfire prediction and mitigation endeavors in natural settings (Li 2018).

### 5.3 Post-Fire Recovery

Post-fire recovery strategies are essential in the aftermath of wildfires, focusing on restoring ecosystems, rehabilitating affected areas, and mitigating the long-term impacts of these destructive events. Typically, these strategies involve a multifaceted approach: Ecological Restoration efforts promote natural regeneration, re-vegetation, and habitat restoration in areas impacted by wildfires, fostering ecosystem recovery and enhancing biodiversity (Scheper et al. 2021). Soil Stabilization measures, such as mulching, re-vegetation, and erosion barriers, are implemented to prevent soil erosion, sedimentation, and degradation in burned areas, safeguarding soil health and stability (Meyer et al. 2021). In numerous regions nationwide, Joint Forest Management Committees (JFMCs) serve as the primary grassroots entities responsible for Forest Fire Prevention and Management (FFPM) within forest-dependent communities. Collaboratively, the forest department often works alongside JFMCs, providing financial support for activities such as clearing fire lines, employing firewatchers, and protecting areas from fire hazards. However, criticism has been directed towards JFMCs for their top-down structure, where decision-making authority and management control predominantly reside within the forest department. Furthermore, there needs to be more concern about declining levels of investment and community participation in JFMC activities (MoEF & CC 2018).

Hazard Mitigation involves assessing post-fire hazards like debris flows, landslides, and flash floods and implementing measures to mitigate risks to public safety, infrastructure, and natural resources in burned areas, enhancing community resilience. Community Support is provided by offering assistance, resources, and support services to communities affected by forest fires, facilitating recovery, rebuilding infrastructure, and addressing social, economic, and psychological impacts (Moreira et al. 2011). Therefore, integrating prevention, suppression, and post-fire recovery strategies into comprehensive forest fire management plans may support stakeholders in effectively reducing wildfire risks, enhancing resilience, and promoting sustainable management of forest ecosystems in the face of increasing fire threats. These integrated efforts foster ecosystem health, safeguard communities, and preserve natural resources after forest fires.

## 6 Conclusion

It is concluded that addressing the challenge of managing forest fires necessitates comprehensive policies and strategies to mitigate risks, protect ecosystems, and safeguard communities. Advancing forest fire management practices requires a holistic approach combining scientific expertise, technological innovation, traditional wisdom, and community participation. Innovation, collaboration, and continuous learning enhance the capacity to manage wildfires effectively, minimize risks to human lives and property, and sustain healthy, fire-resilient landscapes for future generations. This chapter has explored various dimensions of advancing forest fire management practices, highlighting key policies, strategies, and practices that can enhance effectiveness, efficiency, and resilience in managing forest fires. Forest fire management policies integrate prevention and mitigation, preparedness, response, and recovery measures to achieve multiple objectives: reducing the frequency and severity of forest fires, mitigating their impacts on ecosystems and communities, and fostering sustainable fire management practices. Integrating traditional knowledge and indigenous practices enhances cultural resilience, ecological stewardship, and community engagement in fire management efforts. Prescribed fire and landscape restoration programs restore fire-adapted ecosystems, reduce fuel loads, and mitigate wildfire risk. Effective forest fire management also relies on collaborative partnerships and capacity-building initiatives to optimize resource allocation, enhance response capabilities, and foster adaptive management approaches. Engaging with local communities, promoting public awareness, and building social resilience are essential components of forest fire management policies that empower individuals and communities to take ownership of fire safety and preparedness.

## References


Ascoli D, Russo L, Giannino F, Siettos C, Moreira F (2020) Firebreak and fuelbreak. Encyclopedia of wildfires and wildland-urban interface (WUI) fires. Springer International Publishing, Cham, pp 444–452

Bacciu V, Sirca C, Spano D (2022) Towards a systemic approach to fire risk management. Environ Sci Policy 129:37–44

Backer DM, Jensen SE, McPherson GR (2004) Impacts of fire-suppression activities on natural communities. Conserv Biol 18(4):937–946

Barwa S, Singh H (2024) Climate-induced changes in forest ecosystems of Indian Himalayan region: a meta-analysis. In: Singh H (eds) Urban Forests, Climate Change and Environmental Pollution. Springer, Cham. https://doi.org/10.1007/978-3-031-67837-0_39

Blake N (2020) Incident command system. Nurs Manag Pediatr Disaster, 39–83

Bowman DM, Murphy BP, Boer MM, Bradstock RA, Cary GJ, Cochrane MA, Williams, RJ (2013) Forest fire management, climate change, and the risk of catastrophic carbon losses

Carta F, Zidda C, Putzu M, Loru D, Anedda M, Giusto D (2023) Advancements in forest fire prevention: a comprehensive survey. Sensors 23(14):6635

Chang CR (1996) Ecosystem responses to fire and variations in fire regimes. In: Sierra Nevada ecosystem project: final report to Congress, Vol 2, pp 1071–1099

Dombeck MP, Williams JE, Wood CA (2004) Wildfire policy and public lands: integrating scientific understanding with social concerns across landscapes. Conserv Biol 18(4):883–889
Duane A, Castellnou M, Brotons L (2021) Towards a comprehensive look at global drivers of novel extreme wildfire events. Clim Change 165(3–4):43
Flannigan MD, Amiro BD, Logan KA, Stocks BJ, Wotton BM (2006) Forest fires and climate change in the 21st century. Mitig Adapt Strat Glob Change 11:847–859
Francos M, Úbeda X (2021) Prescribed fire management. Curr Opin Environ Sci Health 21:100250
Gebert KM, Black AE (2012) Effect of suppression strategies on federal wildland fire expenditures. J Forest 110(2):65–73
Ghosh A (1993) Forest policy in India. Indian Forestry, a Perspective, 69
Goldammer JG, Fire W (2007) Forest fires global perspective. Global Fire Monitoring Center
Goldammer JG, Scholz C (2004) The global wildland fire network: a global programme for enhancing international cooperation in wildland fire management and preparation of an international wildland fire accord. Floresta, 34(2)
Goldammer JG, Buchholz G, Resch F (2003) The contribution of the global fire monitoring center (GFMC) for early warning and management of wildfires. In: Early warning systems for natural disaster reduction. Berlin, Heidelberg: Springer Berlin Heidelberg. pp 585–592
Husari S, Nichols HT, Sugihara NG, Stephens SL (2006) Fire and fuel management. Fire in California's Ecosystems. University of California Press, Berkeley, CA, pp 444–465
Kalogiannidis S, Chatzitheodoridis F, Kalfas D, Patitsa C, Papagrigoriou A (2023) Socio-psychological, economic and environmental effects of forest fires. Fire 6(7):280
Kaushik P, Singh H (2024) Climate change-induced forest ecosystem functioning and alteration in water-atmosphere-soil–plant (WASP) relationship. In: Singh H (eds) Urban forests, climate change and environmental pollution. Springer, Cham. https://doi.org/10.1007/978-3-031-67837-0_38
Kumar A, Dwivedi GK, Tewari S, Paul J, Anand R, Kumar N, Kumar P, Singh H, Kaushal R (2020) Carbon mineralization and inorganic nitrogen pools under Terminalia Chebula Retz.-Based agroforestry system in Himalayan foothills India. Forest Sci 66(5):634–643
Kumar A, Kumar P, Singh H, Kumar N (2020) Adaptation and mitigation potential of roadside trees with bio-extraction of heavy metals under vehicular emissions and their impact on physiological traits during seasonal regimes. Urb for Urb Green 58:126900
Kumar A, Kumar P, Singh H, Bisht S, Kumar N (2021) Relationship of physiological plant functional traits with soil carbon stock in temperate forest of Garhwal Himalaya. Curr Sci 120(8):1368–1373
Kumar A, Kumar P, Singh H, Kumar N (2021) Modulation of plant functional traits under essential plant nutrients during seasonal regime in natural forests of Garhwal Himalayas. Plant Soil 465:197–212
Kumar N, Singh H, Kumar A, Singh AK, Pandey PK, Kumar A (2024) Forest-fire-induced land degradation. In: Mishra G, Giri K, Singh S, Kumar M. (eds) Sustainable land management in India. Springer Nature, Singapore. https://doi.org/10.1007/978-981-97-5223-2_4
Lelouvier R, Nuijten D, Onida M, Stoof CR (2021) Land-based wildfire prevention: principles and experiences on managing landscapes, forests and woodlands for safety and resilience in Europe. Publications Office of the European Union
Mateus P, Fernandes PM (2014) Forest fires in Portugal: dynamics, causes and policies. Forest context and policies in Portugal: present and future challenges, 97–115
Meyer MD, Long JW, Safford HD, Sawyer SC, North MP, White AM (2021) Principles of post fire restoration. PSW-GTR-270, USDA Forest Service. pp 1–30
Moreira F, Arianoutsou M, Corona P, De las Heras J (2011) Post-fire management and restoration of southern European forests (eds). Springer Science & Business Media. Vol 24
National Disaster Management Authority Report on Forest Fire Management Global Best Practices (2020)
National Institute of Disaster Management, India (2014)

Petrović N, Bošnjak I, Nedeljković S (2017) Disaster risk reduction for sustainable development goals. Eur Project Manag J 7(2):27–35

Ryan RL, Hamin E (2008) Wildfires, communities, and agencies: stakeholders' perceptions of postfire forest restoration and rehabilitation. J Forest 106(7):370–379

Savita, Sharma SK, Singh H, Singh O (2017) Forest fire: a review. Indian for 143(8):719–728

Scheper AC, Verweij PA, van Kuijk M (2021) Post-fire forest restoration in the humid tropics: a synthesis of available strategies and knowledge gaps for effective restoration. Sci Total Environ 771:144647

Schmerbeck J, Fiener P (2015) Wildfires, ecosystem services, and biodiversity in tropical dry forest in India. Environ Manag 56:355–372

Schultz CA, Thompson MP, McCaffrey SM (2019) Forest Service fire management and the elusiveness of change. Fire Ecol 15:1–15

Scott JH, Thompson MP, Calkin DE (2013) A wildfire risk assessment framework for land and resource management

Sett T, Nikam BR, Singh H, Purohit S (2024) Hydrological variability in Indian forest ecosystem: analysis of drought resilience, recovery and water use efficiency in moist and dry deciduous forests. In: Singh H (eds) Urban forests, climate change and environmental pollution. Springer, Cham. https://doi.org/10.1007/978-3-031-67837-0_37

Singh H (2024a) Climate change and forests: biological perspectives. In: Singh H (eds) Forests and climate change. Springer Nature, Singapore. https://doi.org/10.1007/978-981-97-3905-9_1

Singh H (2024c) Climate change and environmental pollution: biological consequences on urban forests. In: Singh H (eds) Urban forests, climate change and environmental pollution. Springer Nature, Cham, Switzerland. https://doi.org/10.1007/978-3-031-67837-0_1

Singh H (2024b) Forests and climate change: biological perspectives on impact, adaptation, and mitigation strategies (ed), Springer Nature Singapore. https://doi.org/10.1007/978-981-97-3905-9

Singh H (2024d) Urban forests, climate change and environmental pollution: physio-biochemical and molecular perspectives to enhance urban resilience (ed), Springer Nature, Cham, Switzerland. https://doi.org/10.1007/978-3-031-67837-0

Smith AM, Kolden CA, Paveglio TB, Cochrane MA, Bowman DM, Moritz MA, Abatzoglou JT (2016) The science of firescapes: achieving fire-resilient communities. Bioscience 66(2):130–146

Spencer AG, Schultz CA, Hoffman CM (2015) Enhancing adaptive capacity for restoring fire-dependent ecosystems: the fire learning network's prescribed fire training exchanges. Ecol Soc, 20(3)

Sreelakshmi PK, Aswin Krishna MV, Maurya LL (2023) Wildfire management and prevention strategies. Adv For Agro-For, 400

Stephens SL, Ruth LW (2005) Federal forest-fire policy in the United States. Ecol Appl 15(2):532–542

Thompson MP, Calkin DE (2011) Uncertainty and risk in wildland fire management: a review. J Environ Manag 92(8):1895–1909

Trumbore S, Brando P, Hartmann H (2015) Forest health and global change. Science 349(6250):814–818

Vose JM, Peterson DL, Fettig CJ, Halofsky JE, Hiers JK, Keane RE, Stambaugh MC (2021) Fire and forests in the 21st century: managing resilience under changing climates and fire regimes in USA forests. Fire ecology and management: Past, present, and future of US forested ecosystems, pp 465–502

Whelan RJ (1995) The ecology of fire. Cambridge University Press

Wise CR, McGuire M (2009) Wildland fire mitigation networks in the western United States. Disasters 33(4):721–746

# Smart Technology and Fire Fighting Systems

**Kishan Kumar and Nishant**

**Abstract** Fires threaten forest ecosystems in the changing climate, demanding innovative solutions for prevention, detection, and suppression. This chapter explores the transformative potential of smart technologies in revolutionizing forest fire management. The alarming rise of forest fires introduces smart technology as a promising game-changer. The current challenges in forest fire management emphasize the need for innovative solutions. The discussion further delves into a range of smart technology solutions, delving into the utilization of sensors and remote sensing for early fire detection, the deployment of drones for monitoring and suppression purposes, and the integration of artificial intelligence for predictive modeling. Additionally, it discusses the integration of IoT and sensor networks and advancements in smart firefighting equipment and tools, all aimed at improving firefighting efficiency and firefighter safety. The chapter illustrates successful implementations of smart technology in forest fire management through real-world case studies while highlighting the associated implementation challenges. Further, it explores the future outlook of smart technology in forest fire management, envisioning further advancements and addressing ethical considerations surrounding its use. It concludes with a hopeful and proactive tone, emphasizing the importance of policy implications and outlining future directions for continued progress in forest fire management.

**Keywords** Remote sensing · Forest fire · IoT (internet of things) · Forest fire management · Artificial intelligence

K. Kumar
Genetics and Tree Improvement Division, Forest Research Institute, Dehradun, India
e-mail: biotechnologistkk@gmail.com

Nishant (✉)
Forest Products Division, Forest Research Institute, Dehradun, India
e-mail: nishantuchiha1@gmail.com



H. Singh and N. Kumar (eds.), *Forest Fire and Climate Change*,
https://doi.org/10.1007/978-3-031-89967-6_16

## 1 Introduction

Over the past few years, the incorporation of smart technology into firefighting systems has transformed the approach to tackling forest fires (Alkhatib 2013). Urgency of Protection explores the transformative potential of smart technology in enhancing fire detection, prevention, and suppression efforts. Leveraging advanced sensors, data analytics, and artificial intelligence may help firefighting agencies deploy more proactive and efficient strategies to mitigate the impact of wildfires (Toledo-Castro et al. 2018).

Adopting smart technology in firefighting systems offers many benefits that significantly enhance the ability to combat wildfires. Firstly, smart sensors and monitoring devices enable early detection of fire outbreaks, allowing for prompt response and containment efforts. Gathering and analysing data in real-time offers crucial insights into fire behaviour, enabling emergency management and firefighters to make more precise and well-informed choices. Moreover, smart technology optimizes resource allocation, ensuring that personnel and equipment are deployed strategically to areas at the highest risk. By prioritizing areas with the greatest potential for fire spread, firefighting agencies can maximize their effectiveness in containing and suppressing wildfires while minimizing resource wastage (Yuan et al. 2015). Integrating artificial intelligence and machine learning algorithms also elevates predictive capabilities, empowering proactive wildfire management strategies (Radovic et al. 2017). Examining past data and environmental elements, these systems can forecast fire behaviour and anticipate future outbreaks, enabling pre-emptive measures to mitigate risk and protect vulnerable communities (Aspragathos et al. 2019).

The growing scale and intensity of wildfires increasingly challenge the traditional methods of forest fire management. Smart technology offers several compelling advantages compared to conventional approaches. Enhanced early warning capabilities provided by smart sensors and monitoring systems enable real-time data collection and prompt alerts of potential outbreaks, facilitating a faster and more effective response. Moreover, advanced data analysis and visualization tools offer firefighters comprehensive situational awareness of the fire's behaviour, enabling them to make informed decisions (Sharma et al. 2020). Additionally, by automating processes and engaging in hazardous activities, smart technologies can potentially improve efficiency and safety, thereby minimizing risks to firefighter safety and optimizing resource allocation. Furthermore, through data-driven decision-making processes facilitated by smart systems, historical data, and current conditions can be analysed to support the development of more effective preventative measures and targeted suppression strategies, ultimately enhancing overall wildfire management efforts (Jemmali et al. 2023) (Fig. 1).

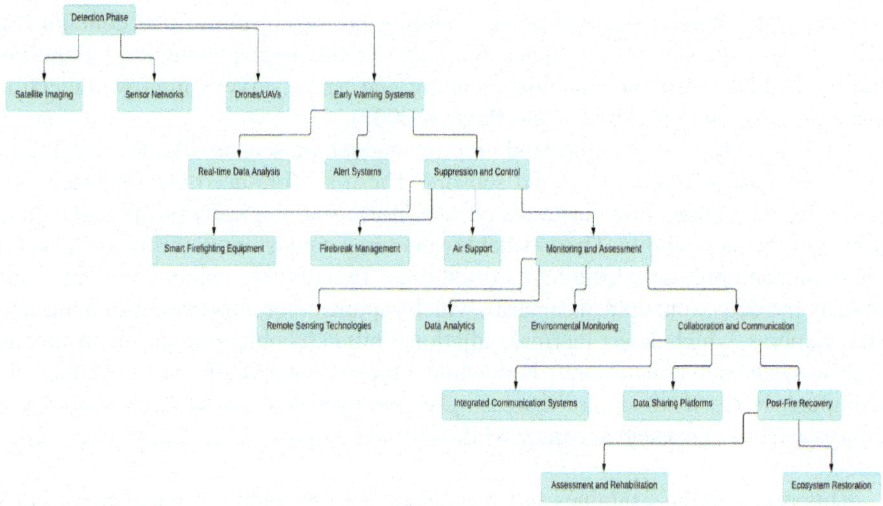


**Fig. 1** Workflow of Smart Technology and Tools in Forest Fire Management

## 2 Smart Technology Solutions for Forest Fires

Historically, developing and developed countries have been focused on suppressing fires, often prohibiting their use altogether. However, it's now recognized that fires are a landscape issue driven by factors like fuel accumulation and continuity rather than just inadequate suppression methods. The impacts of fire are gaining more attention because of their size, intensity, and interaction with changes in land use, demographics, and other factors. Ignoring the ecological, social, cultural, and economic contexts where fires occur has led to landscape alterations and increased population vulnerability (Moore 2019). Despite decades of advancements in firefighting techniques, forest fire management faces significant challenges, exacerbated by the increasing frequency and intensity of wildfires. Traditional methods often struggle to keep pace with the rapid spread of fires, leading to delays in detection, response, and containment efforts. Additionally, unpredictable environmental conditions and terrain complexities further complicate firefighting operations, highlighting the need for innovative solutions to enhance effectiveness and resilience (Martell 2007).

### *2.1 Role of Remote Sensing and Sensors in Early Detection*

Smart sensors and remote sensing technology greatly aid the early monitoring and identification of forest fires. These devices detect subtle temperature, humidity, and smoke density changes, providing real-time data to fire management agencies. By deploying sensor networks across forested areas, authorities can receive early warnings of potential fire outbreaks, enabling prompt response and containment efforts before the situation escalates. Spanning broad geographic regions, gathering thorough spatial data, and satellite imaging and aerial surveys improve detection capacities even more (Hefeeda and Bagheri 2007).

Early detection systems on land employ standalone sensors like fixed, PTZ, or 360° cameras or arrays of ground sensors. These sensors need careful placement for optimal coverage, often installed on watchtowers strategically positioned at high points for better visibility. These watchtowers serve not just for detection but also for validating and precisely locating reported fires. Two primary camera types are used in early fire detection: optical cameras, which capture color information, and infrared (IR) cameras, which detect thermal radiation emitted by objects in the environment. Hybrid systems that incorporate both camera types have recently become prevalent. Computer-based methods are increasingly employed to process large volumes of data to achieve consistent accuracy while minimizing false alarms (Barmpoutis et al. 2020).

This examination examines fire management from three perspectives: pre-fire, post-fire, and detection. Current remote sensing studies highlight a rising trend toward combining many sensors, most notably optical and radar data, with Lidar and optical passive pictures. Many fire products still rely on sensors made for various applications, even if fire sensors have been utilized for several years now. Consequently, fire managers' temporal and spatial requirements often remain inadequately met, with a preference for global scales due to limitations in spatial resolution. Even though future satellite LiDAR systems show promise for operational products, Lidar applications that focus on post-fire regeneration and fuel types tend to be more localized and non-operational at this time. Additionally, a new focus has been on combining regional and worldwide scales, highlighting the importance of upscaling and integrating techniques to reduce uncertainty in global products. According to Chuvieco et al. (2020), validation is a vital stage for any new product based on remote sensing.

## 2.2 *The Application of Drones for Monitoring and Fighting Forest Fires*

The use of drones in managing forest fires is growing. These drones provide countless benefits by entering difficult-to-reach places, tracking the spread of flames, and giving firefighters up-to-date information. Drones with thermal cameras and sensors can identify hotspots, which makes early intervention possible. They can also transport payloads for precise drops, like water or fire retardants. Furthermore, by finding missing people in densely forested areas, drones are indispensable to search and rescue efforts. Drones can obtain precise locations of fire perimeter and high-resolution photos, which can help firefighting teams visualize the size and fire direction. These maps are necessary for planning effective containment and extinguishment measures. Additionally, drones are used to identify forest fires, and manned aircraft are used less frequently, reducing the risk to human pilots. Drones present a reasonably priced substitute for conventional aerial surveillance techniques. (Vallyammai and Manoj 2023).

Drones have emerged as valuable assets in forest fire monitoring and suppression operations. With high-resolution cameras, infrared sensors, and GPS technology, drones can rapidly survey affected areas, assess fire behaviour, and identify hotspots inaccessible to ground personnel. Furthermore, drones equipped with payload delivery systems can rapidly deploy fire retardants, water, or other suppression agents, effectively supporting firefighting efforts in remote or hazardous terrain. The ability to deploy drones quickly and adaptively enhances situational awareness and response capabilities, thereby improving overall firefighting efficiency and effectiveness evaluation of the effects of using a platform to control many drones that can put out forest fires with water or other liquid extinguish ants. Studies examined the impact of drones, assuming they meet certain requirements, on the spread of forest fires. They considered variables like vegetation, moisture content, fire intensity, flame length, wind direction and speed, payload capacity of individual drones, and the time it took to arrive at the firefront (Savita 2017; Singh 2021, 2024a; Singh and Kumar 2022; Sett et al., 2022). Ausonio et al. (2021) calculated the length of an active firefront that may be put out when these characteristics change by using essential water flow estimates based on critical parameters impacting the evolution of fires. They indicated that the drone system could supplement current methods or supply the required water flow to fight forest fires that are limited in scope and severity. However, cutting-edge technology is needed for the platform to be implemented successfully, and further study is required to create a system that satisfies the presumptive requirements. Future research should examine control tactics to identify the ideal fire front portions for drone intervention, potentially using several fire simulation models since the extinguishing liquid is divided into multiple regions, unlike when using airplanes (Alessandri et al. 2021).

I. Synergizing Artificial Intelligence with IoT and Sensor Networks for Enhanced Fire Detection and Monitoring

Artificial intelligence (AI) algorithms are increasingly utilized for fire prediction, modeling, and decision support. AI models can forecast fire behaviour, predict potential ignition points, and assess wildfire risk in real time by analysing vast amounts of historical fire data, environmental factors, and weather patterns. Machine learning techniques enable continuous refinement of predictive models, enhancing their accuracy and reliability. These AI-driven insights empower fire management agencies to proactively allocate resources, implement targeted mitigation strategies, and optimize evacuation plans, thereby minimizing the impact of wildfires on ecosystems and communities. Early warning systems for forest fires are essential to prevent harm to forest ecosystems, with rapid detection being paramount for prompt intervention by authorities. Wahyono et al. (2022) offer a novel approach to solve this, combining machine learning with color-motion-shape data. Beyond being red, fires have other features, such as an uneven shape and continuous movement in certain areas. These characteristics are recorded using color probabilities in the segmentation process, color histograms in the classification process, and picture moments in the verification process. The evaluation uses the intersection over union (IoU) ratio to determine fire localization accuracy and frame-based assessments to quantify fire detection performance. On the VisiFire dataset, experiments show an average processing speed of 21.70 frames per second (FPS) and true-positive rates of 89.97% and 10.03%, respectively. These findings underscore the proposed framework's speed in detection and its ability to maintain accuracy, thus rendering it suitable and dependable for integration into early warning systems.

The potential applications of artificial intelligence (AI) in the forestry and wildlife industries have been greatly expanded by recent developments in data science and digital and satellite technologies. With 7% of the world's forest cover and the eighth-highest biodiverse region, India confronts threats to its diverse biodiversity from urbanization, agricultural growth, and rapid development initiatives (Singh 2024e). Recent advances in data science, digital, and satellite technologies have significantly broadened the potential applications of artificial intelligence (AI) in the forestry and wildlife industries. India, the eighth-highest biodiverse region in the world and 7% of the world's total forest cover faces challenges to its diverse biodiversity from rapid development initiatives, agricultural growth, and urbanization.

Recent advancements in Internet of Things (IoT) technology have made surveillance systems more accessible, emphasizing the importance of energy-efficient smart sensors, particularly in data transmission. Environmental monitoring, for instance, benefits from precise data extracted via smart multimedia sensors, yet transmitting multimedia data can be resource-intensive. Kizilkaya et al. (2022) describe a unique hierarchical system for forest fire detection that uses a hierarchical arrangement of scalar and multimedia sensors to reduce the amount of visual data transmitted. A lightweight deep learning model has also been developed for edge devices to improve detection precision and decrease traffic to the sink. The framework's energy efficiency and detection accuracy are evaluated using real testbeds, network simulations, and tenfold cross-validation. It achieves a validation accuracy of 98.28% and energy savings of 29.94%. The validation accuracy of the proposed deep learning model is nearly as good as that of top-performing architectures, demonstrating its edge computing appropriateness with lower computational needs and smaller model sizes. A study in Indonesia presents a real-time forest fire detection system utilizing solar cell-based IoT technology. Photovoltaic panels power the system and incorporate an Android application for continuous forest monitoring. The Internet of Things uses relays, an Arduino Uno for device control, and an ESP8266 12—F module. Data for temperature and humidity monitoring (DHT22 sensor), water pressure evaluation (pressure transmitter), smoke detection (MQ-2 sensor), and remote tracking of battery charging current and voltage were obtained through laboratory tests. The system can automatically respond to flame detection, enhancing its control capabilities. Laboratory testing indicated that the system effectively monitored and detected fires in real-time, demonstrating functionality across all sensors with satisfactory readings. However, room remains to improve the system's accuracy (Haryo et al. 2022) (Table 1).

**Table 1** Various kinds of sensors are used in the Internet of Things and sensor networks for monitoring and detecting forest fires

| Type of sensor | Features | Advantages | Limitations | Examples of use | References |
|---|---|---|---|---|---|
| Infrared sensors | Detects heat signatures associated with fires | Can detect fires during the day or night | Limited effectiveness in dense smoke | Fire lookout towers | Hendel and Ross (2020), Ya'acob et al. (2021) |
| Optical sensors | Utilizes visual spectrum to identify smoke or flames | Provides real-time visual confirmation | Prone to false alarms from natural phenomena | Unmanned aerial vehicles (UAVs) | Abdikanet al. (2022), Brito et al. (2020) |
| Gas sensors | Detects gases released during combustion | Can identify fires in early stages | Susceptible to interference from other sources | Ground-based sensor networks | Li et al. (2022), Dasari et al. (2020) |
| Weather sensors | Monitors the temperature, humidity, and wind speed of the environment | Provides contextual data for fire behaviour analysis | Limited accuracy in extreme weather conditions | Integrated weather stations | Chandra et al. (2022) |
| Acoustic sensors | Detects sounds associated with fire, such as crackling or roaring | Can supplement other sensor data for enhanced detection accuracy | Prone to false positives from non-fire sources | Acoustic monitoring networks | Aarthi et al. (2020), Duque et al. (2020) |

Forest fires pose significant hazards in wooded areas, with fire monitoring encompassing pre-fire preparation, fire detection and response, and post-fire assessment. Traditionally, manual methods were employed for fire detection, while modern satellite-based systems have limitations in detecting small-scale fires promptly. Research indicates that approximately 80% of forest losses result from delayed fire detection. This study suggests a methodology for early fire detection that uses necessary sensors and Raspberry Pi microcontrollers. A feed-forward, fully connected neural network facilitates prediction while data is evaluated and stored on a centralized server. Alert messages are sent to administrators and those nearby as soon as they are detected (Arisandi et al. 2023). Integrating sensor networks and IoT devices makes comprehensive forest fire warning and monitoring capabilities possible. IoT-enabled sensors positioned throughout forests continuously gather information on the state of the environment, the amount of moisture present, and the health of the vegetation. The real-time data is sent to centralized platforms, where abnormalities suggestive of possible fire activity are found using sophisticated analytics algorithms. Integration with communication networks enables seamless dissemination of alerts to firefighting agencies and local communities, facilitating timely response and coordination efforts. Furthermore, IoT devices can facilitate remote control and monitoring of fire suppression equipment, enhancing operational efficiency and safety (Dubey et al. 2019).

II. Smart Fire Fighting Equipment and Communication and Collaboration Platforms

Smart technology developments have created cutting-edge tools and equipment for combating fires that improve worker safety and operational efficacy. Smart helmets equipped with augmented reality displays provide firefighters with vital information, such as navigation guidance, thermal imaging, and real-time hazard alerts, improving situational awareness in the field. Additionally, sensor-equipped firefighting suits monitor vital signs and environmental conditions, alerting personnel to potential health risks and ensuring timely intervention. Robots are increasingly used in emergencies, especially firefighting operations, thanks to robotic firefighting systems. These systems can autonomously navigate dangerous environments, deploy suppression measures, augment human resources, and reduce exposure to danger previously seen in manufacturing, agriculture, and services.

Preparing vegetation for firefighting requires considerable resources, often facing human and equipment shortages. Robotic technologies offer two significant avenues for improving efficiency in this regard. Firstly, drones can capture aerial imagery essential for planning vegetation preparation tasks. They can identify problem areas, select specific vegetation for removal, and map out optimal routes for extraction (Roldán et al. 2018). Precision agricultural techniques can be applied here, like multispectral imaging, three-dimensional LIDAR scans, and high-resolution image identification of plants and trees (Barrientos et al. 2011). Drones present advantages over satellites due to their greater accessibility, lower costs, and reduced dependence on weather conditions, enhancing their suitability for such tasks. Secondly, ground robots can complement firefighting efforts by assisting in vegetation removal within forests. Designed to fill the gap between firefighters' manual effort and the heavy apparatus they usually use, these robots combine speed, performance, and flexibility in just the right amounts. Novel solutions are needed to effectively deploy agricultural and forestry robotics in vegetation preparation activities due to challenges in both fields, such as planning in uncertain situations, localization and mapping in complex environments, and locomotion in rugged terrain (Aguiar et al. 2020).

Effective communication and collaboration are essential components of successful forest fire management operations. Smart technology solutions facilitate seamless communication and coordination among firefighting agencies, emergency responders, and other stakeholders. Dedicated communication platforms integrate voice, data, and video capabilities, enabling real-time information exchange and decision-making. Outfitting firefighters with smartphones equipped with the built-in ANT protocol presents an opportunity to measure their proximity to one another using low-power communication radios. Subsequently, we employ a clustering approach on the proximity data to identify moving sub-groups among the firefighters. The author conducted experiments during real-life training scenarios involving 16 professional firefighting teams to evaluate the methodology's effectiveness. Six training sessions were manually labelled, involving 51 firefighters, resulting in 79 min of ground truth data. Our algorithm demonstrates an average accuracy of 80% in correctly assigning each group member to their respective ground truth cluster. Furthermore, incorporating height information obtained from atmospheric pressure signals significantly enhances group assignment accuracy to 95% (Feese et al. 2013; Singh et al. 2024; Kumar et al. 2024).

While building amenities, construction rules, and technology have improved occupant evacuation during fires, there is still a great deal of oversight regarding providing a safe environment for firefighters to battle flames and conduct rescue operations. There is a severe flaw in how present fire safety procedures ensure their safety when combating fires. This study presents the idea of Safe Firefighting Time (SFT), which integrates firefighting and evacuation behaviour elements in fire safety design to solve this shortcoming. Detailed calculation methods are established by merging the concepts of Required Safe Firefighting Time (RSFT) and Available Safe Firefighting Time (ASFT). For example, a tunnel's fire safety performance-based design applies the evaluation of SFT, including ASFT and RSFT. SFT can improve intelligent firefighting operations by integrating emerging technology beyond conventional fire safety design. Using clever strategies like IoT infrastructures and cloud services, the framework for future fire safety design and the research needs for smart firefighting with SFT are explored (Zhang et al. 2023).

Resource allocation, incident mapping, and situational investigation are made easier with the help of geographic information systems (GIS), which offer interactive maps and tools for spatial data visualization. Multi-agency coordination is made possible via collaborative platforms, which facilitate sharing of resources, knowledge, and response activities across jurisdictional borders. According to government figures, Turkey has had a noteworthy 58% increase in forest fires during the previous three decades, from 1988 to 2018. Therefore, evaluating the nation's forest fire dangers and creating more potent mitigation measures is critical. Kütahya-Ören'srisk map of forest fire was made in the first part of this study by utilizing geographic information system (GIS) capabilities to analyze various spatial data.

Furthermore, a visibility analysis of the fire towers that are currently in place was done. According to the data, the study area's very-high and low-risk, moderate-risk, and high-risk zones accounted for 2.76%, 36.86%, and 60.39% of the total area. Moreover, the towers provided observation over 82.8% of the area. In the second stage, 4 hectares of the Ören-Çamdibi region were officially reported as burned in October 2001. These areas were marked using remote sensing techniques. The results showed that the actual burned area was 5.6 hectares, with 17% falling into the very-high and high-risk zones and 83% of the burned surfaces designated as moderate-risk areas in the fire risk map (Ozenen Kavlak et al. 2021).

# 3 Case Studies and Implementation Challenges

## *3.1 Present Successful Real-World Case Studies*

California's ALERTWildfire Network.

Natural and human-induced wildfires pose significant threats to the environment and communities. Due to the consequences of climate change and rising global temperatures (Singh 2024b), California alone had horrific wildfires in 2020 that destroyed 10,500 structures, scorched 4.2 million acres of land, and took over 31 lives. Costs associated with combating fires have increased due to the growing size and frequency of wildfires. Present research efforts are mainly focused on early detection of forest fires to stop their spread, reduce damage, and, most importantly, save lives. A robust fire detection system is essential, offering efficiency and accuracy to combat wildfires effectively before they spread uncontrollably. Computer Vision is a promising solution due to its potential for reliability, speed, and widespread applicability. However, existing research in preliminary fire detection encounters challenges, including using inadequate and imbalanced datasets for model training. A suggestion has been made to improve settings, lighting, atmospheric variables, and class distributions to provide more uniform and representative datasets. Upon comprehensive analysis, the outcomes of this research underscore the dataset's efficacy in real-world applications, demonstrating the model's ability to generalize effectively and confirming its practical relevance in machine-learning settings with tangible impacts (Shah 2021).

Australian Bushfire Detection and Response System

The world is grappling with increasingly severe weather phenomena due to climate change, exemplified by the recent catastrophic bushfires in Australia (Singh 2024c). These fires, which have been called the worst in the nation's history, killed or negatively impacted an estimated three billion native species, in addition to causing an economic loss of around one billion Australian dollars. Considering the extent of this devastation, scientists are actively looking for workable ways to anticipate and prevent fires, giving firefighters more time to protect people and property and be ready for battle. This urgency prompted the investigation into a potential solution for this critical issue. Nosouhi et al. (2021) used machine learning (ML) techniques to find abnormalities in spatiotemporal observations of environmental factors like relative humidity and temperature. This process involves training an ML model over a year to identify common spatiotemporal patterns in ecological data. During the detection phase, any deviations from this norm in real-time data collected from field sensors are flagged as anomalies, indicating a potential bushfire situation. They present an additional categorization model based on Moran's I index to validate these anomalies further and rule out sensor failures or security breaches common in Internet of Things (IoT) devices. The datasets were used to train three machine learning models, which were then assessed for comparability and performance.

### 3.2 *Addressing Implementation Challenges: Solutions and Best Practices*

Despite the advancements in smart technology solutions for forest fire management, significant implementation challenges persist. Remote or rugged terrain poses difficulties in deploying sensor networks, drones, and communication systems, limiting their efficacy in certain regions. Integrating data from various sources, such as satellites and ground sensors, presents complexities and requires interoperable systems for seamless information sharing. Additionally, the high initial investment and ongoing maintenance costs associated with smart technology solutions can be prohibitive, especially for smaller agencies with limited budgets. Regulatory constraints, including privacy regulations and airspace restrictions, further impede the deployment of drones for wildfire monitoring. In contrast, concerns about data privacy may arise from collecting and sharing sensitive information. Addressing these challenges will be crucial for maximizing the potential of smart technology in forest fire management and ensuring its widespread adoption and effectiveness.

A comprehensive strategy involving technical innovation, legislative assistance, and stakeholder participation via utilizing traditional knowledge for sustainable development is required to effectively address the implementation issues associated with forest fire control. To facilitate resource-sharing, technology development, and cost-sharing initiatives, government agencies, research institutions, and private sector entities collaborate through public–private partnerships, promoting the broader adoption of smart technology solutions. Standardization and interoperability are essential, requiring the development of standardized protocols and data formats to enable seamless integration and data exchange between different smart technology systems. Investment in infrastructure, particularly in remote areas, necessitates prioritizing resources for broadband networks and sensor deployment frameworks to enhance coverage and connectivity. Clear regulatory frameworks are essential for guiding drone operations, ensuring data privacy, and addressing cyber security concerns to mitigate legal and ethical challenges associated with smart technology implementation. Moreover, capacity building and training programs are essential to enhance the technical skills and awareness of firefighting personnel and stakeholders, fostering greater adoption and utilization of smart technology solutions. Previous strategies of villagers and traditional methods could be used to improve the technological method for forest fire mitigation. By addressing these challenges and implementing best practices, stakeholders can maximize the benefits of smart technology in forest fire management, ultimately improving response capabilities, enhancing firefighter safety, and mitigating the impact of wildfires on communities and ecosystems.

## 4 Future Perspectives

The future of forest fire management holds promising advancements driven by ongoing technological innovation and collaboration. Enhanced integration of artificial intelligence and machine learning algorithms will enable more accurate prediction of fire behaviour and proactive deployment of resources. Advancements in sensor technology will lead to more sophisticated monitoring systems capable of detecting subtle changes in environmental conditions (Singh 2024d). Furthermore, the proliferation of unmanned aerial vehicles (UAVs) and autonomous firefighting robots will revolutionize wildfire response, enabling rapid and targeted suppression efforts in even the most challenging terrain.

Additionally, advancements in communication technology will facilitate real-time coordination and information sharing among firefighting agencies and stakeholders, enhancing overall response capabilities. As we embrace these technological advancements, it is essential to address ethical considerations to ensure responsible and equitable forest fire management practices. To protect individual rights and freedoms, privacy problems in the gathering and using of sensitive data must be adequately controlled. Moreover, deploying autonomous firefighting systems raises ethical questions regarding accountability and decision-making in high-stakes scenarios. Transparency and public engagement are paramount to fostering trust and addressing community concerns regarding deploying emerging technologies in wildfire management. Despite the challenges and ethical considerations, the future of forest fire management is filled with promise and opportunity. We can transform wildfire response and lessen the destructive effects of wildfires on ecosystems and populations by embracing ethical principles and utilizing smart technologies. Through collaboration, innovation, and a proactive approach, we can build resilience and adaptability in the face of evolving fire threats.

Looking ahead, politicians are critical in determining how forest fire control techniques and technology develop. Policy frameworks should prioritize investments in research and development to spur innovation and accelerate the adoption of smart technology solutions. Additionally, regulatory frameworks must be established to ensure the responsible and ethical use of emerging technologies while addressing privacy, safety, and environmental concerns. Collaborative efforts between government agencies, academia, industry, and local communities are essential to develop comprehensive policies that promote resilience and sustainability in wildfire management. Aligning policy objectives with technological advancements and ethical principles forge a path toward a future where wildfires are managed effectively, and the health and well-being of our planet are safeguarded for generations to come.

## 5 Conclusion

Integrating smart technology and advanced tools holds immense promise for revolutionizing forest fire management. This chapter explored the multifaceted applications of these innovations, ranging from early detection and warning systems to efficient suppression and post-fire recovery efforts. Smart technology, including drones, sensor networks, and satellite imaging, enables real-time monitoring and analysis of fire behaviour, facilitating proactive responses and minimizing the impact on ecosystems and communities. Despite significant advancements, challenges persist, such as infrastructure limitations, data integration complexities, and regulatory hurdles. However, collaborative efforts among stakeholders, public–private partnerships, and ongoing research initiatives pave the way for overcoming these obstacles and enhancing the effectiveness of smart technology solutions. Furthermore, the potential of artificial intelligence, machine learning, and IoT integration offers further opportunities for innovation in forest fire management. Leveraging these technologies may enhance predictive capabilities, optimize resource allocation, and improve decision-making processes, ultimately reducing the devastating consequences of wildfires. The adoption of smart tools and technology is more critical than ever in light of the growing effects of climate change and the rising frequency of extreme weather events. Harnessing the power of emerging technology through continuous refinement, interdisciplinary collaboration, and proactive mitigation strategies will help safeguard forests, protect lives and property, and preserve the natural environment for future generations.

## References

Aarthi AVV, Abirami J, Karthika K, Kiruba R (2020) Acoustic wave-based forest fire extinguisher and detection using machine. Int J Eng Appl Sci Technol 4(10):196–202

Abdikan S, Bayik C, Sekertekin A, BektasBalcik F, Karimzadeh S, Matsuoka M, BalikSanli F (2022) Burned area detection using multi-sensor SAR, optical, and thermal data in Mediterranean pine forest. Forests 13(2):347

Aguiar AS, Dos Santos FN, Cunha JB, Sobreira H, Sousa AJ (2020) Localization and mapping for robots in agriculture and forestry: a survey. Robotics 9(4):97

Alessandri A, Bagnerini P, Gaggero M, Mantelli L (2021) Parameter estimation of fire propagation models using level set methods. Appl Math Model 92:731–747

Alkhatib AA (2013) Smart and low cost technique for forest fire detection using wireless sensor network. Int J Comput Appl, 81(11).

Arisandi D, Syamsuadi A, Trisnawati L (2023) Enhancing peatland fire prevention: an incremental LoRa and mobile-based early warning system. Int J Adv Technol Eng Explor 10(108):1368

Aspragathos N, Dogkas E, Konstantinidis P, Koutmos P, Lamprinou N, Moulianitis VC, Xofis P (2019) From pillars to AI technology-based forest fire protection systems. In: Intelligent system and computing. IntechOpen

Ausonio E, Bagnerini P, Ghio M (2021) Drone swarms in fire suppression activities: a conceptual framework. Drones 5(1):17

Barmpoutis P, Papaioannou P, Dimitropoulos K, Grammalidis N (2020) A review on early forest fire detection systems using optical remote sensing. Sensors 20(22):6442

Barrientos A, Colorado J, Cerro JD, Martinez A, Rossi C, Sanz D, Valente J (2011) Aerial remote sensing in agriculture: a practical approach to area coverage and path planning for fleets of mini aerial robots. J Field Robot 28(5):667–689

Brito T, Pereira AI, Lima J, Castro JP, Valente A (2020) Optimal sensors positioning to detect forest fire ignitions. In: Proceedings of the 9th international conference on operations research and enterprise systems, pp 411–418

Chandra R, Agarwal S, Singh N (2022) Semantic sensor network ontology based decision support system for forest fire management. Eco Inform 72:101821

Chuvieco E, Aguado I, Salas J, García M, Yebra M, Oliva P (2020) Satellite remote sensing contributions to wildland fire science and management. Curr for Rep 6:81–96

Dasari P, Reddy GKJ, Gudipalli A (2020) Forest fire detection using wireless sensor networks. Int J Smart Sens Intell Syst 13(1):1–8

Dubey V, Kumar P, Chauhan N (2019) Forest fire detection system using IoT and artificial neural network. In: International conference on innovative computing and communications: proceedings of ICICC 2018, Springer Singapore. Vol 1 pp 323–337

Duque D, Dederichs S, Muhasilovic M (2020) The potential of acoustic sensors to foster the ecological sustainability of a city: a case study in medellin. In: Student engineering conferences, p 1

Feese S, Arnrich B, Tröster G, Burtscher M, Meyer B, Jonas K (2013) Sensing group proximity dynamics of firefighting teams using smartphones. In: Proceedings of the 2013 international symposium on wearable computers, pp 97–104

Haryo RJK, Artono B, Ningrum HNK, Habsari KM, Rahma IM (2022) Forest fire detector and fire fighting monitoring system using solar cell based internet of things (IoT). Telematika 15(1):23–36

Hefeeda M, Bagheri M (2007) Wireless sensor networks for early detection of forest fires. In: 2007 IEEE International conference on mobile adhoc and sensor systems. IEEE. pp 1–6

Hendel IG, Ross GM (2020) Efficacy of remote sensing in early forest fire detection: a thermal sensor comparison. Can J Remote Sens 46(4):414–428

Jemmali M, Kayed BML, Boulila W, Amdouni H, Alharbi MT (2023) Optimizing forest fire prevention: intelligent scheduling algorithms for drone-based surveillance system. Proced Comput Sci 225:1562–1571

Kizilkaya B, Ever E, Yatbaz HY, Yazici A (2022) An effective forest fire detection framework using heterogeneous wireless multimedia sensor networks. ACM Trans Multimed Comput Commun Appl (TOMM) 18(2):1–21

Kumar N, Singh H, Kumar A, Singh AK, Pandey PK, Kumar A (2024) Forest-fire-induced land degradation. In: Mishra G, Giri K, Singh S, Kumar M (eds) sustainable land management in India. Springer Nature, Singapore. https://doi.org/10.1007/978-981-97-5223-2_4

Li Y, Yu L, Zheng C, Ma Z, Yang S, Song F, Tittel FK (2022) Development and field deployment of a mid-infrared CO and $CO_2$ dual-gas sensor system for early fire detection and location. Spectrochim Acta Part A Mol Biomol Spectrosc 270:120834

Manoj S, Valliyammai C (2023) Drone network for early warning of forest fire and dynamic fire quenching plan generation. EURASIP J Wirel Commun Netw 2023(1):112

Martell DL (2007) Forest fire management. Handbook of operations research in natural resources. Springer, US, Boston, MA, pp 489–509

Moore PF (2019) Global wildland fire management research needs. Curr for Rep 5:210–225

Nosouhi MR, Sood K, Kumar N, Wevill T, Thapa C (2021) Bushfire risk detection using internet of things: an application scenario. IEEE Internet Things J 9(7):5266–5274

OzenenKavlak M, Cabuk SN, Cetin M (2021) Development of forest fire risk map using geographical information systems and remote sensing capabilities: Ören case. Environ Sci Pollut Res 28(25):33265–33291

Radovic M, Adarkwa O, Wang Q (2017) Object recognition in aerial images using convolutional neural networks. J Imaging 3(2):21

Roldán JJ, del Cerro J, Garzón-Ramos D, Garcia-Aunon P, Garzón M, De León J, Barrientos A (2018) Robots in agriculture: state of art and practical experiences. Serv Rob, 67–90

Savita, Sharma SK, Singh H, Singh O (2017) Forest fire: a review. Indian for 143(8):719–728

Sett T, Nikam BR, Singh H, Purohit S (2022) Effect of drought stress on forest evapotranspiration—a case study on Indian forests. EGU General Assembly. https://doi.org/10.5194/egusphere-egu22-510

Shah S (2021) Preliminary wildfire detection using state-of-the-art PTZ (Pan, Tilt, Zoom) camera technology and convolutional neural networks. arXiv preprint arXiv:2109.05083

Sharma A, Singh PK, Kumar Y (2020) An integrated fire detection system using IoT and image processing techniques for smart cities. Sustain Cities Soc 61:102332

Singh H (2024) Exploring adaptive modulation in plant functional traits and its impact on the productivity of *Acacia auriculiformis* under $CO_2$-enriched environment. Ind Crops Prod 210:118186. https://doi.org/10.1016/j.indcrop.2024.118186

Singh H, Kumar M (2022) Climate change and its impact on Indian Himalayan forests: current status and research needs. In: Rani S, Kumar R (eds) climate change. Springer Climate. Springer, Cham. pp 223–242, https://doi.org/10.1007/978-3-030-92782-0_11

Singh H, Kumar N, Kumar A (2024) Enhancing resource use efficiency in crops through plant functional traits. Springer, pp 97–117. ISBN: 978-981-97-1509-1

Singh H (2021) Process-based carbon sequestration study in relation to the energy-water-carbon flux in a forest ecosystem. In: Climate impacts on sustainable natural resource management Kumar P, Singh RK, Kumar M, Rani M, Sharma (eds). CRC Press Taylor & Francis Group, United States, pp 336–361

Singh H (2024b) Climate change and forests: biological perspectives. In: Singh H (eds) forests and climate change. Springer Nature, Singapore. https://doi.org/10.1007/978-981-97-3905-9_1

Singh H (2024d) Climate change and environmental pollution: biological consequences on urban forests. In: Singh H (eds) urban forests, climate change and environmental pollution. Springer Nature, Cham, Switzerland. https://doi.org/10.1007/978-3-031-67837-0_1

Singh H (2024c) Forests and climate change: biological perspectives on impact, adaptation, and mitigation strategies (ed), Springer Nature Singapore. https://doi.org/10.1007/978-981-97-3905-9

Singh H (2024e) Urban forests, climate change and environmental pollution: physio-biochemical and molecular perspectives to enhance urban resilience (ed), Springer Nature, Cham, Switzerland. https://doi.org/10.1007/978-3-031-67837-0

Toledo-Castro J, Caballero-Gil P, Rodríguez-Pérez N, Santos-González I, Hernández-Goya C, Aguasca-Colomo R (2018) Forest fire prevention, detection, and fighting based on fuzzy logic and wireless sensor networks. Complexity 2018:1–17

Wahyono, Harjoko A, Dharmawan A, Adhinata FD, Kosala G, Jo KH (2022) Real-time forest fire detection framework based on artificial intelligence using color probability model and motion feature analysis. Fire, 5(1), 23

Ya'acob N, Najib MSM, Tajudin N, Yusof AL, Kassim M (2021) Image processing based forest fire detection using infrared camera. In: Journal of physics: conference series (Vol 1768, No 1, p 012014). IOP Publishing

Yuan C, Zhang Y, Liu Z (2015) A survey on technologies for automatic forest fire monitoring, detection, and fighting using unmanned aerial vehicles and remote sensing techniques. Can J for Res 45(7):783–792

Zhang Y, Zhang X, Huang X (2023) Design a safe firefighting time (SFT) for major fire disaster emergency response. Int J Disaster Risk Reduction 88:103606

# Fire-Resistant Materials and Construction: Designing for Climate Change

**Rishita Pakhira, Subhankar Biswas, and Sukhendu Nandi**

**Abstract** As wildfires become increasingly prevalent and severe due to climate change, there is a growing need for effective fire-resistant materials and construction practices. This chapter explores the importance of designing regarding climate change resilience within the framework of fire-resistant materials and construction. It discusses the characteristics of effective fire-resistant materials, including high heat resistance, non-combustibility, low thermal conductivity, chemical stability, insulation properties, and durability. It also examines various types of natural and synthetic fire-resistant materials and their advantages and limitations. Integrating fire-resistant materials and construction techniques into building design may help communities mitigate the risk of fire damage and enhance resilience to the effects of global warming. Moreover, it highlights the importance of proactive measures in adapting to evolving environmental challenges and underscores the role of fire-resistant materials in creating safer and more sustainable built environments.

**Keywords** Construction design · Fire protection · Fire safety · Wildland-urban interface

## 1 Introduction

A natural occurrence in Earth's systems is fire, which has significantly influenced the landscapes of various biomes over millions of years (Pausas et al. 2017; Shi et al. 2022). Fires cover an area of about 3–5 million square kilometers annually and release approximately $CO_2$ emissions into the atmosphere of 8 billion tons (Chuvieco

R. Pakhira (✉)
TESAF, University of Padua, Padova 35020, Italy
e-mail: rishitapakhira@gmail.com

S. Biswas
Central Botanical Laboratory, Botanical Survey of India, Howrah, West Bengal, India

S. Nandi
ICAR-Indian Agricultural Research Institute, New Delhi, India

 
H. Singh and N. Kumar (eds.), *Forest Fire and Climate Change*,
https://doi.org/10.1007/978-3-031-89967-6_17

et al. 2010; Kumar et al. 2024). These emissions are important parts of the carbon cycle, as is the yearly absorption of about 7 billion tonnes of $CO_2$ through post-fire vegetation regrowth (Yin et al. 2020; Zheng et al. 2021). Fires have a 10% global reduction in the quantity of carbon stored in plants, which has a significant impact on atmospheric $CO_2$ levels and climate (Harrison et al. 2021; Kumar et al. 2020a, 2020b, 2021a; Singh 2024b).

Wildfire patterns vary across ecosystems, particularly concerning their productivity, and are influenced by human activities and broader global changes (Singh 2024c). While rising temperatures and increasing atmospheric $CO_2$ levels are global trends, other factors like population growth, land use practices, precipitation patterns, and invasive species impact wildfire activity differently (Savita 2017; Keeley and Pausas 2019; Kumar et al. 2021b; Singh 2021, 2024a; Singh and Kumar 2022; Sett et al. 2022). The composition of the ecosystem's flora and the kinds of fuel present affect how these drivers of global change behave. Understanding these complexities is essential for effectively managing ecosystems amidst ongoing changes. While global warming is often cited as a major driver of increased wildfire activity, non-climatic human factors may sometimes play equally or even more significant roles (Singh 2024d). Nevertheless, in many ecosystems, the influence of climate change is becoming more pronounced as temperatures continue to rise (Kumar et al. 2017; Williams et al., 2019; Kumar et al. 2019).

The presence of wildfires in an ecosystem depends on the convergence of several elements: ignition sources, available fuel, drought conditions, and suitable meteorological circumstances, such as low humidity, high temperatures, and wind (Singh 2024e; Sett et al. 2024; Kaushik and Singh 2024; Barwa and Singh 2024). These factors interact to initiate wildfires, and global changes can influence all of them. As a result, wildfire patterns are evolving in many ecosystems, often due to various factors. Since fire behavior relies on spatial connections, the relationship between wildfire drivers and activity is unlikely to follow a straightforward, linear path (van Nes et al. 2018).

Urbanization trends are causing urban and wildland areas to merge, giving rise to the Wildland-Urban Interface (WUI), where wildfires pose significant threats. Fires in the WUI often ignite from wildland fuels triggered by natural or human activities. Preventing structural ignitions is crucial in addressing this challenge, primarily achieved by minimizing structures' susceptibility to catching fire (Mell et al. 2010). Utilizing fire-resistant building materials is a crucial strategy, especially in forested areas prone to fires. Because these materials are made to endure heat and flames, there is less chance that a fire may start in a structure and spread. Traditional fire-resistant materials include stone, cement, metal, and brick, while more advanced options like gypsum wallboard and pressure-impregnated lumber offer enhanced protection. Features like smoke detectors, indoor sprinkler systems, and Class A-rated roofing materials further strengthen fire resilience (Caetano et al. 2022; Singh et al. 2024). National building codes provide guidelines for using fire-resistant materials and implementing fire safety measures. Adhering to these guidelines in WUI and other fire-prone residential areas is vital for protecting structures, minimizing property damage, and enhancing overall wildfire resilience.

# 2 Understanding Wildfire Behavior

## *2.1 Factors Influencing Wildfire Ignition and Spread*

Wildfires are caused by several variables that are indirectly related to climate change. With a projected 12% increase in strikes per degree rise in temperature, lightning is linked to global warming and accounts for one-third of forest fires between May and September. In order to comprehend lightning-caused fires, Moris et al. (2020) suggested a strategy for connecting flames to naturally occurring lightning, highlighting the significance of determining what ignites lightning. Changes in soil properties due to fire succession can change how greenhouse gases are consumed and released, impacting carbon sequestration. According to Adkins et al. (2019), wildfires' intensity impacts soil carbon and nitrogen pools, with more burned areas storing less carbon. Fuel moisture plays a vital role in fire incidence, affecting the ignitability of fuel beds and the spread of wildfires. When wildfires occur close to forest-tundra ecotones, where self-pruning of lower branches increases contact with the soil surface, soil organic carbon participation can aid in spreading the fire. Predicting wildfire rate of spread and intensity is essential for sustainable management, with Moinuddin and Sutherland (2020) simulating grassfire and surface-to-crown fire transitions. Drought exacerbates fire risk by affecting vegetation and forest conditions, with reduced precipitation and increased temperatures leading to drier, more flammable fuels. Climate observations and trends in burned areas are used to estimate future temperatures and humidity levels, critical elements in the initiation of wildfires. Davis et al. (2017) projected increased wildfire susceptibility for forests > 40 hectares in the USA due to unsustainable management practices and forest loss. Xu et al. (2020) highlighted the susceptibility of Amazonian forests to fire, with significant loss leading to increased fire risk.

## *2.2 Wildfire Situation Worldwide*

In 2019, Global Forest Watch documented more than 4.5 million fires throughout the globe, with many ignited by human activities like campfires and discarded cigarettes, devastating hectares of forest and vegetation annually. Between 2003 and 2012, an evaluation found an average of every year, 67 million hectares of forest burn, reaching 98 million hectares in 2015, the hottest year of the decade (Van Lierop et al. 2015). Tropical forests in South America and Africa were particularly affected, with about 4% of their total area burned. Notable wildfires in Australia, Brazil, Greece, Russia, and California between 2018 and 2020 resulted in significant loss of life, property, infrastructure, and ecological damage.

Based on Landsat data, the Global Annual Burned Area Mapping (GABAM) project found that there are, on average, 82.09 million hectares of burned forests globally in 2000, 2005, 2010, 2015, and 2018 (Long et al. 2019). Zhang et al. (2020)

reported statistics on burned forests by continent and climate zone from 2000 to 2018, highlighting tropical areas in Africa and South America as hotspots for wildfires.

Wildfires profoundly impact infrastructure, human lives, and wildlife, often caused by human activities encroaching into forested areas, notably in the interface between wildlands and cities (WUI). Wildfires started by people are more likely to occur when WUI expands (Nagy et al. 2018), leading to devastating events globally, including in Australia, Europe, Russia, Portugal, and California. Concerns about future trends and associated damages are growing with increasing wildfire incidence. Studies predict a rise in global wildfire activity based on past trends, with regional variations but consistent biomass-burning patterns (Pechony and Shindell 2010). Government-imposed regulations restricting infrastructure development near forests could mitigate wildfire risk and reduce losses. Following the 2009 Black Saturday Fires in Victoria, Australia, new building codes were introduced to enhance wildfire resilience. Homes were designed with non-combustible materials, ember shields, and metal screens to prevent embers from entering. Additionally, water tanks and sprinkler systems were installed to keep vegetation moist during high-risk periods, reducing property damage in subsequent fires. Similarly, in Brazil's Serra do Mar region, retaining walls, firebreaks, and improved drainage systems were constructed to manage wildfires, prevent erosion, and support firefighting efforts.

# 3 Fire-Resistant Materials

## *3.1 Characteristics of Effective Fire-Resistant Materials*

Effective fire-resistant materials possess several characteristics that enable them to withstand and inhibit fire spread. Here are some key attributes:

1. **High Heat Resistance:** Fire-resistant materials should have a high resistance to heat. They should withstand high temperatures without melting, deforming, or releasing flammable gases.
2. **Non-Combustibility**: These materials should not readily catch fire or support combustion. They should not contribute fuel to a fire.
3. **Low Thermal Conductivity**: Fire-resistant materials should have low thermal conductivity to stop the heat from the fire from spreading to other areas of the building. This helps contain the fire and prevent its spread.
4. **Chemical Stability**: They should be chemically stable, meaning they do not undergo significant chemical changes when exposed to high temperatures. Chemical stability prevents the release of toxic gases or the production of additional fuel for the fire.
5. **Insulation Properties**: Fire-resistant materials should have good insulation properties to protect the underlying structure from heat damage. They should limit the transmission of heat to the protected side.

6. **Durability:** These materials should be durable and resistant to wear and tear, ensuring long-term effectiveness in fire protection.

## 3.2 *Types of Fire-Resistant Materials*

Fire-resistant materials can be categorized into two main types: natural and synthetic. Here's an overview of each:

**Natural Fire-Resistant Materials:**

1. **Mineral-Based Materials:**

**Gypsum**: Gypsum is a naturally occurring mineral often used in building materials like drywall. Because of its low heat conductivity and dehydration characteristics, it provides superior fire resistance.

**Advantages:**

- Because of its low thermal conductivity, it has good fire resistance.
- It is commonly used in drywall construction, providing a barrier against fire spread.
- Affordable and readily available.

**Limitations:**

- Susceptible to water damage, reducing its effectiveness over time.
- Limited structural strength compared to other materials like concrete or steel.

**Clay**: Bricks and clay tiles, for example, are naturally fire-resistant materials. They are resistant to high heat and do not burn.

**Advantages:**

- It is naturally fire-resistant and does not burn.
- Used in various forms, such as bricks and clay tiles for construction.
- Durable and long-lasting.

**Limitations:**

- Heavy and labor-intensive to install.
- Limited design flexibility compared to other materials.

**Mineral wool**: Mineral wool, which is formed from minerals like rock or slag, is an excellent insulator and fire-resistant material because it is non-combustible.

**Advantages:**

- Excellent resistance to fire since it is not flammable.
- Provides effective insulation, enhancing energy efficiency.
- Resistant to mold, mildew, and pests.

**Limitations:**

- Relatively high cost compared to other insulation materials.
- It may settle over time, reducing its insulation effectiveness.

**Synthetic Fire-Resistant Materials**

1. **Fire-retardant treated wood (FRTW):** Wood treated with fire-retardant chemicals becomes fire-resistant. These chemicals reduce the wood's flammability and slow down the spread of flames.

**Advantages:**

- Allows the use of wood in construction while meeting fire safety requirements.
- It can be manufactured from sustainably sourced materials.
- Generally, it is more affordable than non-wood alternatives.

**Limitations:**

- Treatment chemicals may leach over time, posing environmental concerns.
- Requires periodic inspection and maintenance to ensure continued effectiveness.
- **Fire-resistant coatings and paints**: Special coatings and paints can be applied to various surfaces to enhance fire resistance. These coatings form a protective layer that inhibits the ignition and spread of flames.

**Advantages:**

- It can be applied to various surfaces to enhance fire resistance.
- Provides a cost-effective way to upgrade existing materials.
- Some coatings offer additional benefits, such as UV protection and waterproofing.

**Limitations:**

- It may require frequent reapplication to maintain effectiveness.
- Performance can vary depending on the substrate and application method.
- Fire-resistant fabrics and textiles: Fabrics treated with fire-retardant chemicals or made from inherently fire-resistant fibers, such as aramid fibers (e.g., Nomex), can withstand flame exposure and high temperatures.

**Advantages:**

- It is used in furnishings and clothing to reduce flammability and improve safety.
- Available in a wide range of styles and colors.
- Some fabrics are inherently fire-resistant, offering long-lasting protection.

**Limitations:**

- Treatment chemicals may diminish over time, requiring replacement.
- Limited effectiveness against high-intensity fires.
- **Fire-resistant plastics and composites**: Certain types of plastics and composite materials are engineered to be fire-resistant. They may contain additives or be formulated with flame-retardant properties to inhibit combustion.

**Advantages:**

- Lightweight and versatile, it is suitable for various applications.
- It can be engineered to meet specific fire safety standards.
- Resistant to corrosion and degradation.

**Limitations:**

- May emit toxic fumes when burned, posing health risks.
- Performance can degrade over time due to exposure to UV radiation and other factors.
- **Intumescent materials**: When subjected to high temperatures, intumescent materials, such as sealants and coatings, expand and create a protective char layer that shields beneath surfaces from heat and flames.

**Advantages:**

- When exposed to heat, they expand and create a protective covering known as char.
- It can be applied to structural elements to enhance fire resistance.
- Suitable for both interior and exterior applications.

**Limitations:**

- The thickness of the intumescent layer may affect aesthetics and construction costs.
- Performance can be affected by improper installation or substrate preparation.
- **Fire-resistant glass**: Glass treated with fire-resistant coatings or unique formulations can withstand high temperatures and prevent fire from spreading through openings like windows and doors.

**Advantages:**

- It provides transparent fire barriers, allowing natural light transmission and visibility.
- It helps contain fire and smoke in compartmentalization strategies.
- Available in various configurations to meet different performance requirements.

**Limitations:**

- Generally, more expensive than standard glass.
- It may have limited impact resistance compared to laminated or tempered glass.

There are several ways to improve the fire safety of buildings and structures using natural and synthetic fire-resistant materials, and each type has certain benefits and drawbacks. Several variables, including price, availability, performance standards, and building codes, frequently influence the decision between them. Following the 2018 wildfire crisis in Monchique, Portugal, firebreak construction emerged as a key strategy to manage wildfires in the Mediterranean region. Wide strips of vegetation were cleared to prevent fires from spreading to settlements, and gravel roads and walls were built to maintain these firebreaks and ensure access for firefighting vehicles.

Additionally, reservoirs and water points were installed near firebreaks to support fire suppression efforts. This system has since proven effective in containing wildfires in the region.

## 3.3 Innovations in Fire-Resistant Material Technology

In recent years, several notable innovations in fire-resistant material technology have been aimed at enhancing fire safety in various applications. Some of these innovations include:

1. **Nano-engineered materials**: Researchers have developed nano-engineered materials with enhanced fire-resistant properties. These materials often incorporate nanoparticles or nanocoating that improve thermal stability, reduce flammability, and inhibit the spread of flames (Paul et al. 2023).
2. **Self-healing materials**: Self-healing polymers and coatings have been developed that can repair damage caused by fire, such as cracks and charred areas. These materials use mechanisms such as microcapsules containing healing agents released upon damage, restoring the material's integrity (Ming and Cao 2020).
3. **Intumescent coatings**: Intumescent coatings have undergone advancements to improve their performance and versatility. New formulations offer enhanced fire protection, faster expansion rates, and improved adhesion to substrates. Additionally, developments in intumescent technology have led to coatings that can be applied to a broader range of materials, including plastics and composites (Puri and Khanna 2017).
4. **Bio-based fire-resistant materials**: As an alternative to conventional fire-resistant materials, researchers are looking at using bio-based materials such as cellulose fibers, lignin, and bio-based resins. These materials are renewable, sustainable, and biodegradable, offering environmentally friendly solutions for fire protection (Yang et al. 2021).
5. **Flexible fire-resistant textiles**: The development of flexible fire-resistant materials and textiles with enhanced comfort and performance has been made possible by advancements in textile technology. These fabrics incorporate fire-retardant additives or inherently fire-resistant fibers to protect flames and heat while maintaining flexibility and breathability.
6. **Transparent fire-resistant materials**: Transparent fire-resistant materials, such as fire-rated glass and polymers, have been developed for applications where visibility and aesthetics are important, such as building facades, windows, and partitions. These materials offer high levels of fire protection while allowing natural light transmission and unobstructed views.
7. **Smart fire-resistant materials**: Smart materials with sensors and actuators have been designed to detect and respond to fire hazards in real-time. These materials can trigger alarms, activate fire suppression systems, or initiate self-protective mechanisms to prevent fire spread and mitigate damage (Cao et al. 2022).

8. **3D printing of fire-resistant materials**: 3D printing technology has enabled the fabrication of complex geometries and customized structures using fire-resistant materials. This allows for the rapid prototyping and production of fire-resistant components and structures with precise specifications and tailored properties.

# 4 Designing Fire-Resistant Structures

## *4.1 Guidelines for Fire-Resistant Construction:*

Collaboration between fire protection engineers and structural engineers is crucial for designing structures with fire resistance based on performance standards analysis of fire conditions and heat transfer to structural elements is done by fire protection engineers, while structural engineers utilize this analysis to determine structural responses, considering strains caused by heat and changes in material properties at high temperatures. Effective collaboration occurs when fire resistance is integrated early in the design process, allowing flexibility for fire-related considerations to influence structural design. Following the devastating 2018 Camp Fire in Paradise, California, the town was rebuilt with a focus on fire-resistant infrastructure to prevent future disasters. The reconstruction involved using non-flammable materials like concrete and metal roofing, creating defensible spaces around homes, and placing electrical lines underground to reduce fire risks. Roads were also widened, and firebreaks were added to slow the spread of wildfires. This approach serves as a model for building wildfire-resilient communities worldwide.

Several processes are involved in developing structural fire resistance based on performance (Aktan 2007):

1. Using fire modeling to predict the temperature environment surrounding a building structure during a fire occurrence.
2. Conduct heat transfer assessments and convert temperature boundary conditions into heat flux boundary conditions to evaluate the structure's thermal response.
3. Assessing the structural response while considering stresses brought on by thermal expansion and contraction and temperature-dependent material characteristics. Nonlinearities in geometry and material behavior are frequently important factors to consider.

If risk assessment is part of the design, it involves analyzing fire hazards, the structural system's response, and the potential consequences of fire-induced failures.

## *4.2 Designing Concrete Structures to Resist Fires:*

This portion delves into the design principles applicable to various concrete constructions and how reinforced concrete structural elements react to fire exposure, focusing on concrete floor systems. Buchanan & Abu's work in 2001 provides detailed examples of designing concrete components for exposure to fire.

### 4.2.1 Beams for Concrete Floors

In a concrete floor setup, a reinforced concrete beam typically holds positive moment steel near its bottom at midspan, with the top usually integrated with the concrete slab, thus less exposed to fire below. Because concrete below the neutral axis is ignored in conventional flexural design, Its strength reduction does not significantly impact the load capacity in this section. However, steel reinforcement integrity in this area is crucial, with concrete cover as insulation. In a fire, while surface temperatures rise rapidly, interior temperatures increase more slowly due to concrete's heat capacity and density, leading to a nonlinear temperature profile. During the early stages of a fire, intact concrete cover insulates steel. Yet, high temperatures weaken concrete, induce stress from thermal expansion, and may cause spalling, exposing steel directly to fire and reducing its strength even without spalling.

### 4.2.2 Design Fires and Untested Concrete Component Capacity

Designers must perform assessments to justify fire resistance designs when specific designs lack standard fire test results, and informed judgment cannot adapt existing tests. This may entail performing heat transfer analyses to ascertain interior temperature increases over time or using existing tables that depict theoretical concrete component temperatures based on typical exposure fires. The objective is to determine the temperatures of the steel reinforcement as a result of exposure to fire and to pinpoint circumstances and lengths of time that result in decreased concrete component load capacity, usually taking "point-in-time" loads into account.

**Concrete Floors in Continuous Sequence**

Even when midspan positive moment flexural steel is insufficiently strong, continuous concrete floors can support design service loads. Plastic hinges must form at midspan and support points to develop a failure mechanism. Harmful moment resistance steel is embedded close to the top of the slab in structures with continuous beams, frequently shielded from the fire's heat. As a result, positive moment hinges often form first, followed by negative moment hinges resulting from decreased steel capacity.

### 4.2.3 Impact of Thermal Expansion

Strong heat is produced inside the compartment during a fire, but adjacent areas remain relatively cool. As the fire spreads, new areas heat up sequentially. Thermal expansion forces may affect structural components and system survivability. Variations in structural component temperatures can create restraining forces, impacting load capacity. For instance, during initial heating, components expand, pushing up against nearby structures. The development of large compressive forces by heated components may initially counteract the drop in steel reinforcing strength, provided that the structural system remains continuous and the surrounding components can withstand stresses resulting from expansion. However, compressive forces along beam length may generate additional demands on steel reinforcement.

### 4.2.4 Sagging Reinforced Concrete Floor

The warmth of the steel reinforcement can cause floor beams to droop and perhaps break fire barriers. Significant deformations don't always mean failure, especially if preventing collapse is the aim. Large deformations and stiffness loss in systems with continuity can cause tensile forces in beams. Weakened steel can usually sustain applied loads primarily as a tension member as long as the splices work and the surrounding structural detailing is strong and continuous enough.

### 4.2.5 Impacts of the Cooling Phase

While cooling, heated beams may produce extra stresses if they do not collapse during the fire's peak. Surviving elements contract and draw inward on neighboring structures after expanding and sagging in the heat. As contraction reduces thermally induced compressive forces, beams may experience tension, potentially reversing compressive forces on adjacent components. This sequence may induce stresses exceeding typical expectations, especially in the intricacies of the steel reinforcing at the joints.

**Component Design for Reinforced Concrete**

It is common for load capacity to be overlooked in the design of reinforced concrete components after flexural failure. Failure happens at the point where the first plastic hinge emerges in unconstrained, determinate components. Configuration changes or fire protection methods can enhance component fire resistance. Modifications like adding restraint, continuity, increasing concrete cover, or reinforcing steel can bolster resistance, albeit with associated costs. It could be necessary to do heat transfer and structural reaction assessments or consult qualifying fire experiments to evaluate these approaches.

**Designing Steel Structures to Resist Fire:** The design guidelines for typical steel structures are covered in this part, along with how steel structural components and composite floor systems react to fire. The 2001 study by Buchanan & Abu provides thorough illustrations of how to design steel components for fire exposure.

## 4.3 The Properties of Steel at High Temperatures

Steel's thin composition and high heat conductivity make it susceptible to fire effects when unprotected. Steel's yield strength and modulus of elasticity diminish at elevated temperatures (over 400 °C), which might result in plastic and perhaps creep stresses and more significant deflections.

### 4.3.1 The Behavior of Composite Floors at High Temperatures

When a composite floor is heated, thermal expansion occurs, causing compressive loads to be placed on the floor and surrounding framing. In composite sections, heating can cause significant deformation and sagging. Thermal gradients induce thermal bowing, with connections and framing experiencing compressive loads from thermal expansion. Bottom flanges may yield if heated, causing a shift in the neutral axis and rapid loss of moment capacity. If yielding results in plastic hinge formation, the floor may 'hang' between supports, supported through catenary action, with top flanges and slab reinforcement carrying tensile stresses.

**Composite Floor Section Design:** When designing steel components for fire exposure, it is common to assume that both positive and negative moments are constrained to the plastic moment because of the maximum predicted temperatures' effect on yield strength and stiffness. Based on the expected maximum temperatures and the plastic moment across the steel beam and slab, the flexural capacity of a composite floor section is computed. Complete cross-sections of steel beams can function as tensile components if hinges develop in the connections, which can be tailored to carry larger tensile loads.

### 4.3.2 Steel Components' Passive Fire Protection

Fire impacts can be prevented using enclosures, intumescent coatings, and spray-on fire resistance materials (SFRM). Each method provides insulation through mechanisms like low thermal conductivity or expansion. These methods vary in cost, aesthetics, and ease of installation. Traditional methods like gypsum & mineral fiberboard enclosures are simple to set up, while SFRM coatings are lightweight but may require additional finishes for aesthetics. Intumescent coatings offer architectural appeal but may be costly and require specialty application. Regular inspection and maintenance are essential for ensuring the integrity of fire protection systems.

### 4.3.3 Barriers to the Widespread Adoption of Fire-Resistant Materials and Construction Techniques

While fire-resistant materials and construction techniques significantly enhance fire safety in buildings and structures, several barriers hinder their widespread adoption. Some of these barriers include:

1. **Cost**: Fire-resistant materials and construction techniques often incur higher upfront costs than conventional materials. This can be a significant barrier for builders and property developers, especially in markets where cost is a primary consideration.
2. **Absence of consciousness:** Many architects, builders, and property owners may not fully know the benefits and availability of fire-resistant materials and construction techniques. This lack of awareness can result in the continued use of traditional materials and methods, even when safer alternatives exist.
3. **Perception of aesthetics**: Some fire-resistant materials may be perceived as less aesthetically pleasing or less versatile than their conventional counterparts. Builders and property owners may prioritize visual appeal over fire safety, particularly in design-driven projects.
4. **Regulatory challenges**: Building codes and regulations vary widely across regions, and compliance with fire safety standards may not be uniformly enforced or incentivized. This can create inconsistency and ambiguity regarding the requirements for using fire-resistant materials and construction techniques.
5. **Supply chain limitations**: The availability and accessibility of fire-resistant materials may be limited in certain regions or markets. Additionally, specialized expertise and equipment may be required for installation, further complicating the supply chain.
6. **Resistance to change**: The construction industry is traditionally slow to adopt new materials and techniques, particularly when they deviate from established norms. Resistance to change from stakeholders such as builders, contractors, and regulatory agencies can impede the widespread adoption of fire-resistant materials.
7. **Trade-offs with other performance attributes**: Fire-resistant materials may have trade-offs with other performance attributes, such as weight, insulation properties, or environmental impact. Balancing these factors to meet the diverse needs of building projects can be challenging.
8. **Limited research and development**: Despite advancements in fire-resistant material technology, ongoing research and development efforts are needed to address remaining challenges and optimize performance. Limited investment in R&D may slow the pace of innovation and hinder the growth of cost-effective solutions.

### 4.3.4 Practical Examples

Civil construction techniques have been implemented to address wildfire prevention and management in specific locations across the world:

1. Fire-Resistant Infrastructure Development in Paradise, California (USA)
• Location: Paradise, California
• Year: Post-2018 Camp Fire
• After the devastating Camp Fire in 2018, which destroyed much of Paradise, California, civil engineers and architects worked on creating fire-resistant infrastructure. The reconstruction involved.
• Fire-resistant building materials: Use of materials such as concrete, stucco, and metal roofing instead of flammable wood.

  – Defensible space zoning: Houses were built with defensible spaces, which included non-flammable zones around properties.
  – Underground power lines: Electrical wires were placed underground to prevent sparks from igniting wildfires.
  – Fire-resistant roads and barriers: Roads were widened, and firebreaks were constructed to slow down fire spread.

• This civil construction approach is a model for wildfire-resilient communities.
2. Firebreak Construction in the Mediterranean Region (Portugal)
• Location: Monchique, Portugal
• Year: 2018 Wildfire Crisis
• Portugal, a wildfire-prone country, has adopted firebreak construction as a civil engineering solution to manage wildfires. In the Monchique region, firebreaks were created by:

  – Clearing wide strips of vegetation between forests and settlements to stop wildfires from spreading.
  – Constructing gravel roads and walls along these firebreaks to prevent vegetation regrowth and provide access for firefighting vehicles.
  – Building reservoirs and water points near firebreaks to provide water for fire suppression efforts.

• This system proved effective in containing wildfires during subsequent fire seasons.
3. Ember-Resistant Housing in Victoria, Australia
• Location: Victoria, Australia
• Year: Post-2009 Black Saturday Fires
• Following the Black Saturday Fires in 2009, the Australian government introduced new building codes for wildfire-prone areas, requiring:

  – Bushfire-resistant housing designs using non-combustible materials like concrete, metal, and brick.
  – Construction of ember shields and metal mesh screens on windows and vents to prevent embers from entering homes.

  - Water tanks and sprinkler systems installed on properties to keep surrounding vegetation moist during high-risk periods.

- These regulations have helped reduce property damage in subsequent fires.

4. Retaining Walls and Firebreaks in the Serra do Mar Region, Brazil

- Location: Serra do Mar, São Paulo, Brazil
- Year: Ongoing
- The Serra do Mar region, covered by Atlantic rainforest, is vulnerable to wildfires due to climate change and human encroachment. To mitigate this:
  - Civil engineers built retaining walls to prevent soil erosion after wildfires and reduce the chances of landslides.
  - Firebreaks and access roads were constructed along strategic locations in the forest to allow easier access for firefighting teams.
  - Drainage systems were improved to manage water flow and reduce erosion during the rainy season post-fire.
- This project helps protect both the forest ecosystem and nearby communities.

5. Controlled Burning and Firebreaks in the Kruger National Park, South Africa

- Location: Kruger National Park, South Africa
- Year: Ongoing Fire Management
- In Kruger National Park, wildfires are a natural occurrence, but civil engineering plays a critical role in managing them through:
  - Firebreak construction around tourist facilities and vulnerable wildlife habitats.
  - Controlled burning practices supported by fire-resistant pathways and roads to prevent fires from spreading uncontrollably.
  - Construction of watchtowers and communication infrastructure to monitor fire activity and coordinate firefighting efforts.
- These civil measures ensure that wildfires remain part of the natural ecosystem cycle without endangering human lives or infrastructure.
- These examples highlight how civil construction techniques such as firebreaks, fire-resistant materials, and infrastructure upgrades can play a vital role in wildfire prevention and management across different regions and ecosystems.

## 5 Conclusion

Climate change's rising threat of wildfires underscores the critical need for effective fire-resistant materials and construction techniques. As wildfires become more frequent and severe, prioritizing fire safety measures is essential to protect lives, property, and ecosystems. The importance of fire-resistant materials and construction in mitigating wildfire risk cannot be overstated. Understanding the factors influencing wildfire ignition and spread, exacerbated by climate change, highlights the crucial role of proactive fire-resistant design strategies. Effective fire-resistant materials, whether natural or synthetic, are fundamental to developing robust construction solutions. While each material has strengths and limitations, ongoing innovations in fire-resistant technology promise enhanced fire safety. Designing with fire resistance in mind is vital for ensuring the resilience of buildings and infrastructure in wildfire-prone areas. Despite the clear benefits of fire-resistant materials and construction, several barriers to widespread adoption persist, including cost, lack of awareness, and regulatory challenges. Overcoming these obstacles requires a collaborative effort from industry stakeholders, which includes increasing awareness and education, providing policy support and incentives, fostering innovation in material technology, and standardizing fire safety requirements. Embracing fire-resistant design principles and investing in resilient infrastructure may help create a safer and more sustainable future in the face of escalating wildfire threats.

## References

Adkins J, Sanderman J, Miesel J (2019) Soil carbon pools and fluxes vary across a burn severity gradient three years after wildfire in Sierra Nevada mixed-conifer forest. Geod 333:10–22

Aktan AE, Ellingwood BR, Kehoe B (2007) Performance-based engineering of constructed systems. Jstru Eng 133(3):311–323

Barwa S, Singh H (2024) Climate-induced changes in forest ecosystems of Indian Himalayan region: a meta-analysis. In: Singh H (eds) urban forests, climate change and environmental pollution. Springer, Cham. https://doi.org/10.1007/978-3-031-67837-0_39

Buchanan AH, Abu AK (2017) Structural design for fire safety. 2nd Ed.JWS, West Sussex.

Caetano H, Laím L, Santiago A, Durães L, Shahbazian A (2022) Development of passive fire protection mortars. Appl Sci 12(4):2093

Cao CF, Yu B, Huang J, Feng XL, Lv LY, Sun FN, Tang LC, Feng J, Song P, Wang H (2022) Biomimetic, mechanically strong supramolecular nanosystem enabling solvent resistance, reliable fire protection and ultralong fire warning. ACS Nano 16(12):20865–76

Chuvieco E, Aguado I, Salas J, García M, Yebra M, Oliva P (2020) Satellite remote sensing contributions to wildland fire science and management. Curr for Rep 6:81–96

Davis R, Yang Z, Yost A, Belongie C, Cohen W (2017) The normal fire environment—Modeling environmental suitability for large forest wildfires using past, present, and future climate normals. For Ecol Manag 390:173–186

Harrison SP, Prentice IC, Bloomfield KJ, Dong N, Forkel M, Forrest M, Ningthoujam RK, Pellegrini A, Shen Y, Baudena M, Cardoso AW, Huss JC, Joshi J, Oliveras I, Pausas JG, Simpson KJ (2021) Understanding and modelling wildfire regimes: an ecological perspective. Environ Res Lett 16(12):125008

Kaushik P, Singh H (2024) Climate change-induced forest ecosystem functioning and alteration in water-atmosphere-soil–plant (WASP) relationship. In: Singh H (eds) urban forests, climate change and environmental pollution. Springer, Cham. https://doi.org/10.1007/978-3-031-67837-0_38

Keeley JE, Pausas JG (2019) Distinguishing disturbance from perturbations in fire-prone ecosystems. Int J Wildland Fire 28(4):282–287

Kumar A, Dwivedi GK, Tewari S, Paul J, Anand R, Kumar N, Kumar P, Singh H, Kaushal R (2020) Carbon mineralization and inorganic nitrogen pools under *Terminalia chebula* Retz.-Based agroforestry system in Himalayan Foothills India. Forest Sci 66(5):634–643

Kumar A, Kumar P, Singh H, Kumar N (2020) Adaptation and mitigation potential of roadside trees with bio-extraction of heavy metals under vehicular emissions and their impact on physiological traits during seasonal regimes. Urb for Urb Green 58:126900

Kumar A, Kumar P, Singh H, Bisht S, Kumar N (2021) Relationship of physiological plant functional traits with soil carbon stock in temperate forest of Garhwal Himalaya. Curr Sci 120(8):1368–1373

Kumar A, Kumar P, Singh H, Kumar N (2021) Modulation of plant functional traits under essential plant nutrients during seasonal regime in natural forests of Garhwal Himalayas. Plant Soil 465:197–212

Kumar N, Singh H, Kumar A, Singh AK, Pandey PK, Kumar A (2024) Forest-fire-induced land degradation. In: Mishra G, Giri K, Singh S, Kumar M (eds) Sustainable land management in India. Springer Nature, Singapore. https://doi.org/10.1007/978-981-97-5223-2_4

Long T, Zhang Z, He G, Jiao W, Tang C, Wu B, Zhang X, Wang G, Yin R (2019) 30 m resolution global annual burned area mapping based on Landsat images and google earth engine. Remote Sens 11(5):489

Mell WE, Manzello SL, Maranghides A, Butry D, Rehm RG (2010) The wildland–urban interface fire problem–current approaches and research needs. Int J Wildland Fire 19(2):238–251

Ming X, Cao M (2020) Development of eco-efficient cementitious composites with high fire resistance and self-healing abilities-a review. Resour Conserv Recy 162:105017

Moinuddin KA, Sutherland D (2020) Modelling of tree fires and fires transitioning from the forest floor to the canopy with a physics-based model. Math Comput Simul 175:81–95

Moris JV, Conedera M, Nisi L, Bernardi M, Cesti G, Pezzatti GB (2020) Lightning-caused fires in the Alps: Identifying the igniting strokes. Agric for Meteorol 290:107990

Nagy RC, Fusco E, Bradley B, Abatzoglou JT, Balch J (2018) Human-related ignitions increase the number of large wildfires across US ecoregions. Fire 1(1):4

Paul J, Samykano M, Pandey AK, Kadirgama K, Tyagi VV (2023) Nano engineered paraffin-based phase change material for building thermal management. Buildings 13(4):900

Pausas JG, Keeley JE, Schwilk DW (2017) Flammability as an ecological and evolutionary driver. J Ecol 105(2):289–297

Pechony O, Shindell DT (2010) Driving forces of global wildfires over the past millennium and the forthcoming century. Proc Natl Acad Sci 107(45):19167–19170

Puri RG, Khanna AS (2017) Intumescent coatings: a review on recent progress. J Coat Technol Res 14:1–20

Savita, Sharma SK, Singh H, Singh O (2017) Forest fire: a review. Indian for 143(8):719–728

Sett T, Nikam BR, Singh H, Purohit S (2022) Effect of drought stress on forest evapotranspiration—a case study on Indian forests. EGU General Assembly. https://doi.org/10.5194/egusphere-egu22-510

Sett T, Nikam BR, Singh H, Purohit S (2024) Hydrological variability in Indian Forest ecosystem: analysis of drought resilience, recovery and water use efficiency in moist and dry deciduous forests. In: Singh H (eds) Urban forests, climate change and environmental pollution. Springer, Cham. https://doi.org/10.1007/978-3-031-67837-0_37

Shi C, Wang S, Cai HH, Zhang HR, Long XX, Tihelka E, Song WC, Feng Q, Jiang RX, Cai CY, Lombard N, Li X, Yuan J, Zhu J, Yang H, Liu X, Xiang Q, Zhao Z, Long C, Schneider H, Zhang X, Peng H, Li D, Fan Y, Engel MS, Wang Y, Spicer RA (2022) Fire-prone Rhamnaceae with South African affinities in cretaceous Myanmar amber. Nat Plants 8(2):125–35

Singh H (2024) Exploring adaptive modulation in plant functional traits and its impact on the productivity of *Acacia auriculiformis* under $CO_2$-enriched environment. Ind Crops Prod 210(2024):118186. https://doi.org/10.1016/j.indcrop.2024.118186

Singh H, Kumar M (2022) Climate change and its impact on Indian Himalayan forests: current status and research needs. In: Rani S, Kumar R (eds) climate change. Springer Climate. Springer, Cham. pp 223–242, https://doi.org/10.1007/978-3-030-92782-0_11

Singh H, Kumar N, Kumar A (2024) Enhancing resource use efficiency in crops through plant functional traits published by Springer, pp 97–117. ISBN: 978-981-97-1509-1

Singh H (2021) Process-based carbon sequestration study in relation to the energy-water-carbon flux in a forest ecosystem. In: Climate impacts on sustainable natural resource management Kumar P, Singh RK, Kumar M, Rani M, Sharma (eds). CRC Press Taylor & Francis Group, United States, pp 336–361

Singh H (2024b) Climate change and forests: biological perspectives. In: Singh H (eds) Forests and climate change. Springer Nature, Singapore. https://doi.org/10.1007/978-981-97-3905-9_1

Singh H (2024d) Climate change and environmental pollution: biological consequences on urban forests. In: Singh H (eds) Urban forests, climate change and environmental pollution. Springer Nature, Cham, Switzerland. https://doi.org/10.1007/978-3-031-67837-0_1

Singh H (2024c) Forests and climate change: biological perspectives on impact, adaptation, and mitigation strategies (ed), Springer Nature Singapore. https://doi.org/10.1007/978-981-97-3905-9

Singh H (2024e) Urban forests, climate change and environmental pollution: physio-biochemical and molecular perspectives to enhance urban resilience (ed), Springer Nature, Cham, Switzerland. https://doi.org/10.1007/978-3-031-67837-0

van Lierop P, Lindquist E, Sathyapala S, Franceschini G (2015) Global forest area disturbance from fire, insect pests, diseases and severe weather events. For Ecol Manag 352:78–88

van Nes EH, Staal A, Hantson S, Holmgren M, Pueyo S, Bernardi RE, Flores BM, Xu C, Scheffer M (2018) Fire forbids fifty-fifty forest. PLoS ONE 13(1):e0191027

Williams AP, Abatzoglou JT, Gershunov A, Guzman-Morales J, Bishop DA, Balch JK, Lettenmaier DP (2019) Observed impacts of anthropogenic climate change on wildfire in California. Earths Futur 7(8):892–910

Xu X, Jia G, Zhang X, Riley WJ, Xue Y (2020) Climate regime shift and forest loss amplify fire in Amazonian forests. Glob Change Biol 26(10):5874–5885

Yang Y, Haurie L, Wang DY (2021) Bio-based materials for fire-retardant application in construction products: a review. J Therm Anal Calorim 1–20

Yin Y, Bloom AA, Worden J, Saatchi S, Yang Y, Williams M, Liu J, Jiang Z, Worden H, Bowman K, Frankenberg C, Schimel D (2020) Fire decline in dry tropical ecosystems enhances decadal land carbon sink. Nat Commun 11(1): 1900

Zhang Z, Long T, He G, Wei M, Tang C, Wang W, Wang G, She W, Zhang X (2020) Study on global burned forest areas based on Landsat data. Photogramm Eng Remote Sens 86(8):503–508

Zheng B, Ciais P, Chevallier F, Chuvieco E, Chen Y, Yang H (2021) Increasing forest fire emissions despite the decline in global burned area. Sci Adv 7(39):eabh2646

# Sustainable Practices in Forest Fire Management

**Meenakshi Sati and Megha Verma**

**Abstract** Forest fires' increasing frequency and intensity necessitate a paradigm shift in fire management toward sustainable practices. Traditional fire suppression methods have often overlooked the ecological role of fire, leading to unintended consequences such as fuel build-up and increased severity of forest fires. Sustainable fire management aims to strike a balance by integrating controlled burns, ecosystem restoration, and community engagement. This chapter synthesizes the sustainable forest fire management approaches, examining their efficacy in mitigating environmental, social, and economic impacts. The chapter explores fire management strategies' evolution, emphasizing the need for a holistic and adaptive framework. Recent technological advances for early detection, monitoring, and modeling and their implications for enhancing firefighting capabilities are discussed. Indigenous fire knowledge and community engagement are integral to sustainable fire management, promoting resilience and fostering collaborative approaches. The present study also addresses the influence of climate change on fire behavior, underscoring the importance of adaptive strategies. Through a comprehensive analysis of case studies and current research, this review paper aims to provide insights into the effectiveness and challenges of sustainable fire management practices in diverse ecosystems.

**Keywords** Forest fires · Environmental impact · Climate change · Sustainable fire management · Technology

M. Sati (✉)
College of Agricultural Science and Applied Research, Bharatiya Engineering Science and Technology Innovation University, Gownivaripalli, Andhra Pradesh, India
e-mail: minakshi088@gmail.com

M. Verma
Doon P.G. College of Agriculture Science and Technology, Dehradun, India



H. Singh and N. Kumar (eds.), *Forest Fire and Climate Change*,
https://doi.org/10.1007/978-3-031-89967-6_18

# 1 Introduction

It has long been known that fire is an important ecological agent that shapes landscapes, affects biodiversity and maintains the health of ecosystems (Kumar et al. 2024). However, forest fires also pose a severe and intricate threat to land ecosystems and forests, necessitating quick action and preventative measures (Singh 2024a). These fires severely harm forested areas and beyond, with far-reaching ecological, economic, and societal repercussions (EPA US 2016). Forest fires increasing in size, frequency, and intensity bring catastrophic effects to many communities. Nature's resistance to fire is weakened by rising temperatures, longer dry seasons, earlier snowmelts, and stronger winds amidst an increasingly dire climate catastrophe (Singh 2024b). Longer heat waves dry out the fuel, while more intense wet spells encourage rapid vegetation growth. This causes fuel to build. Furthermore, greater lightning strikes mean more chances for ignition (Oris et al. 2014). These days, the fire season begins earlier, lasts longer, and produces more catastrophic forest fires (Singh 2024c). Forest fire susceptibility can be made worse by land use decisions such as deforestation, poor soil management, increased agricultural production, a lack of fire-resistant building techniques, and zoning laws that permit development in high-risk areas. Like previous disasters, fires disproportionately damage communities of color, low-income areas, and Indigenous peoples. They also exacerbate systemic injustices already in place, such as differences in air quality, exposure to chemicals, heat-related illnesses, and hazardous working conditions. (Dillis et al. 2022).Fire management has undergone tremendous change because of the increasing effects of climate change, evolving human views, and technological improvements (Singh 2024d). Sustainable fire management has emerged as a critical framework for balancing fire's ecological, economic, and social dimensions within natural and human-dominated landscapes.

Throughout history, humans have grappled with the dual nature of fire as both a powerful tool and a destructive force. Forest fires have significant ecological effects. They endanger plant and animal species and ruin habitats, contributing to biodiversity loss (Sett et al. 2024; Barwa and Singh 2024). Another important factor is soil degradation, which causes erosion, nutrient loss, and decreased fertility (Kaushik and Singh 2024). In addition, forest fires change succession and regeneration patterns and interfere with critical ecological processes, including the cycling of nutrients.

Additionally, these fires make ecosystems more susceptible to exotic species (Pellegrini et al. 2017; Kumar et al. 2020a). Early civilizations harnessed fire for cooking, warmth, and cultural rituals while experiencing its destructive potential through forest fires. Over time, forest fire suppression became dominant, driven by fears of property loss, resource depletion, and public safety concerns (Davis et al. 2022). However, this suppression-centric paradigm often neglected the ecological benefits of fire, leading to disrupted fire regimes, degraded ecosystems, and increased vulnerability to catastrophic forest fires. The foundation of the argument for sustainable fire management is the understanding that fire is a vital natural phenomenon that shapes ecosystems and preserves biodiversity. The concept of sustainable fire management recognizes the need to balance human requirements, safety concerns, and the ecological roles of fire. Sustainable fire management aims to lower the likelihood of disastrous forest fires, restore ecosystems adapted to fire, and increase ecosystem resilience in the face of shifting environmental conditions. It does this by taking a proactive and adaptive approach.

Sustainable fire management extends beyond fire suppression to encompass a range of strategies and practices to promote ecological health, community resilience, and landscape sustainability. Key objectives include integrating ecological principles into fire management planning, enhancing early detection and response capabilities, fostering collaboration with indigenous communities, and building capacity for adaptive governance. By addressing the complex interactions between fire, climate, and human activities, sustainable fire management seeks to achieve long-term ecological sustainability while safeguarding human lives and livelihoods. Sustainable fire management represents a paradigm shift towards holistic, science-based approaches that recognize the ecological importance of fire and prioritize resilience-building strategies in a changing world. This chapter explores the evolution of fire management, recent advancements in technology and policy, and the challenges and opportunities for promoting sustainable fire management practices in diverse landscapes.

# 2 Evolution of Fire Management

## *2.1 Historical Perspectives*

Larger areas are being consumed by forest fires in many nations, and warmer temperatures are prolonging the length of fire seasons (Jolly et al. 2015). More lives and buildings are at risk of flames due to the growing population in and surrounding forested areas. Approximately 2% of the world's wooded regions, roughly 670,000 km$^2$, burn annually (van Lierop et al. 2015). As a result, billions of tons of $CO_2$ are released into the atmosphere, and it is estimated that illnesses brought on by smoke exposure from forest and landscape fires kill hundreds of thousands of people (Johnston et al. 2012; Sharma et al. 2021). While weather, fuel types, and terrain may affect fire potential and behavior, human activities are responsible for nearly all forest fires globally. Fire plays a role in various conservation issues and biodiversity threats, including inappropriate logging practices, grazing methods, urban expansion into wildland interfaces, alterations in hydrological patterns, and the spread of invasive species (Pausas and Keeley 2019).

Additionally, human negligence, such as careless disposal of cigarettes or inadequate control of burning in agricultural lands, can spark unwanted fires. It is unclear how present forest fire patterns will affect the ecology and the economy in the long run. However, the scientific data that is now available points to a detrimental effect that includes decreased species diversity and impeded natural regeneration (Russell-Smith et al. 2017). But not all flames are destructive; the trick is to maximize the ecological advantages of fire while reducing its adverse effects, understanding that managed fire use can help maintain forests that have evolved to fire. Throughout history, fire has held both reverence and fear in human societies. Early civilizations harnessed fire for cooking, warmth, and land management while witnessing its destructive potential through forest fires. In the United States, the twentieth century saw a significant shift towards fire suppression as federal policies aimed to safeguard lives, property, and natural resources. At its height in the middle of the twentieth century, this suppression-focused strategy was implemented with the "10 a.m. policy," which sought to put out all forest fires by 10 a.m. the day following their report (Deverell 2004). However, this total suppression policy resulted in unintended consequences, including the accumulation of flammable vegetation, altered ecosystem dynamics, and increased forest fire severity.

## 2.2 *Shifting Paradigms: From Suppression to Sustainability*

There has been a notable shift towards adopting more sustainable fire management methodologies in response to the ecological and societal difficulties stemming from years of fire suppression. This transition entails recognizing fire as a natural ecological phenomenon crucial for shaping landscapes and preserving ecosystem well-being. Instead of entirely eradicating fire, modern fire management strategies strive to mimic natural fire patterns through prescribed burns, controlled forest fires, and ecological restoration efforts. Embracing sustainable fire management practices enhances ecosystem resilience and curbs forest fire risks. For instance, prescribed burns can diminish fuel accumulation, lower the likelihood of catastrophic forest fires, and enrich habitat diversity for species adapted to fire. Collaborative partnerships among land stewards, indigenous communities, and stakeholders have also emerged as pivotal forces driving sustainable fire management initiatives, fostering shared learning, cultural exchange, and adaptive governance. The repercussions of forest fires extend beyond the immediate environment. The irreversible damage caused by fires, such as deforestation and habitat loss, poses severe risks to the stability of ecosystems. Significant amounts of carbon emissions are also released by forest fires, which exacerbate global warming and climate change (Kumar et al. 2020b, 2021a; Sousa et al. 2021). Moreover, influences on water resources and hydrological cycles result in deteriorated water quality, decreased availability, and disturbances to watershed functioning. Socially, forest fires cause displaced communities, the loss of livelihoods, and significant financial hardships for the impacted areas. It is essential to implement efficient management techniques to address this hazard completely. Monitoring and surveillance are essential for early detection and quick response systems, which allow for prompt action. To reduce the risk of fire, it is crucial to use fire prevention strategies such as controlled burns, fuel management, and public awareness programs. (Ascoli et al. 2023). In order to confine and put out flames, fire suppression tactics entail deploying infrastructure and firefighting resources. Edgeley et al. (2020) state that active community participation and capacity building strengthen fire-safe practices and community resilience. Utilizing common resources, skills, international collaboration, and knowledge exchange enables cooperative fire management methods. The theoretical framework for forest fire detection and monitoring includes various scientific fields and ideas, such as remote sensing, modeling, data analysis, and fire behavior and spread. Nevertheless, there are a few issues that must be resolved. Adaptive solutions are necessary due to the effects of climate change and the resulting rise in fire hazards. Careful thought and integrated techniques are needed to balance conservation goals with socioeconomic requirements. Traditional ecological knowledge and native fire management techniques can provide insightful information and help develop more potent tactics. Innovative fire suppression techniques and early warning systems are two examples of technological breakthroughs that offer promising opportunities for improvement. Enhancing international cooperation and information sharing is crucial to combat this threat's worldwide scope.

## 2.3 *The Ecological Role of Fire*

Central to the development of fire management practices is a deeper understanding of the ecological function of fire in shaping landscapes and ecosystems. Research has illuminated the intricate relationships among fire, vegetation, and wildlife, underscoring fire's role as a natural disturbance driving ecosystem dynamics and preserving biodiversity (Pausas and Keeley 2009). For instance, plant species adapted to fire have evolved characteristics like serotiny, resprouting, and seed dispersal mechanisms to endure and thrive in fire-prone habitats. Additionally, fire plays a pivotal role in nutrient cycling, soil formation, and carbon sequestration, influencing ecosystem functions at various scales. Evidence suggests that excluding fire can disrupt these processes, leading to soil degradation, biodiversity loss, and heightened vulnerability to invasive species (Pausas and Keeley 2019; Kumar et al. 2021b, 2022).

Conversely, reintroducing fire can stimulate ecosystem recovery, encourage native plant regeneration, and improve wildlife habitat quality. Forest fires, whether caused naturally (e.g., lightning), accidentally (e.g., cigarettes, campfires), or intentionally through arson, are destructive events. Despite this, it may seem paradoxical that fire, which consumes plant life and threatens animals, could foster ecological well-being. Nonetheless, fire is a vital component of natural systems, and regular fires help many ecosystems by removing dead organic matter, which is essential for the survival and procreation of some plant and animal populations. Moreover, nutrients released from burnt material, including dead plants and animals, return to the soil more rapidly than through slow decay, thus enhancing soil fertility, a phenomenon farmers have exploited for centuries (Swetnam et al. 2016). Fire is necessary for the completion of several plant species' life cycles. For instance, pine cones with seeds enclosed in resin require fire to release the seeds.

Similarly, some animals, such as the endangered Karner blue butterfly caterpillar, rely on plants like wild lupine, which need fire to maintain a balanced ecosystem.

These plants do not thrive without fire, affecting caterpillar food sources and the broader food web (Hessburg et al. 2021; Malik et al. 2024). Additionally, fires can aid in controlling invasive species not adapted to regular forest fires, highlighting the ecological necessity of periodic burning. In conclusion, the evolution of fire management reflects a shift from suppression-oriented approaches towards more sustainable practices that acknowledge fire's ecological significance and benefits.

# 3 Technological Advances in Fire Management

## *3.1 Remote Sensing and Early Detection*

Technological advancements have revolutionized fire management practices in recent years, enhancing the early detection, prediction, and monitoring of forest fires. For efficient detection and tracking, it is essential to comprehend the behavior and spread of fires (Andrews 2014). This includes researching fire's thermal dynamics, combustion processes, and other physical aspects. Monitoring and identification of forest fires are greatly aided by remote sensing. In order to gather information regarding fire incidents, smoke plumes, and burned areas, satellite imagery, aerial photography, and other sensor technologies are used (Ding 2023). Forest fires may be located and monitored across broad geographic regions with remote sensing, yielding important resource allocation and decision-making data. (Wooster et al. 2021). Research has indicated that remote sensing is helpful for promptly identifying forest fires and facilitating prompt containment and action. For instance, cutting-edge satellite systems like the Visible Infrared Imaging Radiometer Suite (VIRS) and MODIS (Moderate Resolution Imaging Spectroradiometer) can instantly identify thermal anomalies linked to forest fires and send timely information to fire management agencies (Giuseppe et al. 2022).

Furthermore, the accuracy and effectiveness of forest fire identification from satellite photos have increased by incorporating machine learning algorithms and artificial intelligence techniques. (Kang et al. 2023). Convolutional neural networks (CNNs) and deep learning algorithms can detect forest fires with high precision automatically, reducing false alarms and enhancing early warning systems (Filizzola et al. 2017). These technological advancements in remote sensing have significantly improved the ability to detect forest fires in remote and inaccessible areas, facilitating proactive fire management strategies and reducing the risk of large-scale fire outbreaks.

## *3.2 GIS and Modeling for Fire Behavior Prediction*

Computational modeling methods and geographic information systems (GIS) are essential for forecasting fire behavior, evaluating fire risk, and guiding strategic decision-making in fire control. Recent studies have highlighted the utility of GIS-based fire behavior modeling systems, such as FARSITE (Fire Area Simulator) and BEHAVEPlus, for simulating fire spread, intensity, and direction under different weather and fuel conditions (Andrews 2007, 2014; Finney 1998). These models incorporate spatial data on topography, fuel types, weather patterns, and ignition sources to forecast fire behavior dynamics and inform firefighting tactics. Furthermore, advancements in high-resolution terrain modeling and fuel mapping technologies have improved the accuracy and reliability of fire behavior predictions (Bright et al. 2022). LiDAR (Light Detection and Ranging) data, aerial surveys, and ground-based fuel sampling techniques are integrated into GIS platforms to create detailed spatial representations of fuel loads and vegetation structure (Alonzo et al. 2017). In fire management planning, these integrated modeling approaches are very effective in assessing forest fire risk, prioritizing fuel treatments, and optimizing resource allocation.

### *3.3 Unmanned Aerial Vehicles (UAVs) in Fire Monitoring*

Drones, also referred to as unmanned aerial vehicles (UAVs), are becoming increasingly valuable for situational awareness and fire monitoring during forest fires. UAVs can offer real-time aerial observation of fire behavior, smoke plumes, and hotspots since they are outfitted with thermal sensors, high-resolution cameras, and real-time communication systems (Akhloufi et al. 2021). According to recent studies, unmanned aerial vehicles (UAVs) can improve the effectiveness and safety of firefighting operations by giving ground troops and incident commanders precise and current information. Moreover, integrating UAVs with advanced imaging technologies, such as multispectral and hyperspectral sensors, enables detailed analysis of fire-affected areas and post-fire assessments (Ollero et al. 2006). UAVs with LiDAR and photogrammetry capabilities can generate 3D maps and terrain models of fire-impacted landscapes, aiding post-fire rehabilitation and recovery efforts (Nex et al. 2022). Furthermore, using UAVs for aerial ignitions and prescribed burning operations has shown promise in reducing fuel loads, mitigating forest fire risk, and restoring fire-adapted ecosystems.

# 4 Indigenous Knowledge and Community Engagement

## *4.1 Incorporating Indigenous Fire Practices*

Generation after generation, indigenous societies have long deeply understood fire and its ecological purpose. The value of integrating indigenous fire techniques and encouraging community involvement in fire management initiatives has increasingly come to light in recent years. According to Bowman et al. (2018), Indigenous fire management techniques are essential to preserving the resilience and health of ecosystems. The authors show how traditional burning practices have altered landscapes over millennia, encouraging biodiversity, reducing fuel loads, and lowering the risk of catastrophic forest fires. They do this by drawing on extensive fieldwork and conversations with indigenous tribes.

Similarly, collaborative approaches integrating indigenous fire practices with science offer valuable insights into fire behavior, seasonal patterns, and landscape dynamics, which can inform more effective and culturally sensitive fire management strategies (Smith et al. 2021; Singh et al. 2024a). Hessburg et al. (2019) highlight the potential synergies between Indigenous fire practices and modern techniques such as prescribed burning. Valuing traditional knowledge systems and fostering partnerships with Indigenous communities may help policymakers develop adaptive and sustainable fire management approaches that benefit ecosystems and human societies. Indigenous fire practices, such as cultural burning, patch burning, and selective fuel reduction, are tailored to local ecosystems and promote landscape resilience by maintaining biodiversity, preventing catastrophic forest fires, and rejuvenating fire-adapted plant communities. Collaboration between Indigenous communities and fire management agencies is vital to reintroducing traditional fire practices on Indigenous lands, improving ecosystem health, and reducing fire risk.

## *4.2 Community-Based Fire Management Strategies*

Community-based fire management strategies have emerged promising for enhancing forest fire resilience and fostering greater social cohesion within vulnerable regions. There is a growing emphasis on participatory decision-making processes, collaborative partnerships, and community empowerment in forest fire prevention and response efforts. According to Charnley et al. (2020), involving local populations in the development and execution of fire management strategies can foster a sense of trust, increase communication, and mobilize local knowledge and resources to lower the risk of forest fires and boost ecosystem health. The importance of social networks and community collaborations in improving fire preparedness and response is highlighted by Mockrin et al. (2018). By fostering relationships between residents, emergency responders, and other stakeholders, communities can develop more resilient and adaptive strategies for coping with forest fires and other natural hazards. Community-based fire management approaches are important in decision-making, local capacity-building initiatives, and collaborative partnerships across diverse stakeholder groups (Scott and Carter 2019). Policymakers can increase social justice, strengthen resilience, and create more sustainable landscapes by enabling communities to participate actively in forest fire prevention and response.

### *4.3 Education and Outreach Initiatives*

Education and outreach initiatives are critical in raising awareness, building capacity, and fostering behavior change in forest fire-prone communities. Forest fire education programs significantly enhance public understanding and response to fire risk. The impact of educational materials, workshops, and outreach events in increasing knowledge, changing attitudes, and motivating adaptive behaviors among residents in forest fire-prone areas is multifold (Ager et al. 2017b; Singh et al. 2024b). Similarly, the role of social media and digital technologies in disseminating fire-related information and engaging with diverse stakeholder groups is of paramount importance. Leveraging online platforms, interactive maps, and mobile applications can help educators proactively reach broader audiences, facilitate knowledge exchange, and empower communities to reduce forest fire risk (Ryan et al. 2020). Prior and Eriksen (2013) highlight the importance of tailoring education and outreach initiatives to specific cultural contexts and community needs. Educators can enhance the relevance and effectiveness of fire education programs, fostering greater trust, engagement, and participation among Indigenous and non-Indigenous communities by incorporating Indigenous perspectives, traditional knowledge, and culturally relevant messaging. Hence, education and outreach initiatives are vital in promoting fire awareness, enhancing community resilience, and fostering sustainable fire management practices. Policymakers and practitioners may empower individuals and communities to take proactive measures to decrease forest fire risk, conserve ecosystems, and ensure livelihoods for future generations by utilizing varied communication channels, cultural views, and participatory methodologies.

## 5 Climate Change Impacts on Fire Behavior

### *5.1 Changing Fire Seasons*

One of the most critical concerns of the twenty-first century is climate change, which has a significant impact on ecosystems, biodiversity, and human civilization. Among its many manifestations, altered fire behavior is a critical consequence of changing climatic conditions. One of the most visible effects of climate change on fire behavior is the alteration of fire seasons. According to modeling studies and historical data, there are regular changes in the length, timing, and intensity of fire seasons in different regions. Jolly et al. (2015) analyzed trends in fire season length during the previous four decades and found an average global increase in fire season length of 19%. This extension of the fire season has been attributed to rising temperatures and earlier snowmelt, leading to drier conditions and more conducive forest fire environments.

Similarly, Westerling et al. (2006) looked at patterns in forest fire activity in the western United States. They discovered that, during the 1980s, there was a noticeable rise in the quantity of enormous forest fires and the duration of the fire season. These changes have been linked to warmer temperatures, reduced snowpack, and earlier spring snowmelt, collectively creating conditions favoring fire ignition and spread. In addition to changes in the timing and duration of fire seasons, climate change also influences the spatial distribution of forest fires. Abatzoglou and Williams (2016) utilized climate models to project future changes in forest fire activity across the western United States. Their findings suggest a shift towards more frequent and extensive fires, particularly in regions prone to hot and dry conditions, such as the southwestern United States and the Rocky Mountains.

## 5.2 *Extreme Weather Events*

Heatwaves, droughts, and intense storms are examples of extreme weather events that are becoming more frequent and powerful due to climate change, impacting fire behavior. Heatwaves, in particular, have been identified as a critical driver of increased forest fire risk. Perkins et al. (2017) examined the relationship between heat waves and forest fire activity in Australia and found a significant correlation between extreme heat events and heightened fire danger. Similarly, droughts are critical in amplifying forest fire risk by drying out vegetation and increasing fuel availability. Diffenbaugh et al. (2015) examined how climate change affected California's drought conditions and predicted that drought episodes would occur more frequently and with greater severity in the upcoming decades. These prolonged periods of dryness create conditions ripe for forest fire ignition and spread, posing significant challenges for fire management efforts.

Furthermore, extreme weather events can interact with other factors, such as wind patterns, to exacerbate fire behavior. Lareau et al. (2018b) examined the role of atmospheric conditions in shaping fire behavior during the 2017 California forest fires. Strong winds and atmospheric instability were discovered to contribute to intense fire behavior and rapid fire spread, underscoring the intricate relationships between weather, climate, and fire dynamics.

### 5.3 *Adapting Management Strategies*

In response to the changing fire environment driven by climate change, there is a growing recognition of the need to adapt fire management strategies to enhance resilience and mitigate risks. Traditional approaches to fire management, such as fire suppression, are increasingly complemented by strategies prioritizing ecosystem resilience and community preparedness (Dale et al. 2001). Prescribed burns are one such tactic, which entails purposefully starting fires under controlled circumstances to lower fuel loads and lessen the likelihood of disastrous forest fires. According to Collins et al. (2013), prescribed burning can be an affordable method of controlling ecosystems prone to forest fires. The study assessed the efficacy of prescribed burning in lowering the risk of forest fires in Portugal. Another approach to adapting fire management strategies involves integrating traditional ecological knowledge with scientific expertise to inform decision-making. Indigenous communities have long employed controlled burning practices to manage landscapes and reduce fire risk, drawing on a deep understanding of local ecosystems and fire behavior. The effectiveness and durability of measures can be improved by incorporating this traditional knowledge into fire management planning (Fletcher et al. 2021). Moreover, the significance of community involvement and readiness in lowering the risk of forest fires is becoming increasingly acknowledged. Proactive measures, such as home retrofitting and evacuation planning, can significantly reduce property loss and enhance public safety during forest fire events (Gill and Stephens 2009).

## 6 Future Directions

### 6.1 *Emerging Technologies*

Artificial intelligence (AI) and machine learning (ML) fields have shown great promise in enhancing fire safety procedures by providing new avenues for early detection, prediction, and decision assistance.

#### 6.1.1 Early Detection and Monitoring

The application of AI algorithms for early forest fire identification using a variety of data sources, such as social media feeds, weather reports, and satellite imagery, has been investigated in several studies. For example, convolutional neural networks (CNNs) have been trained to detect smoke plumes and thermal anomalies in satellite images, enabling rapid forest fire detection and alerting systems (Barmpoutis et al. 2020). ML algorithms have also been applied to analyze sensor data from IoT (Internet of Things) devices deployed in fire-prone areas, such as temperature sensors and moisture detectors (Jiao et al. 2019). These smart sensor networks can monitor environmental conditions, fuel moisture levels, and fire danger ratings, aiding in proactive fire management efforts.

### 6.1.2 Fire Behavior Prediction and Modeling

AI and ML techniques are used to develop sophisticated fire behavior prediction models that account for complex interactions between weather, topography, and fuel conditions. Random forests and gradient boosting machines are two ensemble modeling techniques that have demonstrated promise in raising the precision of fire spread forecasts. Researchers are leveraging big data analytics and cloud computing platforms to process large volumes of spatial and temporal data for fire modeling and simulation (Kim and Muminov 2023). By integrating historical fire data, remote sensing imagery, and weather forecasts, ML algorithms can generate probabilistic estimates of fire behavior and potential fire impacts on landscapes.

### 6.1.3 Decision Support Systems

AI-driven decision support systems are being developed to assist fire managers and incident commanders in making informed decisions during forest fire events. These systems utilize ML algorithms to analyze real-time data streams, assess fire risks, and recommend appropriate response actions. Integrating AI technologies with Geographic Information Systems (GIS) and remote sensing platforms enables spatially explicit decision-making in fire management (Bakhtiar et al. 2022). For example, ML algorithms can analyze satellite imagery to identify high-risk areas for forest fire ignition and prioritize resource allocation for prevention and suppression efforts.

### 6.1.4 Post-Fire Assessment and Rehabilitation

AI and ML techniques are being applied to analyze post-fire imagery and assess the extent of fire damage to ecosystems and infrastructure. Image classification algorithms can classify burned areas, estimate vegetation loss, and identify areas at risk of post-fire erosion and debris flows. ML-based predictive models are being developed to prioritize post-fire rehabilitation and restoration activities, such as revegetation, erosion control, and watershed management (Jain 2020). By analyzing environmental data and historical fire patterns, these models can identify areas with the greatest ecological vulnerability and guide targeted restoration efforts.

### 6.2 *Policy Implications*

Policy reforms are essential for supporting and promoting sustainable practices in fire management. Policy-makers play a crucial role in creating enabling environments for sustainable practices and fostering partnerships for collaborative governance. Integrated and adaptive governance frameworks prioritizing ecosystem health, community resilience, and climate change adaptation should be emphasized (Prodyut et al. 2010). Policies related to prescribed burning, land-use planning, and forest fire risk reduction can promote ecosystem resilience and reduce the likelihood of catastrophic forest fires. Collaboration and coordination among government agencies, Indigenous communities, and stakeholders can improve fire management decision-making and facilitate knowledge-sharing and capacity-building efforts (Nurhidayah et al. 2023). Investing in research and innovation and providing financial incentives and technical assistance can support adopting sustainable local, regional, and national fire management practices. Policies with scientific evidence, traditional knowledge, and community values should be adopted to ensure effective and equitable fire management outcomes.

### 6.3 *Research Needs*

Addressing key research gaps and priorities is critical for advancing the science and practice of sustainable fire management. Understanding the long-term ecological impacts of fire suppression and restoration treatments on biodiversity, ecosystem processes, and landscape dynamics is vital (Liu and Wimberly 2016). Incorporating social science perspectives is essential to understand better human perceptions, behaviors, and responses to forest fires and identify strategies for promoting community engagement, resilience, and adaptation. Interdisciplinary research, stakeholder engagement, and adaptive management approaches for addressing complex challenges in fire management are important (Wang et al. 2006). Research and innovation and fostering collaboration between scientists, practitioners, and policymakers are better ways to develop and implement sustainable practices that protect lives, property, and the environment in the face of escalating forest fire risks.

## 7 Conclusion

The escalating frequency and severity of forest fires worldwide necessitate a fundamental shift in fire management practices toward sustainability. While effective in the short term, traditional suppression techniques often overlook the ecological role of fire, leading to unforeseen consequences such as increased fuel accumulation and intensified forest fires. Sustainable fire management addresses these challenges by embracing controlled burns, ecosystem restoration, and active community engagement. Recent technological advancements, including artificial intelligence (AI), the Internet of Things (IoT), and remote sensing, promise to enhance early detection, prediction, and response to forest fires. However, further research and development are required to maximize these technologies' scalability and performance across various environmental conditions. Indigenous fire knowledge and community engagement are integral to sustainable fire management. Recognizing the value of Indigenous wisdom and involving local communities in decision-making processes are essential steps toward building more effective and culturally sensitive fire management strategies. The influence of climate change on fire behavior underscores the urgency of adopting adaptive strategies to cope with changing environmental conditions. Hence, the practices include incorporating Indigenous fire practices, fostering community involvement, considering fire management's ecological, social, and economic dimensions, investing in technological innovations, adopting adaptive strategies, promoting research and knowledge exchange, etc., which may guide policymakers and managers toward sustainable fire management that mitigate the immediate threats of forest fires and build resilience for the future.

## References

Abatzoglou JT, Williams AP (2016) Impact of anthropogenic climate change on wildfire across western US forests. Earth Atmos Planet Sci 113(42):11770–11775

Ager AA, Evers CR, Day MA, Preisler HK, Barros AM, Nielsen-Pincus M (2017) Network analysis of wildfire transmission and implications for risk governance. PLoS ONE 12:e0172867

Akhloufi MA, Couturier A, Castro NA (2021) Unmanned aerial vehicles for wildland fires: sensing, perception. Coop Assist Drones 5(1):15. https://doi.org/10.3390/drones5010015

Alonzo M, Morton DC, Cook BD, Andersen H, Babcock C, Pattison R (2017) Patterns of canopy and surface layer consumption in a boreal forest fire from repeat airborne lidar. Environ Res Lett 12:065004. https://doi.org/10.1088/1748-9326/aa6ade

Andrews PL (2014) Current status and future needs of the behave plus fire modeling system. Int J Wildland Fire 23(1):21. https://doi.org/10.1071/WF12167

Andrews PL (2007) Behave plus fire modeling system: past, present, and future, In: Proceedings of 7th symposium on fire and forest meteorological society, Bar Harbor, Maine pp 1–13

Ascoli D, Plana E, Oggioni SD, Tomao A, Colonico M, Corona P, Giannino F, Moreno M, Xanthopoulos G, Kaoukis K et al (2023) Fire-smart solutions for sustainable wildfire risk prevention: Bottom-up initiatives meet top-down policies under EU green deal. Int J Disaster Risk Reduct 92:103715

Bakhtiar F, Davoud O, Vahid M, Hoda K, Ayyoob S, Bahaoldein K (2022) An integrated approach of artificial intelligence and geoinformation techniques applied to forest fire risk modeling in Gachsaran, Iran. J Environ Plan Manag 65(4). https://doi.org/10.1080/09640568.2022.2027747

Barmpoutis P, Stathaki T, Dimitropoulos K, Grammalidis N (2020) Early fire detection based on aerial 360-degree sensors, deep convolution neural networks, and exploitation of fire dynamic textures. Remote Sens 12:3177

Barwa S, Singh H (2024) Climate-induced changes in forest ecosystems of Indian Himalayan region: a meta-analysis. In: Singh H (eds) Urban forests, climate change and environmental pollution. Springer, Cham. https://doi.org/10.1007/978-3-031-67837-0_39

Bowman D (2018) Wildfire science is at a loss for comprehensive data. Nature 560(7716):07–07

Bright BC, Hudak AT, McCarley TR, Spannuth A, Sánchez-López N, Ottmar RD, Soja AJ (2022) Multitemporallidar captures heterogeneity in fuel loads and consumption on the Kaibab Plateau. Fire Ecol 18(18)

Charnley S, Kelly EC, Fischer AP (2020) Fostering collective action to reduce wildfire risk across property boundaries in the American West. Environ Res Lett 15:025007. https://doi.org/10.1088/1748-9326/ab639a

Collins RD, de Neufville R, Claro J, Oliveira T, Pacheco AP (2013) Forest fire management to avoid unintended consequences: a case study of Portugal using system dynamics. J Environ Manag 130:1–9. https://doi.org/10.1016/j.jenvman.2013.08.033

Dale VH, Joyce LA, McNulty S, Neilson RP, Ayres MP, Flannigan MD, Hanson PJ, Irland LC et al (2001) Climate change and forest disturbances: climate change can affect forests by altering the frequency, intensity, duration, and timing of fire, drought, introduced species, insect and pathogen outbreaks, hurricanes, windstorms, ice storms, or landslides. Bioscience 51:723–734

Davis EJ, Huber-Stearns H, Caggiano M, McAvoy D, Cheng A, Deak A, Evans A (2022) Managed wildfire: a strategy facilitated by civil society partnerships and interagency cooperation. Agric Nat Resour 35:914–932

Deverell W (2004) A companion to the American West Edited. Blackwell Publishing Ltd

Diffenbaugh NS, Swain DL, Touma D (2015) Anthropogenic warming has increased drought risk in California. Proc Natl Acad Sci USA 112(13):3931–3936. https://doi.org/10.1073/pnas.1422385112

Dillis C, Butsic V, Moanga D, Parker-Shames P, Wartenberg A, Grantham TE (2022) The threat of wildfire is unique to cannabis among agricultural sectors in California. Ecosphere 13:e4205

Ding Y, Wang M, Fu Y, Zhang L, Wang X (2023) A wildfire detection algorithm based on the dynamic brightness temperature threshold. Forests 14:477. https://doi.org/10.1038/d41586-018-05840-4

Edgeley CM, Paveglio TB, Williams DR (2020) Support for regulatory and voluntary approaches to wildfire adaptation among unincorporated wild land-urban interface communities. Land Use Policy 91:104394

EPA US (2016) Climate change indicators: wildfires; EPA: Washington. DC, USA, p 2016

Filizzola C, Corrado R, Marchese F, Mazzeo G, Paciello R, Pergola N, Tramutoli V (2017) RST-FIRES, an exportable algorithm for early-fire detection and monitoring: description, implementation, and field validation in the case of the MSG-SEVIRI sensor. Remote Sens Environ 192:2–25

Finney MA (1998) FARSITE: fire area simulator—model development and evaluation, Tech. Rep. RMRS-RP-4, U.S. Department of Agriculture, Forest Service, Rocky Mountain Forest and Range Experiment Station, Ogden, Utah, USA

Fletcher MS, Romano A, Connor S, Mariani M, Maezumi SY (2021) Catastrophic bushfires, indigenous fire knowledge and reframing science in southeast Australia. Fire 4(3):61. https://doi.org/10.3390/fire4030061

Gill AM, Stephens SL (2009) Scientific and social challenges for the management of fire-prone wildland-urban interfaces. Environ Res Lett 4(3) https://doi.org/10.1088/1748-9326/4/3/034014

Giuseppe M, Santis F, Alfredo F, Carolina F, Teodosio L et al (2022) Integrated satellite system for fire detection and prioritization. Remote Sens 14(2):335. https://doi.org/10.3390/rs14020335

Hessburg PF, Miller CL, Parks SA, Povak NA, Taylor AH, Higuera PE, Prichard SJ, North MP et al (2019) Climate, environment, and disturbance history govern resilience of western North American forests. Front Ecol Evol 7:239. https://doi.org/10.3389/fevo.2019.00239

Hessburg PF, Prichard SJ, Hagmann RK, Povak NA, Lake FK (2021) Wildfire and climate change adaptation of western North American forests: a case for intentional management. Ecol Appl 31(8):e02432. https://doi.org/10.1002/eap.2432

Jain P, Coogan CP, Subramanian SG, Crowley M, Taylor S, Flannigan MD (2020) A review of machine learning applications in wildfire science and management. Environ Rev 28(4) https://doi.org/10.1139/er-2020-0019

Jiao Z, Zhang Y, Xin J, Mu L, Yi Y, Liu H, Liu D (2019) A deep learning based forest fire detection approach using UAV and YOLOv3. In: Proceedings of the 1st international conference on industrial artificial intelligence (IAI), Shenyang, China, 23–27, pp 1–5

Johnston FH, Henderson SB, Chen Y, Randerson JT, Marlier M, Defries RS, Kinney P, Bowman DMJS, Brauer M (2012) Estimated global mortality attributable to smoke from landscape fires. Environ Health Perspect 120(5):695–701. https://doi.org/10.1289/ehp.1104422

Jolly W, Cochrane M, Freeborn P, Holden Z, Brown T, Williamson G, Bowman D (2015) Climate-induced variations in global wildfire danger from 1979 to 2013. Nat Commun 6:7537

Kang Y, Sung T, Im J (2023) Toward an adaptable deep-learning model for satellite-based wildfire monitoring with consideration of environmental conditions. Remote Sens Environ 298:113814

Kaushik P, Singh H (2024) Climate change-induced forest ecosystem functioning and alteration in water-atmosphere-soil–plant (WASP) relationship. In: Singh H (eds) Urban Forests, Climate Change and Environmental Pollution. Springer, Cham. https://doi.org/10.1007/978-3-031-67837-0_38

Kim SY, Muminov A (2023) Forest fire smoke detection based on deep learning approaches and unmanned aerial vehicle images. Sensors 23(12):5702

Kumar A, Dwivedi GK, Tewari S, Paul J, Anand R, Kumar N, Kumar P, Singh H, Kaushal R (2020) Carbon mineralization and inorganic nitrogen pools under *Terminalia chebula* Retz.-based agroforestry system in Himalayan foothills India. Forest Sci 66(5):634–643

Kumar A, Kumar P, Singh H, Kumar N (2020) Adaptation and mitigation potential of roadside trees with bio-extraction of heavy metals under vehicular emissions and their impact on physiological traits during seasonal regimes. Urb for Urb Green 58:126900

Kumar A, Kumar P, Singh H, Bisht S, Kumar N (2021) Relationship of physiological plant functional traits with soil carbon stock in temperate forest of Garhwal Himalaya. Curr Sci 120(8):1368–1373

Kumar A, Kumar P, Singh H, Kumar N (2021) Modulation of plant functional traits under essential plant nutrients during seasonal regime in natural forests of Garhwal Himalayas. Plant Soil 465:197–212

Kumar A, Singh H, Kumari G, Bisht S, Malik A, Kumar N, Singh M, Raturi A, Barthwal S, Thakur A, Kaushal R (2022) Adaptive resilience of roadside trees to vehicular emissions via leaf enzymatic, physiological, and anatomical trait modulations. Environ Poll 313:120191

Kumar N, Singh H, Kumar A, Singh AK, Pandey PK, Kumar A (2024) Forest-fire-induced land degradation. In: Mishra G, Giri K, Singh S, Kumar M. (eds) Sustainable Land Management in India. Springer Nature, Singapore. https://doi.org/10.1007/978-981-97-5223-2_4

Lareau NP, Nauslar NJ, Abatzoglou JT (2018) The carr fire vortex: a case of Pyrotornadogenesis. Geophys Res Lett 45:13107–13115

Liu Z, Wimberly MC (2016) Direct and indirect effects of climate change on projected future fire regimes in the western United States. Sci Total Environ 542:65–75

Malik A, Kumar K, Kumari G, Kumar N, Singh H (2024) Application of functional traits in modelling productivity and resilience under climate change published by Springer, pp 77–96. ISBN: 978-981-97-1509-1

Mockrin MH, Fishler HK, Stewart SI (2018) Does wildfire open a policy window? Local government and community adaptation after fire in the United States. Environ Manag 62:210–228

Nex C, Armenakis M, Cramer DA, Cucci M, Gerke E, Honkavaara A, Kukko C, Persello JS (2022) UAV in the advent of the twenties: where we stand and what is next. ISPRS J Photogramm Remote Sens 184:215–242. https://doi.org/10.1016/j.isprsjprs.2021.12.006

Nurhidayah L, Astuti R, Hidayat H, Siburian R (2023) Community-based fire management and peatland restoration in Indonesia. In: Triyanti A, Indrawan M, Nurhidayah L, Marfai MA (eds) Environmental Governance in Indonesia. Environ Policy, Springer, Cham. 61:135–150

Ollero A, de Dios JM, Merino L (2006) Unmanned aerial vehicles as tools for forest-fire fighting. For Ecol Manag 234:S263
Oris F, Asselin H, Ali AA, Finsinger W, Bergeron Y (2014) Effect of increased fire activity on global warming in the boreal forest. Environ Rev 22:206–219
Pausas JG, Keeley JE (2009) A burning story: the role of fire in the history of life. Bioscience 59:593–601
Pausas JG, Keeley JE (2019) Wildfires as an ecosystem service. Front Ecol Environ 17:289–295
Pellegrini AFA, Anderegg WRL, Paine CET, Hoffmann WA, Kartzinel T, Rabin SS, Sheil D, Franco AC, Pacala SW (2017) Convergence of bark investment according to fire and climate structures ecosystem vulnerability to future change. Ecol Lett 20:307–316
Perkins-Kirkpatrick SE, Gibson PB (2017) Changes in regional heatwave characteristics as a function of increasing global temperature. Sci Rep 7:12256
Prior T, Eriksen C (2013) Wildfire preparedness, community cohesion and social–ecological systems. Glob Environ Change 23(6):1575–1586
Prodyut B, Pradhan L, Yadav G (2010) Joint forest management in India: experiences of two decades. Resour Conserv Recycl 54:469–480
Russell-Smith J, Monagle C, Jacobsohn M et al (2017) Can savanna burning projects deliver measurable greenhouse emissions reductions and sustainable livelihood opportunities in fire-prone settings? Clim Chang 140(1):47–61
Ryan B, Johnston KA, Taylor M, McAndrew R (2020) Community engagement for disaster preparedness: a systematic literature review. Int J Disaster Risk Reduct 49:101655
Scott TA, Carter DP (2019) Collaborative governance or private policy making? When consultants matter more than participation in collaborative environmental planning. J Environ Policy Plan 21:153–173
Sett T, Nikam BR, Singh H, Purohit S (2024) Hydrological variability in Indian forest ecosystem: analysis of drought resilience, recovery and water use efficiency in moist and dry deciduous forests. In: Singh H. (eds) Urban forests, climate change and environmental pollution. Springer, Cham. https://doi.org/10.1007/978-3-031-67837-0_37
Singh H, Kumar N, Kumar A (2024a) Enhancing resource use efficiency in crops through plant functional traits. Springer, pp 97–117. ISBN: 978-981-97-1509-1
Singh M, Jyoti, Kumar N, Singh H (2024b) Plant functional traits assisted crop adaptation to abiotic and biotic stress. Springer, pp 239–255. ISBN: 978-981-97-1509-1
Singh H (2024a) Climate change and forests: biological perspectives. In: Singh H. (eds) Forests and climate change. Springer Nature, Singapore. https://doi.org/10.1007/978-981-97-3905-9_1
Singh H (2024c) Climate change and environmental pollution: biological consequences on urban forests. In: Singh H (eds) Urban Forests, Climate Change and Environmental Pollution. Springer Nature, Cham, Switzerland. https://doi.org/10.1007/978-3-031-67837-0_1
Singh H (2024b) Forests and climate change: biological perspectives on impact, adaptation, and mitigation strategies (ed), Springer Nature Singapore. https://doi.org/10.1007/978-981-97-3905-9
Singh H (2024d) Urban forests, climate change and environmental pollution: physio-biochemical and molecular perspectives to enhance urban resilience (ed), Springer Nature, Cham, Switzerland. https://doi.org/10.1007/978-3-031-67837-0
Smith W, Neale T, Weir JK (2021) Persuasion without policies: the work of reviving Indigenous peoples' fire management in southern Australia. Geoforum 120:82–92
Sousa D, Cruz-Jesus F, Sousa A, Painho MA (2021) Multivariate approach to assess the structural determinants of large wildfires: evidence from a Mediterranean country. Int J Wildland Fire 30:241–254
Swetnam TW, Farella J, Roos CI, Liebmann MJ, Falk DA, Allen CD (2016) Multi-scale perspectives of fire, climate and humans in western North America and the Jemez Mountains, USA. Phil Trans R Soc B 371:20150168
vanLierop P, Lindquist E, Sathyapala S, Franceschini G (2015) Global forest area disturbance from fire, insect pests, disease and severe weather events. Forest Ecol Manag 352:78–88

Wang X, He HS, Li X, Chang Y, Hu Y, Xu C, Bu R, Xie F (2006) Simulating the effects of reforestation on a large catastrophic fire burned landscape in Northeastern China. For Ecol Manag 225:82–93

Westerling AL, Hidalgo HG, Cayan DR, Swetnam TR (2006) Warming and earlier spring increase western US forest wildfire activity. Sci 313(5789):940–943

Wooster MJ, Roberts GJ, Giglio L, Roy D, Freeborn P, Boschetti L, Justice C, Ichoku C, Schroeder W, Davies D et al (2021) Satellite remote sensing of active fires: history and current status, applications and future requirements. Remote Sens Environ 267:112694

# Predicting the Future of Forest Fires: Climate Models, Fire Behavior and Fire Spread

**Priyanka Mishra, Hukum Singh, and Narendra Kumar**

**Abstract** Forest fires are becoming an increasing threat to flora and fauna due to anthropogenic activities contributing to global warming and accelerating climate change. These changes significantly influence forest fires' frequency, magnitude, and severity, leading to increased fire risks, longer fire seasons, and more severe fire consequences. To address these challenges, various climate models such as GCM (General Circulation Models), MaxEnt (Maximum Entropy), MMM (Multi-Model Mean) techniques, CNN (Convolutional Neural Networks), BPNN (Backpropagation Neural Networks), LSTM (Long Short-Term Memory), SVM (Support Vector Machines), and ARIMA (AutoRegressive Integrated Moving Average) are utilized to predict and forecast the future behavior of forest fires. Monitoring forest fires' occurrence, behavior, severity, and spread is crucial for effective prevention and mitigation. Reducing uncertainty in baseline data through enhanced monitoring techniques, creating precautionary measures, integrating institutional efforts, implementing legal actions, and funding additional programs are essential steps to improve the current situation. These mitigation measures and methods will contribute to a healthier ecological cycle and more resilient forest ecosystems.

**Keywords** Climate change · Fire · Climate model · MaxEnt · Fire prediction

P. Mishra (✉) · H. Singh
ICFRE-Forest Research Institute, Dehradun, India
e-mail: mishra44pri@gmail.com

P. Mishra
Himalayan Institute of Learning and Leadership, UPES, Dehradun, India

N. Kumar
College of Agriculture, Central Agriculture University (I), Kyrdemkulai, Meghalaya, India



H. Singh and N. Kumar (eds.), *Forest Fire and Climate Change*,
https://doi.org/10.1007/978-3-031-89967-6_19

## 1 Introduction

Forest fire prediction is a major step toward preventing forest fire risks in future scenarios to minimize its impact. Climate change leads to rising global temperature, precipitation, and many other environmental factors that can directly or indirectly accelerate forest fire patterns (Kumar et al. 2024; Singh 2024a). The Intergovernmental Panel on Climate Change's sixth assessment report (AR6) predicted that human activity would raise the average surface temperature by 0.8 °C–1.3 °C globally (IPCC 2021). Because of global warming, there will be more severe precipitation occurrences in the late twenty-first century (Mukherjee et al. 2018; Singh 2024b). Global temperatures are expected to have increased by 5.7 °C, resulting in more volatility and a warmer atmosphere by the end of the century (Singh 2024c). Numerous researches have demonstrated that a rising global mean temperature corresponds with an increasing trend in seasonal monsoonal rainfall (Mall et al. 2021). Extreme precipitation, flash floods, and droughts are examples of hydro-climatic extremes that are extremely dangerous to humans, agriculture, and man-made as well as natural ecosystems (Maurya et al. 2021; Singh 2024d). Based on canopy cover and tree height, the FAO and the UNFCC define three categories of forests: dense, moderately dense, and open.

There are around 76.4 million hectares of forest cover in India, which makes up about 2% of the total area of the country and 1% of the world's primary forest (FSI 2019), which is degrading daily. India sustains 16% of the world's population yet makes up just 2.5% of the world's total land area and forest cover. As per the Forest Survey of India, the country's total area covered by forest cover is 71.22 Mha, or 21.67%, with 3.02% being incredibly thick, 9.39% being moderately dense, and 9.26% being open forest. Between 2003 and 2016, over 380–445 of India's 647 districts experienced fires annually. In 2018, over 37,059 forest fires were recorded, according to MODIS (Moderate Resolution Imaging Spectroradiometer) data. More than 6% of the country's forest land is classified as highly sensitive to fire, and about 4% is extremely vulnerable. Generally, forest fires play a vital role in forest management, whether they are natural or not. In a controlled manner, anthropogenic activities and changing climate have taken imminence and caused uncountable damage to the forest and its ecosystem, which is irreversible. Forest fires result from natural and human-induced events in tropical, temperate, and boreal forests.

Some major factors contribute to forest fires, such as temperatures, low humidity, wind speed, direction, dryness of vegetation and precipitation, etc. However, for a forest fire to occur, three essential elements must be present: oxygen, fuel, and a heat source. Fuel refers to any combustible object nearby of a fire, including trees, grasses, shrubs, and even human settlements (Pourtaghi et al. 2016). The extent of a fire is directly proportional to the total fuel present inside a specific area. Fire necessitates the presence of oxygen, which is supplied by the surrounding air. Heat sources facilitate the ignition of the forest fire by raising the temperature of the fuel.

Several sources, such as lightning, sparks from vehicles, campfires or picnic activities, cigarettes, burning of litter (pine needles), and occasionally direct sunlight, may start forest fires. Three types of forest fires have been categorized: surface fires, ground fires, and crown fires. If a fire occurs at the base and burns litter on the plant's surface, it is considered a surface fire. They are the most manageable fires to cope with and impose the most minor damage to the forest. The second type is ground fires, which occur within humus and piles of litter that are dry enough to burn. This type of fire is dangerous as it is sometimes difficult to manage or entirely extinguish, although such fires spread very slowly. The third, as well as the most destructive fire type, is crown fires, as they burn complete trees. Crown fire is the most harmful forest fire because it can spread quickly across tree tops as winds, tree density, and topography strongly influence it. In addition to consuming the plant, fires also modify the structure, porosity, and water resistivity of soils (Neary et al. 1999). Forest fires may be considered one of the leading causes of areas prone to drought being more vulnerable to forest fires. Joseph et al. (2009) have also highlighted the impact of changing climate patterns on India's forest fires. Their research demonstrated how the diversity of grasslands and woodlands has changed in response to climate change.

In addition to being a major threat to forest communities, timber values, infrastructure, and public safety that supports human habitation, forest fires may also significantly impact ecosystem composition, function, and pattern in both fire-prone and fire-adapted environments. Beverly and Bothwell (2011) state that wildland-urban interface fires can cause evacuations, smoke-related health effects, property destruction, loss of job and company revenue, and even fatalities. Therefore, fire risk assessment, fuel treatment, and fire suppression processes must comprehend and anticipate forest fires and the symptomatic ways they move over, particularly in places where values and communities are susceptible. While more investigation is necessary to evaluate the predictive benefits of estimating fire activity based on daily data instead of monthly, yearly, or decade data, the approach suggested here, at least conceptually, more accurately represents the temporal variability characteristic of massive forest fires emanating from northern forests. Understanding future fires will help us better manage and mitigate in a region prone to forest fires in a changing climate.

## 2 Climate Models Used for Fire Prediction

Climate modeling for forest fires helps forecast the occurrence of forest fires and evaluate the fire risk in response to changing climate conditions using different environmental variables. Scientists examine the dynamics of the climate system and estimate future temperatures and climate change using climate models (Savita et al. 2017; Singh et al. 2020; Singh 2021; Singh and Kumar 2022). In addition to being quantitative, climate models can also be qualitative, meaning they include stories about potential futures that are primarily descriptive (IPCC 2014). Since the 1980s, the research on forest fires has gradually incorporated many innovative models due to the rapid development of computers and artificial intelligence. Researchers have used and developed some models and algorithms to predict future scenarios of forest fires and related parameters, as shown in Table 1, and how different climate models can be used.

The study backed up the advantages of machine learning for forest fire prediction over traditional regression models. The Recurrent Thrifty Attention Network (RTAN) to identify a remote sensing part (Yue et al. 2020). Mohajane et al. (2021) forecasted forest fires in the Mediterranean region using machine learning and remote sensing algorithms. Guo et al. (2010) replicated the incidence of a forest fire using random forest and logistic regression.

The initial machine learning methods were built using linear models, the basic mathematical techniques. Predicting the risk of future fires, Cunningham et al. (1973) fitted Poisson regression models to the Sioux Lookout forest fires. Using logistic regression, Shi et al. (2018) developed a model for estimating line trip probability and assessing forest fire risk. Biu et al. (2016) used logistic regression to identify the area's vulnerability to tropical forest fires. Chang et al. (2013) forecasted the danger of fire using logistic regression. Forest fires were quantitatively analyzed by Qiu et al. (2021) using machine learning and Landsat time series data. An integrated learning-based forest fire detection system was developed by Liu et al. (2021).

**Table 1** Climate models for future projection, fire spread, and fire behavior

| S. No | Models | Input | Output | References |
|---|---|---|---|---|
| 1 | RF model | Bioclimatic data, LULC, forest cover, soil parameters etc | Prediction of fire spread | Hering et al. (2009) |
| 2 | MaxEnt | Bioclimatic data, LULC, forest cover, soil parameters etc | Future projection of fire spread | Merow et al. (2013) |
| 3 | GCMs | Precipitation, minimum and maximum temperature data etc | Anticipate future patterns in fire hazard | Perkins et al. (2007) |
| 4 | WRF-FIRE | Elevation, forest canopy cover, forest canopy height, forest canopy bulk density etc | Fire spread prediction and forecast | Giannaros et al. (2020) |
| 5 | BPNN | Population density and several weather parameters (average relative humidity, wind velocity and daily sunshine data), etc | Forecast the risk of fires | Hu et al. (2021) |
| 6 | CMIP5 | Land use, aerosol, GHGs concentration, etc | Future climate scenarios | Taylor et al. (2012) |
| 7 | CNN | Fire images, smoke sense data, etc | Early forest fire identification | Ghali et al. (2021) |
| 8 | BCC-CSM2-MR climate model | Precipitation, temperature, bioclimatic data, LULC, forest cover, soil parameters, etc | Prediction of future scenario | Liu et al. (2021) |

Mathematical models capable of simulating the interplay of significant climatic drivers are known as numerical climate models, sometimes known as climate system models (Kalantar et al. 2020). The ice, land surface, seas, and atmosphere are these drivers. Early fire forecast and identification are essential to minimize damage and cut down on firefighting operations. The yearly cost of fire management initiatives to prevent or stop the further spread of forest fires is in the millions. For this reason, understanding forest fires and their causes and enhancing forest fire forecasts are crucial in forest fire management domains. Two key steps need to be followed to stop forest fires. The first one is forest fire incidence prediction, which models links between the fire risk and important variables, such as fuel content or weather, and effectively anticipates the possibility of a forest fire eruption earlier in its early initiation (Boer et al. 2008). In contrast, the second step is forest fire detection, which entails detecting and identifying fires that are currently occurring.

The main objective is to provide accurate localization and an early warning system before the fire spreads far and becomes out of control. There are several restrictions on the range of scales at which fire activity may be observed, from seconds to millennia, centimeters to kilometers. For example, in mechanistic models, fire and combustion are historically physical phenomena that can be accurately represented at finer scales. However, such models are often constrained by their inability to resolve significant physical processes and the availability and quality of input data (Hoffman et al. 2016). Moreover, physical models cannot affect research and fire management at larger and longer scales that are occasionally needed in near real-time because of limitations associated with current processing capabilities. As a result, developing empirical and statistical models is crucial to research and forest fire control.

The capabilities and higher accessibility of remote sensing data technology have allowed major advancements in observing and tracking forest fires. However, the complexities of these events can occasionally provide modeling challenges. AVHRR, VIIRS, and MODIS are onboard satellite fire detection sensors, including NASA, TERRA, and AQUA. These sensors track vegetation distributions and changes regularly. Additionally, advances in climate models and numerical weather prediction offer longer lead times for forecasts with less geographical resolutions (Bauer et al. 2015), which may enhance the forecasting of intense fire weather events. Given sufficient data, these developments make a data-centric approach to forest fire modeling a logical next step for many research problems. As a result, using machine learning techniques in research and forest fire control has recently received much more attention.

Taylor et al. (2012) describe the Coupled Model Inter-comparison Project (CMIP5). Using a series of Representative Concentration Pathways (RCPs), combined land use, aerosol, and greenhouse gas concentration scenarios involve CMIP5 for modelling future climate scenarios (Wu et al. 2022). Various combinations of RCPs lead to varying radiative forcing levels (Van Vuuren et al. 2011). RCP 2.6, RCP 4.5, RCP 6.0, and RCP 8.5 are the modeled end-of-century radiative forcing levels corresponding to forcing in W m^2.

Quantifying the effects of fire severity scenarios on secondary succession patterns, landscape forest composition, and future fire dynamics using the ALFRESCO model

(Rupp et al. 2007). Earlier researchers used the multi-model mean technique to lower forecast uncertainty (Katzenberger et al. 2021). However, MMM, which depicts daily rainfall uncertainty, has been applied in many research. Numerous studies use the random forest model to forecast the potential frequency of forest fires. The RF modeling was used to identify areas of fire danger (Comert and Matc 2019). The RF worked effectively, using all available parameters (topography, vegetation kinds, infrastructure, meteorological and social-economic aspects) to anticipate the occurrence of fires with a high prediction accuracy (Hering et al. 2009).

A common feed-forward neural network model in deep learning is the CNN (Convolutional Neural Network). Because of its excellent feature learning skills and ability to drastically reduce the number of model parameters, it is primarily used in image recognition. CNN is currently widely utilized for time series processing (Xu et al. 2021).

MaxEnt (Maximum Entropy modelling) fits mathematical functions of environmental predictor factors to presence locations using a machine learning technique and the maximum entropy principle (Phillips and Dudík 2008). According to Merow et al. (2013), it does this by optimizing the probability ratio of average presence values to average values from a sizable random sample of the background environment. The application of machine learning techniques to experimentally simulate interactions between fire and the environment is growing (West et al. 2016). In order to improve model accuracy, identify uncertainties, and establish its credibility, these approaches vary from process-based methods in that they enable model calibration and assessment using real-world fire observational data (Alexander et al. 2013).

A machine learning-based multi-factor forest fire prediction model was developed in a different study. Zheng et al. (2000) produced a high-precision forest fire prediction model based on an improved dynamic neural network and created a forest fire sensitivity prediction model with GIS, while Ghali et al. (2021) created a deep convolutional neural network-based forest fire prediction model and developed the UAV forest fire detection model by using deep learning. In order to identify forest fire smoke, Lin et al. (2023) devised a technique that merged the double-stream model, which is made up of double short-term memory (BLSTM) and visual geometric group network (VGG), with the time domain robust principal component analysis (TRPCA) (TSVB). LSTM was applied by Natekar et al. (2021) to forecast forest fires in India.

Hu et al. (2021) developed a backpropagation neural network (BPNN) prediction model. They compared it to the LSTM approach to forecast the risk of past fires and fire incidence based on each contributing component. The RNN (Recurrent Neural Networks) model, the LSTNet model, and the Long Short-Term Memory (LSTM) model are successful for data with sequence characteristics (Zaremba et al. 2014). It can extract semantic and temporal sequence information from the data, retain prior knowledge, and utilize it to compute the current output model and Support Vector Regression model, demonstrating that LSTNet's prediction results for high-risk area prediction are smaller than those of other models (Kong et al. 2019).

Decision-makers, forest managers, and other stakeholders can benefit substantially from these models' insightful advice on reducing the frequency of forest fires and their adverse effects on the ecosystem. According to Giorgi et al. (2008), future model projections show that there will likely be more days with maximum temperatures above and an increase in these days overall each year. These alterations might raise the immensity of forest fires in a given area. As a result of these potential future modifications to forest fire patterns, there will be a higher chance of forest fire-related risks and vulnerabilities, including altered microclimates, infrastructure destruction, floods, and human casualties. These post-forest fire conditions will impact other changes in the natural environment and lead to modified succession, land degradation, vegetation patterns, desertification, and disruption of the hydrological cycle.

Moreover, emissions of greenhouse gases, including carbon, from forest fires result in climatic feedback loops (Scheffer et al. 2001; Kumar et al. 2020a, 2020b). Deep learning has advanced to the point where it can more successfully and independently extract useful feature information from vast amounts of data than standard machine learning methods. More experts and scholars in forest fire prediction are attempting deep learning.

## 3 Understanding Behaviors of Fire and Severity

A study to understand the consequences of fire on individuals, landscapes, and ecosystems is based on research on fire behavior and heat transmission processes (Miyanishi et al. 2001). Variations in the biophysical environment, such as the topography, weather, and vegetation structure, spatially and temporally influence the behavior of fires (Catchpole et al. 2002). Slope, aspect, elevation, and soil changes impact the possible plant community and the site's water and energy budgets. Because these topographical characteristics affect moisture conditions and fuel flammability on hourly, seasonal, and succession time scales, they also significantly impact fire-induced disruption (Wotton et al. 2010).

The intensity and aboveground impacts of a fire are greatly influenced by weather variations, particularly those related to wind and relative humidity, over fine time scales ranging from minutes to hours. Normal seasonal variations in dryness and occasional droughts primarily affect the burn depth and belowground effects, affecting the intensity and severity of fires. Changes in the vegetative structure over decades impact the mass of fuel that may burn and, consequently, the potential energy that can be produced during a fire. The biophysical environment and the fire's perimeter position influence the fire's intensity, which fluctuates over time and place. Variations in fire intensity in space and time can have long-term effects on the species composition and organization of post-fire communities and the likelihood of further disturbances. Fire severity varies with variation in vegetation structure (Kafka et al. 2001), whereas in many ecosystems, fire severity has a significant role in determining the post-fire trajectories of vegetation recovery (Keeley et al. 2005; Malik et al. 2024).

According to Davies et al. (2016), fire severity evaluations are related to how much biomass and organic matter are burned during combustion (Oliva 2013). Significant variations in the amount of fuel used throughout the fire are associated with changes in the intensity of the fire. Thus, improving emissions estimates requires understanding the amount of pre-fire fuel burned and the degree of fire damage (Kumar et al., 2021a, 2021b). During a forest fire, fuel consumption can occur at varying rates. Therefore, it is crucial to evaluate the proportion of fuel that is burnt as a geographical and temporal dynamic variable connected to the severity level (Oliva et al. 2013). Since fire emissions are typically estimated as the product of the burned area, fuel load, and burned fraction. A specific emission factor (Urbanski 2014), an inaccurate estimation of this burned fraction can lead to large errors in the quantification of emissions from forest fires (De Santis et al. 2010; Singh et al. 2024).

Changes in successional routes may result in alterations in the forest cover caused by a rise in the intensity of surface fires (Johnstone et al. 2010). Accordingly, in addition to considering the shifting distribution of stand ages and successional stages, projections of the future effects of changing climatic and fire regimes on forest composition also need to consider the impact of fire on successional trajectories (Suffling et al. 1995). Suppose changes in forest cover have a knock-on impact on the likelihood of future disturbances. In that case, interactions between fire severity and successional trajectories may be crucial in controlling the effects of climate change (Higuera et al. 2009).

Important implications for forest fuel types and probable fire behavior arise from the onset of deciduous or mixed deciduous-conifer successional trajectories on areas previously dominated by black spruce self-replacement cycles (Krawchuk et al. 2006). Characteristic temperature histories of surface, ground, and crown fires are employed to show differences in fire severity. The influence of different burn depths on the depth at which different fire effects occur in the soil profile is demonstrated using a soil-heating model. A conceptual model represents the impact of fire intensity on the interactions between fire and plant regeneration. Restoration ecologists can develop and carry out fire treatments to preserve biodiversity by using the conceptual model to assess the differences in impacts between controlled or managed burns and forest fires. (Ryan 2002).

## 4 Forecasting and Fire Spread

Research indicates that the following factors are important for forecasting fire spots and fire danger levels: temperature, wind speed, relative humidity, soil moisture, precipitation, drought index, vegetation indices, and pressure. In order to anticipate future patterns in fire hazards brought on by climate change, global evaluations of fire weather indicators using General Circulation Models (GCMs) are helpful. These evaluations highlight the requirement of taking systematic model mistakes and biases into account for accurate forecasts. Three distinct General Circulation Models (GCMs) were used for the study: CanESM2, HadGEM2-ES, and CSIROMk3-6-0, with RCP 2.6, RCP 4.5, and RCP 8.5. Using a best-performance selection technique like that mentioned by Perkins et al. (2007), these three GCMs were used. By entering data, traditional machine learning models like SVM can forecast a forest fire's vulnerability.

Around the world, efforts have been undertaken to create information, alarm, and rating systems for fire hazards to provide various stakeholders with timely information about possible fire danger. Fuel type, terrain, and weather are inputs used by the National Fire Risk Rating System (NFDRS) of the United States of America to assess fire hazards for the next twenty-four hours (USDA 2022). Daily maps of fire hazard and fire weather indices are provided by the Canadian Forest Fire Weather Index (FWI) system. In the European region, EFFIS offers a "fire danger" forecast that is good for nine days. On a near real-time basis, the Forest Survey of India tracks and detects major forest fires in India by using satellite data SNPP (Suomi-National Polar-orbiting Partnership)-VIIRS (FSI 2022).

Furthermore, advanced numerical two-way coupled models, like WRF-FIRE, have been created and evaluated for operational fire spread simulation in various nations, including Greece (Giannaros et al. 2020), Israel (Mandel et al. 2014) and the United States. These models allow for the hourly prediction of forest fire spread after identifying the ignition location. It is sometimes tricky to swiftly deploy the resources to contain a forest fire at short notice, even when various fire early warning technologies are available to offer the earliest possible alerts regarding the fire's severity, threat, and location. Complex terrain causes an additional delay in response time. In order to pre-allocate resources at potentially susceptible places throughout the fire season and hence reduce the reaction time of the forest fire fighting authority, mid to long-term (monthly and annual) fire predictions must be used.

The forest fire's time series analysis (TSA) can be used to anticipate fire episodes at particular periods. Forecasts might be based on the fire time series alone or on several independent factors. The dynamic interactions between forest fires and related independent factors concerning time are the foundation of fire predictions considering independent variables. Some studies have found multivariate time series models superior to univariate models. Others have suggested this model be preferred over multivariate models (Aboagye-Sarfo et al. 2015; Iwok and Okpe 2016).

Sethi and Mittal (2020) discovered that the univariate ARIMA model outperformed the multivariate vector autoregression model (VAR) in predicting the air quality index. The TSA is a data-intensive procedure requiring years' worth of data to produce insightful insights. It requires less data than the multivariate model. Using the ARIMA (Autoregressive Integrated Moving Average) model (Kadir et al. 2020); the Probit model (Albertson et al. 2009); longitudinal negative bionomial (NB) and zero-inflated negative bionomial (ZNIB) mixed models (Viedma et al. 2018); the Poisson's regression method (Graff et al. 2020) to predict fire activity for one to five days in the future using MODIS satellite data active fire count and ERAinterim reanalysis-based weather data; and mapping the fire risk using machine learning and satellite vegetation index time series (Michael et al. 2021) and an analytical hierarchical process (Sivrikaya and Kucuk 2021).

## 5 Mitigation, Challenges, and Future Direction

Forest fires are one of the many adverse effects caused by climate change. Climate change due to anthropogenic activities is trying to inflict increased stress on many forest species and impair their resistance to forest fires (Hawbaker et al. 2013). Within some parts of the globe, more intense and recurring forest fires are projected to occur due to climate change (Pitman et al. 2007).

The National Master Plan for Forest Fire Control, the National Forest Policy, and the Guidelines, developed in 1988, are a few of the major steps the Indian government has taken to control fires. In India, fire prevention techniques commonly involve clearing fire lines, managing fires to reduce fuel loads, implementing fire-adapted tree species planting and selective thinning in high-risk regions, utilizing early warning systems, and implementing fire hazard rating systems.

Another important milestone in India's micro-level control of forest fires was the establishment of Joint Forest Control (JFM) village-level committees. Short-term understanding about systematic forest fire suppression is lacking. The review has flaws in everything from the coordination of preventative measures to the detection of fires. Developing an adaptive management strategy is crucial right now. Clear guidelines are provided for the sector-specific execution of a comprehensive fire prevention strategy in light of climate change and short- and long-term aspirations.

Preventing forest fires is the critical initiatives for the containment and prevention of forest fires are mapping and identifying regions that are prone to fire, collecting information on the effects of forest fires, implementing forest risk grading instruments, and creating and implementing forecasting systems for forest fires to guarantee budgetary allocation, forest protection-related activities should be included in scheduled activities. Precautions must be implemented far before the fire season, and a fire control committee is to be established for monitoring and controlling the fire at the block, district, and state headquarters levels.

Plenty of resources should be given to them to combat fire incidents. A communication network should be established to guarantee the prompt delivery of supplies, labor, and information to fire locations. Groups such as Joint Forest Management should be organized, trained, and involved in managing and controlling forest fires. Particularly, those communities that reside close to the forest must participate actively in these committees. Forest staff employees and other members of fire protection committees should get regular training to keep them informed about new knowledge and resources for preventing forest fires. Raising public awareness plays an essential role in preventing forest fires. Strict legal measures must be taken to avoid forest fires.

An automated early warning system for forest fires, called firewatch, is a computerized, remote, terrestrial surveillance system that monitors more expansive woody areas and analyzes, assesses, links, and stores the collected data. The system's sensitivity, precision, and dependability allow for the early detection of forest fires using a radio-acoustic sounding method for fire detection. A radio-acoustic sounding system for the early detection of surface and crown fires, but not ground fires. That technology may also be used in conjunction with current ones to improve the effectiveness of fire detection for all kinds of forest fires. There must be some educational measures to raise public awareness of the danger and damage caused by forest fires. For example, posters with messages about fire prevention or fire risk meters installed by the project's divisions have been put up to raise awareness of the severity of seasonal fires.

Furthermore, for controlling forest fires, several methods exist, such as airborne firefighting in which airplanes and other airborne resources are used for fighting forest fires. Fixed-wing and helicopter aircraft are among the types of aircraft in use. Aerial firefighters include smoke jumpers and rappelers, who are sent to the fire by rappelling from helicopters or jumping from various fixed-wing aircraft. Air tankers, sometimes known as water bombers, are fixed-wing aircraft equipped with tanks that may be refilled on the ground at an air tanker base or by skimming water from lakes, reservoirs, or big rivers in the case of flying boats and amphibious aircraft. Air tankers are big aircraft fitted with tanks that carry and release water or fire retardants to help extinguish a fire and support ground-based fire line-building activities. The main applications for air tankers are the defense of structures and first attacks. Chemical retardants are usually paired with a red dye to provide pilots with visual cues about the precision of the drop. Although feasibility studies for bigger capacities have been carried out, capacities typically range from 2,000 to 3,000 gallons. Because they are more mobile and have a lower capacity, single-engine air tankers are advantageous in hilly terrain.

The establishment of patches throughout an area with a potential for fire spread, known as fire lines, must be eradicated yearly. As the developed vegetation may generate continuity. Chemicals like foams, gels, and specifically made fire retardants can all be used to put out flames. Borate salts worked as a fire retardant and were once utilized by forest firefighters, but it was discovered that they sterilized the soil, were poisonous to animals, and were thus outlawed. Ammonium sulfate, ammonium polyphosphate, ammonium phosphate, and a thickener derived from guar gum are the ingredients of more recent retardants. These are not only less harmful, but they also serve as fertilizers to aid in the plant's post-fire regeneration.

In addition to being colored red with ferric oxide or another fugitive color to indicate where they have been dumped, fire retardants frequently contain wetting agents, preservatives, and rust inhibitors. Chemical fire control, like detergent, prolongs the usable life of water for cleaning, and retardants and foams increase the effectiveness of water as a suppression by dispersing it across a surface and allowing it to penetrate deeper into porous fuels. Retardants can also reduce ignition temperatures, speed up fuel conversion, and expedite extinction by preventing flames. Ground control operations involving fire engines utilize water and e-version 33 water enhancers, such as foams and fire retardants. Water makes up 85% of the retardant that air tankers and helicopters drop, followed by diammonium phosphate or sulphates (10%) and small components (5%). The combination weighs around 9 lb per gallon. Diammonium phosphate is the active component that stops the fire from spreading and fertilizes the ground when it is extinguished. Gums and clay thickeners are minor components, as is iron oxide. Because airborne retardants may be hazardous to aquatic life, their use may be limited in the vicinity of creeks and streams.

Evaluation is an essential component for academics and resource managers to embrace a model and give it legitimacy. It is challenging to get the kind of high-quality data required for a sufficient evaluation of fire behaviour models due to the nature and features of free-burning wildland fires (Alexander et al. 2013). Although using models to forecast fire behaviour in a changing climate provides insightful information, conceding their limits and potential policy significance is equally critical. Accurate input data, including vegetation type, fuel moisture, and meteorological conditions, is crucial for models. Predictions can be profoundly inaccurate if these inputs contain errors or any uncertainty in data. Some models make oversimplified assumptions that might not adequately represent the complexity of fire dynamics, particularly in diverse environments or during catastrophic weather events. A possible drawback of the fire behaviour datasets currently available for use in assessing physics-based models is the possibility of undocumented input variables, which would require adopting assumptions (Mell et al. 2007; Linn et al. 2012). Models frequently have trouble striking a balance between local specificity and generalizability. Without contextual changes, policymakers may rely on model results, leading to misaligned management methods. Using predictive models too often could cause concern when taking preemptive steps like community involvement initiatives or controlled burning. In the name of fire prevention, policymakers can exploit the outcomes of models to support contentious choices like excessive logging.

## 6 Conclusion

Forest fire prevention is significantly enhanced by modeling and forecasting the potential for such fires. Recognized as a global natural hazard, forest fires have prompted several nations to emphasize the importance of mountain fire forecasts in their forest fire prevention and control efforts. Numerous studies have projected the geographical distribution of forest fire occurrences by utilizing MODIS fire point data, historical meteorological, topographical, vegetation, social data, and temperature and precipitation data from future scenarios. These studies have revealed a substantial correlation between precipitation, elevation, temperature, and the frequency of forest fires. Temperature-related relative humidity changes accelerate water evaporation from flammable materials in forests. Rising temperatures also increase the likelihood of strong winds, elevating the risk of forest fires and the rapid spread of flames. Precipitation directly influences the water content of flammable materials, with higher plant water content reducing the likelihood of forest fires. Projections using the BCC-CSM2-MR climate model indicate that severe precipitation events will become more frequent and intense under the higher carbon emission scenario SSP585 compared to the lower emission scenario SSP126. Future research should focus on how large-scale climatic variations impact the frequency of forest fires and carbon emissions. Also, studies should explore forest fires' environmental, ecological, and evolutionary consequences. It is crucial to consider how different climatic scenarios will affect forest fire frequency and how forest age and structure will change over the next century. While decadal data on forest fires, climate, and topographic traits provide valuable predictions, short-term climate data may not fully capture the long-term variability in fire-climate relationships, potentially affecting the comparability of present and future fires. Evaluating the climate conditions before future fires will become increasingly important. The prolonged growing season caused by global warming is likely to increase the incidence of forest fires. Researchers can better identify forest fire occurrences under future climate change scenarios by combining active fire data and multiple factors. These data serve as essential references and support for developing sustainable forest management practices, preventing and mitigating future forest fires, and predicting future occurrences. The correlations observed through climate models will aid in predicting the effects of climate change and its consequences in future scenarios.

## References


Aboagye-Sarfo P, Mai Q, Sanfilippo FM, Preen DB, Stewart LM, Fatovich DM (2015) A comparison of multivariate and univariate time series approaches to modelling and forecasting emergency department demand in Western Australia. J Biomed Inform 57:62–73

Albertson K, Aylen J, Cavan G, Mcmorrow J (2009) Forecasting the outbreak of moorland wildfires in the English Peak district. J Environ Manag 90:2642–2651. https://doi.org/10.1016/j.jenvman.2009.02.011

Alexander ME, Cruz MG (2013) Are the applications of wildland fire behaviour models getting ahead of their evaluation again? Environ Model Softw 41:65–71

Bauer P, Thorpe A, Brunet G (2015) The quiet revolution of numerical weather prediction. Nature 525:47–55

Beverly JL, Bothwell P (2011) Wildfire evacuations in Canada 1980–2007. Nat Hazards 59:571–96

Boer MM, Sadler RJ, Bradstock RA, Gill AM, Grierson PF (2008) Spatial scale invariance of southern Australian forest fires mirrors the scaling behaviour of weather events. Landsca Ecol 23:899–913

Bui DT, Le K, Nguyen VC, Le HD, Revhaug I (2016) Tropical forest fire susceptibility mapping at the Cat Ba National Park Area, Hai Phong City, Vietnam, using GIS-based kernel logistic regression. Remote Sens 8:347

Catchpole W (2002) Fire properties and burn patterns in heterogeneous landscapes. Flammable Australia: the fire regimes and biodiversity of a continent (ed by Bradstock RA, Williams JE, Gill AM), pp 50–75. Cambridge University Press, Cambridge

Chang Y, Zhu Z, Bu R, Chen H, Feng Y-T, Li Y, Hu Y, Wang Z (2013) Predicting fire occurrence patterns with logistic regression in Heilongjiang Province China. Landsc Ecol 28:1989–2004

Comert R, Matc DK, Avdan U (2019) Object based burned area mapping with random forest algorithm. Int J Eng Geosci 4(2):078–087

Conard SG, Solomon AM (2008) Chapter 5 Effects of wildland fire on regional and global carbon stocks in a changing environment. Dev Environ Sci

Cunningham AA, Martell DL (1973) A stochastic model for the occurrence of man-caused forest fires. Can J for Res 3(2):282–287

Davies GM, Domènech R, Gray A, Johnson PCD (2016) Vegetation structure and fire weather influence variation in burn severity and fuel consumption during peatland wildfires. Biogeosciences 13:389–398

De Santis A, Asner GP, Vaughan PJ, Knapp DE (2010) Mapping burn severity and burning efficiency in California using simulation models and Landsat imagery. Remote Sens Environ 114:1535–1545. https://doi.org/10.1016/j.rse.2010.02.008

FSI [Forest Survey of India] (2019) India state of forest report 2019. Dehradun: Forest Survey of India, Ministry of Environment and Forests. Available online at https://fsi.nic.in/isfr2019/isfr-fsi-vol2.pdf

FSI (2022) FSI forest fire alert system (FAST 3.0). Available online at: http://fsi.nic.in/uploads/documents/technical_information_series_vol1_no2.pdf. Accessed January 14, 2022

Ghali R, Akhloufi MA, Jamal M, Mseddi WS, Attia R (2021) Forest fires segmentation using deep convolutional neural networks. In: 2021 IEEE International conference on systems, man, and cybernetics (SMC). IEEE. pp 2109–2114

Giannaros TM, Lagouvardos K, Kotroni V (2020) Performance evaluation of an operational rapid response fire spread forecasting system in the southeast mediterranean (Greece). Atmosphere 11:1264. https://doi.org/10.3390/atmos11111264

Giorgi F, Lionello P (2008) Climate change projections for the Mediterranean region. Glob Planet Chang 63:90–104

Graff CA, Coffield SR, Chen Y, Georgiou EF, Randerson JT, Smyth P (2020) Forecasting daily wildfire activity using poisson regression. IEEE Trans Geosci Remote Sens 58:7. https://doi.org/10.1109/TGRS.2020.2968029

Guo FT, Hu HQ, Ma ZH, Zhang Y (2010) Applicability of different models in simulating the relationships between forest fire occurrence and weather factors in Daxing'an Mountains. J Appl Ecol 21:159–164

Hawbaker TJ, Radelof VC, Stewart SI, Hammer RB, Keuler NS, Clay ton MK (2013) Human and biophysical infuences on fire occurrence in the United States. Ecol Appl 23(3):565–582

Hering AS, Bell CL, Genton MG (2009) Modeling spatio-temporal wildfire ignition point patterns. Environ Ecol Stat 16:225–250

Higuera PE, Brubaker LB, Anderson PM, Hu FS, Brown TA (2009) Vegetation mediated the impacts of postglacial climate change on fire regimes in the south-central Brooks Range Alaska. Ecol Monogr 79(2):201–219

Hoffman CM, Canfield J, Linn RR, Mell W, Sieg CH, Pimont F, Ziegler J (2016) Evaluating crown fire rate of spread predictions from physics-based models. Fire Technol 52:221–237

Hu Y, Zhao FJ, Chen F, Shu LF (2021) Impacts of global warming and large-scale climate fluctuation on forest fires and forest carbon emissions. Terr Ecosyst Conserv 1(01):75–81

IPCC (2014) AR5 Synthesis report—climate change 2014. Contribution of working groups I, II and III to the fifth assessment report of the intergovernmental panel on climate change

IPCC (2021) In climate change 2021 the physical science basis summary for policymakers working group I contribution to the sixth assessment report of the intergovernmental panel on climate change

Iwok IA, Okpe AS (2016) A comparative study between univariate and multivariate linear stationary time series models. Am J Math Stat 6:203–212

Johnstone JF, Hollingsworth TN, Chapin FS III, Mack MC (2010) Changes in fire regime break the legacy lock on successional trajectories in Alaskan boreal forest. Glob Change Biol 16:1281–1295

Joseph S, Anitha K, Murthy MSR (2009) Forest fire in India: a review of the knowledge base. J For Res 14(3)127–134

Kadir EA, Dayana NE, Rosa SL, Othman M, Saian R (2020) Prediction of hotspots in Riau Province, Indonesia using the autoregressive integrated moving average (ARIMA) model. SAR J 3:101–110. https://doi.org/10.18421/SAR33-03

Kafka V, Gauthier S, Bergeron Y (2001) Fire impacts and crowning in the boreal forest: study of a large wildfire in western Quebec. Int J Wild Land Fire 10(2):119–127

Kalantar B, Ueda N, Idrees MO, Janizadeh S, Ahmadi K, Shabani F (2020) Forest fire susceptibility prediction based on machine learning models with resampling algorithms on remote sensing data. Remote Sens 12:3682

Katzenberger A, Schewe J, Pongratz J, Levermann A (2021) Robust increase of Indian monsoon rainfall and its variability under future warming in CMIP6 models. Earth Syst Dyn 12(2):367–386. https://doi.org/10.5194/esd-12-367-2021

Keeley JE, Fotheringham CJ, Baer-Keeley M (2005) Determinants of post-fire recovery and succession in mediterranean-climate shrub lands of California. Ecol Appl 15(5):1515–1534

Kong W, Dong ZY, Jia Y, Hill DJ, Xu Y, Zhang Y (2019) Short-term residential load forecasting based on LSTM recurrent neural network. IEEE Trans Smart Grid 10:841–851

Krawchuk MA, Cumming SG, Flannigan MD, Wein RW (2006) Biotic and abiotic regulation of lightning fire initiation in the mixed wood boreal forest. Ecology 87(2):458–468

Kumar A, Dwivedi GK, Tewari S, Paul J, Anand R, Kumar N, Kumar P, Singh H, Kaushal R (2020) Carbon mineralization and inorganic nitrogen pools under *Terminalia chebula* Retz.-based agroforestry system in Himalayan Foothills India. Forest Sci 66(5):634–643

Kumar A, Kumar P, Singh H, Kumar N (2020) Adaptation and mitigation potential of roadside trees with bio-extraction of heavy metals under vehicular emissions and their impact on physiological traits during seasonal regimes. Urb for Urb Green 58:126900

Kumar A, Kumar P, Singh H, Bisht S, Kumar N (2021) Relationship of physiological plant functional traits with soil carbon stock in temperate forest of Garhwal Himalaya. Curr Sci 120(8):1368–1373

Kumar A, Kumar P, Singh H, Kumar N (2021) Modulation of plant functional traits under essential plant nutrients during seasonal regime in natural forests of Garhwal Himalayas. Plant Soil 465:197–212

Kumar N, Singh H, Kumar A, Singh AK, Pandey PK, Kumar A (2024) Forest-fire-induced land degradation. In: Mishra G, Giri K, Singh S, Kumar M. (eds) Sustainable Land Management in India. Springer Nature, Singapore. https://doi.org/10.1007/978-981-97-5223-2_4

Lin X, Li Z, Chen W, Sun X, Gao D (2023) Forest fire prediction based on long-and short-term time-series network. Forests 14(4):778

Linn R, Anderson K, Winterkamp J, Brooks A, Wotton M, Dupuy JL, Pimont F, Edminster C (2012) Incorporating field wind data into FIRETEC simulations of the international crown fire modeling experiment (ICFME): preliminary lessons learned. Can J for Res 42:879e898

Liu C, Zhang H, Wu J (2021) Impact assessment of extreme precipitation in China under SSPs scenario. Environ Prot 49(08):29–34

Lucas C, Hennessy KJ, Mills GA, Bathols JM (2007) Bushfire weather in southeast Australia: recent trends and projected climate change impacts. Report to the Climate Institute of Australia Bushfire Cooperative Research Centre, Australian Bureau of Meteorology and CSIRO Marine and Atmospheric Research, Canberra

Malik A, Kumar K, Kumari G, Kumar N, Singh H (2024) Application of functional traits in modelling productivity and resilience under climate change published by Springer, pp 77–96. ISBN: 978-981-97-1509-1

Mall RK, Chaturvedi M, Singh N, Bhatla R, Singh RS, Gupta A, Niyogi D (2021) Evidence of asymmetric change in diurnal temperature range in recent decades over different agro-climatic zones of India. Int J Climatol 41(4):2597–2610. https://doi.org/10.1002/joc.6978

Mandel J, Amram S, Beezley JD, Kelman G, Kochanski AK, Kondratenko VY et al (2014) New features in WRF-SFIRE and the wildfire forecasting and danger system in Israel. Nat Hazards Earth Syst Sci Discuss 2:1759–1797. https://doi.org/10.5194/nhessd-2-1759-2014

Maurya S, Srivastava PK, Yaduvanshi A, Anand A, Petropoulos GP, Zhuo L, Mall RK (2021) Soil erosion in future scenario using CMIP5 models and Earth observation datasets. J Hydrol 594:125851. https://doi.org/10.1016/j.jhydrol.2020.125851

Mell W, Jenkins MA, Gould J, Cheney P (2007) A physics-based approach to modelling grassland fires. Int J Wildland Fire 16:1e22

Merow C, Smith MJ Jr, Silander JA (2013) A practical guide to MaxEnt for modeling species' distributions: what it does, and why inputs and settings matter. Ecography 36:1058–1069
Michael Y, Helman D, Glickman O, Gabay D, Brenner S, Lensky IM (2021) Forecasting fire risk with machine learning and dynamic informationderived from satellite vegetation index time-series. Sci Total Envion 764:142844. https://doi.org/10.1016/j.scitotenv.2020.142844
Miyanishi K (2001) Forest fires, behavior, and ecological effects. Acad Press San Francisco CA 594:p4
Mohajane M, Costache R, Karimi F, Pham QB, Essahlaoui A, Nguyen H, Oudija F (2021) Application of remote sensing and machine learning algorithms for forest fire mapping in a Mediterranean area. Ecol Ind 129:107869
Mukherjee S, Aadhar S, Stone D, Mishra V (2018) Increase in extreme precipitation events under anthropogenic warming in India. Weather Climate Extremes 20:45–53. https://doi.org/10.1016/j.wace.2018.03.005
Natekar S, Patil S, Nair A, Roychowdhury S (2021) Forest fire prediction using LSTM. In: Proceedings of the 2021 2nd international conference for emerging technology (INCET), Belgaum, India, 21–23, pp 1–5
Neary GD, Klopatek CC, DeBano LF, Ffolliott PF (1999) Fire effects on belowground sustainability: a review and synthesis. For Ecol Manage 122:51–71
Oliva P (2013) FEMM—fire effects modeling and mapping: an approach to estimate the spatial variability of burning efficiency. In: en Fernández DY Sabia R (Coords.): Remote sensing advances for system science the ESA science network: project, pp 93–102. Springer, Berlin, pp 2009–2011
Perkins SE, Pitman AJ, Holbrook NJ, Mcaneney J (2007) Evaluation of the AR4 climate models' simulated daily maximum temperature, minimum temperature, and precipitation over Australia using probability density functions. J Clim 20:4356–76
Phillips SJ, Dudík M (2008) Modeling of species distributions with MaxEnt: new extensions and a comprehensive evaluation. Ecography 31(2):161–175
Pitman AJ, Narisma GT, McAneney J (2007) The impact of climate change on the risk of forest and grassland fires in Australia. Clim Change 84(3–4):383–401. https://doi.org/10.1007/s10584-007-9243-6
Pourtaghi ZS, Pourghasemi HR, Aretano R, Semeraro T (2016) Investigation of general indicators influencing on forest fire and its susceptibility modelling using different data mining techniques. Ecol Ind 64:72–84. https://doi.org/10.1016/j.ecolind.2015.12.030
Qiu Z, Feng Z, Song Y, Li M, Zhang P (2020) Carbon sequestration potential of forest vegetation in China from 2003 to 2050: predicting forest vegetation growth based on climate and the environment. J Clean Prod 252:119715
Qiu J, Wang H, Shen W, Zhang Y, Su H, Li M (2021) Quantifying forest fire and post-fire vegetation recovery in the Daxin'anling area of Northeastern China using Landsat time-series data and machine learning. Remote Sens 13:792
Rupp TS, Chen X, Olson M (2007) Sensitivity of simulated boreal fire dynamics to uncertainties in climate drivers. Earth Interact 11(3):1–21
Ryan KC (2002) Dynamic interactions between forest structure and fire behaviour in boreal ecosystems. Silva Fennica 36(1):13–39
Savita, Sharma SK, Singh H, Singh O (2017) Forest fire: a review. Indian for 143(8):719–728
Scheffer M, Carpenter S, Foley J, Folke C, Walker B (2001) Catastrophic shifts in ecosystems. Nature 413:591–596
Sethi JK, Mittal M (2020) Analysis of air quality using univariate and multivariate time series models, In: Proceedings of the 10th international conference on cloud computing, data science & engineering (Confluence), Noida, pp 823–827. https://doi.org/10.1109/Confluence47617.2020.9058303
Shi S, Yao C, Wang S, Han WA (2018) Model design for risk assessment of line tripping caused by wildfires. Sensors 18:1941
Singh H, Kumar N, Kumar M, Singh R (2020) Modelling habitat suitability of western tragopan (*Tragopan melanocephalus*) a range-restricted vulnerable bird species of the Himalayan region

in response to climate change. Climate Risk Manag 29:100241. https://doi.org/10.1016/j.crm.2020.100241
Singh H, Kumar N, Singh R, Kumar M (2023) Assessing the climate change impact on the habitat suitability of the range-restricted bird species (*Catreus wallichii*) in the Indian Himalayan ecosystem. Environ Sci Pollut Res. https://doi.org/10.1007/s11356-023-30789-x
Singh H, Kumar M (2022) Climate change and its impact on Indian Himalayan forests: current status and research needs. In: Rani S, Kumar R (eds) Climate change. Springer Climate. Springer, Cham. pp 223–242, https://doi.org/10.1007/978-3-030-92782-0_11
Singh H, Kumar N, Kumar A (2024) Enhancing resource use efficiency in crops through plant functional traits. Springer, pp 97–117. ISBN: 978-981-97-1509-1
Singh H (2021) Process-based carbon sequestration study in relation to the energy-water-carbon flux in a forest ecosystem. In: Climate impacts on sustainable natural resource management (Kumar P, Singh RK, Kumar M, Rani M, Sharma (eds). CRC Press Taylor & Francis Group, United States, pp 336–361
Singh H (2024a) Climate change and forests: biological perspectives. In: Singh H (eds) Forests and climate change. Springer Nature, Singapore. https://doi.org/10.1007/978-981-97-3905-9_1
Singh H (2024c) Climate change and environmental pollution: biological consequences on urban forests. In: Singh H (eds) Urban forests, climate change and environmental pollution. Springer Nature, Cham, Switzerland. https://doi.org/10.1007/978-3-031-67837-0_1
Singh H (2024b) Forests and climate change: biological perspectives on impact, adaptation, and mitigation strategies (ed), Springer Nature Singapore. https://doi.org/10.1007/978-981-97-3905-9
Singh H (2024d) Urban forests, climate change and environmental pollution: physio-biochemical and molecular perspectives to enhance urban resilience, Springer Nature (ed), Cham, Switzerland. https://doi.org/10.1007/978-3-031-67837-0
Sivrikaya F, Kucuk O (2021) Modeling forest fire risk based on GIS-based analytical hierarchy process and statistical analysis in Mediterranean region. Ecol. Inf. 68:101537. https://doi.org/10.1016/j.ecoinf.2021.101537
Suffling R (1995) Can disturbance determine vegetation distribution during climate warming? A boreal test. J Biogeogr 22:501–508
Taylor KE, Stouffer RJ, Meehl GA (2012) An overview of CMIP5 and the experimental design Bull. Am Meteorol Soc 93:485–98
Urbanski S (2014) Wild land fire emissions, carbon, and climate: emission factors. For Ecol Manag 317:51–60
USDA (2022) National fire danger rating system. Available online at:https://www.fs.usda.gov/detail/cibola/landmanagement/resourcemanagement/?cid=stelprdb5368839. Accessed February 18, 2022
Van Vuuren DP et al (2011) The representative concentration pathways: an overview Clim. Change 109:5–31
Viedma O, Urbieta IR, Moreno JM (2018) Wildfres and the role of their drivers are changing over time in a large rural area of west-central Spain. Sci Rep 8:17797. https://doi.org/10.1038/s41598-018-36134-4
West AM, Kumar S, Jarnevich CS (2016) Regional modeling of large wildfires under current and potential future climates in Colorado and Wyoming, USA.Clim. Change 134:565–577
Wotton M, Nock C, Flannigan M (2010) Forest fire occurrence and climate change in Canada. Int J Wild Land Fire 19:253–271
Wu H, Lei H, Lu W, Liu Z (2022) Future changes in precipitation over the upper Yangtze River basin based on bias correction spatial downscaling of models from CMIP6. Environ Res Commun 4(4):045002. https://doi.org/10.1088/2515-7620/ac620e
Xu R, Lin HX, Lu K, Cao L, Liu Y (2021) A forest fire detection system based on ensemble learning. Forests 12:217
Yue C, Luo C, Shu L, Shen Z (2020) A review on wildfire studies in the context of global change. Acta Ecol Sin 40(02):385–401

Zaremba W, Sutskever I, Vinyals O (2014) Recurrent neural network regularization. arXiv, arXiv: 1409.2329

Zhang S, Zhu C, Chen Z (2000) Research on forest fire meteorological environmental elements and large forest fires. J Nat Disasters 02:111–117

# Innovations on the Horizon: Technology Advancements for Fire Management

**Yashasvi Thakur**

**Abstract** In recent years, the intensity and frequency of forest fires have escalated significantly, primarily due to rising global temperatures, resulting in severe environmental consequences. The environmental and economic damage caused by forest fires underscores the importance of early prediction and detection for effective management strategies. While manual patrols remain essential for fire monitoring, deploying advanced technologies, including remote sensing and fire detectors, can substantially enhance the effectiveness and efficiency of these efforts. Researchers worldwide have introduced various technologies and techniques for anticipating and detecting forest fires. Despite these advancements, there is an ongoing need for innovative technologies and integrated management systems to accurately forecast fire likelihood, assess the risk of fire-induced damage, and manage future fire events. This chapter explores a range of innovations and new technologies for forest fire management, covering aspects from prediction and prevention to the restoration of burnt forest areas. It also discusses emerging technologies such as Artificial Intelligence (AI), geospatial technologies, 5G and enhanced connectivity, the Internet of Things (IoT), the Internet of Behaviours (IoB), virtual and augmented reality, and robotics and their applications in forest fire management.

**Keyword** Forest fire management

## 1 Introduction

Forest fire is one of the biggest and most common disasters that take place in many forest systems throughout the world. It is generally caused either naturally through volcanic eruptions, lightning events, and spontaneous ignition or anthropogenically due to arson, negligence, unscientific agricultural practices, and changes in land use.

Y. Thakur (✉)
Department of Environmental Science, Dr YS Parmar University of Horticulture and Forestry, Nauni, Himachal Pradesh, India
e-mail: yashasvi.thakur@gmail.com

 
H. Singh and N. Kumar (eds.), *Forest Fire and Climate Change*,
https://doi.org/10.1007/978-3-031-89967-6_20

However, they have dire ecological, economic, environmental, and social ramifications, such as biodiversity loss, soil degradation, disruption of nutrient and hydrological cycles, alteration in succession and regeneration patterns, and increased ecosystems' vulnerability to invasive species (Savita et al. 2017; Singh and Kumar 2022; Singh 2021). Changing climate, higher temperatures, early melting of winter snow, and extended drought periods have amplified the vegetation flammability, significantly increasing the frequency and intensity of the forest of late (Kaushik and Singh 2024; Kumar et al. 2024). The carbon emissions from forest fires further exacerbate this destructive cycle, accelerating climate change (Singh et al. 2024; Malik et al. 2024; Barwa and Singh 2024).

Early fire prediction and detection and rapid response systems can be instrumental in formulating effective fire management strategies for monitoring and surveillance, enabling timely action (Singh 2024a, b, c, d). For effective detection and tracking, it is crucial to understand the fire behavior, how they ignite, their spread, and environmental interactions in a better way, which has further enhanced the capabilities of the scientific and departmental officials to develop precise strategies for early detection and monitoring.

## 2 Emerging Technologies in Changing Climate

The fast-evolving progressive innovations of interdisciplinary technologies converging to harness the competitive advantage of diverse distinct technologies constitute emerging technologies. Besides significantly transforming industry and human life, technological evolution equips us to tackle the menace of fire events more efficiently. Historically, 'manual patrols' and fire lookout towers built on mountain tops housing 'fire spotters' during the early 1900s, after the Great Fire of the US, became the first organized fire alert network. These spotters were entrusted with observing any fire event or signs in the surroundings and communicating through heliographs. With scientific advancements, better technologies were made available. These watchtowers and human monitors were retired by the twentieth century in many parts of the world (Kumar et al. 2020a, b).

There are various practical implications linked with the traditionally adopted models and methods, which are efficient in terms of cost, reliability, and capabilities, besides being governed through human input and judgment for decision-making. The inefficiencies and limitations can be significantly reduced with emergent technologies and by improving the efficacy of the strategies already involved for suppressing forest fires, assessing and reducing loss in the burnt area, and reducing its ecological and environmental impacts. Characteristically, these emergent technologies are radically novel, fast-evolving, and coherent with a notable impact in addition to being more precise. An account of such technological options and opportunities with a special focus on appropriate fire handling in the forest is presented in this chapter. Satellite imaging, wireless sensor networks (WSNs), fire and smoke detectors, and unmanned remotely operated vehicles are already employed worldwide in the surveillance of forest fires. Many emerging technologies, such as deep and machine learning, robotics, and IoT, are being integrated to address this threat effectively (Alkhatib et al. 2023).

## 3 Geospatial Technologies

Geospatial technologies constitute various tools used for geographic mapping and earth analysis. These technologies can be traced back to prehistoric times when maps were drawn for the first time. They evolved further in the nineteenth century with aerial photography by balloon and pigeon-mounted cameras and then aircraft used during the twentieth century. The era of satellites and computers accelerated these technologies, particularly during the Second World War and then during the Cold War. It was now possible to take images of Earth's surface and monitor human activities to some extent. The geographical information systems (GIS) utilize computers as a repository and transmit the imageries augmented with the development of analytical software, maps, and data sets related to socio-economic, physiographic features, and environmental attributes for processing and interpretation. Using GIS tools, the geospatial data can be converted into an assemblage of layered information on maps wherein a wide range of themes are effectively presented, which otherwise have interdisciplinary complexities. The geospatial data, which refers to the precise location of specific features on Earth's surface, are digitized to prepare feature-specific layers using appropriate software. Arrays of such geospatial technologies are now available that find potential usage to deal with the hazardous fire events in the forest, which include the following:

Remote Sensing: Various public sector agencies and private enterprises at national and international levels are actively engaged in acquiring imageries with a resolution of one meter or more through satellites, aerial photography, sensing platforms, etc., which constitute remote sensing, thus offering opportunities to detect and monitor forest fires more effectively (Gitas et al. 2014). The remote sensing data generated through imageries and photographs acquired from space and airborne platforms and sensors confers precision in detecting fire events, smoke fumes, and destruction caused in the burnt area (Khan et al. 2022).

Geographic Information Systems (GIS): A variety of open source and paid software offers a wide range of applications to map and analyze geo-referenced data concerning particular locations of time-spaced events in relation to their position on the Earth's surface.

Global Positioning System (GPS): A set of satellites synchronizes to precisely transmit the geographic coordinates of locations to a receiver with the user interface.

Internet Mapping Technologies: The visualization and exchange of geospatial data is relatively simpler with user-friendly softwares like Google Earth Engine and

Microsoft Virtual Earth, which has enhanced its utilization manifold for a wide range of masses and the public. However, tech-savvy specialists rely on complex GIS software that requires time to draw interpretations.

Predicting the site of ignition events and their furtherance in the forest area is done by mapping fire risk areas. This helps generate information to anticipate the quantum of damage such an event may cause. The set of variables responsible for spiking any fire incident's probability provides valuable inputs to map fire risk and associated hazards. There are two basic methods to accomplish this:

- Operating systems that are based on point-wise meteorological systems
- Remote sensing technologies and Geographical information systems (RS and GIS).

In the first method, four major operating systems find an application across the world: the Fire Weather Index (FWI), Wild and Fire Assessment System (WFAS), the Russian Nesterov Index, and McArthur's Forest Fire Danger Rating System (FFDRS). These systems use meteorological data comprising temperature, precipitation, wind speed, and humidity to assess fire risk spanned over a large geographic area. However, this method has many shortfalls, as the input data correspond to a few point locations where weather data collection points or observation stations are positioned, and fire risk mapping has to rely on interpolating this point data distributed over a wide area. On the contrary, the RS and GIS approach is more applicable for generating fire risk maps because it can acquire continuous data about a particular area.

The meteorological component does not assess fire hazards but differentiates them from fire dangers. The fire hazard maps are based on quantifying environmental conditions, including factors like topography state of fuel, and considering the human settlements and infrastructural features present in the vicinity to find out the sites posing a risk of fire occurrence. Mapping of fire risk resultant with fire hazard maps with augmentation of variables used to assess vulnerability.

Fire hazard maps are prepared by integrating the variables used to determine vulnerability and fire risk mapping. Some dynamic parameters of weather and vegetation conditions may also be used to prepare fire risk maps. As a result, we get a map where we may find a range of fire risk from low to very high with varying degrees, and such maps can be categorized into two: long-term and short-term. The risk mapping using factors that do not show high temporal variations, like the type of vegetation, human habitations, and topographical conditions, generates long-term (seasonal) maps, whereas short-period (days–weeks) estimates are generated using short-term fire risk maps (Rasel Rahman et al. 2023). Long-term risk maps are not updated so frequently; hence, high temporal resolution is not necessary. The input data required to prepare long-term fire risk maps can be acquired from moderate to high resolution (< 100 m) sensors like those onboard Landsat Series, Sentinel-2, and ASTER (Advanced Space-borne Thermal Emission and Reflection Radiometer). Although < 30 m spatial resolution is considered appropriate for long-term risk maps where inputs like land cover/land use can be classified. Many inputs, such as those of long-term risk maps, apply to short-term risk maps besides other dynamic variables like moisture content in fuel, weather parameters, and vegetation conditions. A higher temporal resolution is required for short-term maps, which can be acquired from the sensors mentioned earlier. However, there is a limitation due to the inability of the user to update the dynamic variables rapidly.

Thus, sensors like MODIS or those mounted on Sentinel-3 have a coarse spatial resolution (> 100 m) and shorter revisit intervals, which are five days for Sentinel-2, thus enabling recording weekly updates at a resolution of 10–20 m. The sensors with moderate to high spatial resolution are used for mapping under static conditions, whereas sensors recording data at a coarser resolution have applicability to update under dynamic conditions daily. Such maps are instrumental for field-level workers in diminishing the eventuality of fire occurrence through controlled fires, thinking mechanically, and working out evacuation measures to be taken by firefighting functionaries.

In recent fire risk assessment studies, remote sensing has been primarily used to evaluate vegetation conditions, categorize land cover, supply elevation data, and verify various proposed risk assessments. As vegetation and land cover data are associated with the characteristics and type of fuel available, they play a pivotal role in assessing the fire risk of a given area, and an accurate and precise land cover classification is necessary to eliminate high-risk areas. The data associated with the vegetation indices, such as the Normalised Difference Moisture Index (NDMI) and Normalised Difference Vegetation Index (NDVI), provide the current status of the potential fuel sources and, hence, for making an effective fire risk model, land cover classification data needs to be supplemented with vegetation conditions. The susceptibility of vegetation to ignition and a spreading event are accurately identified and modeled for fuel loads by these vegetation indices. Since these conditions evolve rapidly, frequent updates are required for dynamic fire risk mapping. Data regarding elevation slope and aspect, or topographic data, highly influence an area's fire ignition and/or spread risk. Data regarding active fires and burn severity are commonly used to test the accuracy of these fire risk assessments, assuming that forest fire events generally occur in these high-risk areas.

Advanced Very High-Resolution Radiometer (AVHRR), Geostationary Operational Environmental Satellite (GOES), Himawari-8, Meteosat Second Generation Spinning Enhanced Visible and Infra-red Imager (MSG-SEVIRI), Moderate Resolution Imaging Spectroradiometer (MODIS), and Visible Infrared Imaging Radiometer Suite (VIIRS) are some of the many remote sensing systems being used these days for active fire detection. Owing to their high temporal resolution and channels specifically designed for the detection of fire and its monitoring, MODIS Terra and Aqua have been used as primary sensors for wildfire management over the last two decades. In recent studies regarding forest fire detection and active fire monitoring, Sentinel-2 and Landsat 8 Operational Land Imager (OLI) sensor data have also been used despite their poor temporal resolution and inappropriate for this purpose. However, due to the increased number of sensors (moderate to high spatial resolution < 100 m), their use for short temporal periods in fire detection and tracking is becoming more practical.

The wide applications of satellite optical imagery come with considerable constraints in the case of wildfire management. It only illustrates and visualizes the canopy and ascertains the total extent of the forest fire spread. A system that can penetrate the canopy or the foliage gaps to capture vegetation underneath and on the forest floor is required. For example, in pine forests in the north-western Himalayas, the fuel for these wildfires is generally the dried pine needles on the forest floor, which not only act as kindling but also the fuel, but satellite imagery is less efficient in its detection and thus inaccurate in fire risk mapping and fuel load modeling in such cases. To resolve this, a remote sensing method, known as LiDAR, stands for Light Detection and Ranging, in which light in the form of a pulsed laser is used to measure variable distances or ranges. Since the LiDAR measures the three-dimensional structure of vegetation and ground elevation, it helps understand the ecosystem attributes pre- and post-fire. Then, these data derivatives are linked to other drivers related to burn severity, fire spread, etc., which are generally not observed using other traditional optical imagery techniques. Due to its effectiveness in approaches like the evaluation of forest fuel loads and fire severity, estimations of base height, and simulation of wildfire, LiDAR has been widely used.

Parameters such as canopy height model, bulk density, canopy cover, etc., can be measured easily using the LiDAR technique. Geometric characterization of various fuel parameters to generate 3D models for simulations of wildfire behaviours and progressions and fuel characterization maps for predicting scenarios of fire and their prevention requirements is also being done using LiDAR technology.

In recent years, drone-mounted LiDAR systems and LiDAR-equipped Unmanned Aerial Vehicles (UAV platforms), flown across survey sites at a low altitude and a slow speed, enabled the acquisition of high-resolution data and in greater detail for the characterization of forest fuels at district or precinct level. The vertical profile of the forest and individual trees can also be surveyed.

## 4 Unmanned System Technology: UAV, UAS, RPAS

Unmanned vehicles or aircraft are emerging technologies that find applications in many fields and are not limited to personnel, businesses, and governments, particularly in the military. Even though remote sensing systems, especially satellite-based optical imagery systems, are finding an increased application in forest fire detection and monitoring, the cost aspect and high-flying altitudes can make identifying young fires extremely difficult. Also, due to hostile environmental conditions during a forest fire, such as high temperature, low visibility, smoke, and sometimes nighttime, piloting an aircraft is highly hazardous, especially at lower altitudes. Therefore, these unmanned, remotely piloted, or autonomous aircraft can be a practical solution in these adverse scenarios.

Unmanned System Technology refers to vehicles and associated equipment that lift the vehicle using aerodynamics, can be piloted autonomously or remotely, and are expendable or recoverable. UAS are commonly referred to as unmanned aerial systems (UAS), unmanned aerial vehicles (UAV), remotely piloted aircraft systems (RPAS), and drones. This Unmanned System Technology was initially developed for military applications as it can easily carry a payload specific to the intended purpose in hostile environments without the involvement or risking of human life. Its payload often includes a specialized sensor or a camera that collects data in near real-time and transmits it for analysis.

The mobility of these UAVs and their accessibility in dangerous environments gives them a unique vantage point over other traditional means of data collection. They gather data in real-time, are highly precise, and allow continuous and rapid-fire movement and progression monitoring.

UAVs were first designed for wildfire fighting purposes and operated within a 1000–4000 ft. range, but with the advancement of technology over the recent year, this range of gas improved to 30,000 ft. To gather forest intelligence, these remotely piloted systems are paired with sensors such as infrared, etc., to identify forest fire hot spots quickly. This can help decision-makers make an informed firefighting plan and response strategy. This unmanned aerial system technology collects data from GPS locations, video feeds, images, sensor node readings, etc. It can monitor vegetation regrowth at temporal intervals (independent of satellite orbits) and high spatial resolutions. Some UAVs can also carry additional devices onboard, such as a water tank for fighting, a chemical fire extinguishing agent or a specialized fire sensor. For example, the IGNIS drone system by Drone Amplified can utilize several such firefighting features onboard. It can combat active wildfires through the release of chemical payloads. It can be actively controlled by preset instructions sent by a mobile app and/or a remote controller.

Based on their size, application, design, weight, autonomy, and flying mechanisms, UAVs can be classified into fixed-wing, rotary-wing, and hybrid fixed/rotary wing. Out of these three, a fixed-wing UAV, capable of vertical takeoff/landing UAV (VTOL), is generally operated at higher altitudes to survey a wider area rapidly. In contrast, rotary-wing UAVs, capable of horizontal takeoff/landing (HTOL-UAVs), generally hover at lower altitudes to collect high-resolution data and are also called observer UAVs.

The structures of fixed-winged UAVs are generally less complicated with more superficial structures, require low maintenance and repair, and therefore can operate for more extended periods carrying larger payloads at higher speeds over longer distances while using less power, as they have natural gliding capabilities with no power requirement in comparison to a rotary-wing UAV. On the other hand, a hybrid drone UAV combines the advantages of both fixed- and rotary-wing UAVs. However, all three are effectively used in different forest fire management opérations.

## 5 Wireless Sensor Networks (WSNs)

A wireless, infrastructure-less network positioned in many wireless sensors in an adhoc manner is a Wireless Sensor Network (WSN). It monitors a system's physical or environmental conditions and controls a specific phenomenon or event. It has many advantages over stand-alone traditional sensors and controllers. Some of them are:

- *Site specificity*: Sensors can be deployed in the vicinity of vulnerable forests.
- *Target specificity*: Customize network nodes to monitor only the variables of interest. It reduces the total number of sensors used and, therefore, the cost of the overall network.
- *High spatial resolution*: By using a multiple number of nodes, the number of controllers and sensors per unit area can be increased.

WSNs consist mainly of nodes, routers, and a gateway. Nodes can be of two types: sensor nodes (SNs) and actuator nodes (ANs). A router is used to extend or increase the communication range. It can also be used to circumvent an obstacle. A gateway device aggregates the information received from nodes and transmits it to the user platform in real-time or near real-time, managing and controlling the network. A microcontroller operates each node, and for long-term operability, it is typically powered by rechargeable batteries and solar energy.

Wireless sensor networks can be used to develop an affordable and real-time technique for early fire detection. To observe and gather physical information and then convert it into electrical signals to transmit it to a remote location for analysis, using the results in various applications, wireless sensor network technologies use many low-cost, small, and densely deployed sensors. The type of sensors, network topology, communication approaches, and data processing techniques diversify WSNs. Physical parameters, including but not limited to temperature, pressure, moisture, and humidity, and chemical parameters, such as carbon dioxide, carbon monoxide, and nitrogen dioxide, are detected by the wireless nodes by using various types of sensors (Sett et al. 2024). Therefore, sensor networks enables quicker fire detection and high accuracy even under smoky and cloudy conditions, larger scanning intervals and higher resolutions over traditional satellite imagery.

Enhanced fire detection capability of wireless sensor networks in early fire detection, along with their deployment in inaccessible and remote places, enables a timely intervention in firefighting operations, as it can be controlled at initial stages before escalating to havoc.

The average distance between sensor nodes and the distribution model for efficient communication between these nodes needs to be carefully considered to enhance its capabilities. For example, choosing a layout out of hexagonal or a square as it impacts the number of nodes present in each cluster. In the case of a hexagonal layout, six nodes are present compared to the four nodes of a square design. Still, it effectively enhances the system's robustness and leads to lesser congestion than the other.

An alternative to wireless networks is using optical fiber to connect various sensors. It is called the fibre optic sensor network (FOSN). It comes with all the features of a traditional wireless sensor network but with additional benefits of fibre optics, such as highly efficient signal transmission and lesser electromagnetic interference. To cover a large geographical area with a broader bandwidth. Lower transmission losses are crucial, and this fibre optics allows the position of these sensors inside the structure so that the monitoring in flammable environs without any power supply can be done effectively.

**Internet of Things (IoT)**

The term IoT, or Internet of Things, refers to a collection of network-connected devices and the technology that facilitates the communication between these devices and the cloud. IoT data transfers over a network do not require human-to-computer or human-to-human interactions. Smart devices that are web-enabled and use embedded system technologies such as sensors, processors and computer hardware for collection, transmission and action on data acquired from the environments constitute an IoT ecosystem. It is one of the best, cheapest, and readily available and scalable forest fire detection systems. To detect forest fires in real-time, IoT sensors can detect fire-related variables such as carbon dioxide, carbon monoxide, temperature, relative humidity, and smoke. A wireless sensor network is then used to send data to the Cloud Server via an Internet Gateway, where after data analysis, fire alerts are instantaneously disseminated to the concerned parties (firefighters, government agencies, local population and, and other stakeholders). A wholesome forest fire management system can be consolidated by integrating an IoT system with unmanned aerial vehicles, drones, global positioning systems, satellites, etc.

IoT device hardware or embedded devices can have a connectivity module with a printed circuit board with sensors for heat, smoke, flame, moisture, etc. embedded in a circuit board with the connectivity modules. Forests can be entirely covered with these sensors since forest fire mainly depends upon the organic matter's temperature, smoke, humidity and dryness or moisture content. And these embedded board clusters accurately detect wildfire. For prediction models the data collected from these sensors in the cloud servers can be trained on machine learning algorithms, and a model using temperature and humidity variables can be built with the help and intervention of artificial intelligence.

## 6 Artificial Intelligence: Machine Learning and Deep Learning

Artificial intelligence uses an algorithm or a set of rules to incorporate machines with human intelligence. AI comprises two words: "Artificial," meaning anything of non-natural origin or something manmade, and "Intelligence," which is the ability to think, comprehend or, understand, and learn. The human mind's capabilities in solving complex problems and making difficult decisions are mimicked by software and computers, which use artificial intelligence to give simple solutions to complex problems. AI allows the machine to play with comprehensive data of a heterogeneous nature by aggregating, extracting, integrating, and then analyzing it to derive solutions. For maximum efficiency, AI focuses on three important skills. These are learning, reasoning and self-correction. The performance of AI gradually improves over time as it learns continuously with each interaction. With the passage of time and the advancement of technology, artificial intelligence techniques such as neural networks, machine learning, predictive analytics, and natural learning processing are being developed.

**Machine Learning**

Machine Learning is a subset of artificial intelligence that develops programs to access data, allowing system learning without human intervention. It is a process when a system or computer, without being programmed learns through its own experiences and makes improvements automatically. Machine learning identifies patterns by making observations on data to make future decisions. Machine learning can be classified into three main categories depending on human involvement. It can either be supervised, unsupervised, or reinforced by machine learning. In supervised machine learning, humans help machines solve problems by identifying and gathering necessary data and feeding it to the system. Here, a machine is fed with specific data to examine, classify and detect a pattern to ascertain whether a prediction is correct. In unsupervised machine learning, machines can spot trends and patterns without human intervention. This type of learning is primarily concerned with categorizing unlabelled, unorganized data. Reinforced machine learning is a type of machine learning where machines are made to solve problems in a confined foreign setting through mistakes and trials and errors. In this learning method, machines find the best answer using the reward and punishment methods. As in the scenario in several games, machines are rewarded for their good trials and punished for committing errors.

It is quite a challenge to develop a complex deterministic algorithm to precisely predict the occurrence of wildfires based on a certain set of features. On the other hand, machine learning can be quite valuable in such cases, as it can directly utilize data to gain insights and extract information using powerful models. Several techniques for automatic fire detection are already employed, such as videos and images and classification models trained with environmental data.

**Deep Learning**

In Deep Learning, neural networks mimic human brain-like behavior. It is a subset of machine learning where the algorithm focuses on the mechanism of information processing patterns to identify patterns like those in the human brain and then classify this information for intended purposes accordingly. Unlike Machine Learning, Deep Learning works on a larger data set, and the machines then self-administer the reduction mechanism.

Deep Learning models are a promising prospect in wildfire detection. Integrating forest fire detection systems allows for better accuracy than traditional computer vision techniques, as it has better semantic segmentation, picture classification, and object detection. Since neural networks in Deep Learning applications are similar to neurons working in our brains and are not programmed instead trained, they can take advantage of the available video and image data without needing fine tunings or expert analysis.

Significant improvements, especially in detecting minor fire incidents, are still required to make the Deep Learning systems more significant in forest fire detection and monitoring systems.

Improvements by utilizing additional prediction models and then integrating them to make a hybridized model can have huge potential in the whole firefighting scenario as they will enhance both the complexity of the system and its accuracy.

## 7 Internet of Behaviours (IoB)

Forest fires are generally manmade. The leading cause of the forest fire is human beings. Although much time, money, effort, energy, and other resources are allocated yearly to create mass awareness regarding forest fires and their ignition, man remains the leading cause.

IoB, or the Internet of Behaviours, is a term coined by Professor Gote Nyman in 2012, and it refers to the interconnected network of digital and physical objects that collect information and exchange it over the internet and links this data to specific human inferred or measured behaviours. It aims at data collection and interpretation from a human sociological and psychological point of view, and it uses it to understand how it influences or changes human behaviour for various interests. As each device has an IP address in IoT, similar behavioural patterns in a network would have an IoB.

To reduce human-induced forest fires, IoB can monitor masses near the forest areas, especially those in high-risk areas, and develop programs to educate and make those directly involved aware.

## 8 5G and Enhanced Connectivity

Modern and futuristic forest fire management shall rely on advanced machinery, precision of sensors, and advanced devices that generate a large amount of data. Such digital data has to be transmitted through a high throughput communication network capable of handling data transfer efficiently. The efficiency of a communication network appropriate for futuristic fire management systems and strategies is primarily determined by its high speed, low lag or latency, large bandwidth, or network capacity, which is readily available and reliable. The latest 5th-generation telecommunication technology, which offers a theoretical peak speed of 20 Gbps, is viable in keeping with the data transfer requirements of a compelling fire management system. The promptness of 5G-based telecommunication technology inherently has features, i.e., low latency and exceptionally high speed for handling large volumes of real-time data acquired from digital devices, advanced sensors, and other machines integrating to constitute a comprehensive fire management system.

A crucial factor is the timely responsive mechanism for controlling the spread of fire and seizing such an event in a conclusive environment where n dry and windy conditions can play havoc. Besides this, fire control without direct human exposure to fire seems to be workable nowadays with unmanned vehicles and robots, which again require a high-performance network like 5G. Fire detection using a 5G network synchronized with computing devices having multiple user interfaces has the advantage of fast transmission, quick analysis, and significantly reduced response time.

Efforts are underway to make fire suppression and control strategies even more efficient using artificial intelligence tools like Pano AI, which utilizes 5G technology to record, deliver, and tactically analyze the real-time imageries acquired by the UAVs.

## 9 Virtual and Augmented Reality

With the advancement of technology, computer graphics allow the operator to interact with the computer-generated imagery in both virtual and even real environmental scenarios. This is known as eXtended Reality (XR), which constitutes Augmented Reality (AR), Virtual Reality (VR), and Mixed Reality (MR).

The fire events in the forest mostly create a highly hostile, unpredictable, dynamic environment and require specialized firefighting vehicles and equipment. The design and specifications of such vehicles are of great concern to researchers, who increasingly employ VR to create simulations to integrate the motion and optimize suspension systems in forest fire vehicles.

Real-time integration of precisely related items visualized from an aircraft position with three three-dimensional (3-D) terrain visualization systems is possible with augmented reality solutions at a low cost. Such simulations make it convenient for the pilot to coordinate the fire control activities effectively and help achieve more effective suppression.

A simulator based on virtual reality can integrate fire dynamics data comprising of and calculate heat, poisonous gases, smoke, and flame, which mocks any fire event or emanating situations, simplifies training the firefighters, and acquires a virtually real experience. Such training simulators provide effective training opportunities for threatening situations while handling fire in forests, such as falling trees, direct flames, ergonomic risks, rough and unknown terrain, etc. The learning acquired from the experience and erroneous practices used in controlled virtual reality conditions allow them to realize the gravity of the situation, make decisions, and act accordingly.

## 10 Robotics

Robotics is designing, manufacturing, and operating intelligent machines to assist humans in various activities. It is a beautiful amalgamation of science (computer science), engineering (mechanical and electrical), and technology. The machines are called robots and can be programmed to undertake various tasks with or without human supervision. These days, with the advent of Artificial Intelligence, integrated robots are finding applications in forest fire management.

During a forest fire, immovable, traditional sensors find a limited application as their safety is a major concern in a harsh environment. To overcome these limitations, movable, intelligent robots can be deployed with fire sensors and also as fire extinguishers and firefighters. Robots' mobility on the ground and in the air gives them an added advantage over the other sensor systems. It can significantly reduce the firefighting casualties and make it a less risky operation.

Robotics finds application in all three phases of wildfire management, viz. prevention. It includes surveillance by aerial robots and clearing of kindlings and fuel on the forest floor (by grinding it into mulch) by ground-based robotics systems; suppression by monitoring and detecting early-stage fires; and restoration of the ecosystem after forest fire (Fernández-Álvarez et al. 2019; Carta et al. 2023).

Robots are developed for forest firefighting to have real-time field applications. Forest-fighting robots equipped with advanced AI-assisted imaging technology are being commercialized by many companies to withstand dangerous smoky conditions with high temperatures and low visibility while effectively managing forest fires. Fire extinguishment is the foremost objective of these firefighting robots. Traditional fire extinction techniques use water, chemicals, foam, and $CO_2$.

In hostile senarios, firefighting robots can assist human firefighters. However, since human intellect, decision-making ability, and quick response to changing conditions surpass that of robots, they are still a less preferred choice. However, in case of a fire hazard, instead of jeopardizing human life, using robots is a much safer option. With the continuous advancement and the advent of new and emerging technology, robots can be promising and highly effective firefighting tools.

## 11 Conclusion

Humans have coexisted with forest ecosystems for millennia, inevitably encountering forest fires that have shaped ecosystems and species. However, human activities have disrupted these delicate relationships, increasing the frequency and intensity of forest fires. Anthropogenic factors, the primary drivers of climate change, have contributed to rising global temperatures, thereby increasing the likelihood and severity of forest fires. Such conditions threaten the entirety of Earth's environment, adversely affecting human life and the broader socio-economic fabric. The vulnerabilities and risks associated with forest fires extend beyond the loss of flora and fauna within the forest ecosystem; they also threaten nearby human settlements, putting lives in danger. In combination with global warming, hot and dry weather conditions significantly enhance the potential for fires, making the development of a dynamic and efficient forest response and management system essential. Early detection of fires is complicated by factors such as fuel/plant types, the spectrum of emissions, color differences, and visual appearance. We found various advancements and emerging tools and technologies that hold promise for innovating and potentially reducing or managing forest fires. The Internet of Behaviors (IoB) plays a crucial role by analyzing the profiles and needs of local populations and formulating prevention strategies tailored to specific demographic setups. While IoT devices offer considerable potential for fire management, they present challenges—especially for engineers—when deployed in remote forests. Similarly, their energy needs limit the use of sensors and devices in inaccessible and challenging forest terrains. Continuous long-term usage requires extending battery life or relying on alternative energy sources, which is critical to their success. Another significant challenge is establishing fast and reliable network connectivity over large areas. No system, tool, or technology will effectively prevent, detect, and manage wildfires. Success depends on developing strategies based on ground realities supported by technological advancements. Emerging technological interventions include remote sensing, GIS, VR, AR, LiDAR, Artificial Intelligence, and 5G networking, equipping firefighters with UAVs and robots and reducing the need for direct human involvement in combating fires. Addressing training needs through VR and AR-generated simulation systems, categorizing forest fuels, modeling forest fires with geospatial tools and remote sensing, placing sensors in remote forest locations for fire detection, synchronizing and transferring data through high-throughput 5G networks, integrating all data inputs with AI interfaces, and employing machine and deep learning for active fire identification and prediction collectively provide a comprehensive suite of functionalities. These technologies form the backbone of future strategies and tactical responses to forest fire events, reducing human lives' exposure to fire dangers.

## References

Alkhatib R, Sahwan W, Alkhatieb A, Schütt B (2023) A brief review of machine learning algorithms in forest fires science. Appl Sci 13:8275. https://doi.org/10.3390/app13148275

Barwa S, Singh H (2024) Climate-induced changes in forest ecosystems of Indian Himalayan region: a meta-analysis. In: Singh H (ed) Urban forests, climate change and environmental pollution. Springer, Cham. https://doi.org/10.1007/978-3-031-67837-0_39

Carta F, Zidda C, Putzu M et al (2023) Advancements in forest fire prevention: a comprehensive survey. Sensors 23:6635. https://doi.org/10.3390/s23146635

Fernández-Álvarez M, Armesto J, Picos J (2019) LiDAR-based wildfire prevention in WUI: the automatic detection, measurement and evaluation of forest fuels. Forests 10:148. https://doi.org/10.3390/f10020148

Gitas IZ, San-Miguel-Ayanz J, Chuvieco E, Camia A (2014) Advances in remote sensing and GIS applications in support of forest fire management. Int J Wildl Fire 23:603. https://doi.org/10.1071/wf14117

Kaushik P, Singh H (2024) Climate change-induced forest ecosystem functioning and alteration in water–atmosphere–soil–plant (WASP) relationship. In: Singh H (ed) Urban forests, climate change and environmental pollution. Springer, Cham. https://doi.org/10.1007/978-3-031-67837-0_38
Khan F, Xu Z, Sun J et al (2022) Recent advances in sensors for fire detection. Sensors 22:3310. https://doi.org/10.3390/s22093310
Kumar A, Dwivedi GK, Tewari S, Paul J, Anand R, Kumar N, Kumar P, Singh H, Kaushal R (2020a) Carbon mineralization and inorganic nitrogen pools under *Terminalia chebula* Retz.-based agroforestry system in Himalayan foothills, India. For Sci 66(5):634–643
Kumar A, Kumar P, Singh H, Kumar N (2020b) Adaptation and mitigation potential of roadside trees with bio-extraction of heavy metals under vehicular emissions and their impact on physiological traits during seasonal regimes. Urb For Urb Green 58:126900
Kumar N, Singh H, Kumar A, Singh AK, Pandey PK, Kumar A (2024) Forest-fire-induced land degradation. In: Mishra G, Giri K, Singh S, Kumar M (eds) Sustainable land management in India. Springer Nature, Singapore. https://doi.org/10.1007/978-981-97-5223-2_4
Malik A, Kumar K, Kumari G, Kumar N, Singh H (2024) Application of functional traits in modelling productivity and resilience under climate change. Springer, pp 77–96. ISBN: 978-981-97-1509-1
Rasel Rahman A, Nabil Sakif SM, Sikder N et al (2023) Unmanned aerial vehicle assisted forest fire detection using deep convolutional neural network. Intell Autom Soft Comput 35:3259–3277. https://doi.org/10.32604/iasc.2023.030142
Savita, Sharma SK, Singh H, Singh O (2017) Forest fire: a review. Indian For 143(8):719–728
Sett T, Nikam BR, Singh H, Purohit S (2024) Hydrological variability in Indian forest ecosystem: analysis of drought resilience, recovery and water use efficiency in moist and dry deciduous forests. In: Singh H (ed) Urban forests, climate change and environmental pollution. Springer, Cham. https://doi.org/10.1007/978-3-031-67837-0_37
Singh H (2021) Process-based carbon sequestration study in relation to the energy-water-carbon flux in a forest ecosystem. In: Kumar P, Singh RK, Kumar M, Rani M, Sharma P (eds) Climate impacts on sustainable natural resource management. CRC Press, Taylor & Francis Group, United States, pp 336–361
Singh H (2024a) Climate change and forests: biological perspectives. In: Singh H (ed) Forests and climate change. Springer Nature, Singapore. https://doi.org/10.1007/978-981-97-3905-9_1
Singh H (ed) (2024b) Forests and climate change: biological perspectives on impact, adaptation, and mitigation strategies. Springer Nature, Singapore. https://doi.org/10.1007/978-981-97-3905-9
Singh H (2024c) Climate change and environmental pollution: biological consequences on urban forests. In: Singh H (ed) Urban forests, climate change and environmental pollution. Springer Nature, Cham, Switzerland. https://doi.org/10.1007/978-3-031-67837-0_1
Singh H (ed) (2024d) Urban forests, climate change and environmental pollution: physio-biochemical and molecular perspectives to enhance urban resilience. Springer Nature, Cham, Switzerland. https://doi.org/10.1007/978-3-031-67837-0
Singh H, Kumar M (2022) Climate change and its impact on Indian Himalayan forests: current status and research needs. In: Rani S, Kumar R (eds) Climate change. Springer, Cham, pp 223–242. https://doi.org/10.1007/978-3-030-92782-0_11
Singh H, Kumar N, Kumar A (2024) Enhancing resource use efficiency in crops through plant functional traits. Springer, pp 97–117. ISBN: 978-981-97-1509-1

# The Road Ahead: Shaping Policies and Strategies for Climate Resilience in India

**Tsenbeni N. Lotha and Latonglila Jamir**

**Abstract** This chapter addresses critical issues and techniques for creating policies and implementing measures to enhance climate resilience. It emphasizes the urgent need to address climate change and build resilience to its impacts. It begins by defining climate resilience and highlighting its importance in the context of ongoing climate change. The chapter then examines current climate change trends, vulnerabilities, and threats, laying the groundwork for understanding the need for action. A comprehensive policy framework for climate resilience is explored, including analyzing national policy landscapes and identifying critical components of successful resilience programs. The chapter discusses various options for increasing climate resilience, such as adaptation, mitigation, and integrated approaches. Case studies and best practices illustrate real-world implementations of climate resilience methods, providing insights into effective strategies and lessons learned. Additionally, the chapter covers the challenges and opportunities associated with implementing resilience measures, addressing obstacles to action while emphasizing the growing potential for progress. It highlights the necessity for collaborative efforts in developing climate resilience policies and strategies, advocating for innovation and immediate action to address the consequences of a changing climate.

**Keywords** Adaptation · Climate change · IPCC · Mitigation · Vulnerability

## 1 Introduction

The global society is looking for innovative ways to handle climate change adaptation and developmental challenges such as food, water, and energy security (Singh 2024a; Barwa and Singh 2024). 'The Future We Want' highlights the significance of incorporating economic, social, and environmental factors into development efforts. Among the most significant trials of our time is finding ways to address global warming and minimize the negative impacts on both social and natural environments (Debnath

T. N. Lotha · L. Jamir (✉)
Department of Environmental Science, Nagaland University Lumami, Lumami, Nagaland, India
e-mail: latongli.jamir@gmail.com



H. Singh and N. Kumar (eds.), *Forest Fire and Climate Change*,
https://doi.org/10.1007/978-3-031-89967-6_21

et al. 2022; Singh 2024b; Kaushik and Singh 2024). The severity of existing hazards and the possibility of extreme weather occurrences is expected to rise. It is also anticipated that new dangers and vulnerabilities will be introduced to affect different areas and socioeconomic groups. This will likely erode the resilience of impoverished and vulnerable communities in many Indian cities. As it begins to impact India's robust economic growth and the well-being of millions, it is poised to become a critical strategic, economic, and political issue. Anticipated climate change will likely affect biological diversity at various scales, from local to global (Sett et al. 2024; Singh 2024c). This impact may manifest in many ways, such as direct effects on species performance, variations, and also the increasing frequency and intensity of abiotic and biotic disturbances driven by changing climate, which may hasten disturbances in ecosystems, thereby compromising the benefits they provide to human well-being (Essl et al. 2012; Singh 2024d).

Adapting to environmental changes is a historical practice seen in both traditional and industrial societies, where they have adjusted to the environments to mitigate risks related to climate variability. This adaptation has been a constant feature throughout human history, and variations in the world's climate and observable impacts on physical systems have been reported. Economies and societies have probably initiated some adaptation measures (Tompkins et al. 2010). India faces vulnerability to various natural hazards, including droughts, floods, and cyclones. Throughout history, the population has developed coping mechanisms to deal with these hazards, i.e., natural and human-induced. The rapid growth of the population, coupled with high population significance, has heightened vulnerability, particularly in India's urban areas, over recent decades (Gupta et al. 2024).

Urban climate action is becoming increasingly significant across multiple disciplines, including development policy, disaster mitigation, and environmental governance. It is anticipated that 68% world's inhabitants will reside in urban regions by 2050 (Ogwu 2019). Increasing worries exist regarding how quickly expanding and financially restricted communities will address climate issues (Dulal 2019). Climate change has an especially major effect on Asia, where urban populations are widely concentrated, will be highly exposed to natural elements, and have limited capacities to cope (Sethi and de Oliveira 2018). In the past decade, cities have played an increasingly important role in global energy, environmental, and sustainability policies. However, the successful execution of resilient city solutions in underdeveloped countries continues to lack the requisite scientific precision and political coherence. Although an initial emphasis in science-policy discussions regarding urban systems was on mitigation strategies (Hunt and Watkiss 2011), recent bibliometric analyses of urban climate change literature indicate a significant increase in attention to adaptation strategies (Sharifi 2021). Local climate plans typically outline adaptation strategies tailored to specific climate hazards, such as floods, within urban areas (Sethi 2018). Resilience and adaptation contribute to just 13.1% of global urban developments (Castán Broto and Bulkeley 2013). However, there is a rising emphasis on the need for a scientifically based policy strategy to identify urban risks better and design suitable remedies (Andersson-Sköld et al. 2015). Thus, this chapter aims to overcome these difficulties by proposing a set of policy frameworks and strategies to enhance climate resilience. The framework would include measures to mobilize resources effectively, streamline governance structures, and enhance community engagement. Additionally, it would explore the part of technology and invention in building climate resilience, highlighting best practices and case studies. Overall, the chapter would provide useful ideas and suggestions for policymakers, and scholars working on climate resilience.

## 2 Understanding the Impacts of Climate Change

The lack of a comprehensive national urbanization policy, as well as worries about the local execution of the antiquated national climate plan, add to the complexity and urgency of this investigation. For instance, in 2008, India presented the National Action Plan on Climate Change (NAPCC) (Sharma and Chauhan 2011), wherein it outlines eight missions aimed at addressing mitigation and adaptation strategies, covering areas like sustainable habitats, water management, the Himalayan ecology, afforestation, and environmentally friendly agriculture. However, when it comes to execution, the method requires an understanding of local climate possibilities and their respective effects. Furthermore, sectoral policies linked to the surroundings, drinking water, hygiene, vitality, and transportation are not specifically targeted to address climate change issues. Amid India's fast urbanization, cities must improve their response and resistance to the changing climate, mainly in critical urban areas like water management, infrastructure development, building construction, urban planning, and public health. However, despite its importance for resilience, energy management is not under municipal jurisdiction but is addressed within specific sectors like infrastructure and buildings. India's diverse climatic features, ranging from harsh winters in the north to tropical temperatures in the south, emphasize the need for specialized resilience tactics across different regions (García Sánchez and Govindarajulu 2023). The northeastern regions of India experience hot and humid conditions, while arid climates characterize the northwestern regions. During India's southwest monsoon season, which runs from June to September, rainfall is a major part of agricultural productivity. However, extreme weather phenomena, including heatwaves, cold waves, flash floods, cyclones, and hailstorms, also have a major impact (Subramanian et al. 2023). The climate also impacts various plant processes, photosynthesis, and phenology, ultimately affecting quality and crop yield. Major crops like rice and wheat are less productive in the Indo-Gangetic Plains (IGP). Maize in the Southern Plateau and mid-IGP is also vulnerable to these climate-related issues, which might lower crop yields in India by an estimated 4.5–9.0% (Pathak 2023). It is predicted to have significant economic impacts on India, with agriculture, which presently subsidizes about 17% of the country's GDP, projected to suffer losses of approximately 1.0% of GDP from the crop sector (Table 1).

**Table 1** A summary of climate change's wide-ranging physical, spatial, and socio-economic effects

| | Variables | Impacts | References |
|---|---|---|---|
| Physical impacts | Increased carbon dioxide ($CO_2$) concentration, crop development and yield | Enhance photosynthetic and water-use efficacy in crops, increasing biological and economic yields. Additionally, it can decrease transpiration and stomatal conductance | Sekhar et al. (2021) and Kumar et al. (2021b, c) |
| | Rise in temperature, crop development and yield | Early maturation of crops and increased respiration rates under elevated temperatures. Deterioration in biomass, grain profits, and grain quality, particularly with increased nighttime temperatures | Kumar et al. (2015, 2016, 2017, 2019, 2021) and Mamrutha et al. (2020) |
| | Increased temperature and soil fertility | Decreased soil organic carbon levels, soil fertility, increased nitrogen mineralization and denitrification | Andry et al. (2019) and Mondal (2019) |
| | Increased temperature and water for irrigation | Water availability for irrigation in the Brahmaputra, Ganges, and Indus River basins is projected to decrease | |
| | Temperature, rainfall, $CO_2$, and agricultural biodiversity | Genetic erosion of crops may increase due to climate change | Göçmez and Aydın (2022) |
| | Temperature, rainfall, $CO_2$, and pest population | Pests may become more severe, leading to decreased crop production and increased losses | Skendžić et al. (2021) |

(continued)

**Table 1** (continued)

| | Variables | Impacts | References |
|---|---|---|---|
| | Increased $CO_2$, temperature, and weed population | Weeds may exhibit increased rhizome growth and higher seed production, attributed to their enhanced photosynthetic efficiency under these conditions | Malhi et al. (2021) |
| Socio-economic impacts | Unemployment, rural migration, and poverty | It could also affect food prices and consumer utility, potentially leading to a decline | Shuaibu (2014) |
| | Farm income | In a scenario where the temperature is 1 °C higher and rain is deficient by 101 mm compared to the kharif season decreases by 4.4 and 13.8%, and in the rabi season by 4.2 and 5.6% | Singh et al. (2018) |
| Spatial impacts | Region | Impacts | |
| | North East | Climate change adversely affects rice yields in Northeast India | Dkhar et al. (2017) |
| | Assam | Forced migration of rural people to urban centers due to food shortages poses risks of land erosion to farmlands | Sarma (2015) |
| | North-West India | The $CO_2$ fertilization effect on rice cannot counterbalance the negative impacts of higher temperatures resulting from the changing climate | Singh et al. (2020) |
| | Central India | The increasing temperatures in central India are having severe effects on wheat productivity | Daloz et al. (2021) |

(continued)

**Table 1** (continued)

| | Variables | Impacts | References |
|---|---|---|---|
| | Indo-Gangetic Plains | Severe weather conditions might have a discernible impact on crop yields | Singh et al. (2019) |
| | Kerala | Rice yield could drop by 7% for every 1 °C increase in temperature | Jan and Ajith (2023) |
| | Punjab | The fluctuations in climate experienced in Punjab over the last couple of years have resulted in reduced yields of rice and wheat | Gaydon et al. (2023) |

## 3 Assessing Trends of Climate Change, Vulnerabilities, and Risks

Within the Indian context, evaluating climate change trends, vulnerabilities, and risks is vital, given the country's varied climate patterns, dense population, and extensive socio-economic diversity. Analysis of historical climate data in India indicates a consistent warming trend, alongside notable fluctuations in rainfall across various regions. Vulnerability is the susceptibility to harm, encompassing risks such as injury, loss of life, livelihood disruptions, and other adverse effects following extreme events. It also refers to the challenges of recovering from such impacts (Wisner 2002). Poverty, rapid urbanization, and inadequate infrastructure in numerous regions compound India's susceptibility to climate change. Forecasts from climate models propose that temperatures in India will continue to climb, resulting in more frequent and intense heat waves. Fluctuations in precipitation are expected to become more pronounced, leading to heightened variability, including more intense rainfall in some areas and prolonged dry periods in others (Kumar et al. 2024).

Furthermore, shifts in precipitation patterns can strain water resources, impacting irrigation, drinking water, and industrial needs. Coastal regions face escalating threats from rising sea levels and more frequent and severe storm surges, endangering infrastructure and livelihoods (Zhuang et al. 2020). India must adopt an integrated strategy that involves sustainable land-use practices, advancement of renewable energy, improved water management, and bolstered disaster preparedness and response capabilities. Engaging stakeholders, including local communities, governmental bodies, and civil society organizations, is crucial for developing and executing successful climate resilience plans. By taking proactive steps to tackle climate change, India can diminish its vulnerabilities, enhance resilience, and secure a sustainable future for its populace.

## 4 Policy Framework and Key Components of Effective Climate Resilience Policies

Policy frameworks for climate resilience include a wide range of tactics and initiatives aimed at reducing vulnerability and increasing the ability to adapt to the effects of climate change. These frameworks typically incorporate a blend of policies, regulations, incentives, and programs targeting different facets of resilience. Gaps in functional mandates, plans, policies, and institutional capabilities in Indian cities are shown by the multi-level governance of the environment and climate in that country. In 2010, the Government of India's Ministry of Agriculture and Farmers' Welfare initiated the National Mission for Sustainable Agriculture (NMSA) to tackle challenges related to sustainable agriculture, develop suitable adaptation strategies to improve livelihoods and food security and help financial constancy. The assignment is centered around ten critical dimensions, including the enhancement of crop seeds, livestock, and fish farming practices, water efficiency, pest control, enhanced agricultural techniques, nutrient management, agronomic assurance, financial provision, market access, and information availability (https://nmsa.dac.gov.in). Introduced in 2011, the National Initiative on Climate Resilient Agriculture (NICRA) is a pioneering endeavor by the Indian Council of Agricultural Research (ICAR). It aims to bolster the flexibility of Indian farming. With focused research and skill demonstrations, NICRA seeks to fortify the agricultural sector, ensuring its sustainability in changing climatic conditions (http://nicra-icar.in). Significant programs like the National Food Security Mission (NFSM) (https://nfsm.gov.in), the Mission for Integrated Development of Horticulture (MIDH) (https://midh.gov.in), National Mission on Oilseed and Oil Palm (NMOOP) (https://nmoop.gov.in), Pradhan Mantri Krishi Sinchayee Yojana (PMKSY) (https://pmksy.gov.in), Rashtriya Krishi Vikas Yojana (RKVY) (https://rkvy.nic.in), and Paramparagat Krishi Vikas Yojana (PKVY) (https://pgsindiancof.gov.in) stress the importance of integrating climate-smart skills into agricultural practices, a specific program focusing on On-Farm Water Management (OFWM) has been implemented on National Mission for Sustainable Agriculture (NMSA). This initiative intends to advance effective water management techniques among farmers, emphasizing the need for sustainable practices in water usage.

Additionally, the National Agroforestry Policy (NAP) was created to support the planting of integrated trees alongside crops and animals. The policy aims to enhance resilience in agriculture, which is critical as changing rainfall and temperature patterns threaten global agricultural production. The term "climate-smart agriculture" describes how agricultural methods are being changed and adapted to meet the problems brought on by climate change while maintaining food supply. Vulnerable communities need to anticipate climatic risks and be prepared to respond effectively during emergencies (Taylor 2018). Bhalerao et al. (2020) in Northeast India agreed on essential skills for each stakeholder group using a modified Delphi approach with multiple iterations. In Northeastern India, strategies that promote climate-resilient and sustainable agriculture are greatly aided by the Krishi Vigyan Kendras. They are uniquely positioned to spearhead such projects because they are the institutional link to farming communities and rural villages. The research is aligned with SDG Goal 13, which emphasizes the need for climate action. Management strategies for climate-resilient agriculture include crop rotation, diversification, preserving and utilizing regional genetic resources, organic management, and modern irrigation methods like drip and sprinkler irrigation to combat water scarcity (Molotoks et al. 2021).

India is currently actively pursuing many mission-driven urban development initiatives, such as the Swachh Bharat Mission (SBM), the Atal Mission for Rejuvenation and Urban Transformation (AMRUT), and the Smart Cities Mission (SCM). These missions include capacity-building components, enhancing city infrastructure functionality, and adopting novel technologies to enhance efficacy and improve urban living standards. They haven't, however, yet included the advancement of scientific capabilities for producing regional climate change estimates, accurately determining the susceptibility and effects of climate change in urban areas, and putting mitigation plans into action (Table 2).

Scaling up climate initiatives and shifting focus from mitigation to resilience action is crucial. Effective policies should include comprehensive risk assessments to identify vulnerable areas and populations and adaptation strategies. Integrating climate considerations into development planning and decision-making is vital to attaining sustainable land use practices, improved water management, and resilient infrastructure. Developing early warning systems and emergency plans is also vital. Governance structures, capacity-building, public awareness, and international cooperation are essential for promoting climate resilience globally.

## 5 Strategies for Enhancing Climate Resilience

Owing to rapid population and industrial expansion, the demand for cereals will likely grasp 476 million tonnes by 2025, nearly double the 2000 figure of 241 million tonnes (FAO 2011). Nonetheless, higher agricultural output must be achieved using the same or potentially even less land, given that rising populations, development, and commercialization drive competing land uses. For example, China, although having just 7% of the world's farmed land area, feeds 23% of the worldwide population (Rasul and Sharma 2016). Economic growth has led to shifts in dietary habits, with a greater preference for meat and other resource-intensive foods. This growth has raised the need for water, energy, and grains for livestock feed (Liu et al. 2007). The issues remain magnified in the Himalayan region, characterized by steep and delicate terrain, where agricultural and grazing activities have expanded to marginal lands. Increased agricultural intensity and land use can quickly degrade land resources and diminish production capacity (Rasul and Sharma 2016). Various strategies have been proposed to conserve biodiversity, such as assisted colonization, utilizing invasive species for ecosystem restoration, expanding and creating protected areas and networks, and enhancing resource efficiency. The core elements of sustainable development are often seen as ensuring access to affordable energy and effectively managing to prevent

**Table 2** Gaps in integrating climate resilience components within the government's urban missions

| Mission | Infrastructure | Gaps |
|---|---|---|
| AMRUT | Supply of water, sewage systems, stormwater drains, parks, recreational facilities, and parking areas | The way that open spaces and transportation infrastructure are provided does not take into account climate change concerns like heat waves, landslides, flooding, and sea level rise. Rebuilding or adapting to expected climate impacts is not included in expanding new infrastructure or renovating existing infrastructure. Furthermore, there are no water harvesting and recycling requirements in the drainage, sewage, or water supply systems |
| Housing for All (Urban) | Infrastructure and public facilities | Currently, no laws are required to protect civic infrastructure and structures from natural disasters brought on by climate change |
| HRIDAY | Parking lots, street infrastructure, solid waste management, availability of clean drinking water, traffic control, and sanitation facilities | Pre-audits for engineering resilience against future climate impacts are absent from newly constructed facilities, making them susceptible to issues brought on by climate change |
| Smart City | Essential services such as water supply, cleanliness, waste management, urban mobility, public transportation, health, education, power, affordable housing, and strong IT and digitalization infrastructure | In developing smart city characteristics, the existing methodology fails to consider urban climate vulnerability, future consequences, and resilience strategies. This gap is critical for designing infrastructure, buildings, and health policies, among other things |
| Swachh Bharat Mission | Urban sanitation infrastructure includes toilets in residences, communal facilities, public places, urinals, and solid waste management | New facilities are developed without undergoing pre-audits to ensure they are engineered to withstand future climate impacts, thus lacking in resilience planning |

*Source* MoHUA (2020)

further biodiversity loss. The interconnectedness of water, energy, and food, along with their relationship to adaptation, is illustrated in Fig. 1. Policymakers must grasp these connections when formulating sustainable adaptation plans.

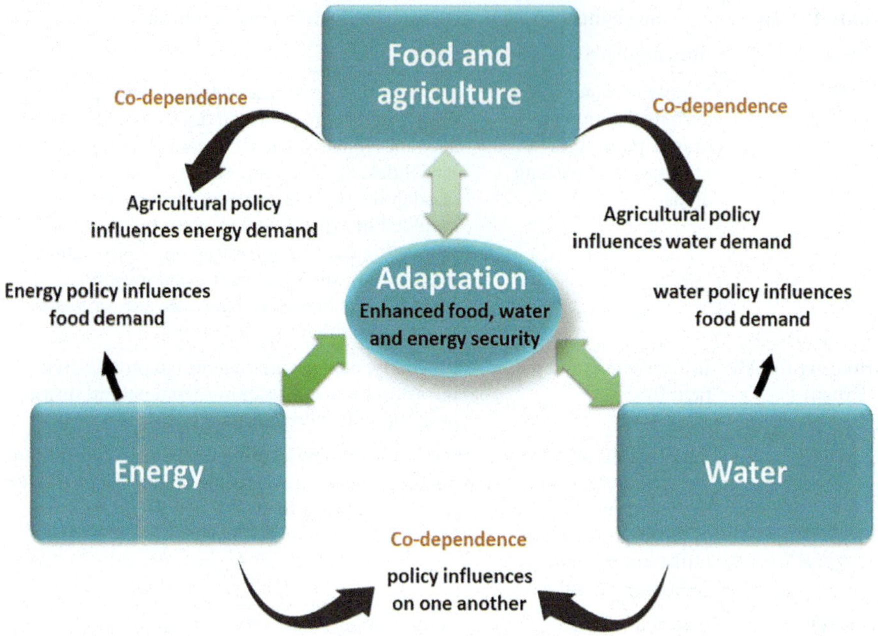


**Fig. 1** Interconnection among water, energy, and food with their relationship to adaptation

## *5.1 Adaptation, Mitigation, and Integrated Strategies to Climate Change*

Crop varieties resistant to various environmental stressors, including drought, heat, flood, salt, pests, and disease, while maintaining high yield potential are crucial for mitigating the adverse impact of climate-related risks on agriculture. This approach can aid in the availability of food and sustainable agriculture in the midst of altering environmental circumstances. Enhancing the efficiency of roots in absorbing water and nutrients from the soil is another crucial aspect. In India, varieties like CR Dhan 801 and 802, tolerant to multiple abiotic stresses, including rice, are being developed and promoted as part of these efforts (Pathak et al. 2018). The successful inter-specific transplanting of crops for flood resistance has been proven, such as grafting plants of tomatoes (cv. Arka Rakshak) on the brinjal plant (cv. Arka Neelkanth) rootstocks (Bahadur et al. 2016). Diversifying crops can mitigate the negative impacts of higher temperatures, reduce risks from unpredictable monsoons, and often lead to higher yields. Extensive portions of the upper Indo-Gangetic Plains (IGP) that are now reliant on rice agriculture that requires substantial groundwater consumption might be converted to fewer water-intensive crops such as maize and millets.

Similarly, the fallow fields following the harvest of rice in the lower IGP, which retain sufficient moisture in the soil, could be diversified with crops such as dragon fruit, offering substantial potential for climate-stressed areas (Nangare et al. 2020).

Globally, India's North East Climate Change Adaptation Program stands out as one of the pioneers in scientifically projecting the potential consequences of changes in the climate and has leveraged these assessments to secure funding for implementing adaptation measures at the grassroots level and aims to improve the adaptive capacity of local communities. For example, climate change has had an unfavorable impact on potato agriculture in Meghalaya, prompting the implementation of farm-level adaption methods. Challenges hindering the implementation of these practices include the limited availability of high-quality seeds, market price fluctuations, insufficient technical knowledge of enhanced potato cultivation techniques, and inadequate transportation and marketing infrastructure. Including traditional farmer adaption practices into scientific research is vital to develop efficient, farmer-centric, affordable, and sustainable adaptation approaches (Kharumnuid et al. 2017).

Rainwater collection provides considerable benefits for adaptation and mitigation by utilizing on-farm water conservation strategies, adopting crop-specific irrigation practices, and applying technologies such as alternating wetting and drying (Pathak et al. 2013). The Jalkund technique, a low-cost method for gathering rainwater, has proven beneficial in conserving water in the dry season (Panwar et al. 2023). Optimal nutrient management can offset the detrimental impacts of climate change on agricultural yields. For instance, decreased yield caused by late rice planting due to delayed monsoons can be offset by increasing Nitrogen (N) application. Thus, proper nutrient management may strengthen the agricultural system by limiting the effects of harsh weather occurrences. Nutrient-smart methods include using nutrient-efficient crop varieties, tailoring fertilizers, and applying highly efficient Nitrogen (N) fertilizers. India has also effectively utilized techniques like as neem-coated urea, soil health cards, and color of leaves charts to increase fertilizer effectiveness (Pathak et al. 2016). Conservation agriculture (CA) includes measures such as zero or low tillage, direct planting, residual cover, and crop rotation, which holds promise for enhancing climate resilience in agriculture (Pathak et al. 2021; Pathak 2023). Mitigating climate change requires collaborative efforts from governments and the public. This can be accomplished by implementing integrated farming techniques and watershed development, which involve using animals, fisheries, and hedge row cropping to conserve soil and moisture while recycling nutrients.

Additionally, promoting technologies that enhance biological nitrogen fixation and nutrient and water usage efficiency can help reduce nitrous oxide ($N_2O$) emissions and dependence on non-renewable energy sources (Grogan et al. 2012). Given the shifting trends in rainfall and rising temperatures in North East India, it is critical to reconsider planting dates and crop kinds. For instance, in Meghalaya's mid-altitude regions, around 950 m above sea level, where previously it was not viable to cultivate a second rice crop after the kharif (monsoon) season due to early winters, double cropping has become feasible, at least in experimental fields, by adjusting planting dates and using suitable varieties (Das et al. 2008).

The IPCC (2014) identified several approaches for lowering emissions of greenhouse gases in the agricultural sector, and the use of improved crop varieties, diversified crop rotations, efficient nutrient management practices using nitrification inhibitors and efficient fertilizers, CA with minimal tillage and residue retention including agroforestry (IPCC 2014). Studies by Somasundaram and co-workers (2020) have demonstrated how CA decreases the carbon footprint of agricultural production systems and enhances flexibility by managing soil temperature and moisture. Pathak and co-workers (Kumar et al. 2020a, b; Pathak et al. 2021) evaluated that adopting various CA practices in India mitigates approximately 25 million tonnes of $CO_2$ equivalent in terms of global warming potential annually. India's agricultural sector emits 408 million tonnes of $CO_2$ equivalent, with 56% of emissions stemming from the fermentation process, 20% via soil sources, and 17% from rice fields. Scholars propose adopting alternating drying and wetting methods to reduce $CH_4$ emissions from rice cultivation (Bhatia et al. 2012). However, occasional flooding may result in a small increase in $N_2O$ emissions, estimated at around 7%. Direct seeding of rice (DSR) is a viable solution to reduce methane emissions. Implementing a complete set of greenhouse gas (GHG) mitigation techniques in rice, such as alternating soaking and drying, producing shorter duration rice varieties, and applying neem-coated urea, could lower methane emissions by almost 60% (Pathak 2015). Enhancing nitrogen (N) use by precisely applying the appropriate type and amount of fertilizer at the correct time, place, and method is crucial for reducing nitrous oxide emissions. Implementing a demand-driven approach for N use, guided by tools like the leaf color chart (LCC), can lead to an 11–14% reduction in nitrous oxide emission (Jain et al. 2014).

## 6 Case Studies and Best Practices for Successful Climate Resilience Initiatives

In the Indian context, numerous case studies and exemplary practices showcase successful climate resilience endeavors. For instance, the Sikkim Organic Mission stands out, with Sikkim attaining 100% organic farming status, reducing reliance on chemical fertilizers and pesticides. This endeavor enhanced soil health and biodiversity. Another illustration of effective climate resilience initiatives is the implementation of Watershed Development Projects in states like Maharashtra and Rajasthan. These projects prioritize water conservation, afforestation, and soil conservation, leading to recharged groundwater levels, reduced soil erosion, and enhanced resilience to droughts and floods. Community-driven adaptation efforts have also demonstrated significant success, such as restoring traditional water harvesting structures like 'Johads' in Rajasthan or 'Kulams' in Tamil Nadu. These structures are pivotal in storing rainwater, replenishing groundwater, and providing irrigation water, particularly during periods of low rainfall.

Azhoni and Kumar (2018) investigated the viewpoints of different stakeholders on the implications of climate change, existing adaptation approaches, knowledge deficiencies, and adaptation obstacles. This was accomplished through semi-structured interviews with Sikkim State Government officials, scientists, specialists, aid groups, and agriculturalists, emphasizing executing the Sikkim State Action Plan on Climate Change. Their study employed stakeholder network Analysis tools to delve into the challenges related to understanding climate change implications, finding suitable approaches, and executing adaptative measures. Farmers may not be well-informed about the implications of climate change on water availability. Still, their familiarity with local circumstances and direct involvement with the State Government Agriculture Department provides opportunities for action. While it has made significant progress, it has yet to actualize successful adaptation implementation completely. Therefore, enhancing the coordination networks among the various implementing agencies is crucial. As highlighted by key stakeholders in this study, the identified knowledge gaps and the necessity for capacity building are pertinent for the research community and policymakers.

Utilizing traditional and genetically modified (GMO) crops, oilseed, sugarcane, the seeds of rap, and the Jatropha for biofuel production could help mitigate the adverse effects of vehicular $CO_2$ emissions (Sarin et al. 2007). Sasmal and co-workers identified *Areca catheu*, *Albizia lucida* and *Ziziphus rugosa* as promising biomass sources for biofuel invention in Northeast India (Sasmal et al. 2012). Consequently, energy-efficient agriculture will involve machinery powered by bio-diesel and bio-ethanol rather than traditional fossil fuels (Jaradat 2018). It's evident that reduced rainfall, increased prevalence of weeds, pests, and diseases due to microbial activity and climate change collectively contribute to decreased agricultural yields (Lin et al. 2008).

Kattumuria et al. (2015) studied the present adaption mechanisms used by rural families in two desert villages in Bagepalli Block, Chikkaballapur district, Karnataka, southern India. They evaluated the effectiveness of these adaptation strategies. It revealed that rural households, especially farmers, have embraced various practices to manage present climate risks. These include irrigation from groundwater sources, transitioning to further resilient but economically less valuable harvests, implementing agroforestry, maintaining varied livestock assets, and being dependent on government efforts. It also meant leaving croplands fallow, selling assets like animals and trees, and resorting to relocation. However, current approaches to climate-related agricultural difficulties are insufficient to address existing threats. This underlines that rural households may struggle to manage escalating climate variability and variation. There's an urgent need for a greater understanding of present adaptation measures and attempts to strengthen resilience and establish systematic plans to manage present-day environmental hazards and long-term global warming.

Kamruzzaman et al. (2023) conducted a study to assess farmers' perceptions of the changing climate in the Rajshahi and Bogra Districts of Bangladesh and compare them to climatic records. The study also looked at how farmers used various adaptation tactics to deal with climate change and its effects, where the Mann–Kendall (MK/mMK) test revealed a declining trend in annual rainfall at most stations, consistent with seasonal trends. Conversely, most stations' yearly and seasonal temperatures increased, aligning with farmers' perceptions. Farmers have employed various strategies to address climate change, including technological advancements, government initiatives, farm management practices, and financial strategies. However, the most prevalent strategy included adjusting planting dates, crop rotation, supplementary irrigation, and variations in fertilizer use.

Due to its coastal location, Gujarat is susceptible to changing climate, including rising sea levels and increased salinity. These factors contribute to uncertainties in regional agricultural production. Pandey (2023) conducted a study as part of the NPCC project and examined climate trends and extremes using records from many locations in Gujarat. Crop simulation algorithms were also tested using experimental information from various crops in Gujarat to see how they respond to climate variability. The findings indicated a general temperature increase over time, although significant temperature rises were mainly observed in specific locations, particularly at night. Rainfall in Gujarat has shown a marginal increase, and future temperature projections indicate varying magnitudes of increase across different regions under RCP4.5 and RCP8.5 scenarios. Rainfall is also forecast to rise, while sunlight hours will also decrease. Despite a rise in $CO_2$ levels, these changes are likely to lead to a significant reduction in yields, suggesting that present crop types could fail to be sufficient to sustain crop output levels in the face of climate change.

Sabar and Midya (2022) investigated the conventional agriculture ways of the Chuktia Bhunjia tribe in Odisha, India, to see how they interact with their surroundings to preserve agricultural productivity and livelihoods. Records obtained via interviews and observations show that the tribe's agricultural techniques are strongly entrenched in the local environment, ideas, practices, and wisdom. Their agroforestry-based practices, belief systems, rituals, and technology contribute to cost-effective agriculture and ecological conservation. The study also highlights how climate change influences agricultural decisions, ensuring sustainable food production and livelihoods.

## 7 Challenges and Future Prospects

As the changing climate poses issues, it is critical to emphasize research and development on new technologies and ways to strengthen resilience. This includes investing in energy efficiency, sustainable agriculture, water conservation, and urban planning. International cooperation and partnerships will also be vital in addressing global climate issues. Countries may collaborate by sharing their expertise, assets, and efficient methods in building more resilient communities and ecosystems. Additionally, efforts to raise awareness and educate the public will be key in fostering a culture of resilience. Adaptive governance structures that enable flexibility and responsiveness to changing climate conditions will be essential. This includes integrating climate considerations into policy-making processes at all levels of government and ensuring that decisions are based on the best available science. Overall, the future of climate resilience depends on our ability to adapt, innovate, and work together to build a more sustainable and resilient world for future generations.

Shaping policies and strategies for climate resilience include many obstacles that must be handled to ensure their efficacy. Among the main issues is the availability of resources, particularly in developing countries and vulnerable communities, to fund and implement resilience initiatives. Additionally, significant data and information gaps related to climate change impacts pose challenges in developing evidence-based policies and strategies. It is challenging to assess risks and develop targeted interventions without comprehensive data accurately. Climate change is a multi-faceted issue with interconnected impacts on ecosystems, economies, and societies. This complexity makes it challenging to develop policies that effectively address all aspects of climate resilience. Implementing climate resilience policies requires strong political commitment, which may be lacking in some regions or among certain stakeholders. Lacking approval from politicians may be challenging to organize finances to execute successful programs. Building institutional capacity is also a challenge. Developing and implementing climate resilience policies requires robust institutional structures and processes, which may be lacking in areas with weak governance systems. Community engagement is essential for the success of resilience policies, but it can be challenging to engage communities effectively. Communities often have diverse interests and priorities, making it difficult to develop policies that meet their needs and build resilience. Finally, technological and infrastructure constraints can hinder the implementation of resilience strategies. Implementing climate resilience often requires new technologies and infrastructure, which may be costly or difficult to deploy in some contexts. Addressing these challenges will require a coordinated and multi-faceted approach involving governments, communities, the private sector, and civil society. It will also require innovative solutions and partnerships to overcome these challenges and build climate resilience effectively.

## 8 Conclusion

Shaping effective policies and strategies for climate resilience is critical amid current and predicted climate change implications. The road ahead requires a holistic approach integrating scientific knowledge, traditional ecological knowledge, and local community participation. Key elements for successful climate resilience policies include robust risk assessment and planning, investment in resilient infrastructure, sustainable natural resource management, community engagement, capacity building, and education. Additionally, early warning systems, financial mechanisms, policy coherence, and monitoring and evaluation are vital components of a comprehensive resilience strategy and essential to distinguish that changing climate affects different regions and communities differently, and solutions must be tailored to local contexts. Collaboration between governments, communities, and academia is paramount to ensure the effectiveness of resilience efforts. Implementing these strategies and policies may direct us to work towards building a more resilient future where communities are better equipped to deal with and cope with the effects of the changing climate.

## References

Andersson-Sköld Y, Thorsson S, Rayner D et al (2015) An integrated method for assessing climate-related risks and adaptation alternatives in urban areas. Clim Risk Manag 7(1):31–50. https://doi.org/10.1016/j.crm.2015.01.003

Andry A, Tiphaine C, Dominique M et al (2019) Land management modifies the temperature sensitivity of soil organic carbon, nitrogen, and phosphorus dynamics in a Ferralsol. Appl Soil Ecol 138(6):112–122

Azhoni A, Kumar M (2018) Diagnosing climate change impacts and identifying adaptation strategies by involving key stakeholder organisations and farmers in Sikkim, India: challenges and opportunities. Sci Total Environ 626(6):468–477. https://doi.org/10.1016/j.scitotenv.2018.01.112

Bahadur A, Rai N, Kumar R et al (2016) Grafting tomato on eggplant as a potential tool to improve waterlogging tolerance in hybrid tomato. Veg Sci 42(2):82–87

Barwa S, Singh H (2024) Climate-induced changes in forest ecosystems of Indian Himalayan region: a meta-analysis. In: Singh H (ed) Urban forests, climate change and environmental pollution. Springer, Cham. https://doi.org/10.1007/978-3-031-67837-0_39

Bhalerao AK, Rasche L, Schneider UA (2020) Preparing for a better future: Delphi forecasts on competency development to enhance climate-resilient farming in Northeastern India. Int J Sustain Dev World Ecol 28(3):255–266. https://doi.org/10.1080/13504509.2020.1765043

Bhatia A, Pathak H, Jain N (2012) Greenhouse gas mitigation in rice–wheat system with leaf color chart-based urea application. Environ Monit Assess 184(5):3095–3107. https://doi.org/10.1007/s10661-011-2174-8

Castán Broto V, Bulkeley H (2013) A survey of urban climate change experiments in 100 cities. Glob Environ Change 23(1):92–102. https://doi.org/10.1016/j.gloenvcha.2012.07.005

Daloz AS, Rydsaa JH, Hodnebrog et al (2021) Direct and indirect impacts of climate change on wheat yield in the Indo-Gangetic plain in India. J Agric Food Res 4(1):100132. https://doi.org/10.1016/j.jafr.2021.100132

Das A, Patel DP, Munda GC et al (2008) Nutrient recycling potential in rice-vegetable cropping sequences under in situ residue management at mid-altitude subtropical Meghalaya. Nutr Cycl Agroecosyst 82(3):251–258. https://doi.org/10.1007/s10705-008-9184-0

Debnath J, Das N, Sahariah D (2022) Impacts of temperature–rainfall and land use/land cover changes on the hydrological regime in the Muhuri River basin, Northeast India. Sustain Water Resour Manag 8(5):1–21. https://doi.org/10.1007/s40899-022-00738-6

Dkhar DK, Feroze SM, Singh R, Ray LIP (2017) Effect of rainfall variability on rice yield in north eastern hills of India: a case study. Agric Res 6(4):341–346. https://doi.org/10.1007/s40003-017-0276-4

Dulal HB (2019) Cities in Asia: how are they adapting to climate change? J Environ Stud Sci 9(1):13–24. https://doi.org/10.1007/s13412-018-0534-1

Essl F, Dullinger S, Moser D et al (2012) Vulnerability of mires under climate change: implications for nature conservation and climate change adaptation. Biodivers Conserv 21(3):655–669. https://doi.org/10.1007/s10531-011-0206-x

FAO (2011) The state of the world's land and water resources for food and agriculture (SOLAW)—managing systems at risk

García Sánchez F, Govindarajulu D (2023) Integrating blue-green infrastructure in urban planning for climate adaptation: lessons from Chennai and Kochi, India. Land Use Policy 124(11):1–10. https://doi.org/10.1016/j.landusepol.2022.106455

Gaydon DS, Khaliq T, Ahmad MUD, Cheema MJM (2023) How will future climates in the Pakistani Punjab rice-wheat system affect the optimal agronomic settings, and can adaptation offset losses? Field Crops Res 302(6):109037. https://doi.org/10.1016/j.fcr.2023.109037

Göçmez S, Aydın GB (2022) Different tillage effects on some soil biochemical properties and soil organic carbon

Grogan P, Lalnunmawia F, Tripathi SK (2012) Shifting cultivation in steeply sloped regions: a review of management options and research priorities for Mizoram state, Northeast India. Agroforest Syst 84(2):163–177. https://doi.org/10.1007/s10457-011-9469-1

Gupta AK, Bhardwaj S, Goyal MK, Gupta A (2024) Disaster risk reduction through climate adaptive development: strategies and road ahead

Hunt A, Watkiss P (2011) Climate change impacts and adaptation in cities: a review of the literature. Clim Change 104(1):13–49. https://doi.org/10.1007/s10584-010-9975-6

IPCC (2014) Summary for policymakers. In: Climate change 2014: impacts, adaptation, and vulnerability. Part A: global and sectoral aspects. Contribution of working group II to the fifth assessment report of the Intergovernmental Panel on Climate Change

Jain N, Dubey R, Dubey DS et al (2014) Mitigation of greenhouse gas emission with system of rice intensification in the Indo-Gangetic Plains. Paddy Water Environ 12(7):355–363. https://doi.org/10.1007/s10333-013-0390-2

Jan PS, Ajith K (2023) Projections of rice yield in the 21st century: a study on the below sea level farming region of Kerala, India. Res J Agric Sci 14(5):1492–1496

Jaradat DMM (2018) Thirteen decades of peptide synthesis: key developments in solid phase peptide synthesis and amide bond formation utilized in peptide ligation. Amino Acids 50(1):39–68. https://doi.org/10.1007/s00726-017-2516-0

Kamruzzaman M, Rahman ATMS, Basak A et al (2023) Assessment and adaptation strategies of climate change through the prism of farmers' perception: a case study. Int J Environ Sci Technol 20(5):5609–5628. https://doi.org/10.1007/s13762-022-04254-0

Kattumuria R, Ravindranatha D, Esteves T (2015) Local adaptation strategies in semi-arid regions: study of two villages in Karnataka, India. Clim Dev 9(1):36–49. https://doi.org/10.1080/17565529.2015.1067179

Kaushik P, Singh H (2024) Climate change-induced forest ecosystem functioning and alteration in water–atmosphere–soil–plant (WASP) relationship. In: Singh H (ed) Urban forests, climate change and environmental pollution. Springer, Cham. https://doi.org/10.1007/978-3-031-67837-0_38

Kharumnuid P, Rao IS, Sudharani V, Kumar S (2017) Farm level adaptation practices of potato growing farmers in East Khasi Hills district of Meghalaya, India. J Environ Biol 39(9):575–580. https://doi.org/10.22438/jeb/39/5/MRN-639

Kumar N, Kumar N, Shukla A, Shankhdhar SC, Shankhdhar D (2015) Impact of terminal heat stress on pollen viability and yield attributes of rice (*Oryza sativa* L.). Cer Res Commun 43(4):616–626

Kumar N, Shankhdhar SC, Shankhdhar D (2016) Impact of elevated temperature on antioxidant activity and membrane stability in different genotypes of rice (*Oryza sativa* L.). Indian J Plant Physiol 21(1):37–43

Kumar N, Suyal DC, Sharma IP, Verma A, Singh H (2017) Elucidating stress proteins in rice (*Oryza sativa* L.) genotype under elevated temperature: a proteomic approach to understand heat stress response. 3 Biotech 7:205

Kumar N, Jeena N, Singh H (2019) Elevated temperature modulates rice pollen structure: a study from foothill Himalayan agro-ecosystem in India. 3Biotech 9:175

Kumar A, Dwivedi GK, Tewari S, Paul J, Anand R, Kumar N, Kumar P, Singh H, Kaushal R (2020a) Carbon mineralization and inorganic nitrogen pools under *Terminalia chebula* Retz.-based agroforestry system in Himalayan foothills, India. For Sci 66(5):634–643

Kumar A, Kumar P, Singh H, Kumar N (2020b) Adaptation and mitigation potential of roadside trees with bio-extraction of heavy metals under vehicular emissions and their impact on physiological traits during seasonal regimes. Urb For Urb Green 58:126900

Kumar N, Jeena N, Kumar A, Khairakpam R, Singh H (2021a) Comparative response of rice cultivars to elevated air temperature in Bhabar region of Indian Himalaya: status on yield attributes. Heliyon 7:e07474

Kumar A, Kumar P, Singh H, Bisht S, Kumar N (2021b) Relationship of physiological plant functional traits with soil carbon stock in temperate forest of Garhwal Himalaya. Curr Sci 120(8):1368–1373

Kumar A, Kumar P, Singh H, Kumar N (2021c) Modulation of plant functional traits under essential plant nutrients during seasonal regime in natural forests of Garhwal Himalayas. Plant Soil 465:197–212

Kumar N, Singh H, Kumar A, Singh AK, Pandey PK, Kumar A (2024) Forest-fire-induced land degradation. In: Mishra G, Giri K, Singh S, Kumar M (eds) Sustainable land management in India. Springer Nature, Singapore. https://doi.org/10.1007/978-981-97-5223-2_4

Lin BB, Perfecto I, Vandermeer J (2008) Synergies between agricultural intensification and climate change could create surprising vulnerabilities for crops. Bioscience 58(9):847–854

Liu J, Dietz T, Carpenter SR et al (2007) Complexity of coupled human and natural systems. Science (80-) 317(10):1513–1516

Malhi GS, Kaur M, Kaushik P (2021) Impact of climate change on agriculture and its mitigation strategies: a review. Sustainability 13(3):1–21. https://doi.org/10.3390/su13031318

Mamrutha HM, Rinki K, Venkatesh K et al (2020) Impact of high night temperature stress on different growth stages of wheat. Plant Physiol Rep 25(4):707–715. https://doi.org/10.1007/s40502-020-00558-w

MoHUA (2020) Schemes/programmes. Ministry of Housing and Urban Affairs, Government of India, New Delhi

Molotoks A, Smith P, Dawson TP (2021) Impacts of land use, population, and climate change on global food security. Food Energy Secur 10(1):1–20. https://doi.org/10.1002/fes3.261

Mondal S (2019) Impact of climate change on soil fertility

Nangare D, Taware P, Singh Y et al (2020) Dragon fruit: a potential crop for abiotic stressed areas. Tech Bull 46(10):24

Ogwu MC (2019) Towards sustainable development in Africa: the challenge of urbanization and climate change adaptation. Geogr Clim Change Adapt Urb Afr 1:1–548. https://doi.org/10.1007/978-3-030-04873-0

Pandey V (2023) Climate variability, trends, projections and their impact on different crops: a case study of Gujarat, India. J Agrometeorol 1665(25):224–238

Panwar P, Machiwal D, Kumari V et al (2023) Sustainable water harvesting for improving food security and livelihoods of smallholders under different climatic conditions of India. Sustainability 15(12):9230

Pathak H (2015) Greenhouse gas emission from indian agriculture: trends, drivers and mitigation strategies. Proc Indian Natl Sci Acad 81(5):1133–1149. https://doi.org/10.16943/ptinsa/2015/v81i5/48333

Pathak H (2023) Impact, adaptation, and mitigation of climate change in Indian agriculture. Environ Monit Assess 195(1):1–22. https://doi.org/10.1007/s10661-022-10537-3

Pathak H, Sankhyan S, Dubey DS et al (2013) Dry direct-seeding of rice for mitigating greenhouse gas emission: field experimentation and simulation. Paddy Water Environ 11(1):593–601. https://doi.org/10.1007/s10333-012-0352-0

Pathak H, Jain N, Bhatia A et al (2016) Improved nitrogen management: a key to climate change adaptation and mitigation. Indian J Fertil 12(11):151–162

Pathak H, Nayak A, Jena M et al (2018) Rice research for enhancing productivity, profitability and climate resilience

Pathak H, Srinivasarao CH, Jat ML (2021) Conservation agriculture for climate change adaptation and mitigation in India. J Agric Phys 21(1):182–196

Rasul G, Sharma B (2016) The nexus approach to water–energy–food security: an option for adaptation to climate change an option for adaptation to climate change. Clim Policy 16(6):682–702. https://doi.org/10.1080/14693062.2015.1029865

Sabar B, Midya DK (2022) Intersecting knowledge with landscape: indigenous agriculture, sustainable food production and response to climate change—a case study of Chuktia Bhunjia tribe of Odisha, India. J Asian Afr Stud 59(1):123–141

Sarin R, Sharma M, Sinharay S, Malhotra RK (2007) Jatropha–palm biodiesel blends: an optimum mix for Asia. Fuel 86(12):1365–1371. https://doi.org/10.1016/j.fuel.2006.11.040

Sarma M (2015) A study of migration from Bangladesh to Assam, India and its impact

Sasmal S, Goud VV, Mohanty K (2012) Characterization of biomasses available in the region of North-East India for production of biofuels. Biomass Bioenergy 45(7):212–220. https://doi.org/10.1016/j.biombioe.2012.06.008

Sekhar KM, Kota VR, Reddy TP et al (2021) Amelioration of plant responses to drought under elevated $CO_2$ by rejuvenating photosynthesis and nitrogen use efficiency: implications for future climate-resilient crops. Photosynth Res 150(1–3):21–40. https://doi.org/10.1007/s11120-020-00772-5

Sethi M (2018) Co-benefits assessment tools and research gaps

Sethi M, de Oliveira JAP (2018) Cities and climate co-benefits

Sett T, Nikam BR, Singh H, Purohit S (2024) Hydrological variability in Indian forest ecosystem: analysis of drought resilience, recovery and water use efficiency in moist and dry deciduous forests. In: Singh H (ed) Urban forests, climate change and environmental pollution. Springer, Cham. https://doi.org/10.1007/978-3-031-67837-0_37

Sharifi A (2021) Co-benefits and synergies between urban climate change mitigation and adaptation measures: a literature review. Sci Total Environ 750(8):141642. https://doi.org/10.1016/j.scitotenv.2020.141642

Sharma SK, Chauhan R (2011) Climate change research initiative: Indian network for climate change assessment. Curr Sci 101(3):308–311

Shuaibu H (2014) Farm households' coping strategies to climate change: a review. Br J Appl Sci Technol 4(20):2864–2877. https://doi.org/10.9734/bjast/2014/4402

Singh H (2024a) Climate change and forests: biological perspectives. In: Singh H (ed) Forests and climate change. Springer Nature, Singapore. https://doi.org/10.1007/978-981-97-3905-9_1

Singh H (ed) (2024b) Forests and climate change: biological perspectives on impact, adaptation, and mitigation strategies. Springer Nature, Singapore. https://doi.org/10.1007/978-981-97-3905-9

Singh H (2024c) Climate change and environmental pollution: biological consequences on urban forests. In: Singh H (ed) Urban forests, climate change and environmental pollution. Springer Nature, Cham, Switzerland. https://doi.org/10.1007/978-3-031-67837-0_1

Singh H (ed) (2024d) Urban forests, climate change and environmental pollution: physio-biochemical and molecular perspectives to enhance urban resilience. Springer Nature, Cham, Switzerland. https://doi.org/10.1007/978-3-031-67837-0

Singh NP, Anand B, Khan MA (2018) Micro-level perception to climate change and adaptation issues: a prelude to mainstreaming climate adaptation into developmental landscape in India. Nat Hazards 92(3):1287–1304. https://doi.org/10.1007/s11069-018-3250-y

Singh NP, Anand B, Singh S, Khan A (2019) Mainstreaming climate adaptation in Indian rural developmental agenda: a micro-macro convergence. Clim Risk Manag 24(7):30–41. https://doi.org/10.1016/j.crm.2019.04.003

Singh Y, Jat HS, Jat SL (2020) Wheat productivity enhancement through climate smart practices

Skendžić S, Zovko M, Živković IP et al (2021) The impact of climate change on agricultural insect pests. Insects 12(5):1–31. https://doi.org/10.3390/insects12050440

Somasundaram J, Sinha NK, Dalal RC et al (2020) No-till farming and conservation agriculture in South Asia—issues, challenges, prospects and benefits. Crit Rev Plant Sci 39(3):236–279

Subramanian A, Nagarajan AM, Vinod S et al (2023) Long-term impacts of climate change on coastal and transitional eco-systems in India: an overview of its current status, future projections, solutions, and policies. RSC Adv 13(18):12204–12228. https://doi.org/10.1039/d2ra07448f

Taylor M (2018) Climate-smart agriculture: what is it good for? J Peasant Stud 45(1):89–107. https://doi.org/10.1080/03066150.2017.1312355

Tompkins EL, Adger WN, Boyd E et al (2010) Observed adaptation to climate change: UK evidence of transition to a well-adapting society. Glob Environ Change 20(4):627–635. https://doi.org/10.1016/j.gloenvcha.2010.05.001

Wisner B (2002) Who? What? Where? When? in an emergency: notes on possible indicators of vulnerability and resilience: by phase of the disaster management cycle and social actor. In: Environment and human security: contributions to a workshop in Bonn, vol 12, no 10, pp 7–12

Zhuang M, Sanganyado E, Xu L et al (2020) High throughput sediment DNA sequencing reveals azo dye degrading bacteria inhabit nearshore sediments. Microorganisms 8(2):1–15. https://doi.org/10.3390/microorganisms8020233

# Evaluation of Forest Fire Susceptibility in Mizoram State Utilizing Analytical Hierarchy Processes and Frequency Ratios

**K. V. Suresh Babu and Rama Chandra Prasad Pillutla**

**Abstract** The forest cover in Mizoram State is essential for the region's socio-economic development and environmental sustainability. This state features abundant natural resources and diverse biodiversity, predominantly harbored within its forests. However, forest fires pose a significant risk, especially during dry seasons. Factors such as human activities, weather conditions, and traditional bamboo and Jhum cultivation practices contribute to the occurrence of forest fires in Mizoram. It's crucial to map forest fire susceptibility to manage this risk effectively. This proactive approach enables better fire management, reduces risks, and facilitates informed decision-making. By identifying areas vulnerable to wildfires, it's possible to promote sustainable land use practices, preserve ecosystems, and safeguard people and property. Statistical techniques such as frequency ratio (FR) and analytic hierarchy process (AHP) are used to generate the fire susceptibility maps for Mizoram based on satellite datasets. The MODIS Terra and Aqua active fire points (MCD14) are the basis, divided into training and testing datasets. AHP and FR techniques establish relationships between the training dataset and fourteen key factors, including slope, aspect, curvature, elevation, Normalized Difference Vegetation Index (NDVI), Normalized Multiband Drought Index (NMDI), rainfall, temperature, wind speed, and proximity to settlements and roads. Various datasets such as MODIS Terra (surface reflectance, land surface temperature, vegetation indices), SRTM Digital Elevation Model, ERA-5, and CHRS are utilized in this study. The accuracy of the susceptibility maps generated by FR and AHP is validated against active fire test datasets, ensuring reliability in predicting fire-prone areas.

**Keywords** Fire susceptibility map · AHP · FR · MODIS · SRTM

K. V. Suresh Babu · R. C. P. Pillutla (✉)
Lab for Spatial Informatics, International Institute of Information Technology, Hyderabad, India
e-mail: rcprasad@iiit.ac.in

 
H. Singh and N. Kumar (eds.), *Forest Fire and Climate Change*,
https://doi.org/10.1007/978-3-031-89967-6_22

# 1 Introduction

Mizoram, nestled in the northeastern region of India, possesses expansive forested landscapes renowned for their significance. These woodlands serve as linchpins for the state's socio-economic progress and are pivotal in upholding its environmental equilibrium (Department of Forests, Mizoram 2020). Within their verdant embrace lie treasures of biodiversity and natural wealth, rendering them indispensable ecological assets (Singh 2024a). Yet, the menace of forest fires, particularly prevalent during arid spells (Darlong 2002; Kumar et al. 2024), looms large, jeopardizing their longevity. The genesis of these fires in Mizoram can be ascribed to a confluence of human actions and natural phenomena. Traditional practices like jhum cultivation, entailing controlled burning of vegetation for land preparation, are significant contributors to these conflagrations (Vanlalruata 2011; Malik et al. 2024). Additionally, haphazard waste disposal practices and unintentional human activities such as carelessly discarded cigarettes or abandoned campfires inadvertently stoke the flames (Singh 2015, 2024b).

Climate conditions also wield significant influence (Singh 2024c). The dry season, spanning typically from February to April, brings forth low humidity, elevated temperatures, and robust winds, fostering ideal circumstances for fire ignition and rapid propagation (Yokomizo et al. 2005). While factors like lightning strikes, spontaneous combustion under specific conditions, and volcanic activity, where applicable, also contribute to fire genesis, their incidences pale compared to human-induced causes (Singh 2024d). Forest fires exact severe ecological and socio-economic tolls. The incineration of vegetation precipitates habitat loss and biodiversity decline (Jain et al. 2005a, b).

Furthermore, fires exacerbate soil erosion and degradation, impairing soil fertility and curtailing ecosystem services (Kumar et al. 2024; Yadav et al. 2017). Moreover, forest fires pose a substantial threat in the context of climate change by releasing copious amounts of carbon dioxide, a potent greenhouse gas, into the atmosphere, thereby intensifying global warming (Yadav et al. 2017; Singh et al. 2024a). These conflagrations imperil lives and property in neighboring settlements and disrupt agricultural activities, resulting in diminished crop yields and adversely affecting the livelihoods of agriculture-dependent communities (Reddy et al. 2006; Sett et al. 2024; Kaushik and Singh 2024).

Precisely delineating the spatial distribution of forest fire susceptibility has emerged as an imperative tool at the national level for proficient fire management (Chuvieco and Congalton 1989). The amalgamation of on-ground data collection, remote sensing (RS) insights, Geographic Information Systems (GIS) methodologies, and statistical analyses empowers researchers to construct dependable spatial forecasts of potential forest fire hazard zones across varied terrains (Keane and Karau 2010; Singh et al. 2024b). Forest fire risk zoning maps, susceptibility maps, and fire danger assessments are pivotal to proactive fire management strategies (Pourtaghi et al. 2016). These tools furnish essential spatial insights for evaluating forest fires' probability, severity, and potential ramifications (Chuvieco et al. 2010; Jain et al.

2005a, b; Yıldız et al. 2023). Typically derived from satellite remote sensing data, these resources offer a comprehensive understanding of diverse factors influencing fire dynamics, encompassing vegetation composition, terrain characteristics, and atmospheric conditions (Jain et al. 2005a, b; Mallick et al. 2021a, b).

### *1.1 Objective of the Study*

The primary objective of this study is to evaluate forest fire susceptibility within Mizoram, India, through a comparative analysis of two established methodologies: the Frequency Ratio (FR) and the Analytic Hierarchy Process (AHP). Although both techniques harness geospatial data to pinpoint regions with elevated fire risk, they operate on distinct underlying principles. Both methodologies will be deployed utilizing an extensive dataset comprising 11 critical factors acknowledged to influence forest fire incidences within the region. Subsequently, the resultant fire susceptibility maps will be juxtaposed against the active fire points sourced from MODIS (Moderate Resolution Imaging Spectroradiometer) and VIIRS (Visible Infrared Imaging Radiometer Suite). This comparative scrutiny will facilitate the assessment of each methodology's efficacy in accurately delineating forest fire susceptibility within the unique context of Mizoram.

## 2 Study Area

Located in the northeastern stretch of India, Mizoram stands out for its varied natural panoramas and vibrant cultural legacy (Fig. 1). The study zone, Mizoram, sprawls across 21,081 km$^2$, between 21° 56′ to 24° 35′ North latitude and 92° 15′ to 93° 26′ East longitude (DoF Mizoram 2020), bordered by Myanmar to the east and south, and neighboured by the states of Tripura, Assam, and Manipur to the north. Characterized by undulating hills and mountains, Mizoram's terrain forms a segment of the eastern extension of the Himalayan range. Its enchanting vista displays steep inclines, deep ravines, and a plethora of rivers and cascades, amplifying its scenic allure (Barwa and Singh 2024). Towering over all is the regal Phawngpui Tlang, also renowned as the Blue Mountain, rising majestically to a height of 2210 m (DoF Mizoram 2020).

A considerable stretch of Mizoram is cloaked in dense forests, constituting a cornerstone in the state's ecological integrity and economic framework. These forested realms encompass a blend of tropical moist deciduous, subtropical broadleaf, bamboo, and evergreen species, fostering a haven for a rich diversity of flora and fauna, inclusive of numerous indigenous species (Department of Environment and Forests, Government of Mizoram 2010). Bamboo resources thrive abundantly in Mizoram and are integral to its forest ecosystems. Mizoram experiences a humid subtropical climate, characterized by temperate to warm summers with temperatures ranging between 20 and 30 °C. The monsoon season, from June to September, brings

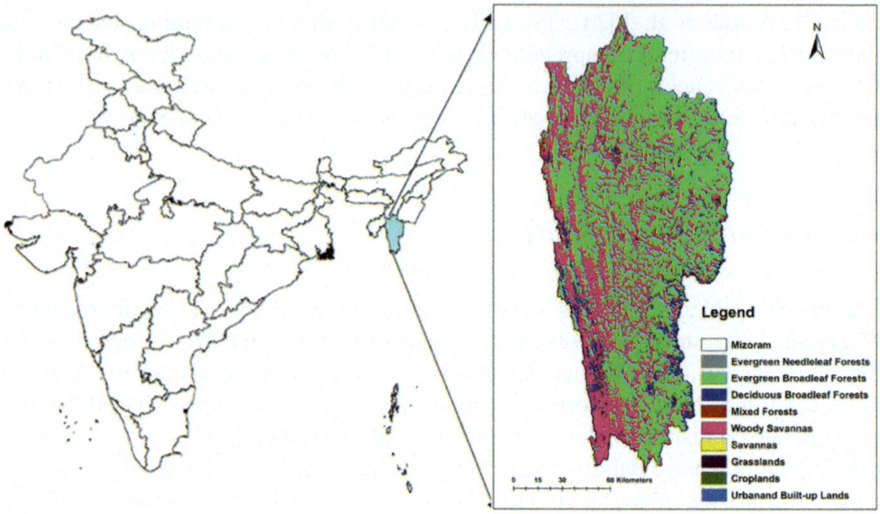


**Fig. 1** Location of the study area (MODIS IGBP land cover map)

heavy rainfall and high humidity, sustaining its lush forests. Winters are marked by a cool and arid ambiance, with temperatures typically fluctuating between 10 and 20 °C.

## 3 Materials and Methods

The dataset encompasses various components, notably the MODIS Terra Land Surface Reflectance data, which furnishes insights into land surface reflectance captured by the MODIS instrument aboard the Terra satellite. This dataset serves as the foundation for computing the Normalized Difference Moisture Index (NDMI), offering estimations of vegetation moisture content are downloaded from the Earth data website (Earth data). Complementing this is the MODIS Terra Vegetation Indices dataset, offering pre-calculated vegetation indices derived from land surface reflectance data (LPDAAC). Among these, the Normalized Difference Vegetation Index (NDVI) stands out, gauging vegetation greenness and photosynthetic activity. Augmenting these datasets is the SRTM DEM dataset, furnishing detailed topographic information, including elevation, slope, aspect, and curvature data concerning the Earth's terrain (Earth Explorer). Ancillary data supplements these primary datasets, featuring details on proximity to roads and settlements derived from spatial data on road networks and human settlements. The ERA-5 Climate Data dataset also provides atmospheric information such as temperature, wind speed, and precipitation. Table 1 shows the datasets utilized in this study. These calculated parameters will be integrated with additional datasets to formulate fire susceptibility maps employing both the Frequency Ratio and Analytic Hierarchy Process methodologies within the study. The entire method is presented in Fig. 2.

Forest fire susceptibility in Mizoram, India, is shaped by various environmental elements that interact to determine the probability of fire occurrence. Among these factors, elevation emerges as a critical influencer of fire risk (Pourtaghi et al. 2016).

**Table 1** Datasets used in the study

| S. No. | Datasets | Parameters |
|---|---|---|
| 1 | MODIS Terra Land surface reflectance (MOD0A1) | NMDI |
| 2 | MODIS Terra vegetation indices (MOD13A2) | NDVI |
| 3 | SRTM DEM | Elevation, slope, aspect, curvature |
| 4 | Ancillary data | Distance from roads, settlements |
| 5 | ERA-5 climate data | Temperature, wind speed, rainfall |

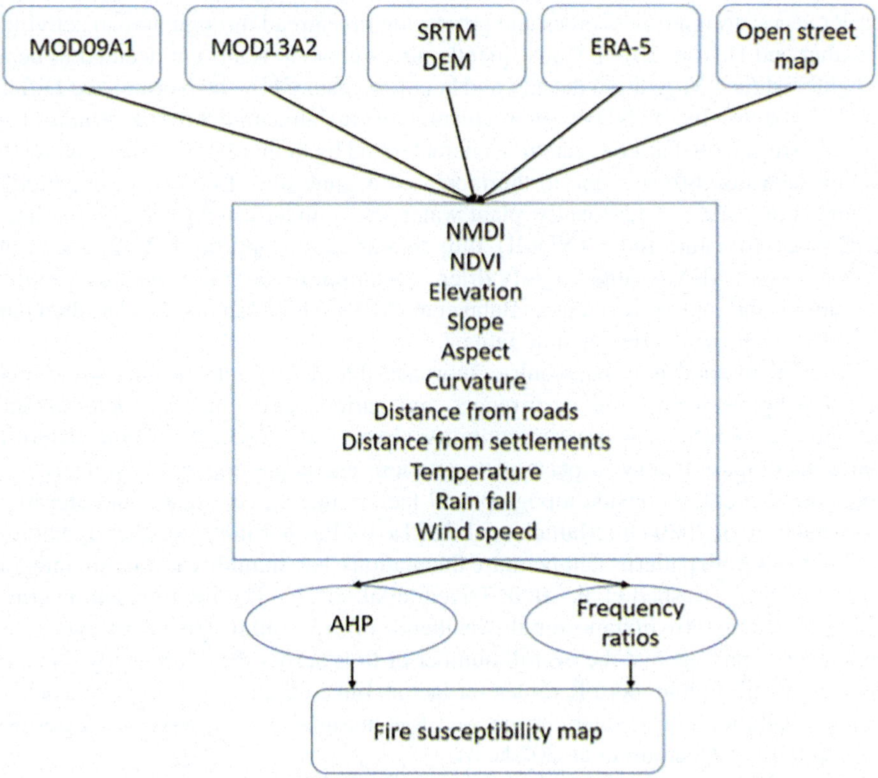


**Fig. 2** Schematic diagram shows the methodology

Elevated regions typically exhibit cooler temperatures, higher humidity levels, and reduced wind velocities (Chuvieco and Congalton 1989), creating less conducive fire ignition and spread conditions. Additionally, the curvature of terrain, delineating slope shapes such as concave, flat, or convex, plays a crucial role in fire propagation dynamics (Adab et al. 2013). Concave slopes, characterized by moisture retention, impede fire spread (Babu et al. 2016), whereas convex slopes, conducive to drier conditions, facilitate faster fire propagation (Yokomizo et al. 2005). To account for this curvature effect, the study area's slopes are scrutinized and categorized accordingly. Furthermore, denoting the direction a slope faces impacts fire risk by influencing temperature and moisture levels (Mitri et al. 2017). South-facing slopes, receiving more sunlight, typically experience higher temperatures and potentially drier conditions, increasing fire risk compared to north-facing slopes (Yadav et al. 2017).

Human activities significantly contribute to forest fires, especially in areas proximate to settlements, intensifying fire risk (Zote et al. 2019). Common practices such as cooking, improper disposal of cigarettes, and deliberate burning are more prevalent in close vicinity to settlements, amplifying the likelihood of fire outbreaks (Vanlalruata 2011). Moreover, road networks can exacerbate fire risk by facilitating easier access to ignition sources and promoting fire spread through human activities (Akther and Hassan 2011). Consequently, areas close to roads are deemed to be at heightened risk. Vegetation density and health, as gauged by the Normalized Difference Vegetation Index (NDVI), play a pivotal role in delineating fire risk (Zhang et al. 2019). Areas with denser vegetation, characterized by higher NDVI values, typically exhibit elevated fire risk due to the increased availability of combustible material (Babu et al. 2016). Additionally, plant water stress, manifested by the Normalized Difference Moisture Index (NDMI), directly correlates with fire risk (Zhang et al. 2019). Lower NDMI values signify drier vegetation, rendering it more susceptible to ignition and rapid propagation (Babu et al. 2019). These factors are classified into distinct categories, as depicted in Table 2.

The FR method is a data-centric approach that delves into the statistical relationship between past fire occurrences and various environmental determinants recognized to influence fire susceptibility (Jain et al. 2005a, b). These determinants encompass diverse aspects of vegetation, encompassing type, biomass, and moisture content, alongside topographical factors like slope, aspect, and elevation (Khanduri et al. 2020a, b; Mallick et al. 2021a, b). Furthermore, weather dynamics, including rainfall patterns, temperature fluctuations, and humidity levels, are integral to the analysis (Rashed et al. 2020; Trivedi et al. 2021). The FR for each determinant is calculated by juxtaposing the incidence of fires within specific categories of that determinant against the overall number of fires across the entire study domain. Regions exhibiting higher FR values for factors linked to increased fire risk, such as dense vegetation cover, steep terrain, and low humidity levels, possess greater fire susceptibility (Khanduri et al. 2020a, b).

The AHP is a multi-criteria decision-making (MCDM) technique, drawing upon expert knowledge and judgment to quantify fire susceptibility (Saaty 1988). It follows a structured framework involving several fundamental steps. Initially, a hierarchical

**Table 2** Categorization of factors

| S. No. | Factors | Class |
|---|---|---|
| 1 | Slope (°) | < 15<br>15–25<br>25–35<br>35–45<br>> 45 |
| 2 | Elevation (m) | < 400<br>400–800<br>800–1200<br>1200–1600<br>> 1600 |
| 3 | Aspect | Flat (− 1)<br>North (0–22.5)<br>Northeast (22.5–67.5)<br>East (67.5–112.5)<br>Southeast (112.5–157.5)<br>South (157.5–202.5)<br>Southwest (202.5–247.5)<br>West (247.5–292.5)<br>Northwest (292.5–337.5)<br>North (337.5–360) |
| 4 | Curvature | Concave<br>Flat<br>Convex |
| 5 | NDVI | < 0.1<br>0.1–0.3<br>0.3–0.6<br>0.6–0.7<br>> 0.7 |
| 6 | NDMI | < 0.05<br>0.05–0.1<br>0.1–0.2<br>> 0.2 |
| 7 | Distance from settlements (m) | < 100<br>100–200<br>200–300<br>300–400<br>> 400 |
| 8 | Distance from roads (m) | < 100<br>100–200<br>200–300<br>300–400<br>> 400 |
| 9 | Rainfall (mm) | < 1200<br>1200–1500<br>1500–1800<br>1800–2100<br>> 2100 |

(continued)

**Table 2** (continued)

| S. No. | Factors | Class |
|---|---|---|
| 10 | Temperature (°C) | < 19<br>19–21<br>21–22<br>22–23<br>> 24 |
| 11 | Wind speed | < 3.5<br>3.5–4<br>4–4.5<br>4.5–5<br>> 5 |

structure is delineated, with the uppermost tier representing the overarching objective of assessing fire risk. Subsequent tiers hold pertinent criteria impacting fire susceptibility, such as topography, vegetation, and meteorological conditions, further elaborated by sub-criteria such as slope, aspect, and vegetation type. Experts then compare each criterion and sub-criterion pairwise, assigning values reflecting their relative significance in influencing fire risk. Eigenvectors and weights are subsequently computed based on these comparisons, delineating the relative importance of each criterion and sub-criterion. Individual spatial models are crafted for each weighted factor utilizing appropriate analytical methodologies like logistic regression or random forest. These weighted spatial models are then amalgamated to formulate a composite fire susceptibility map, comprehensively depicting fire risk distribution across the landscape (Khanduri et al. 2020a, b). Employing FR and AHP methodologies and juxtaposing the resultant fire susceptibility maps with extant fire occurrence data in Mizoram can assess the efficacy of each technique in capturing spatial fire risk patterns. This comparative analysis promises valuable insights into each method's strengths, limitations, and applicability in informing fire management strategies within the region.

## 4 Results and Discussion

The fire susceptibility map is generated by utilizing the FR method, which involves analyzing historical fire occurrence data alongside diverse environmental variables to delineate regions with higher or lower susceptibility to fires. The map shown in Fig. 3 is segmented into five distinct danger classes: very low, low, moderate, high, and very high, reflecting varying degrees of susceptibility. To measure the map's accuracy, active fire points sourced from satellites such as MODIS Terra and Aqua and SNPP VIIRS (FIRMS website) are overlaid onto the map. The precision of the map can be assessed by comparing the identified fire points with the corresponding fire danger classes depicted on the map.

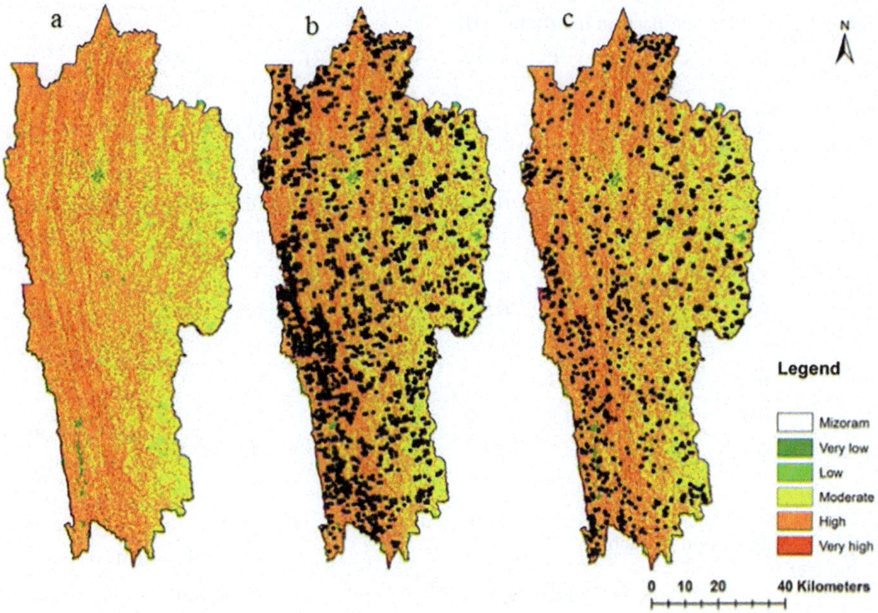


**Fig. 3** **a** Fire susceptibility map generated from frequency ratio method, **b** overlaid with MODIS, **c** VIIRS fire points

The Analytic Hierarchy Process (AHP) method assesses the relative significance of factors influencing forest fire danger. This method systematically compares each pair of factors, such as temperature and wind speed, to ascertain their relative contributions to the likelihood of forest fires. Expert judgment is employed to determine which factor holds greater importance in influencing fire risk within each pair. The outcomes of these comparisons are subsequently transformed into final weights, representing each factor's relative importance in the overall assessment. These weights serve as a quantitative measure of the influence and significance of each factor in determining forest fire risk. This approach facilitates informed decision-making and prioritization of resources in fire management and prevention efforts by attributing appropriate weights to the factors driving forest fire occurrences. The pair-wise comparison matrix is presented in Table 3, while Table 4 displays the weights assigned to each parameter. The fire susceptibility map presented in Fig. 4 is generated using the AHP method.

The studies conducted by Akther and Hassan (2011), Adab et al. (2013), Babu et al. (2016, 2019), and Mitri et al. (2017) employed accuracy calculations based on the frequency of fire hotspots across distinct categories within fire danger maps. These studies operated under the assumption that fires occurring within the very low to moderate danger classes were unanticipated, while those beyond these classes were considered as predicted fires. The enumeration of fire points (from MODIS and VIIRS) within each fire danger class for the FR and AHP methods is tabulated

**Table 3** Pair-wise comparison matrix by AHP

| Criteria | E | S | A | C | NDVI | NMDI | Rf | T | WS | DS | DR |
|---|---|---|---|---|---|---|---|---|---|---|---|
| E | 1 | 3 | 2 | 4 | 2 | 1 | 2 | 1 | 2 | 3 | 2 |
| S | 0.33 | 1 | 0.5 | 3 | 0.5 | 0.33 | 0.5 | 0.33 | 0.5 | 2 | 1 |
| A | 2 | 2 | 1 | 2 | 1 | 0.5 | 1 | 0.5 | 1 | 2 | 1 |
| C | 0.25 | 0.33 | 0.5 | 1 | 0.33 | 0.25 | 0.5 | 0.25 | 0.33 | 0.5 | 0.33 |
| NDVI | 2 | 2 | 1 | 3 | 1 | 0.5 | 2 | 1 | 2 | 3 | 2 |
| NMDI | 1 | 3 | 2 | 4 | 2 | 1 | 2 | 1 | 2 | 3 | 2 |
| RF | 0.5 | 2 | 1 | 2 | 0.5 | 0.5 | 1 | 0.5 | 0.5 | 1 | 0.5 |
| T | 1 | 3 | 2 | 4 | 1 | 1 | 2 | 1 | 2 | 3 | 2 |
| WS | 0.5 | 2 | 1 | 3 | 0.5 | 0.5 | 2 | 0.5 | 1 | 2 | 1 |
| D S | 0.33 | 0.5 | 0.5 | 2 | 0.33 | 0.33 | 1 | 0.5 | 0.5 | 1 | 0.5 |
| D R | 0.5 | 1 | 1 | 3 | 0.5 | 0.5 | 2 | 0.5 | 1 | 2 | 1 |

**Table 4** AHP weight of each parameter

| Criteria | AHP weight |
|---|---|
| Elevation | 0.149 |
| Slope | 0.053 |
| Aspect | 0.089 |
| Curvature | 0.03 |
| NDVI | 0.124 |
| NMDI | 0.149 |
| Rainfall | 0.061 |
| Temperature | 0.14 |
| Wind speed | 0.082 |
| Distance to settlements | 0.045 |
| Distance to roads | 0.077 |

in Table 5, with accuracy determined utilizing Eq. 1 as per Babu et al. (2016). This methodology facilitated an evaluation of the efficacy of fire danger maps in forecasting fire occurrences.

$$Accuracy = \frac{Total\ number\ of\ fire\ incidences\ in\ High\ and\ very\ high\ danger\ classes}{Total\ number\ of\ fire\ incidents\ in\ all\ classes} * 100 \tag{1}$$

The accuracy of both the FR and AHP methods, utilizing fire points from MODIS and VIIRS, is determined by employing Eq. 1. This equation computes the ratio of correctly predicted fire points to the total number of fire points. The accuracy for each method is calculated independently based on fire points obtained from both MODIS and VIIRS data sources, offering a quantitative assessment of the reliability and

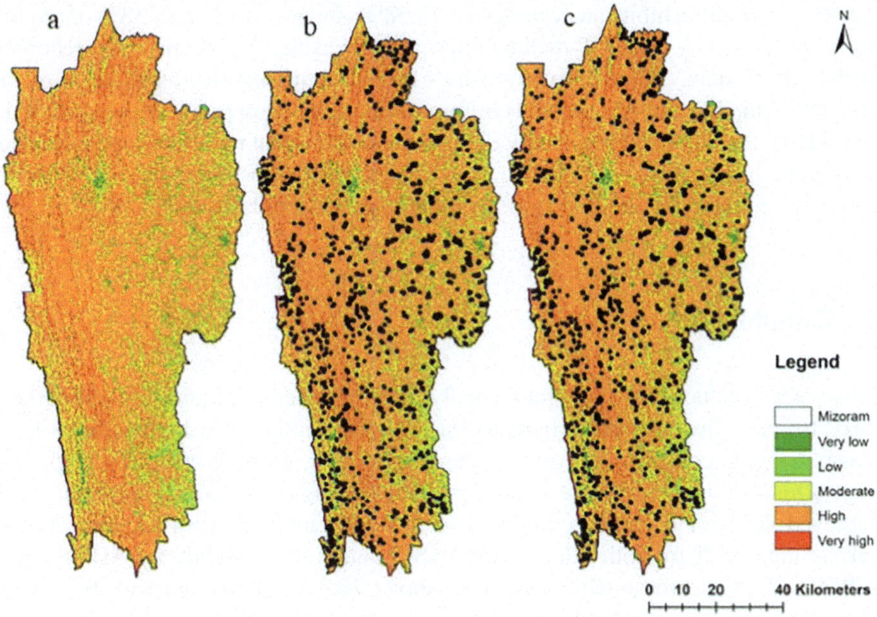


**Fig. 4** **a** Fire susceptibility map generated from AHP method, **b** overlaid with MODIS, **c** VIIRS fire points

**Table 5** Number of fire points in each fire danger class

| Fire danger class | Number of fire points | | | |
|---|---|---|---|---|
| | Frequency ratio | | AHP | |
| | MODIS | VIIRS | MODIS | VIIRS |
| Very low | 9 | 10 | 9 | 10 |
| Low | 62 | 49 | 121 | 104 |
| Moderate | 288 | 284 | 320 | 306 |
| High | 2094 | 1564 | 2135 | 1454 |
| Very high | 241 | 54 | 109 | 87 |

effectiveness of these methods in forecasting fire occurrences. For MODIS data, the FR method demonstrates an accuracy of 86.67%, signifying that 86.67% of the fire points predicted by this method correspond with the actual fire occurrences observed in the MODIS data. Similarly, for VIIRS data, the FR method exhibits an accuracy of 82.51%, indicating that 82.51% of the fire points predicted align with the actual fire occurrences recorded in the VIIRS data.

For MODIS data, the AHP method demonstrates an accuracy of 83.30%. This indicates that 83.30% of the fire points predicted by the AHP method coincide with the actual fire occurrences observed in the MODIS data. Similarly, for VIIRS data, the AHP method exhibits an accuracy of 78.58%, suggesting that 78.58% of the fire points predicted by the AHP method align with the actual fire occurrences recorded in the VIIRS data. These accuracy values furnish a quantitative assessment of the efficacy of the FR and AH methods in forecasting fire occurrences utilizing MODIS and VIIRS data. A higher accuracy denotes more excellent reliability and effectiveness in predicting fire points. Hence, the FR method demonstrates superior accuracy compared to the AHP method in generating a fire susceptibility map.

# 5 Conclusion

This study conducted a comparative analysis of two methodologies, the FR and the AHP, to create fire susceptibility maps for Mizoram, India. Both approaches utilized geospatial data concerning diverse factors influencing fire risk, such as topography, vegetation, and meteorological conditions. The findings revealed that the FR method achieved superior accuracy compared to the AHP method, with accuracy values surpassing 82.5% for both MODIS and VIIRS data sources. While the AHP method still demonstrated reasonable accuracy (above 78.5%), it fell short of the performance achieved by the FR method. This emphasizes that the FR method is more effective in generating fire susceptibility maps within the studied region. This superiority can be attributed to the data-centric nature of the FR method, which directly incorporates historical fire occurrence data to establish correlations between fire events and environmental variables. On the other hand, the AHP method, despite incorporating expert knowledge, may be vulnerable to subjective interpretations and may not fully capture the intricate relationships between various factors and fire risk. Future research avenues could explore the performance of these methodologies in diverse geographical contexts characterized by varying fire regimes and environmental attributes. Additionally, investigating the integration of advanced machine learning algorithms with these methodologies can further enhance their accuracy and predictive capabilities. While this study focused on Mizoram, India, the methodologies and findings have broader implications. The FR method's data-driven approach can be potentially applied to other regions with sufficient historical fire data and readily available geospatial information. However, the specific factors influencing fire risk and their relative importance may vary significantly across different landscapes. Therefore, careful consideration of the unique environmental and socio-economic conditions of each region is crucial for effective application of these methodologies.

**Acknowledgements** The authors acknowledge the MODIS Science team for the Science Algorithms, the Processing Team for producing MODIS data, and the GES DAAC MODIS Data Support Team for making MODIS data available to the user community, the NASA Earthdata team, and FIRMS websites for free MODIS TERRA datasets and active fire data.

## References

Adab H, Kanniah KD, Solaimani K (2013) Modeling forest fire risk in the northeast of Iran using remote sensing and GIS techniques. Nat Hazards 65(3):1723–1743

Akther MS, Hassan QK (2011) Remote sensing-based assessment of fire danger conditions over boreal forest. IEEE J Sel Top Appl Earth Obs Remote Sens 4(4):992–999

Babu KVS, Roy A, Prasad PR (2016) Forest fire risk modeling in Uttarakhand Himalaya using TERRA satellite datasets. Eur J Remote Sens 49(1):381–395

Babu KVS, Kabdulova G, Kabzhanovam G (2019) Developing the forest fire danger index for the country Kazakhstan by using geospatial techniques. J Environ Inform Lett 1:48–59

Barwa S, Singh H (2024) Climate-induced changes in forest ecosystems of Indian Himalayan region: a meta-analysis. In: Singh H (ed) Urban forests, climate change and environmental pollution. Springer, Cham. https://doi.org/10.1007/978-3-031-67837-0_39

Chuvieco E, Congalton RG (1989) Application of remote sensing and geographic information systems to forest fire hazard mapping. Remote Sens Environ 29(2):147–159

Chuvieco E, Aguado I, Yebra M, Nieto H, Salas J, Martín MP et al (2010) Development of a framework for fire risk assessment using remote sensing and geographic information system technologies. Ecol Model 221(1):46–58

Darlong V (2002) Forest fire disasters in Mizoram. J Ind Soc Remote Sens 30(3):185–189

Department of Environment and Forests, Government of Mizoram (2010) State of environment report—Mizoram. Accessed 2 Feb 2024. https://forest.mizoram.gov.in/

Department of Forests, Government of Mizoram (2020) Forest fire management in Mizoram. Accessed 29 Feb 2024. https://forest.mizoram.gov.in/post/forest-fire-management-in-mizoram

Earth data. Accessed 2 Jan 2024. https://search.earthdata.nasa.gov/search

FIRMS website. Accessed 10 Jan 2024. https://firms.modaps.eosdis.nasa.gov/download/

Jain AK, Murthy MSR, Nagaraju V (2005a) Assessment of forest fire risk using remote sensing and GIS. Int J Remote Sens 26(11):2217–2232

Jain TB, Goldstein G, Asner GP (2005b) Forest fire response to early logging and slash-and-burn agriculture in a tropical deciduous forest of western Lesser Himalaya, India. For Ecol Manag 213(1–3):342–354

Kaushik P, Singh H (2024) Climate change-induced forest ecosystem functioning and alteration in water–atmosphere–soil–plant (WASP) relationship. In: Singh H (ed) Urban forests, climate change and environmental pollution. Springer, Cham. https://doi.org/10.1007/978-3-031-67837-0_38

Keane RE, Karau E (2010) Evaluating the ecological benefits of wildfire by integrating fire and ecosystem simulation models. Ecol Modell 221(8):1162–1172

Khanduri DK, Pandey A, Sharma LK, Verma HC (2020a) Assessment of forest fire susceptibility using remote sensing and geographic information system in the Shivalik hills of Himachal Pradesh, India. J Mt Sci 17(8):1–12

Khanduri S, Gupta RP, Sharma DP (2020b) Forest fire susceptibility mapping using analytical hierarchy process and remote sensing data in Pauri Garhwal district, northwest Himalaya, India. Environ Monit Assess 192(8):1–17

Kumar N, Singh H, Kumar A, Singh AK, Pandey PK, Kumar A (2024) Forest-fire-induced land degradation. In: Mishra G, Giri K, Singh S, Kumar M (eds) Sustainable land management in India. Springer Nature, Singapore. https://doi.org/10.1007/978-981-97-5223-2_4

Malik A, Kumar K, Kumari G, Kumar N, Singh H (2024) Application of functional traits in modelling productivity and resilience under climate change. Springer, pp 77–96. ISBN: 978-981-97-1509-1

Mallick J, Zhao L, Bhotika A, Islam T, Rahman MM, Rahman A (2021a) Application of remote sensing and GIS in forest fire risk assessment using multi-criteria decision analysis (MCDA) techniques. Int J Remote Sens 42(19):7666–7692

Mallick J, Zhao Y, Bhotika R, Kumar R, Saran S (2021b) Forest fire susceptibility assessment using machine learning and geospatial data in the hilly terrain of Central Himalaya, India. Arab J Geosci 14(1)

Mitri G, Saba S, Nader M, McWethy D (2017) Developing Lebanon's fire danger forecast. Int J Disaster Risk Reduct 24:332–339

Pourtaghi ZS, Pourghasemi HR, Aretano R, Semeraro T (2016) Investigation of general indicators influencing on forest fire and its susceptibility modeling using different data mining techniques. Ecol Indic 64:72–84

Rashed TA, Prakash A, Ghosh S (2020) Forest fire risk mapping using remote sensing and geospatial techniques in the Chittagong Hill Tracts, Bangladesh. Remote Sens Appl Soc Environ 5:100121

Reddy CS, Kumar BM, Parthasarathy N, Reddy MR (2006) Community-based assessment, monitoring and planning of forest fires in the Eastern Ghats of Andhra Pradesh, India. Environ Manage 37(5):705–713

Saaty TL (1988) What is the analytic hierarchy process? Springer Berlin Heidelberg, Berlin, Heidelberg, pp 109–121

Sett T, Nikam BR, Singh H, Purohit S (2024) Hydrological variability in Indian forest ecosystem: analysis of drought resilience, recovery and water use efficiency in moist and dry deciduous forests. In: Singh H (ed) Urban forests, climate change and environmental pollution. Springer, Cham. https://doi.org/10.1007/978-3-031-67837-0_37

Singh V (2015) Human-induced forest fires: causes and preventive measures. Int J For Res 789432

Singh H (2024a) Climate change and forests: biological perspectives. In: Singh H (ed) Forests and climate change. Springer Nature, Singapore. https://doi.org/10.1007/978-981-97-3905-9_1

Singh H (ed) (2024b) Forests and climate change: biological perspectives on impact, adaptation, and mitigation strategies. Springer Nature, Singapore. https://doi.org/10.1007/978-981-97-3905-9

Singh H (2024c) Climate change and environmental pollution: biological consequences on urban forests. In: Singh H (ed) Urban forests, climate change and environmental pollution. Springer Nature, Cham, Switzerland. https://doi.org/10.1007/978-3-031-67837-0_1

Singh H (ed) (2024d) Urban forests, climate change and environmental pollution: physio-biochemical and molecular perspectives to enhance urban resilience. Springer Nature, Cham, Switzerland. https://doi.org/10.1007/978-3-031-67837-0

Singh H, Kumar N, Kumar A (2024a) Enhancing resource use efficiency in crops through plant functional traits. Springer, pp 97–117. ISBN: 978-981-97-1509-1

Singh M, Jyoti, Kumar N, Singh H (2024b) Plant functional traits assisted crop adaptation to abiotic and biotic stress. Springer, pp 239–255. ISBN: 978-981-97-1509-1

Trivedi AK, Kumar V, Pandey R (2021) Assessment of forest fire susceptibility using GIS and remote sensing in the hilly terrain of Pauri Garhwal district, northwest Himalaya, India. Ecol Indic 130:108250

Vanlalruata C (2011) Shifting cultivation and community development programs: a study in Mizoram, north-east India. J South Asian Dev 6(1):103–126

Yadav U, Chauhan NK, Sharma U (2017) Forest fires in north-eastern states of India: vegetation types, burned areas and probable causes. Ind For 143(7):721–734

Yıldız C, Çömert R, Tanyaş H, Yılmaz A, Akbaş A, Akay SS et al (2023) The effect of post-wildfire management practices on vegetation recovery: insights from the Sapadere fire, Antalya, Türkiye. Front Earth Sci 11:1174155

Yokomizo H, Murayama T, Ushio M, Numata M (2005) Relationships between weather conditions and forest fire occurrences in Mizoram, northeast India. Int J Wildland Fire 14(3):309–317

Zhang Y, Wang J, Li X, Yang J, Huang L (2019) Forest fire susceptibility mapping using frequency ratio and logistic regression models in Daxing'an Mountains, China. Forests 10(9):821

Zote L, Lalrammawia, Prasad SN (2019) Forest fires in Mizoram: causes, impacts, and management strategies. Int J Curr Microbiol Appl Sci 8(9):139–147

# A Comprehensive Review on Controlling Forest Fires Using AI Technologies

**Chandan Kumar and Seema Sharma**

**Abstract** Conservation of our natural resources, mainly forests and woodlands, has been addressed numerously in many ways. This literature review explores innovative fire safety and detection approaches using Artificial Intelligence (AI) based techniques across various contexts. This chapter comprehensively analyzes existing methods and technologies for fire detection and management. Several contributions are examined where the integration of artificial intelligence and advanced sensing technologies exists. The Intelligent Fire Engineering Tool (IFETool) offers efficient fire safety analysis, demonstrating 97% accuracy in predicting key parameters. AI-based forest fire detection utilizes Convolutional Neural Networks (CNN) and transfer learning for early detection. An improved method using Detectron2 achieves higher precision in forest fire detection. The Smart Fire Detection System (SFDS) for smart cities leverages the YOLOv8 algorithm, achieving high precision. An IoT-based sensor solution addresses peat reserve fires, emphasizing early detection and monitoring. Transfer learning from UAV images achieves 99.32% accuracy in forest fire detection. Residential fire detection explores optimal sensor combinations and AI algorithms for accurate and efficient detection. Collectively, these contributions showcase advancements in fire safety through the fusion of AI and cutting-edge technologies.

**Keywords** Forest fire · AI · Deep learning · Machine learning · Smart fire detection system

C. Kumar
Department of Computer Science and Engineering, Career Point University, Hamirpur, Himachal Pradesh, India

S. Sharma (✉)
Department of Forest Products, College of Horticulture and Forestry, Neri, Hamirpur, Dr Y S Parmar University of Horticulture and Forestry, Nauni, Solan, Himachal Pradesh, India
e-mail: sharsee06@gmail.com

 
H. Singh and N. Kumar (eds.), *Forest Fire and Climate Change*,
https://doi.org/10.1007/978-3-031-89967-6_23

# 1 Introduction

Forests play a crucial role in preserving the fragile ecological equilibrium of the planet (Barwa and Singh 2024; Kaushik and Singh 2024). However, the threat of forest fires is substantial (Kumar et al. 2024). Detecting forest fires promptly is crucial, as by the time they're spotted, they may have already spread widely, complicating containment efforts and sometimes rendering them futile. The consequences of forest fires are devastating, leading to widespread loss and irreparable damage to the environment and atmosphere (Singh 2024b). Forest fires contribute to approximately 30% of $CO_2$ emissions into the atmosphere, exacerbating global warming and disrupting local weather patterns (Singh 2024a).

Additionally, these fires can result in the extinction of rare plant and animal species (Alkhatib 2014; Kumar et al. 2020a, b; Singh et al. 2024). Forests' remote and often unmanaged nature makes them susceptible to fire outbreaks. Dry foliage, branches, and other combustible materials fuel the rapid spread of fires. Ignition can occur due to various factors, including human activities like smoking or natural phenomena like intense heat. Once ignited, fires can quickly escalate from surface to crown fires, engulfing vast areas of forestland and becoming uncontrollable. Millions of hectares of forest are lost to fires yearly, emitting large amounts of carbon monoxide into the atmosphere. Early detection and monitoring of potential fire risk areas are essential for reducing the damage caused by forest fires. Various detection and monitoring systems are utilized by authorities, including patrols, monitoring towers, aerial surveillance, and satellite monitoring.

Additionally, optical camera sensors and specialized detection sensors are increasingly being employed to enhance fire detection capabilities. Improving fire detection systems aims to shorten response times and minimize damage and firefighting costs. The objective is to detect fires at their earliest stages and promptly notify firefighting units for swift intervention. Utilizing detection and monitoring technologies is essential for reducing the impact of forest fires and protecting these critical ecosystems (Attri et al. 2020; Kumar et al. 2021a, b).

Forest fires are among the most common hazards forest ecosystems face, posing significant threats to biodiversity and ecology (Singh 2024c). Recent reports from the Brazilian National Institute of Space Research (INPE) highlight the severity of the issue, with alarming statistics indicating a substantial increase in burned trees in regions like Amazonas. Similarly, devastating wildfires in Australia have resulted in the loss of millions of hectares of forested areas. The importance of forests cannot be overstated, as they are crucial in sustaining life on Earth. Forests serve as habitats for diverse species and offer ecosystem services such as regulating climate, preventing soil erosion, and maintaining water cycles (Sett et al. 2024; Singh 2024d).

Furthermore, forests contribute to oxygen production and serve as carbon sinks, helping mitigate the effects of climate change. As technology grows, there is interest in leveraging image processing technology for forest fire detection. Computer vision systems, mainly based on deep learning techniques such as Convolutional Neural

Networks (CNN), have emerged as powerful tools. These systems utilize sophisticated algorithms to analyze visual data and detect signs of fire in images with high accuracy (Xu et al. 2021). Forest fires threaten ecosystems, biodiversity, and human communities worldwide (Savita et al. 2017; Singh 2021, 2024e; Singh and Kumar 2022; Sett et al. 2022). These events degrade the environment and threaten public health, infrastructure, and economies. Timely detection and effective management of forest fires are crucial to mitigating their adverse impacts and minimizing loss of life and property (Flannigan et al. 2006). In recent years, the integration of artificial intelligence (AI) techniques has emerged as a promising approach to enhance the capabilities of forest fire detection systems. AI-driven solutions leverage advanced algorithms and sensing technologies to improve detection accuracy, reduce false alarms, and enable proactive fire management strategies. AI algorithms can identify potential fire incidents and alert authorities in real time by analyzing various data sources such as satellite imagery, wireless sensor networks, and infrared sensors. Additionally, AI techniques enable the analysis of historical fire data to identify patterns and trends, facilitating predictive modeling and early warning systems (Ai et al. 2023).

## *1.1 Artificial Intelligence and Machine Learning*

Technological progress has revolutionized many aspects of our lives in recent years, making processes more efficient and precise. One rapidly expanding field is artificial intelligence (AI), which involves using machines to perform tasks previously thought to require human intelligence. Within AI, machine learning (ML) is a critical application that enables machines to learn from data and make predictions. ML has become popular due to advancements in computer technology, allowing us to gather and analyze large amounts of data effectively. ML encompasses various methods, tools, and algorithms that teach machines to improve their performance over time by learning from data. One of the main methods in ML is supervised learning, where machines are trained using labeled data with input from human experts, guiding the learning process (Gaur et al. 2020). Machines examine certain aspects of data to identify patterns, categorize items, and evaluate their predictions' accuracy.

Conversely, unsupervised learning focuses on clustering or grouping uncategorized data without human intervention. Machines autonomously identify patterns and trends in unlabeled data, allowing for the discovery of valuable insights. Reinforcement learning involves machines learning to solve problems through trial and error, similar to how humans learn from experience. Machines receive feedback through rewards or punishments based on their actions, enabling them to learn and refine their decision-making processes. The ML process typically involves several steps, including data gathering and preprocessing. High-quality data is essential for developing accurate ML models, as insufficient or low-quality data can lead to biased predictions and inaccurate interpretations (Abid 2021). Data cleaning and preparation are critical stages that involve formatting, modifying, and integrating datasets to ensure their suitability for analysis. Feature engineering and data cleaning further refine the dataset, allowing for extracting relevant features and attributes. The modeling stage focuses on selecting appropriate ML algorithms, optimizing hyperparameters, and training the model using the prepared dataset.

Deep learning is a type of machine learning inspired by how the human brain works. It uses artificial neural networks to learn from data. Unlike regular machine learning, deep learning has many processing layers, which helps it understand complex patterns better. Some common deep learning techniques include convolutional neural networks (CNNs), recurrent neural networks (RNNs), and generative adversarial networks (GANs). CNNs, in particular, have found extensive applications in fire detection systems, where they excel at detecting the presence of fires in images. ML and DL technologies offer powerful tools for analyzing data, making predictions, and solving complex problems across various domains. Such technologies can develop innovative solutions for addressing critical challenges such as forest fire detection and prediction, ultimately contributing to enhanced safety and environmental protection (Hossain et al. 2020; Kumar et al. 2021a, b).

Artificial Intelligence (AI) has revolutionized fire prevention and fighting techniques worldwide, introducing various innovative products and technologies to enhance safety and efficiency. These AI products encompass various aspects of fire safety monitoring, analysis, forecasting, and intervention, offering advanced solutions to address the challenges of fires. One significant application of AI in fire prevention is the development of sound and exit warning systems. These systems utilize AI technology to enhance traditional fire alarms, enabling more sensitive operation and accurate analysis. In the event of a fire, AI-based sound systems provide specific instructions to individuals, guiding them on how to respond and evacuate safely. Equipped with cameras, AI-based fire alarms can distinguish between different types of smoke, ensuring more accurate and timely detection of fires. Remote sensing satellite technology represents another groundbreaking achievement in fire prevention. NASA's Form EOSDIS data, developed in collaboration with the University of Maryland, utilizes satellite imagery to monitor forests and detect fires. By analyzing smoke, hot spots, and temperature variations, this technology facilitates early detection and rapid response to wildfires, enabling firefighting units to mobilize quickly and efficiently.

AI-driven tools developed by NASA enable firefighters to collect and analyze on-scene data, such as temperature and gas levels, to optimize firefighting strategies and minimize exposure to hazardous environments. Additionally, AI products are employed to store and analyze information related to fire incidents, enabling authorities to assess the situation, forecast potential risks, and adjust fire management policies accordingly. Regarding firefighting vehicles and equipment, AI enhances capabilities and effectiveness. Fire trucks equipped with AI technology can easily navigate congested roads, utilizing special mechanisms to maneuver through narrow spaces. Helicopters and aircraft with AI systems can conduct precision firefighting operations and efficiently deliver water or flame retardant to affected areas. Robotic systems designed for firefighting are equipped with advanced sensors and cameras, enabling them to operate in hazardous environments and effectively support firefighting efforts. These robots can pump water, extinguish fires, and provide surveillance in areas inaccessible to human firefighters, enhancing overall safety and operational efficiency.

Furthermore, drones equipped with firefighting capabilities are being developed to combat fires from the air. These drones can carry and deploy fire extinguishing balls or spray water/foam onto fires, offering a rapid and effective response to fire incidents, especially in high-rise buildings or remote areas. AI-driven products and technologies are transforming the landscape of fire prevention and fighting, offering innovative solutions to mitigate risks, enhance response capabilities, and save lives in the face of fire emergencies (Hossain et al. 2020). The most common method to detect flame and smoke is represented by Fig. 1 (Hossain et al. 2020). The forest fire detection technique using a convolution neural network (CNN) is depicted in Fig. 2 (Zheng et al. 2022).

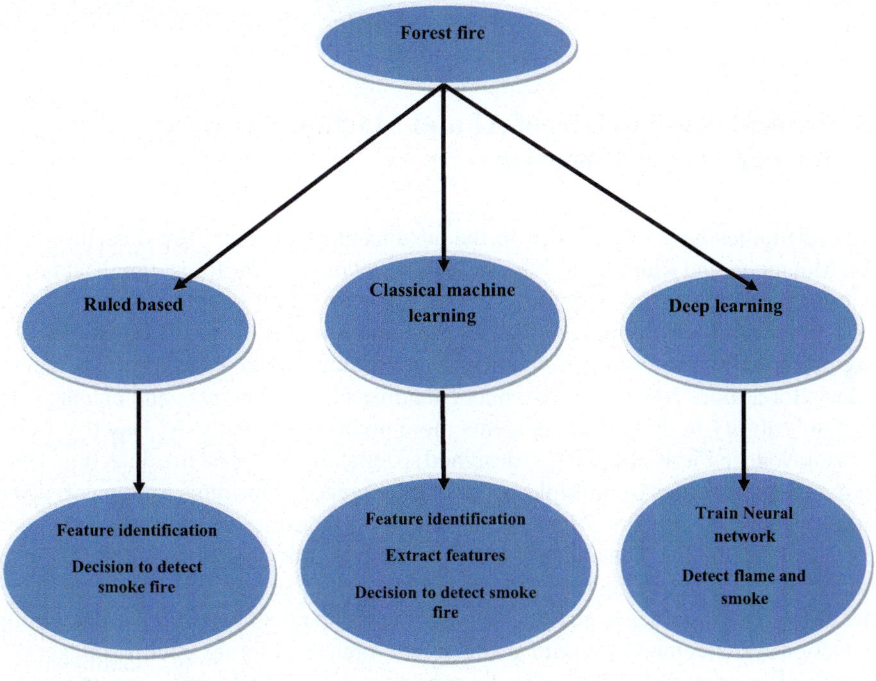


**Fig. 1** Common method to detect flame and smoke (Hossain et al. 2020)

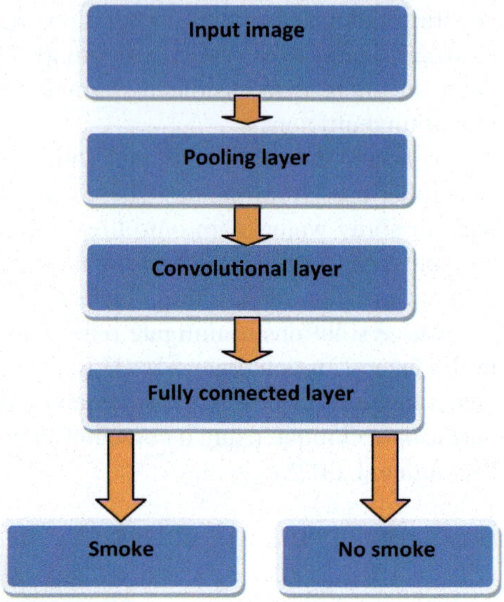


**Fig. 2** The forest fire smoke detection algorithm utilizing convolutional neural networks (CNN) (Zheng et al. 2022)

## 2 Several Applications of AI and Machine Learning for Forest Fire Detection

Several studies have contributed to the advancement of forest fire detection using AI. Mahmoud and Ren (2018) introduced a novel method for forest fire detection by combining a rule-based image processing algorithm with temporal variation analysis. Their innovative approach focuses on improving the accuracy and efficiency of fire detection by examining temporal changes in image data over time. This method allows for a more comprehensive understanding of fire dynamics and enhances the system's ability to detect fire incidents more precisely. A Multi-AI Based Wireless Sensor Node in Pieris et al. (2022) designed specifically for forest fire detection, leveraging the capabilities of multiple artificial intelligence algorithms to improve detection accuracy and response time. Arrue et al. (2000) developed an intelligent system focused on reducing false alarms in infrared forest fire detection, highlighting the importance of minimizing unnecessary alerts to optimize resource allocation during fire emergencies. James et al. (2023) contributed to the field with an efficient wildfire detection system embedded with artificial intelligence, utilizing satellite imagery for rapid and accurate fire detection in diverse environments. Advancements in wireless sensor networks and image compression techniques have also contributed significantly to forest fire detection. Bouakkaz et al. (2021) investigated using wireless multimedia sensor networks and image compression techniques for efficient fire detection in remote forested regions. Kizilkaya et al. (2022) introduced a robust forest fire detection framework that leverages heterogeneous wireless multimedia sensor networks, showcasing improved capabilities in identifying and mitigating fire hazards.

Additionally, researchers have capitalized on the potential of deep learning and convolutional neural networks to enhance forest fire detection accuracy. Dutta and Ghosh (2021) introduced an innovative method that merges separable convolution with image processing techniques to improve fire detection accuracy, highlighting the effectiveness of deep learning in this domain. Chanthiya and Kalaivani (2021) employed support vector machine (SVM) algorithms to detect forest fires in LANDSAT images, showcasing the capability of machine learning in analyzing satellite data for fire detection. Tailored convolutional neural networks (CNNs) have also been crafted to detect forest fires using satellite imagery.

A novel CNN architecture tailored for satellite image analysis, achieving improved performance in detecting forest fires, was introduced by Kalaivani and Chanthiya (2022). Real-time detection of full-scale forest fire smoke has been enabled through deep convolutional neural networks (CNNs), as demonstrated by Zheng et al. (2022), highlighting the potential of AI-driven solutions in monitoring and early warning systems. In addition, feature extraction techniques and multi-view image analysis advancements have contributed to forest fire recognition and detection. Wu et al. (2021) introduced a feature extraction-based method for forest fire recognition from multi-view images, showcasing the importance of data preprocessing techniques in enhancing detection accuracy. Hu et al. (2022) developed a fast forest fire smoke detection system using MVMNet, emphasizing the role of efficient algorithms in real-time fire detection applications. Moreover, pixel-level automatic annotation techniques and ensemble-pertained residual networks have been employed for automated and accurate forest fire detection. Yang et al. (2021) presented an innovative approach for annotating forest fire images at the pixel level, showcasing the effectiveness of AI-powered annotation techniques in optimizing the detection process. Dogan et al. (2022) introduced an automated fire detection system employing ensemble retrained residual networks, underscoring the efficacy of ensemble learning techniques in enhancing detection precision. Additionally, deep learning-based computer vision algorithms have streamlined the early identification of wildfires using unmanned aerial vehicles (UAVs). Bouguettaya et al. (2022) conducted a comprehensive review of early wildfire detection methodologies, emphasizing the role of deep learning algorithms in UAV-based forest fire detection. Li et al. (2022) introduced a novel deep-learning approach utilizing SRN-YOLO for forest fire detection, highlighting the promise of innovative deep-learning architectures in enhancing detection efficacy. Moreover, transfer learning methodologies have been leveraged to recognize forest fires in UAV images, facilitating the effective transfer of knowledge from pre-existing models to boost detection precision. Zhang et al. (2022) demonstrated a forest fire recognition method based on transfer learning, highlighting the importance of leveraging pre-existing knowledge to improve detection performance in UAV-based systems. Tian et al. (2022) introduced a method for monitoring forest fire spread and vegetation changes using remote sensing images. This comprehensive approach integrates diverse data sources for effective fire management. AI-driven solutions enable swift and precise fire detection, enhancing forest fire management strategies and mitigating wildfire impacts. Avazov et al. (2023) employed Internet of Things (IoT) devices equipped with sensors to gather up-to-date environmental data, including temperature and humidity levels, in real-time. AI algorithms then processed data and trained to detect patterns indicative of potential forest fires. When a fire risk was detected, the system notified relevant authorities for prompt intervention. El Madafri et al. (2023) employed a dataset comprising images of forest landscapes and fires to train CNNs for fire detection. The CNN models were trained using supervised learning techniques, where they learned to distinguish between normal forest scenes and images containing signs of fire. The trained models were then integrated into an inference system for automated fire detection. Sathishkumar et al. (2023) assembled a dataset comprising images recorded by surveillance cameras stationed in forested regions. These images were annotated to pinpoint areas of significance like flames and smoke. Deep learning models, such as convolutional neural networks (CNNs) and recurrent neural networks (RNNs), underwent training on this annotated dataset to swiftly and accurately detect forest fires and smoke signals as they occur. Paidipati et al. (2023) established a network of wireless sensors strategically positioned throughout forested areas to capture environmental data, encompassing temperature, humidity, and air quality. This data was then relayed to a central server for analysis. Utilizing deep learning models like convolutional neural networks (CNNs) and recurrent neural networks (RNNs), they were trained on this dataset to detect anomalies indicative of forest fires. Carta et al. (2023) extensively reviewed existing literature concerning forest fire prevention strategies. James et al. (2023) curated satellite imagery depicting forested regions and meticulously annotated these images to pinpoint areas affected by wildfires. They then employed deep learning models, specifically convolutional neural networks (CNNs) and recurrent neural networks (RNNs), trained on this annotated dataset to accurately classify and detect wildfires in satellite imagery. Ahmad et al. (2023) introduced FireXnet, a specialized deep-learning model crafted for wildfire detection on devices with limited resources. Their approach focused on developing a lightweight neural network architecture optimized for efficient operation on low-power devices. By employing transfer learning techniques, the model was trained on a dataset of wildfire images, enabling adaptation of pre-trained models to the specific task of wildfire detection. Almeida et al. (2023) introduced EdgeFireSmoke, a novel lightweight algorithm for real-time forest fire detection through an IoT-human–machine interface. Their methodology centered on crafting an edge computing framework capable of processing sensor data locally to identify fire and smoke patterns. This localized approach minimized latency, facilitating swift responses to fire incidents without over-reliance on cloud-based processing. Peruzzi et al. (2023) explored the application of ensemble learning techniques in forest fire detection. Their approach entailed training various AI models on diverse datasets, including decision trees, random forests, and gradient-boosting machines. Subsequently, they combined the predictions from these models to enhance the overall accuracy and resilience of the fire detection system. Almasoud (2023) devised an intelligent deep learning-enabled forest fire detection system, employing a vast dataset of annotated forest images to train convolutional neural networks (CNNs) and recurrent neural networks (RNNs). This system leveraged UAV resources for immediate fire detection, utilizing typical convolution layers to analyze aerial imagery and pinpoint potential fire hotspots.

Rao et al. (2023) developed a neural fuzzy system for real-time forest fire detection, combining neural networks and fuzzy logic to enhance early detection and response capabilities, thereby contributing to improved forest fire management and mitigation efforts. Reis and Turk (2023) explored forest fire detection using deep convolutional neural networks (CNNs) with a transfer learning approach, demonstrating the efficacy of AI-driven solutions in fire management by achieving satisfactory performance in fire detection and response. Abdusalomov et al. (2023) proposed an enhanced forest fire detection method based on the detectron2 model and deep learning techniques, outperforming traditional methods with a remarkable accuracy rate of 99.3% in fire detection. Their approach showcased superior capabilities for accurate and efficient fire detection by integrating deep learning techniques. The forest fire smoke detection algorithm utilizing Convolutional Neural Networks (CNN) is depicted in Wahyono Harjoko et al. (2022). The study emphasized the importance of maintaining high performance while ensuring real-time operation of the integrated fire detection module, addressing the critical need for timely fire detection to prevent disasters. Fouda et al. (2022) introduced a lightweight hierarchical AI model for UAV-enabled edge computing specifically designed for forest fire detection. Utilizing hierarchical AI frameworks, the study aimed to optimize image classification tasks, such as detecting forest fires, to enhance early warning systems and mitigate fire risks. Khan et al. (2022) developed DeepFire, a novel dataset and deep transfer learning benchmark for forest fire detection, aiming to advance fire detection systems by providing a comprehensive dataset and benchmark for evaluating deep learning models. This initiative facilitates the design of more accurate and reliable fire detection algorithms. The proposed forest fire detection method (Rahman et al. 2023) utilizes a Convolutional Neural Network (CNN) architecture with separable convolution layers for real-time detection. Trained on a new fire detection dataset, the method achieves high accuracy (97.63%), F1 Score (98.00%), and Kappa coefficient (80%), making it a valuable tool for quickly detecting forest fires and enabling prompt response from authorities to minimize damage. This approach represents a significant advancement in UAV-based fire detection systems, leveraging deep learning and image processing algorithms for improved disaster preparedness and response. Kumar and Akhlaq (2022) presented a study on forest fire detection using AI techniques, focusing on relapse procedures and AI archive development. The research aimed to propose AI-based solutions for analyzing environmental and geographical factors to improve forest fire detection and prediction accuracy, facilitating proactive fire management measures.

Seydi et al. (2022) introduced Fire-Net, a deep learning framework for active forest fire detection, aiming to develop an efficient and reliable fire detection system capable of detecting forest fires in real time. The study aimed to enable timely response and mitigation efforts by leveraging deep learning techniques. Kang et al. (2022) created a deep learning model for forest fire detection by harnessing geostationary satellite data. They aimed to minimize detection latency and enhance accuracy. The research showcased the efficacy of artificial intelligence in promptly identifying forest fires through satellite imagery analysis, thereby bolstering early warning systems and firefighting capabilities. Khan and Khan (2022) proposed FFireNet, a deep learning-based forest fire classification and detection method for smart cities. The study focused on leveraging artificial intelligence to develop an effective forest fire detection system tailored for urban environments, aiming to enhance the safety and resilience of smart city infrastructures.

Grari et al. (2022) conducted a comprehensive review focusing on integrating the Internet of Things (IoT) and machine learning (ML) for forest fire detection, monitoring, and prediction. The review emphasized the importance of utilizing IoT and ML technologies to devise proactive fire management solutions. It underscored the value of data-driven approaches in improving forest fire prevention and response strategies, thereby contributing to more effective forest conservation efforts. Authors (Xu et al. 2021) proposed a forest fire detection system based on ensemble learning, combining the YOLOv5 and EfficientDet object detectors within an ensemble learning framework. The detectors were trained on labeled datasets containing forest images with and without fires, and ensemble techniques were used to enhance detection accuracy. Thapa et al. (2021) conducted a comprehensive review of forest fire detection and monitoring systems, synthesizing information on remote sensing, GIS, and fire-danger forecasting techniques to enhance forest management frameworks. Dampage et al. (2022) developed a forest fire detection system using wireless sensor networks (WSNs) and machine learning algorithms, deploying sensor-equipped WSN nodes to monitor environmental parameters associated with fire risk. Guede-Fernández et al. (2021) developed an innovative object identification system using deep learning for forest fire detection. Their approach involved gathering and preprocessing extensive image datasets depicting forests under varying fire conditions. By training convolutional neural network (CNN) models like ResNet and VGG on these datasets, the system learned to recognize key features indicative of fire and smoke. Transfer learning methods were applied to fine-tune the pre-trained models to enhance detection accuracy. The system's effectiveness was assessed using performance metrics, including precision, recall, and F1-score, demonstrating its promising capabilities in fire detection and mitigation efforts. Ya'acob et al. (2021) devised an infrared camera-based forest fire detection system, processing captured infrared images to enhance detection. The method involved preprocessing images, segmenting fire regions, and utilizing color models to differentiate fire pixels from background elements.

Field experiments validated the system's effectiveness in detecting forest fires, addressing the challenge of early detection and mitigation of these devastating events. However, the rise of Unmanned Aerial Vehicle (UAV) image-based fire detection systems offers a hopeful solution. Xue et al. (2022) introduced a refined YOLOv5 model for detecting small-target forest fires. Their approach involved optimizing the YOLOv5 architecture through adjustments in hyperparameters, data augmentation, and loss functions. The modified model demonstrated improved performance in detecting small fires, which was evaluated using metrics like mean average precision (mAP). Alkhatib et al. (2021) developed a wireless sensor network (WSN) tailored for forest fire detection and behavior analysis. Sensor nodes with environmental sensors monitor temperature, humidity, and smoke levels in forested areas. Machine learning algorithms were trained on collected data to identify patterns indicative of fire events. Field trials and simulations validated the system's effectiveness. Kukuk and Kilimci (2021) thoroughly analyzed forest fire detection methods by evaluating the difference between conventional machine learning algorithms and deep learning models. They collected and preprocessed datasets containing forest images with and without fire events. Various deep learning architectures and traditional ML methods were evaluated (Table 1).

## 3 Conclusion

Fire prevention and firefighting are essential for protecting lives and properties and safeguarding the interests of nations, organizations, and individuals. With the rapid advancement of technology, AI products have emerged as promising tools to enhance fire safety measures. However, despite their growing popularity and sophistication, the widespread adoption of AI in fire prevention and firefighting remains limited in many countries. Various factors contribute to this limitation, including economic constraints, scientific and technical development disparities, and differing policy priorities among nations. While AI offers numerous benefits, its application in fire safety also presents challenges and obstacles that require careful consideration. Based on our research, we propose several recommendations and solutions to enhance the effectiveness and efficiency of AI in fire prevention and firefighting activities. These measures include investing in research and development to improve AI technologies tailored to fire safety, promoting international collaboration and knowledge sharing, and implementing policies to incentivize the adoption of AI solutions in fire prevention and firefighting efforts. Addressing these challenges and implementing targeted strategies may harness the full potential of AI to mitigate fire risks, enhance emergency response capabilities, and ultimately save lives and protect communities from the devastating impact of fires.

**Table 1** Number of techniques used and their findings

| S. No. | Technique used | Findings | References |
|---|---|---|---|
| 1 | Rule-based image processing algorithm combined with temporal variation analysis | Aimed to enhance accuracy and efficiency of fire detection by analyzing temporal variations in image data | Mahmoud and Ren (2018) |
| 2 | Multi-AI-based wireless sensor node | Leveraged multiple artificial intelligence algorithms to improve detection accuracy and response time | Pieris et al. (2022) |
| 3 | Intelligent system for reducing false alarms in infrared forest fire detection | Focused on minimizing unnecessary alerts to optimize resource allocation during fire emergencies | Arrue et al. (2000) |
| 4 | Wildfire detection system embedded with artificial intelligence | Utilized satellite imagery for rapid and accurate fire detection in diverse environments | James et al. (2023) |
| 5 | Wireless multimedia sensor networks combined with image compression | Explored efficient fire detection in remote forested areas | Bouakkaz et al. (2021) |
| 6 | Heterogeneous wireless multimedia sensor networks | Demonstrated enhanced performance in detecting and mitigating fire risks | Kizilkaya et al. (2022) |
| 7 | Combination of separable convolution and image processing techniques | Showcased the potential of deep learning in improving detection accuracy | Dutta and Ghosh (2021) |
| 8 | Support vector machine (SVM) algorithms on LANDSAT images | Demonstrated effectiveness of machine learning in analyzing satellite data for fire detection | Chanthiya and Kalaivani (2021) |
| 9 | Custom-optimized convolutional neural network (CNN) architecture | Achieved improved performance in detecting forest fires from satellite images | Kalaivani and Chanthiya (2022) |
| 10 | Deep convolutional neural networks (CNNs) for real-time detection of full-scale forest fire smoke | Enabled real-time monitoring and early warning systems for forest fires | Zheng et al. (2022) |
| 11 | Feature extraction-based approach for forest fire recognition from multi-view images | Highlighted the importance of data preprocessing techniques in enhancing detection accuracy | Wu et al. (2021) |
| 12 | Fast forest fire smoke detection system using MVMNet | Emphasized the role of efficient algorithms in real-time fire detection | Hu et al. (2022) |
| 13 | Pixel-level automatic annotation method for forest fire images | Demonstrated the potential of AI-driven annotation techniques in streamlining the detection process | Yang et al. (2021) |

(continued)

**Table 1** (continued)

| S. No. | Technique used | Findings | References |
|---|---|---|---|
| 14 | Automated fire detection system using ensemble pre-trained residual networks | Showcased the effectiveness of ensemble learning approaches in improving detection accuracy | Dogan et al. (2022) |
| 15 | Early wildfire detection from unmanned aerial vehicles (UAVs) using deep learning-based computer vision algorithms | Reviewed methodologies for UAV-based forest fire detection, highlighting the role of deep learning | Bouguettaya et al. (2022) |
| 16 | Deep learning method based on SRN-YOLO for forest fire detection | Showcased the potential of novel deep learning architectures in improving detection performance | Li et al. (2022) |
| 17 | Forest fire recognition method based on transfer learning | Highlighted the importance of leveraging pre-existing knowledge for improving detection performance in UAV-based systems | Zhang et al. (2022) |
| 18 | Forest fire spread monitoring and vegetation dynamics detection using multi-source remote sensing images | Demonstrated a holistic approach to integrating various data sources for comprehensive fire management | Tian et al. (2022) |
| 19 | IoT devices equipped with sensors for real-time data collection on environmental conditions | Utilized AI algorithms to detect patterns indicative of potential forest fires, triggering notifications for prompt intervention | Avazov et al. (2023) |
| 20 | CNNs trained on forest landscapes and fire images for automated fire detection | Integrated CNN models into the inference system for automated fire detection | El Madafri et al. (2023) |
| 21 | Deep learning models trained on surveillance camera images for real-time forest fire detection | Collected annotated datasets and trained CNNs and RNNs for accurate detection | Sathishkumar et al. (2023) |
| 22 | Network of wireless sensors transmitting environmental data for analysis | Deep learning models trained on sensor data for anomaly detection indicative of forest fires | Paidipati et al. (2023) |
| 23 | Comprehensive review of forest fire prevention methods | Analyzed various approaches, including machine learning and AI, to develop fire detection algorithms | Carta et al. (2023) |
| 24 | Deep learning models trained on satellite images for wildfire detection | Manually annotated satellite images are used to train CNNs and RNNs for accurate detection | James et al. (2023) |

(continued)

**Table 1** (continued)

| S. No. | Technique used | Findings | References |
|---|---|---|---|
| 25 | Adaptable deep learning solution for wildfire detection on resource-limited devices | FireXnet is designed to run efficiently on low-power devices and is trained using transfer learning | Ahmad et al. (2023) |
| 26 | Efficient IoT-enabled human–machine interface for real-time forest fire detection | EdgeFireSmoke processed sensor data locally to detect fire and smoke patterns, reducing latency | Almeida et al. (2023) |
| 27 | Utilizing ensemble learning for forest fire detection | Combined predictions from multiple AI models to improve detection accuracy and robustness | Peruzzi et al. (2023) |
| 28 | Forest fire detection system empowered by deep learning | Trained CNNs and RNNs on annotated datasets for accurate detection and classification | Almasoud (2023) |
| 29 | Real-time forest fire detection system enhanced by neural fuzzy technology | Combined neural networks and fuzzy logic for improved early detection and response | Rao et al. (2023) |
| 30 | Utilizing transfer learning in deep convolutional neural networks for forest fire detection | Achieved satisfactory performance in detecting and responding to forest fires | Reis and Turk (2023) |
| 31 | Enhanced forest fire detection method employing deep learning strategy | Outperformed traditional methods with a high accuracy rate in fire detection | Abdusalomov et al. (2023) |
| 32 | AI-driven framework for rapid forest fire detection in real-time | The color probability model is utilized, and motion feature analysis for timely fire detection | Wahyono Harjoko et al. (2022) |
| 33 | Compact hierarchical AI model tailored for edge computing on UAVs | Optimized image classification tasks, such as fire detection, for early warning systems | Fouda et al. (2022) |
| 34 | DeepFire dataset and deep transfer learning benchmark for forest fire detection | Provided comprehensive dataset and benchmark for evaluating deep learning models for fire detection | Khan et al. (2022) |
| 35 | Wireless sensor node employing multiple AI techniques for forest fire detection | Explored various AI architectures to enhance detection accuracy and efficiency | Rahman et al. (2023) |
| 36 | AI techniques for forest fire detection and prediction | Proposed AI-based solutions for analyzing environmental factors to improve detection accuracy | Kumar and Akhlaq (2022) |
| 37 | Deep learning framework (Fire-Net) for active forest fire detection | Developed an efficient and reliable system for real-time fire detection | Seydi et al. (2022) |

(continued)

**Table 1** (continued)

| S. No. | Technique used | Findings | References |
|---|---|---|---|
| 38 | Deep learning approach for detecting forest fires with geostationary satellite data | Reduced detection latency and improved accuracy in analyzing satellite imagery for fire detection | Kang et al. (2022) |
| 39 | Forest fire detection and classification method tailored for smart cities using a deep learning-based approach | Leveraged AI to develop an effective detection system tailored for urban environments | Khan and Khan (2022) |

## References


Abdusalomov AB, Islam BMS, Nasimov R, Mukhiddinov M, Whangbo TK (2023) An improved forest fire detection method based on the detectron2 model and a deep learning approach. Sensors 23(3):1512

Abid F (2021) A survey of machine learning algorithms based forest fires prediction and detection systems. Fire Technol 57(2):559–590

Ahmad K, Khan MS, Ahmed F, Driss M, Boulila W, Alazeb A, Ahmad J (2023) FireXnet: an explainable AI-based tailored deep learning model for wildfire detection on resource-constrained devices. Fire Ecol 19(1):54

Ai W, Wu Y, Wang C, Liang L (2023) Impacts of climate change on forest fires. In: Annual report on actions to address climate change (2019) climate risk prevention. Springer Nature Singapore, Singapore, pp 195–209

Alkhatib AA (2014) A review on forest fire detection techniques. Int J Distrib Sens Netw 10(3):597368

Alkhatib AA, Abdelal Q, Kanan T (2021) Wireless sensor network for forest fire detection and behavior analysis. Int J Adv Soft Comput Appl 13(1)

Almasoud AS (2023) Intelligent deep learning enabled wild forest fire detection system. Comput Syst Sci Eng 44(2)

Almeida JS, Jagatheesaperumal SK, Nogueira FG, Albuquerque VHC (2023) EdgeFireSmoke++: a novel lightweight algorithm for real-time forest fire detection and visualization using internet of things-human machine interface. Expert Syst Appl 221:119747

Arrue BC, Ollero A, De Dios JM (2000) An intelligent system for false alarm reduction in infrared forest-fire detection. IEEE Intell Syst Appl 15(3):64–73

Attri V, Dhiman R, Sarvade S (2020) A review on status, implications and recent trends of forest fire management. Arch Agric Environ Sci 5(4):592–602

Avazov K, Hyun AE, Sami SAA, Khaitov A, Abdusalomov AB, Cho YI (2023) Forest fire detection and notification method based on AI and IoT approaches. Future Internet 15(2):61

Barwa S, Singh H (2024) Climate-induced changes in forest ecosystems of Indian Himalayan region: a meta-analysis. In: Singh H (ed) Urban forests, climate change and environmental pollution. Springer, Cham. https://doi.org/10.1007/978-3-031-67837-0_39

Bouakkaz F, Ali W, Derdour M (2021) Forest fire detection using wireless multimedia sensor networks and image compression. Instrum Mes Métrol 20:57–63

Bouguettaya A, Zarzour H, Taberkit AM, Kechida A (2022) A review on early wildfire detection from unmanned aerial vehicles using deep learning-based computer vision algorithms. Signal Process 190:108309

Carta F, Zidda C, Putzu M, Loru D, Anedda M, Giusto D (2023) Advancements in forest fire prevention: a comprehensive survey. Sensors 23(14):6635

Chanthiya P, Kalaivani V (2021) Forest fire detection on LANDSAT images using support vector machine. Concurr Comput Pract Exp 33(16):e6280

Dampage U, Bandaranayake L, Wanasinghe R, Kottahachchi K, Jayasanka B (2022) Forest fire detection system using wireless sensor networks and machine learning. Sci Rep 12(1):46

Dogan S, Barua PD, Kutlu H, Baygin M, Fujita H, Tuncer T, Acharya UR (2022) Automated accurate fire detection system using ensemble pretrained residual network. Expert Syst Appl 203:117407

Dutta S, Ghosh S (2021) Forest fire detection using combined architecture of separable convolution and image processing. In: 2021 1st international conference on artificial intelligence and data analytics (CAIDA), Apr 2021. IEEE, pp 36–41

Flannigan MD, Amiro BD, Logan KA, Stocks BJ, Wotton BM (2006) Forest fires and climate change in the 21st century. Mitig Adapt Strateg Glob Change 11:847–859

Fouda MM, Sakib S, Fadlullah ZM, Nasser N, Guizani M (2022) A lightweight hierarchical AI model for UAV-enabled edge computing with forest-fire detection use-case. IEEE Netw 36(6):38–45

Gaur A, Singh A, Kumar A, Kumar A, Kapoor K (2020) Video flame and smoke based fire detection algorithms: a literature review. Fire Technol 56:1943–1980

Grari M, Yandouzi M, Idrissi I, Boukabous M, Moussaoui O, Azizi M, Moussaoui M (2022) Using IoT and ML for forest fire detection, monitoring, and prediction: a literature review. J Theor Appl Inf Technol 100(19):5445–5461

Guede-Fernández F, Martins L, de Almeida RV, Gamboa H, Vieira P (2021) A deep learning based object identification system for forest fire detection. Fire 4(4):75

Hossain FA, Zhang YM, Tonima MA (2020) Forest fire flame and smoke detection from UAV-captured images using fire-specific color features and multi-color space local binary pattern. J Unmanned Veh Syst 8(4):285–309

Hu Y, Zhan J, Zhou G, Chen A, Cai W, Guo K, Li L (2022) Fast forest fire smoke detection using MVMNet. Knowl-Based Syst 241:108219

James GL, Ansaf RB, Al Samahi SS, Parker RD, Cutler JM, Gachette RV, Ansaf BI (2023) An efficient wildfire detection system for AI-embedded applications using satellite imagery. Fire 6(4):169

Kalaivani V, Chanthiya P (2022) A novel custom optimized convolutional neural network for a satellite image by using forest fire detection. Earth Sci Inform 15(2):1285–1295

Kang Y, Jang E, Im J, Kwon C (2022) A deep learning model using geostationary satellite data for forest fire detection with reduced detection latency. GISci Remote Sens 59(1):2019–2035

Kaushik P, Singh H (2024) Climate change-induced forest ecosystem functioning and alteration in water–atmosphere–soil–plant (WASP) relationship. In: Singh H (ed) Urban forests, climate change and environmental pollution. Springer, Cham. https://doi.org/10.1007/978-3-031-67837-0_38

Khan S, Khan A (2022) Ffirenet: deep learning based forest fire classification and detection in smart cities. Symmetry 14(10):2155

Khan A, Hassan B, Khan S, Ahmed R, Abuassba A (2022) DeepFire: a novel dataset and deep transfer learning benchmark for forest fire detection. Mob Inf Syst 2022

Kizilkaya B, Ever E, Yatbaz HY, Yazici A (2022) An effective forest fire detection framework using heterogeneous wireless multimedia sensor networks. ACM Trans Multimed Comput Commun Appl (TOMM) 18(2):1–21

Kukuk SB, Kilimci ZH (2021) Comprehensive analysis of forest fire detection using deep learning models and conventional machine learning algorithms. Int J Comput Exp Sci Eng 7(2):84–94

Kumar A, Akhlaq A (2022) Forest fire detection. Int J Health Sci II:7504–7510

Kumar A, Dwivedi GK, Tewari S, Paul J, Anand R, Kumar N, Kumar P, Singh H, Kaushal R (2020a) Carbon mineralization and inorganic nitrogen pools under *Terminalia chebula* Retz.-based agroforestry system in Himalayan foothills, India. For Sci 66(5):634–643

Kumar A, Kumar P, Singh H, Kumar N (2020b) Adaptation and mitigation potential of roadside trees with bio-extraction of heavy metals under vehicular emissions and their impact on physiological traits during seasonal regimes. Urb For Urb Green 58:126900

Kumar A, Kumar P, Singh H, Bisht S, Kumar N (2021a) Relationship of physiological plant functional traits with soil carbon stock in temperate forest of Garhwal Himalaya. Curr Sci 120(8):1368–1373

Kumar A, Kumar P, Singh H, Kumar N (2021b) Modulation of plant functional traits under essential plant nutrients during seasonal regime in natural forests of Garhwal Himalayas. Plant Soil 465:197–212

Kumar N, Singh H, Kumar A, Singh AK, Pandey PK, Kumar A (2024) Forest-fire-induced land degradation. In: Mishra G, Giri K, Singh S, Kumar M (eds) Sustainable land management in India. Springer Nature, Singapore. https://doi.org/10.1007/978-981-97-5223-2_4

Li Y, Shen Z, Li J, Xu Z (2022) A deep learning method based on SRN-YOLO for forest fire detection. In: 2022 5th international symposium on autonomous systems (ISAS), Apr 2022. IEEE, pp 1–6

Madafri I, Peña Carrera M, Olmedo Torre N (2023) Applying artificial intelligence models for the automatic forest fire detection. In: Avenços en recerca i desenvolupament del Departament d'Enginyeria Gràfica i de Disseny. OmniaScience, pp 95–104

Mahmoud MA, Ren H (2018) Forest fire detection using a rule-based image processing algorithm and temporal variation. Math Probl Eng 2018

Paidipati KK, Kurangi C, Reddy ASK, Kadiravan G, Shah NH (2023) Wireless sensor network assisted automated forest fire detection using deep learning and computer vision model. Multimed Tools Appl 1–18

Peruzzi G, Pozzebon A, Van Der Meer M (2023) Fight fire with fire: detecting forest fires with embedded machine learning models dealing with audio and images on low power IoT devices. Sensors 23(2):783

Pieris TPD, Kulasekera AL, Chathuranga KVDS (2022) Multi-AI based wireless sensor node for forest fire detection. In: 2022 7th international conference on information technology research (ICITR), Dec 2022. IEEE, pp 1–6

Rahman AKZR, Sakif S, Sikder N, Masud M, Aljuaid H, Bairagi AK (2023) Unmanned aerial vehicle assisted forest fire detection using deep convolutional neural network. Intell Autom Soft Comput 35(3)

Rao GN, Rao PJ, Duvvuru R, Beulah K, Lydia EL, Rathnala P, Motru VR (2023) Neural fuzzy system design in forest fire detection. Microsyst Technol 1–13

Reis HC, Turk V (2023) Detection of forest fire using deep convolutional neural networks with transfer learning approach. Appl Soft Comput 143:110362

Sathishkumar VE, Cho J, Subramanian M, Naren OS (2023) Forest fire and smoke detection using deep learning-based learning without forgetting. Fire Ecol 19(1):1–17

Savita, Sharma SK, Singh H, Singh O (2017) Forest fire: a review. Indian For 143(8):719–728

Sett T, Nikam BR, Singh H, Purohit S (2022) Effect of drought stress on forest evapotranspiration—a case study on Indian forests. In: EGU general assembly 2022. https://doi.org/10.5194/egusphere-egu22-510

Sett T, Nikam BR, Singh H, Purohit S (2024) Hydrological variability in Indian forest ecosystem: analysis of drought resilience, recovery and water use efficiency in moist and dry deciduous forests. In: Singh H (ed) Urban forests, climate change and environmental pollution. Springer, Cham. https://doi.org/10.1007/978-3-031-67837-0_37

Seydi ST, Saeidi V, Kalantar B, Ueda N, Halin AA (2022) Fire-Net: a deep learning framework for active forest fire detection. J Sens 2022:1–14

Singh H (2021) Process-based carbon sequestration study in relation to the energy-water-carbon flux in a forest ecosystem. In: Kumar P, Singh RK, Kumar M, Rani M, Sharma P (eds) Climate impacts on sustainable natural resource management. CRC Press, Taylor & Francis Group, United States, pp 336–361

Singh H (2024a) Climate change and forests: biological perspectives. In: Singh H (ed) Forests and climate change. Springer Nature, Singapore. https://doi.org/10.1007/978-981-97-3905-9_1

Singh H (ed) (2024b) Forests and climate change: biological perspectives on impact, adaptation, and mitigation strategies. Springer Nature, Singapore. https://doi.org/10.1007/978-981-97-3905-9

Singh H (2024c) Climate change and environmental pollution: biological consequences on urban forests. In: Singh H (ed) Urban forests, climate change and environmental pollution. Springer Nature, Cham, Switzerland. https://doi.org/10.1007/978-3-031-67837-0_1

Singh H (ed) (2024d) Urban forests, climate change and environmental pollution: physio-biochemical and molecular perspectives to enhance urban resilience. Springer Nature, Cham, Switzerland. https://doi.org/10.1007/978-3-031-67837-0

Singh H (2024e) Exploring adaptive modulation in plant functional traits and its impact on the productivity of *Acacia auriculiformis* under $CO_2$-enriched environment. Ind Crops Prod 210(2024):118186. https://doi.org/10.1016/j.indcrop.2024.118186

Singh H, Kumar M (2022) Climate change and its impact on Indian Himalayan forests: current status and research needs. In: Rani S, Kumar R (eds) Climate change. Springer, Cham, pp 223–242. https://doi.org/10.1007/978-3-030-92782-0_11

Singh H, Kumar N, Kumar A (2024) Enhancing resource use efficiency in crops through plant functional traits. Springer, pp 97–117. ISBN: 978-981-97-1509-1

Tian Y, Wu Z, Li M, Wang B, Zhang X (2022) Forest fire spread monitoring and vegetation dynamics detection based on multi-source remote sensing images. Remote Sens 14(18):4431

Thapa S, Chitale VS, Pradhan S, Shakya B, Sharma S, Regmi S, Dangol GS (2021) Forest fire detection and monitoring. In: Earth observation science and applications for risk reduction and enhanced resilience in Hindu Kush Himalaya region: a decade of experience from SERVIR, pp 147–167

Wahyono Harjoko A, Dharmawan A, Adhinata FD, Kosala G, Jo KH (2022) Real-time forest fire detection framework based on artificial intelligence using color probability model and motion feature analysis. Fire 5(1):23

Wu D, Zhang C, Ji L, Ran R, Wu H, Xu Y (2021) Forest fire recognition based on feature extraction from multi-view images. Trait Signal 38(3)

Xu R, Lin H, Lu K, Cao L, Liu Y (2021) A forest fire detection system based on ensemble learning. Forests 12(2):217

Yang X, Chen R, Zhang F, Zhang L, Fan X, Ye Q, Fu L (2021) Pixel-level automatic annotation for forest fire image. Eng Appl Artif Intell 104:104353

Ya'acob N, Najib MSM, Tajudin N, Yusof AL, Kassim M (2021) Image processing based forest fire detection using infrared camera. In: Journal of physics: conference series, vol 1768, no 1. IOP Publishing, p 012014

Zhang L, Wang M, Fu Y, Ding Y (2022) A forest fire recognition method using UAV images based on transfer learning. Forests 13(7):975

Zheng X, Chen F, Lou L, Cheng P, Huang Y (2022) Real-time detection of full-scale forest fire smoke based on deep convolution neural network. Remote Sens 14(3):536

# Policies for Advancing Forest Fire Management Practices and Adaptation Strategies Concerning Climate Change

**Ramesh Kumar, Prity Singh Pippal, Rupendra Singh, Varsha Tanu, Kumari Gouri, Pankaj Kumar, and Rajesh Kumar**

**Abstract** The escalating threat of forest fires due to climate change necessitates robust policies for advancing forest fire management practices and implementing adaptation strategies. This paper examines current policies and explores potential frameworks to enhance resilience in the face of increasing fire risks. It delves into strategies for prevention, early detection, rapid response, and recovery measures, emphasizing the integration of climate change adaptation into forest management practices. The analysis considers international best practices, technological innovations, and community engagement approaches to develop a comprehensive policy framework for effective forest fire management. Forests are vital ecosystems globally, supporting millions of people and playing a crucial role in biodiversity preservation and climate regulation. However, the integrity and resilience of these ecosystems are under serious threat from the increasing frequency and severity of forest fires, impacting the environment, biodiversity, economy, and society. Factors contributing to this trend include climate change, land-use changes, invasive species, and human activities. Effective forest fire management is essential to mitigate these destructive impacts, with integrating traditional acquaintance into modern strategies proving

R. Kumar
Department of Civil Engineering, Marwadi University Research Centre, Marwadi University, Rajkot, Gujarat, India

P. S. Pippal · R. Kumar (✉)
Department of Environmental Science, Central University of Rajasthan, Bandar Sindri, Ajmer, Rajasthan, India
e-mail: rajesh.kumar@curaj.ac.in

R. Singh
CPS, School of Social Sciences, Jawaharlal Nehru University, Delhi, India

V. Tanu
IIHMR University, Jaipur, Rajasthan, India

K. Gouri
Magadh University, Bodhgaya, Gaya, Bihar, India

P. Kumar
Department of Environmental Science, Parul Institute of Applied Sciences, Parul University, Vadodara, Gujarat, India

 
H. Singh and N. Kumar (eds.), *Forest Fire and Climate Change*,
https://doi.org/10.1007/978-3-031-89967-6_24

beneficial, especially in areas with indigenous populations. Internationally, treaties such as the UN Framework Convention on Climate Change (UNFCCC) and the Sendai framework for disaster risk reduction provide essential frameworks for addressing wildfires within the climate change framework. These agreements stress the importance of global cooperation, research, and capacity building to improve fire management and disaster resilience. Countries develop tailored policies and strategies to mitigate wildfire risks at the regional level, focusing on community preparedness, fuel management, fire suppression, and land-use planning. Regional wildland fire networks facilitate collaboration and knowledge exchange among stakeholders across specific geographical areas.

**Keywords** Wildfire · Fire management · Policy initiatives · Climate change · Resilience

## 1 Introduction

The critical environmental policy and sustainability components are advancing forest fire management practices and developing adaptation strategies for climate change. The escalation in the frequency and intensity of forest fires attributed to climate change has spurred governments and organizations worldwide to acknowledge the pressing need for comprehensive policies. These policies aim to enhance forest fire prevention, early detection, rapid response, and recovery measures. Moreover, they stress integrating climate change adaptation strategies into forest management practices to bolster ecosystem resilience and mitigate the vulnerability of communities and biodiversity to fire-related risks (Savita et al. 2017; Singh 2021, 2024; Singh and Kumar 2022; Sett et al. 2022). This introduction lays the groundwork for delving into the diverse policies and initiatives geared towards advancing forest fire management and adaptation strategies to the changing climate.

Changing climate is a significant catalyst for global forest fires, impacting ecosystems, economies, and communities. As temperatures climb and weather patterns shift, forests are increasingly prone to extended droughts, higher temperatures, and heightened pest infestations, creating optimal conditions for wildfires to spread and intensify rapidly (Kumar et al. 2016). These fires devastate landscapes and contribute to environmental pollution. Understanding the intricate interplay between changing climate change and forest fires is essential for formulating effective mitigation and adaptation strategies to safeguard natural ecosystems and human livelihoods (Kaushik and Singh 2024).

Forests are among the Earth's most essential ecosystems as biodiversity reservoirs, climate regulators, and income streams for millions of people worldwide (Brack 2019; Sett et al. 2024; Barwa and Singh 2024). However, forest fires' fostering frequency and intensity severely threaten the ecosystem's integrity and resilience, with worldwide ecological, economic, and societal consequences (Vilà-Vilardell et al. 2020; Pereira et al. 2020). While forest ecosystems have shown resilience and historical adaptability to changing conditions, there is concern that future changes may surpass the natural adaptive capacity of ecosystems (Seppala 2009; Keenan 2015). In the last few years, the world has witnessed a stark increase in forest fires' frequency, intensity, and spatial extent, igniting concerns about their environmental, societal, and economic impacts (Garcia et al. 2021; Jones et al. 2022). Amazon rainforests in South America, North America, Eurasia boreal forests, and tropical/temperate forests in India and Africa are essential to natural heritage and give significant ecological, economic, and socioeconomic advantages. The forest cover supports a broad array of wildlife, acts as a carbon sink, and sustains the livelihoods of millions of people. However, this essential resource is facing a growing threat from forest fires. These forest fires have become progressively frequent and severe, exacerbating habitat loss, biodiversity decline, and carbon emissions (Millar et al. 2007; Keenan 2015). A forest fire, commonly called wildfire, is an uncontrolled fire in forested areas or other vegetated landscapes. These fires can vary in size, intensity, and behaviour, ranging from minor, low-intensity burns to large, intense conflagrations that spread rapidly across vast areas. Forest fires typically result from the ignition and combustion of vegetation, often fueled by dry conditions, high temperatures, and strong winds. They might be caused by natural phenomena such as lightning strikes, human activities such as campfires left unattended, discarded cigarettes, or intentional burning such as fuel, industrial, and crop residue (Saikanth et al. 2023). The increasing incidence of forest fires has surged across diverse biomes and regions, fueled by climate change, land-use change, invasive species, and human activities (Duane et al. 2021; Loehman et al. 2021). Forests are increasingly confronted with more frequent and severe disturbance (Dale et al. 2001; Bolte et al. 2009; Lorz et al. 2010; Abdullah et al. 2019). In European forests, approximately 33.4 billion tons of forest biomass faces substantial vulnerability to disturbances, particularly with heightened relative losses of 40% in wind throws and 34% in fires compared to a 26% impact from insect outbreaks during the period 1979–2018 (Forzieri et al. 2021). The susceptibility of forests to calamities is influenced by climate change, which can alter the frequency, strength, period, and timing of these events (Stanturf et al. 2014). The increased risk posed by climate-induced factors, such as reduced precipitation leading to more frequent and prolonged droughts, poses a significant threat to changing climate (Brázdil et al. 2009; Kolström et al. 2011). Rising annual temperatures and alterations in precipitation patterns that involve reduced seasonal rainfall in South Europe and increased levels in the north of Central Europe are anticipated to bring about substantial summer droughts/floods. These climatic shifts are expected to impact growth processes within forest environments considerably (Seidl et al. 2011; Kumar et al. 2019; Christidis and Stott 2022). Notably, the temperature in Europe has increased more rapidly, surpassing the global mean, registering a 2 °C rise from pre-industrial levels from 2012 to 2021. The most pronounced warming has been observed in Northeast Europe and Scandinavia (IPCC 2022).

Climate change is one of the most significant challenges of the twenty-first century and carries extensive consequences for the sustainability and resilience of metropolitan locations globally (IPCC 2021, 2023). Human-induced climate change introduces substantial threats to the planet, manifesting in elevated temperatures, severe weather events, and rising sea levels. This urban development modifies local climates, positioning urban areas at the forefront of climate-related consequences (Srivastava and Maity 2023). To assess the impact of climate change, scientists employ diverse methodologies and approaches (Kelman 2017; Reckien et al. 2017; Toukabri and Youssef 2022). Addressing the complexities of climate change involves a multifaceted approach, incorporating various methodologies and tools.

Additionally, the establishment of dedicated climate observatories plays a critical role in monitoring and learning the evolving dynamics of climate change. In this context, wildfires share similarities with climate change as they are not purely natural disasters but involve anthropogenic hazards that can be mitigated through human intervention. Using the climate change framework, "mitigation" is employed to reduce the magnitude of wildfire hazards through strategic fuel management (Sharifi 2021). Similarly, within this study, the term "adaptation" describes actions that aim to decrease the exposure and vulnerability of people and assets to wildfires. It is important to note that the mitigation-adaptation dichotomy remains robust, as specific actions can contribute to both mitigation and adaptation. Notably, mitigation and adaptation play crucial roles in diminishing wildfire damage. Therefore, determining an effective strategy for damage reduction involves identifying the optimal balance of mitigation and adaptation measures tailored to specific contexts. Mitigation minimizes forest damage by directly reducing the extent of burnt areas. This underscores the importance of a comprehensive approach integrating mitigation and adaptation strategies for effective wildfire management.

Therefore, this chapter investigates forest fire management's status in India, critically evaluates existing policies and practices, showcases best practices and case studies, and outlines recommendations for better policies. Through this comprehensive exploration, we hope to provide valuable insights and actionable recommendations for policymakers, practitioners, researchers, and other stakeholders involved in the endeavour to protect and sustainably manage India's forests in the face of escalating fire risks. Recognizing the urgency of addressing the global challenge of forest fires, policymakers, researchers, and practitioners have been actively formulating and implementing policies, strategies, and initiatives to enhance forest fire management practices. These efforts span national boundaries and encompass various approaches, from fire suppression and prevention to community-based forest management and landscape restoration. Through a comprehensive review of policy responses, technological innovations, and stakeholder engagement strategies, this chapter aims to contribute to developing holistic and adaptive approaches to forest fire management that integrate ecological, social, and economic considerations.

## *1.1 Factors Contributing to Forest Fires*

Forest fires are impacted by various elements that need a comprehensive understanding to formulate successful management strategies. These factors include climate conditions, human activities, fuel availability, topography, lightning, forest management practices, climate change, and insect infestations. Climate conditions, such as dry and hot weather, low humidity, and high winds, create ideal circumstances for fire ignition and spread. Additionally, the accumulation of dry and combustible vegetation, exacerbated by inadequate forest management and land-use practices, contributes significantly to fire risks. Human activities, including agricultural burning, industrial operations, campfires, and arson, also play a pivotal role in igniting forest fires.

Furthermore, topographic features such as slope steepness and terrain can influence fire behaviour by affecting wind patterns and fuel distribution. Natural ignition sources like lightning strikes further contribute to fires, often presenting challenges in remote areas. Forest management practices, such as fire suppression policies and prescribed burning, impact fire dynamics, with suppression policies sometimes leading to more severe fires due to fuel accumulation. Climate change exacerbates fire risks by altering weather patterns, increasing temperatures, and prolonging drought conditions. However, insect infestations and disease outbreaks weaken trees, increasing fuel loads and fire susceptibility. An integrated approach considering these factors is essential for mitigating fire risks and promoting resilient forest ecosystems. Understanding these factors and their interactions is critical for developing effective forest fire management strategies that address the underlying drivers of wildfires and reduce their impacts on communities, ecosystems, and economies. Integrated approaches considering climate projections, landscape characteristics, human activities, and ecological processes are necessary to mitigate fire risks and promote resilient forest ecosystems.

### *1.2 Impact of Climate Change on Forest Fires*

The changing climate is indisputably reshaping the landscape of forest fires worldwide, and an era of heightened frequency, intensity, and duration of these devastating events accompanies it. This reality, far from being a speculative future scenario, is unfolding before our eyes across various regions. The increase in fire seasons is one of climate change's most conspicuous effects. As global temperatures rise, even a modest 1 °C increase can extend the fire season by several weeks, providing ample time for ignitions and existing fires to intensify (Abatzoglou et al. 2018). Moreover, earlier snowmelt, documented by Mote et al. (2005), exposed dry vegetation sooner, extending the window of vulnerability to fires. In addition to prolonging fire seasons, climate change amplifies the behaviour of wildfires. The escalation of drought severity and soil moisture reduction fosters drier vegetation that ignites faster and burns hotter, unleashing more formidable fires (Trenberth et al. 2018). However, altered wind patterns, including heightened speeds, exacerbate fire spread by fanning flames and hindering containment efforts (Diffenbaugh et al. 2015). Beyond immediate impacts, climate change is reshaping fire regimes and ecosystems. Changes in temperature and precipitation patterns drive shifts in vegetation composition, potentially expanding the range of fire-prone species and fundamentally altering fire dynamics in affected regions (Kelly and Goulden 2008). The warming temperatures fuel insect outbreaks, contributing to increased wildfire risk by weakening trees and enhancing fuel loads (Kurz et al. 2008). The interplay between wildfires and climate change forms a dangerous feedback loop. Wildfires release copious amounts of greenhouse gases, exacerbating global warming and fueling further fire activity (Carmichael et al. 2003). Therefore, burned forests lose their ability to sequester carbon dioxide, perpetuating atmospheric $CO_2$ levels and intensifying warming (Randerson et al. 2006). Addressing the complex faces shown by climate change-driven wildfires requires a multifaceted approach.

## 2 Adaptation Strategies for Climate-Resilient Forest Fire Management

Climate change significantly challenges forest ecosystems, amplifying wildfires' frequency, intensity, and duration. To effectively adapt and manage this escalating threat, a comprehensive set of strategies can be implemented to bolster resilience in our forests. Forest management practices such as fuel reduction through prescribed burns, mechanical thinning, and forest restoration are vital in mitigating wildfire risks (Stephens 2018). Landscape management approaches, including creating firebreaks and fuel breaks, also comprehend fires and limit their spread (Agee 1993). Enhanced monitoring systems and improved forecasting models enable early detection and faster response to wildfires, utilizing real-time data on fire risk factors (Eidenshink et al. 2007; Abatzoglou et al. 2019). Investing in firefighting resources and community preparedness further strengthens wildfire response capabilities, emphasizing the importance of adequate funding, personnel, and public education (Headley 2016). Policy measures, including land-use planning and international collaboration, are essential for reducing risks to people and property and improving global preparedness for fire management challenges (Radeloff 2005; FAO 2015). Forest managers traditionally prioritize long-term thinking, weighing the enduring consequences and incorporating uncertainty and unknowns into their management approach (Guariguata et al. 2008). Moreover, investing in research and innovation, such as studying fire behaviour and leveraging technological advancements, holds promise for enhancing fire management capabilities in changing climates (Ince et al. 2011). Through collaborative efforts and long-term planning, we can work towards a future where forests are better equipped to withstand the challenges posed by a changing climate.

## 3 Community Engagement and Traditional Methods of Forest Fire Management

Community engagement is crucial in leveraging local knowledge and fostering collective responsibility. Communities possess invaluable insights into local fire behaviour, historical fire patterns, and fuel types (Gavin et al. 2013). Engaging communities in fire management planning and decision-making tap into this wealth of knowledge and cultivates a sense of ownership and accountability (McCarry 2010). Collaborative platforms facilitate effective communication about fire risks, prevention strategies, and response plans, enhancing overall preparedness and resilience (The Nature Conservancy 2023). However, various strategies to enhance community engagement include community fire planning workshops, Fire wise communities' initiatives, and indigenous knowledge exchange programs. These initiatives empower communities to actively participate in fire management efforts, tailor solutions to local needs and concerns, and preserve traditional knowledge and practices that have sustained ecosystems for generations. The traditional methods of fire management, rooted in indigenous knowledge and practices, offer additional layers of resilience and effectiveness. For centuries, indigenous communities have employed controlled burns, fuel reduction techniques, and fire monitoring systems to manage vegetation growth, reduce fuel loads, and monitor fire risks (Gavin et al. 2003). These methods have cultural significance and provide ecological benefits by promoting biodiversity, ecosystem health, and resilience.

Further, combining modern and traditional approaches to fire management requires integration, respect, and collaboration. Integrating community knowledge and traditional practices with modern scientific understanding and technology can lead to more effective and culturally sensitive fire management strategies. However, challenges such as lack of recognition, power imbalances, and cultural sensitivity must be addressed through respectful partnerships and dialogue between communities, authorities, and researchers. Implementing community-based fire management practices, including establishing fire management committees, volunteer firefighter groups, and community patrols to monitor and respond to fire incidents. Therefore, by integrating community engagement and traditional methods into forest fire management strategies, we can enhance the effectiveness, sustainability, and resilience of efforts to prevent and mitigate the impacts of forest fires while fostering greater community empowerment and resilience.

## 4 Current Policies and Challenges in Forest Fire Management

Climate change significantly challenges forest ecosystems, amplifying wildfires' frequency, intensity, and duration. To effectively adapt and manage this rising threat, a comprehensive set of strategies can be implemented to bolster resilience in our forests. Forest management practices such as fuel reduction through prescribed burns, mechanical thinning, and forest restoration are vital in mitigating wildfire risks (Stephens 2018; National Park Service 2023). Landscape management approaches, including creating firebreaks and fuel breaks, also contain fires and limit their spread (Agee 1993). Enhanced monitoring systems and improved forecasting models enable early detection and faster response to wildfires, utilizing real-time data on fire risk factors (Eidenshink et al. 2007; Abatzoglou et al. 2019). Investing in firefighting resources and community preparedness further strengthens wildfire response capabilities, emphasizing the importance of adequate funding, personnel, and public education (Headley 2016; National Fire Protection Association 2023). Policy measures, including land-use planning and international collaboration, are essential for reducing risks to people and property and improving global preparedness for fire management challenges (Radeloff 2005; FAO 2015). Moreover, investing in research and innovation, such as studying fire behaviour and leveraging technological advancements, holds promise for enhancing fire management capabilities in a changing climate. Adapting to climate change requires a coordinated, multifaceted approach that integrates traditional knowledge with modern scientific understanding, fostering climate-resilient forest fire management practices.

## *4.1 Global Initiatives and Frameworks*

The UNFCCC and the Sendai Framework for Disaster Risk Reduction are pivotal international agreements addressing the challenges posed by wildfires. The UNFCCC focuses on climate change mitigation and adaptation, recognizing the increasing significance of wildfires due to global warming. It emphasizes the importance of international cooperation in research and technology to enhance fire management effectiveness. On the other hand, the Sendai Framework for Disaster Risk Reduction, spanning from 2015 to 2030, takes an inclusive approach to risk reduction, encompassing wildfires as one of its key focal points. Together, these frameworks provide a critical foundation for international collaboration and action toward the complex problem of wildfires within the broader context of climate change and disaster risk management. Globally unified policy for forest fire management, several vital approaches and frameworks guide fire management efforts worldwide, often implemented at regional or national levels. These policies provide a foundation for forest fire management in India, addressing various aspects such as prevention, early detection, suppression, and community participation.

### 4.2 *Regional Approaches for Sustainable Forest Fire Management and Resilience*

Regional methods for sustainable forest fire management and resilience involve coordinated efforts among multiple stakeholders to mitigate the risk of wildfires, protect ecosystems and communities, and endorse long-term resilience. Collaborative planning among regional stakeholders, including government agencies, local communities, indigenous groups, non-governmental organizations (NGOs), and industry representatives, is essential for developing comprehensive wildfire management plans. These plans should consider ecological conditions, land use patterns, community vulnerabilities, and available resources. Conducting comprehensive risk assessments at the regional level helps identify areas with the highest wildfire risk, vulnerable ecosystems, critical infrastructure, and communities at risk. This information guides prioritizing mitigation efforts and allocating resources to needy areas. Investing in ecosystem restoration initiatives, such as reforestation, watershed protection, and habitat conservation, is crucial in enhancing ecosystem resilience to wildfires and promoting long-term sustainability. Adopting a holistic, collaborative approach to wildfire management at the regional level, stakeholders can enhance resilience to wildfires, protect communities and natural resources, and promote sustainable forest management practices. This integrated approach acknowledges the interconnectedness of ecosystems, communities, and wildfire dynamics, leading to more effective and resilient wildfire management strategies.

These initiatives aim to proactively prevent, mitigate, and respond to forest fires, fostering enduring ecological and community well-being. These approaches underscore the necessity for joint strategies and inclusive planning on a regional scale. By embracing such regional frameworks, stakeholders can consolidate resources, exchange knowledge, and cultivate a more robust environment capable of enduring and rebounding from the impacts of forest fires. The core focus is achieving sustainability by harmonizing ecological health, community safety, and economic considerations. Some approaches have been elaborated in Table 1 and discussed briefly.

**Table 1** Regional approaches for sustainable forest fire management and resilience to climate change

| Regional approaches for sustainable forest fire management | | | | | |
|---|---|---|---|---|---|
| United States National Fire (2000) | European Union Forest Strategy (2020) | Forest Conservation Act, India (1980) | National Forest Policy of India (1988) | National Action Plan on Forest Fire in India (2000) | National Disaster Management Framework (2005) |

### 4.2.1 United States National Fire Plan (2000)

The United States National Fire Plan (NFP) was initiated in response to severe wildfires in the late 1990s and established in August 2000. It responded to severe wildfires and aimed to enhance wildfire response, restore fire-affected ecosystems, empower communities, and ensure firefighting capacity. It focuses on reducing wildfire risks and protecting communities, natural resources, and infrastructure. The plan emphasizes collaborative efforts among federal, state, local, and tribal agencies, with prime components including fuel reduction projects, community protection plans, and technology development. It prioritized safe firefighting, ecosystem restoration, hazardous fuels reduction, community assistance, and accountability. The NFP remains a cornerstone of wildland fire management, fostering initiatives like the Healthy Forests Initiative. Over the past five years (2004–2008), 44,000 fuel treatments were executed across the western United States (Schoennagel et al. 2009). This concerted effort reflects a commitment to proactive measures to manage and mitigate the threat of wildfires, emphasizing the importance of strategic interventions to enhance landscape resilience and protect ecosystems and communities. To evaluate the effectiveness of wildfire risk reduction, focusing on areas where such measures would have the most significant potential to mitigate fire risks for nearby homes and communities. Despite federal policies emphasizing substantial investment in the WUI, the findings revealed that only 3% of the treated area fell within the WUI. This discrepancy underscores challenges in aligning policy objectives with on-the-ground implementation, particularly in areas critical for community fire safety. At the same time, treatments conducted in areas distant from the wildland-urban interface (WUI) may provide certain fire mitigation benefits. This strategic emphasis is essential in the National Fire Plan (NFP) for effectively reducing fire risk. However, achieving this goal may necessitate a shift in management and policy focus, moving from an exclusive emphasis on public lands to a more inclusive approach encompassing private lands (Schoennagel et al. 2009). This adjustment acknowledges the importance of cooperative efforts across diverse landownership types, recognizing the shared responsibility in bolstering the overall resilience of communities against the threat of wildfires. The National Fire Plan has improved firefighting capabilities, enhanced community preparedness, and restored fire-affected landscapes in the US. In India, similar policies have helped reduce wildfire risks, protect biodiversity, and involve communities in fire management activities.

### 4.2.2 European Union Forest Strategy (2020)

The European Union Forest Strategy 2020 promotes sustainable forest management and resilience to climate change, pests, and diseases. While not explicitly focused on fire management, it addresses the importance of preventing and mitigating forest fires. The strategy emphasizes risk-based approaches, early warning systems, and cross-border cooperation. Similar strategies are adopted in India under national and state-level forest policies. The EU Forest Strategy has improved forest management practices and resilience to forest fires in Europe, fostering collaboration, knowledge exchange, and public awareness. In India, similar policies have enhanced forest protection, community engagement, and technological advancements in fire management.

### 4.2.3 Forest Conservation Act (FCA), India (1980)

The Forest Conservation Act (FCA) in India, sanctioned in 1980 by the Government of India, plays a pivotal role in regulating and standardizing procedures related to the alteration of forest areas for non-forest purposes. This legislation establishes a framework for managing the forest land for activities such as mining, industries, and the construction of dams. The Forest Conservation Act of 1980 is a central legislation aimed at conserving and protecting forests in India. While the act primarily focuses on regulating the change of forest land for non-forest purposes, it also plays a role in forest fire management. The act empowers the central government to declare areas as reserved forests or protected forests, which receive legal protection against unauthorized activities, including deliberate setting of fires. By safeguarding forest areas from encroachment and exploitation, the act indirectly contributes to reducing the occurrence of forest fires and preserving forest ecosystems. The State of Forests Report (SFR 1987) indicates that India's most significant loss of forests occurred between 1950 and 1980, just before implementing the FCA (Table 2). In the above-mentioned period, substantial forest areas were allocated to various sectors for developmental activities. The conversion of forest areas for non-forestry purposes during this timeframe encompassed uses such as agriculture, industries/townships, and road construction. This historical context underscores the need for regulatory measures like the Forest Conservation Act to address and manage the allocation of forested land for development while emphasizing the importance of sustainable forestry practices and environmental conservation. The Forest Conservation Act (FCA) incorporates a built-in mechanism to address forest loss resulting from diversion through compensatory afforestation (CA), which involves planting trees in non-forest or degraded forest sites near the project area. The National Forest Policy of 1988 underscores the significance of safeguarding forests as state property, managed with the expertise of specialists (Kohli and Menon 2011). A Central Empowered Committee (CEC) report highlighted that by 2002, less than 60% of the total funds recovered from user agencies had been spent, covering only 61% of the total target area for CA (Kohli et al. 2011). This structured approach aims to ensure more effective utilization of funds and streamline afforestation efforts to offset the environmental impact of forest diversions under the Forest Conservation Act.

### 4.2.4 National Forest Policy of India (1988)

The National Forest Policy (NFP) of 1988 is a comprehensive policy framework aimed at conserving and managing India's forest resources sustainably. While the

**Table 2** Conversion of forested areas for non-agricultural uses from 1951 to 1980 based on the State of Forests Report (SFR 1987)

| Purpose | Area covered (million hectares) |
|---|---|
| Farming | 2.623 |
| Projects in river valleys | 0.502 |
| Industrial facilities and residential areas | 0.134 |
| Electricity transmission infrastructure and road networks | 0.061 |
| Others | 1.008 |
| Total | 4.328 |

policy does not explicitly focus solely on forest fire management, it emphasizes the need to protect forests from various threats, including fires. The policy advocates for measures such as afforestation, regeneration of degraded forests, and promoting community participation in forest management. These practices indirectly contribute to mitigating forest fire risks. Additionally, the NFP underscores the importance of scientific management practices, including fire prevention and suppression techniques, to ensure forests' ecological integrity and productivity. This change in perspective toward forest resources has brought about substantial socioeconomic, cultural, and ecological transformations. The NFP, initiated in 1952, has consistently advocated for achieving a 33% forest cover nationwide, with a heightened target of 60% forest cover in mountainous regions (Joshi et al. 2011). The continuity of this goal reflects a sustained commitment to fostering environmental sustainability and acknowledging the multifaceted benefits of forest conservation and responsible land management.

The National Forest Policy (NFP) examines forest cover trends at the state level to meet its policy objectives. Currently, only three Indian states meet these targets, with another three having the potential to achieve them through the afforestation of their wasteland areas. Several other states could reach the 33% target if they have adequate possessions attainable in the states (FSI 2009).

Recognizing the importance of incorporating additional natural ecosystems, the National Forest Policy (NFP) aligns with the Planning Commission's focus during the XI Five-year Plan (2007–12). This paper evaluates the targets above in light of the Planning Commission's suggestions. It argues that the NFP should undergo reassessment and revision to align with current realities, establishing more feasible and achievable goals for forest and tree cover in India. This proactive approach reflects a commitment to addressing challenges and establishing a more efficient and sustainable framework for forest and land management in the country.

### 4.2.5 National Action Plan on Forest Fire in India (2000)

According to the Forest Survey of India report (FSI 2009), 54.40% of India's forests face rare fires, 7.49% experience frequent fires, 2.40% encounter frequent fires, and

35.71% have not yet been exposed to fire incidents. The forests in northeastern India and the Deccan Plateau are the most severely affected regions. Extended periods of drought have increased forest susceptibility to fires, a vulnerability further compounded by climate change. Many of India's forests are fragmented and interspersed with settlements of varying sizes, leading to high population densities within forested areas and increased fire risk.

In response to these challenges, the National Action Plan on Forest Fire (NAPFF) was established in 2000 to address India's escalating threat of forest fires. This plan outlines a strategic framework for improving forest fire management practices nationwide, aiming to reduce forest vulnerabilities to fire hazards across diverse ecosystems in the country. Implementing such measures is expected to enhance the capabilities of forest personnel and other relevant institutions in effectively combating fires and expediting recovery efforts following fire incidents.

Key components of the NAPFF include early detection and monitoring of fires, enhancing fire suppression capabilities, raising awareness among stakeholders, and promoting research and development in fire management technologies. The plan also underscores the importance of collaborative efforts involving government agencies, local communities, non-governmental organizations (NGOs), and research institutions to ensure effective forest fire management.

### 4.2.6 National Disaster Management Framework (NDMF)

The National Disaster Management Act of 2005 (NDMA) is a comprehensive law administered by the Ministry of Home Affairs, serving as India's central coordinator for responses to natural and man-made disasters. Established following the 2001 Gujarat earthquake, it sets up a structured disaster management framework nationwide, assigning specific roles and tasks to different departments or ministries to handle emergencies efficiently. It takes a holistic approach to disaster management, including natural calamities like forest fires, and emphasizes integrating disaster risk reduction measures into forest management practices.

The National Disaster Management Plan (NDMP), launched in 2016, is India's strategic blueprint to enhance the country's resilience to disasters, whether natural or human-induced. Its objective is to strengthen India's disaster resilience, reduce disaster risks significantly, and minimize losses across various aspects of life and resources—economic, physical, social, cultural, and environmental. This is achieved by enhancing disaster coping capacities at all administrative levels and within communities. The NDMP is aligned with the Sendai Framework for Disaster Risk Reduction 2015–2030, a United Nations initiative to substantially reduce global disaster risks and losses by 2030. Additionally, it adheres to the Disaster Management Act of 2005 and the National Policy on Disaster Management of 2009. The framework promotes proactive approaches like conducting risk assessments, developing contingency plans, and enhancing capacity through training to improve preparedness and response during forest fire incidents. It stresses the importance of coordinated

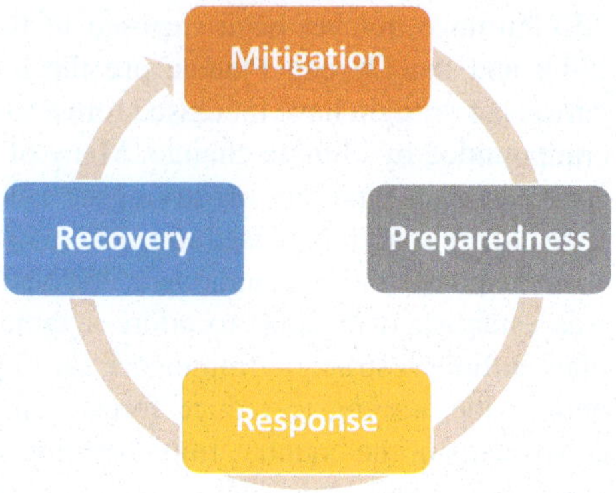


**Fig. 1** The emergency management cycle signifies the role of all communities in at least one phase of emergency management at any time

efforts among national, state, and local stakeholders to manage forest fire emergencies effectively.

Emergency management considers disasters in four phases: mitigation, preparedness, response, and recovery, which can be illustrated with the diagram relating the four phases of emergency management (Fig. 1).

## 5 Limitations and Recommendations

Despite these forest fire management efforts, challenges such as inadequate resources, limited capacity, and climate change continue to pose challenges to effective forest fire management in India. Addressing these challenges requires sustained commitment, enhanced coordination among stakeholders, and integration of scientific knowledge, traditional practices, and community participation into forest fire management strategies. Effective forest fire management requires a holistic and multifaceted approach encompassing prevention, mitigation, and adaptation strategies. By adopting proactive measures and fostering collaboration among governments, communities, and stakeholders, we can mitigate the impacts of forest fires and build resilience to future fire hazards in a changing climate.

## 6 Conclusion

The analysis of the national and international policies emphasized the need for comprehensive frameworks addressing prevention, early detection, suppression, and post-fire recovery. Successful forest fire management relies on collaboration and partnerships among government agencies, local communities, NGOs, research institutions, and other stakeholders. Fostering cooperation and leveraging collective expertise, resources, and technologies can enhance efforts and achieve greater resilience to wildfires. Community engagement is crucial, empowering local communities through education, capacity building, and participatory decision-making processes. Integrating traditional knowledge and practices of local communities into modern fire management strategies can further improve effectiveness and sustainability. While conventional methods like controlled burning, firebreaks, and early warning systems have proven valuable, the success of these policies underlines the importance of adaptive management and continuous evaluation. Proactive policies, collaborative efforts, community engagement, and integrating traditional knowledge with modern approaches are essential for effective forest fire management. Adopting these principles and building on successful global experiences may help policymakers enhance wildfire resilience and safeguard valuable forest ecosystems for future generations. Leveraging technological advancements, such as early warning systems and fire monitoring tools, further strengthens fire management capabilities by aiding prediction and response efforts. However, the dynamic nature of the wildfire landscape, driven by climate change, necessitates ongoing adaptation and innovation to address the frequency, intensity, and unpredictability of wildfires. Integrating international cooperation, regional collaboration, community engagement, and technological advancements lays the groundwork for holistic and sustainable forest fire management practices. This comprehensive approach will contribute to a future where forests are better equipped to withstand the impacts of climate change and wildfires, ensuring the protection of ecosystems and livelihoods.

## References

Abatzoglou JT, Williams AP, Moritz MA (2018) Climate change and shifting fire regimes in California and the Pacific Northwest. Nat Clim Change 8(11):917–922, 1028–1032. https://doi.org/10.1038/s41558-018-0280-3

Abatzoglou JT, Williams AP, Moritz MA (2019) Climate change and shifting fire regimes in California and the Pacific Northwest. Environ Res Lett 14(3):034005. https://doi.org/10.1088/1748-9326/aafe9c

Abdullah MM, Assi AT, Asadalla NB (2019) Integrated ecosystem sustainability approach: toward a holistic system of thinking of managing arid ecosystems. Open J Ecol 9(11):493

Agee JK (1993) Fire ecology & management: principles & applications. Sinauer Associates, Inc. Publishers

Barwa S, Singh H (2024) Climate-induced changes in forest ecosystems of Indian Himalayan region: a meta-analysis. In: Singh H (ed) Urban forests, climate change and environmental pollution. Springer, Cham. https://doi.org/10.1007/978-3-031-67837-0_39

Bolte A, Ammer C, Löf M, Nabuurs GJ, Schall P, Spathelf P (2009) Adaptive forest management: a prerequisite for sustainable forestry in the face of climate change. In: Sustainable forest management in a changing world: a European perspective, pp 115–139

Brack D (2019) Forests and climate change. In: Proceedings of background study prepared for the fourteenth session of the United Nations forum on forests. United Nations Forum on Forests, New York, NY, USA

Brázdil R, Trnka M, Dobrovolný P, Chromá K, Hlavinka P, Žalud Z (2009) Variability of droughts in the Czech Republic, 1881–2006. Theor Appl Climatol 97(3):297–315

Carmichael GR, Wong DC, Ye Q (2003) Tropospheric ozone formation at continental scales: a satellite analysis. Glob Biogeochem Cycl 17(4):1027. https://doi.org/10.1029/2002GB001801

Christidis N, Stott PA (2022) Anthropogenic climate change and the record-high temperature of May 2020 in Western Europe. Bull Am Meteorol Soc 103(3):S33–S37

Dale VH, Joyce LA, McNulty S, Neilson RP, Ayres MP, Flannigan MD, Hanson PJ, Irland LC, Lugo AE, Peterson CJ, Simberloff D (2001) Climate change and forest disturbances: climate change can affect forests by altering the frequency, intensity, duration, and timing of fire, drought, introduced species, insect and pathogen outbreaks, hurricanes, windstorms, ice storms, or landslides. Bioscience 51(9):723–734

Diffenbaugh NS, Swain DL, Touma D (2015) A tale of two California droughts: differences in climatic drivers and socioeconomic impacts. Proc Natl Acad Sci 112(14):3981–3986. https://doi.org/10.1073/pnas.1421983112

Duane A, Castellnou M, Brotons L (2021) Towards a comprehensive look at global drivers of novel extreme wildfire events. Clim Change 165(3–4):43

Eidenshink J, Schwind B, Held A, Kollmeyer W, Mukhorta A (2007) A 13-year record of annual burned area in boreal and temperate Eurasia derived from the earth observing system MODIS moderate resolution imaging spectroradiometer (MODIS). Remote Sens Environ 111(3):362–378. https://doi.org/10.1016/j.rse.2007.01.024

FAO (Food and Agriculture Organization of the United Nations) (2015) Building resilience for adaptation to climate change through sustainable forest management

Forzieri G, Girardello M, Ceccherini G, Spinoni J, Feyen L, Hartmann H, Beck PS, Camps-Valls G, Chirici G, Mauri A, Cescatti A (2021) Emergent vulnerability to climate-driven disturbances in European forests. Nat Commun 12(1):1081

FSI (2009) Indian state of forest report (ISFR) 2009. Forest Survey of India, MoEF, GoI, Dehradun, p 199

Garcia LC, Szabo JK, de Oliveira RF, Pereira AD, da Cunha CN, Damasceno-Júnior GA, Morato RG, Tomas WM, Libonati R, Ribeiro DB (2021) Record-breaking wildfires in the world's largest continuous tropical wetland: integrative fire management is urgently needed for both biodiversity and humans. J Environ Manage 293:112870

Gavin DG, Brubaker LB, Lertzman KP (2003) Holocene fire history of a coastal temperate rain forest based on soil charcoal radiocarbon dates. Ecology 84(1):186–201

Gavin MC, McCarry MK, Flannigan PN (2013) Can local communities help reduce fire risk? Exploring the potential of community-based fire management in boreal forests. J Appl Ecol 50(1):102–112. https://doi.org/10.1111/1365-2664.12034

Guariguata MR, Cornelius JP, Locatelli B, Forner C, Sánchez-Azofeifa GA (2008) Mitigation needs adaptation: tropical forestry and climate change. Mitig Adapt Strateg Glob Change 13:793–808. https://doi.org/10.1007/s11027-007-9141-2

Headley S (2016) Resource constraints and fire management: a wicked problem. J Environ Manage 173:113–121

Ince PJ, Kramp AD, Skog KE, Yoo DI, Sample VA (2011) Modeling future US forest sector market and trade impacts of expansion in wood energy consumption. J for Econ 17:142–156. https://doi.org/10.1016/j.jfe.2011.02.007

IPCC (2022) Climate change 2022: mitigation of climate change. Cambridge University Press, Cambridge, UK; New York, NY, pp 783–2042

IPCC (2023) Climate change 2023: synthesis report. In: Core Writing Team, Lee H, Romero J (eds) Contribution of working groups I, II and III to the sixth assessment report of the Intergovernmental Panel on Climate Change. IPCC, Geneva, Switzerland, pp 35–115. Accessed 28 Nov 2023. Available online: https://www.ipcc.ch/report/ar6/syr/downloads/report/IPCC_AR6_SYR_FullVolume.pdf

Jones MW, Abatzoglou JT, Veraverbeke S, Andela N, Lasslop G, Forkel M, Smith AJ, Burton C, Betts RA, van der Werf GR, Sitch S (2022) Global and regional trends and drivers of fire under climate change. Rev Geophys

Joshi AK, Pant P, Kumar P, Giriraj A, Joshi PK (2011) National forest policy in India: critique of targets and implementation. Small-Scale for 10:83–96

Kaushik P, Singh H (2024) Climate change-induced forest ecosystem functioning and alteration in water–atmosphere–soil–plant (WASP) relationship. In: Singh H (ed) Urban forests, climate change and environmental pollution. Springer, Cham. https://doi.org/10.1007/978-3-031-67837-0_38

Keenan RJ (2015) Climate change impacts and adaptation in forest management: a review. Ann for Sci 72:145–167. https://doi.org/10.1007/s13595-014-0446-5

Kelly AE, Goulden ML (2008) Regional variations in hydrologically driven $CO_2$ exchange from California forests. Ecol Lett 11(8):831–840. https://doi.org/10.1111/j.1461-024

Kelman I (2017) Linking disaster risk reduction, climate change, and the sustainable development goals. Disaster Prev Manag Int J 26:254–258

Kohli K, Menon M (2011) Banking on forests: assets for a climate cure? Kalpavriksh and Heinrich Boll Foundation, New Delhi

Kohli KM, Menon V, Samdariya S, Guptabhaya (2011) Pocketful of forests: legal debates on valuating and compensating forest loss in India. Kalpavriksh & WWF-India, New Delhi

Kolström M, Lindner M, Vilén T, Maroschek M, Seidl R, Lexer MJ, Netherer S, Kremer A, Delzon S, Barbati A, Marchetti M, Corona P (2011) Reviewing the science and implementation of climate change adaptation measures in European Forestry. Forests 2(4):961–982

Kumar M, Singh H, Pandey R, Singh M, Ravindranath NH, Kalra N (2019) Assessing vulnerability of forest ecosystem in the Indian Western Himalayan region using trends of net primary productivity. Biodivers Conserv 28(8):2163–2182

Kumar R, Singh S, Kumar R, Singh A, Bhardwaj A, Bhardwaj A, Sam L, Randhawa SS, Gupta A (2016) Development of a glacio-hydrological model for discharge and mass balance reconstruction. Water Resour Manag 30(10):3475–3492

Kurz WA, Stinson G, Rampley GJ, Dymond CC, Neilson ET (2008) Risk of natural disturbances makes future contribution of Canada's forests to the global carbon cycle highly uncertain. Proc Natl Acad Sci USA 105(5):1551–1555

Loehman RA, Friggens MM, Sherriff RL, Keyser AR, Riley KL (2021) Impacts of climate changes and amplified natural disturbance on global ecosystems. In: The Routledge handbook of landscape ecology, pp 175–198

Lorz C, Fürst C, Galic Z, Matijasic D, Podrazky V, Potocic N, Simoncic P, Strauch M, Vacik H, Makeschin F (2010) GIS-based probability assessment of natural hazards in forested landscapes of central and south-eastern Europe. Environ Manage 46:920–930

Masson-Delmotte V, Zhai P, Pirani A, Connors SL, Péan C, Berger S, Caud N, Chen Y, Goldfarb L, Gomis MI (eds) (2021) Climate change 2021 the physical science basis. Contribution of working group I to the sixth assessment report of the Intergovernmental Panel on Climate Change. Cambridge University Press, Cambridge, UK, New York, NY, USA

McCarry MK (2010) Public participation in fire management: a literature review. Environ Pract 12(3):239–253. https://doi.org/10.1007/s10026-010-9100-y

Millar CI, Stephenson NL, Stephens SL (2007) Climate change and forests of the future: managing in the face of uncertainty. Ecol Appl 17(8):2145–2151. https://doi.org/10.1073/pnas.1617464114

Mote PW, Hamlet AF, Clark MP, Lettenmaier DP (2005) Declining mountain snowpack in western north america. Bull Am Meteorol Soc 86(1):39–49

National Fire Protection Association (2023) Firewise USA. https://www.nfpa.org/education-and-research/wildfire/firewise-usa

National Park Service (2023) How we manage fire. https://www.nps.gov/subjects/fire/index.htm

Pereira SC, Marta-Almeida M, Carvalho AC, Rocha A (2020) Extreme precipitation events under climate change in the Iberian Peninsula. Int J Climatol 40(2):1255–1278. https://doi.org/10.1002/joc.6269

Radeloff VC (2005) The impact of urban sprawl on fire risk and fire suppression costs in California. Ecol Appl 15(3):1760–1775

Randerson JT, Liu H, Flanner M, Chambers SD, Jin Y, Hess PG, Pfister G, Mack MC, Treseder KK, Welp LR, Chapin FS, Harden JW, Goulden ML, Lyons E, Neff JC, Schuur EAG, Zender CS (2006) The impact of boreal forest fire on climate warming. Science 314(5802):1130–1132

Reckien D, Creutzig F, Fernandez B, Lwasa S, Tovar-Restrepo M, Mcevoy D, Satterthwaite D (2017) Climate change, equity and the sustainable development goals: an urban perspective. Environ Urb 29:159–182

Saikanth DR, Kumar S, Rani M, Sharma A, Srivastava S, Vyas D, Singh GA, Kumar S (2023) A comprehensive review on climate change adaptation strategies and challenges in agriculture. Int J Environ Clim Change 13(11):10–19

Savita, Sharma SK, Singh H, Singh O (2017) Forest fire: a review. Indian for 143(8):719–728

Schoennagel T, Nelson CR, Theobald DM, Carnwath GC, Chapman TB (2009) Implementation of national fire plan treatments near the wildland–urban interface in the western United States. PNAS 106(26):10706–10711

Seidl R, Seidl R, Schelhaas M-J, Lexer MJ (2011) Unraveling the drivers of intensifying forest disturbance regimes in Europe. Glob Chang Biol 17(9):2842–2852

Seppala R (2009) A global assessment on adaptation of forests to climate change. Scand J for Res 24:469–472. https://doi.org/10.1080/02827580903378626

Sett T, Nikam BR, Singh H, Purohit S (2022) Effect of drought stress on forest evapotranspiration—a case study on Indian forests. In: EGU general assembly 2022. https://doi.org/10.5194/egusphere-egu22-510

Sett T, Nikam BR, Singh H, Purohit S (2024) Hydrological variability in Indian forest ecosystem: analysis of drought resilience, recovery and water use efficiency in moist and dry deciduous forests. In: Singh H (ed) Urban forests, climate change and environmental pollution. Springer, Cham. https://doi.org/10.1007/978-3-031-67837-0_37

Sharifi A (2021) Co-benefits and synergies between urban climate change mitigation and adaptation measures: a literature review. Sci Total Environ 750:141642

Singh H (2021) Process-based carbon sequestration study in relation to the energy-water-carbon flux in a forest ecosystem. In: Kumar P, Singh RK, Kumar M, Rani M, Sharma P (eds) Climate impacts on sustainable natural resource management. CRC Press, Taylor & Francis Group, United States, pp 336–361

Singh H (2024) Exploring adaptive modulation in plant functional traits and its impact on the productivity of *Acacia auriculiformis* under $CO_2$-enriched environment. Ind Crops Prod 210(2024):118186. https://doi.org/10.1016/j.indcrop.2024.118186

Singh H, Kumar M (2022) Climate change and its impact on Indian Himalayan forests: current status and research needs. In: Rani S, Kumar R (eds) Climate change. Springer, Cham, pp 223–242. https://doi.org/10.1007/978-3-030-92782-0_11

Srivastava A, Maity R (2023) Assessing the potential of AI–ML in urban climate change adaptation and sustainable development. Sustainability 15(23):16461

Stanturf JA, Palik BJ, Williams MI, Dumroese RK (2014) Forest restoration paradigms. J Sustain For 33:161–194

Stephens SL (2018) Fire and fire surrogates for reducing forest fire severity in the western United States. Fire 1(1):3

The Nature Conservancy (2023) Working fire. https://www.nature.org/en-us/what-we-do/our-priorities/protect-water-and-land/land-and-water-stories/why-we-work-with-fire/

Toukabri M, Youssef MAM (2022) Climate change disclosure and sustainable development goals (SDGs) of the 2030 agenda: the moderating role of corporate governance. J Inf Commun Ethics Soc 21:30–62

Trenberth KE, Cheng L, Jacobs P, Zhang Y, Fasullo J (2018) Hurricane Harvey links to ocean heat content and climate change adaptation. Earth's Futur 6:730–744

Vilà-Vilardell L, Vilà-Vilardell L, Keeton WS, Thom D, Gyeltshen C, Tshering K, Gratzer G (2020) Climate change effects on wildfire hazards in the wildland-urban-interface—Blue pine forests of Bhutan. For Ecol Manag 461:117927

# Early Warning Systems: Enhancing Fire Prediction and Response

**Unshani Daryal, Anmol Giri, Sanjivani Karki, Passangkit Lepcha, Shehanaz Alam, and Aditya Pratap Singh**

**Abstract** Forest fires, typically uncontrolled by humans and stemming from natural or accidental causes, significantly damage wildlife and natural resources. In 2015, approximately 3% of the global forest area was affected by fires, with India facing the vulnerability of over 36% of its forest cover to frequent fires. The increasing frequency of forest fires is attributed to human activities, climate change, and other contributing factors. However, traditional detection methods have proven inefficient in keeping pace with this escalation. Recent advancements in machine learning, computer vision, and remote sensing technologies offer promising avenues for identifying and monitoring forest fires. Nonetheless, a precise and coordinated prevention, early warning, and response system is imperative to effectively mitigate these fires' impacts. Various factors influence fire behavior, including weather conditions, fuel moisture, climate change patterns, lightning occurrences, and vegetation types. Understanding these factors is essential for accurate fire prediction and effective response strategies. Modern technologies such as geospatial data analysis, satellite

U. Daryal
Department of Silviculture and Agroforestry, Sher-e-Kashmir University of Agricultural Sciences and Technology, Chatha, Jammu and Kashmir, India

A. Giri
Department of Agricultural Economics, Bidhan Chandra Krishi Viswavidyalaya, Mohanpur, West Bengal, India

S. Karki
Department of Agronomy, Uttar Banga Krishi Viswavidyalaya, Pundibari, Cooch Behar, West Bengal, India

P. Lepcha
Systematic Botany Discipline, Forest Botany Division, Forest Research Institute, Dehradun, Uttarakhand, India

S. Alam
Department of Agricultural Extension, Bidhan Chandra Krishi Viswavidyalaya, Mohanpur, Nadia, West Bengal, India

A. P. Singh (✉)
Department of Genetics and Plant Breeding, Bidhan Chandra Krishi Viswavidyalaya, Mohanpur, West Bengal, India
e-mail: adityapratapbckv@gmail.com

 
H. Singh and N. Kumar (eds.), *Forest Fire and Climate Change*,
https://doi.org/10.1007/978-3-031-89967-6_25

imagery interpretation, remote sensing techniques, Geographic Information Systems (GIS), Artificial Intelligence (AI), Machine Learning (ML), and Deep Learning (DL) play a pivotal role in comprehending fire behavior dynamics. The integration of these technologies has revolutionized wildfire management by enhancing prediction accuracy and response timeliness. Furthermore, utilizing remote sensing and GIS in disaster management encompasses mitigation, preparedness, response, and recovery phases. Looking ahead, there is a growing need to develop integrated, multi-hazard early warning systems that prioritize community needs and encompass the entire disaster management cycle. Investment in research and development is essential to address technological limitations and adapt to the evolving environmental landscape.

**Keywords** Behavior · Detection methods · Fire · Forest · Management · Remote sensing

## 1 Introduction

Forest fires are unwanted events that threaten forest ecosystems (Kumar et al. 2024). It is either natural or man-made and has serious consequences, claiming many human lives and causing significant damage to wildlife and natural resources (Singh 2024a). In 2015, about 98 Mha, or 3% of the global forest area, were affected by fire (GFRA 2020). In India, over 36% of forest cover is prone to frequent fires, with nearly 4% significantly fire-prone and 6% highly prone (ISFR 2019). Over the years, the frequency of forest fires has increased dramatically due to human activities, climate change, and other factors (Roohi 2015; Singh 2024b). According to the 2020 report of WWF International, around 75% of wildfires are due to anthropogenic activities. Early identification of forest fires is the most effective and efficient way to reduce or mitigate their impacts.

Since traditional forest fire detection methods have become inefficient, several techniques are currently employed to avoid or mitigate forest fire incidences. Although it is impossible to eliminate forest fires, it is crucial to detect them on time to minimize the impact of more significant damages and losses. The primary methods for early fire detection include Ground-based, aerial, and space-borne (Barmpoutis et al. 2020). Recent developments in machine learning, computer vision, and remote sensing technologies have introduced new tools for identifying and monitoring forest fires.

Additionally, progress in microelectronics has made sensors more efficient at detecting active fires (Gaur et al. 2019). Adopting a precise and closely coordinated effective prevention, early warning, and response system that will warrant related communities and agencies to initiate fire-preventative measures and mitigate fire impacts efficiently is necessary. Early warning (EW) is "the provision of timely and effective information, through identified institutions, that allow individuals exposed to hazard to take action to avoid or reduce their risk and prepare for

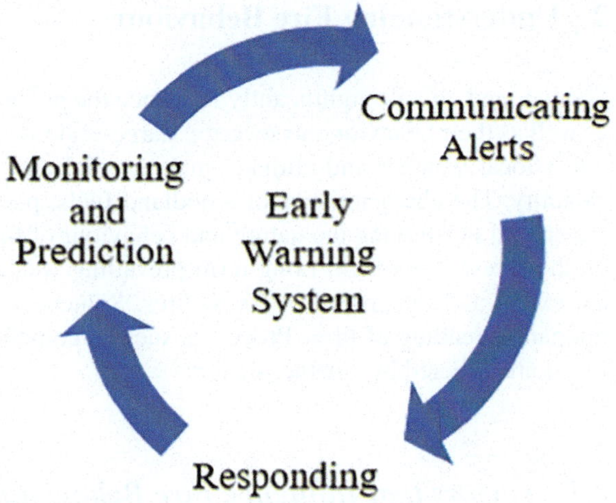


**Fig. 1** Early warning system operational aspects. *Source* Grasso (2016, draft report)

effective response". It amalgamates four main elements: Risk Knowledge, Monitoring and Predicting, Disseminating Information, and Response (ISDR UN 2006; Fig. 1). The whole system will fail if any part of the system fails.

For early warning systems to be effective, local governments, institutions, and communities must be actively involved in the entire policymaking process. This ensures they are fully informed and prepared to implement short- and long-term action plans. In April 2020, global fire alerts increased by 13% compared to the previous year (WWF, BCG 2020). Forest Survey of India (FSI) analysis shows that 10.66% of India's forest cover is extremely fire-prone. Early warnings are crucial for preventive measures. FSI is developing a system like the Canadian Fire Danger Rating System (CFFDRS), using the Fire Weather Index (FWI), along with Forest Type Layer and Forest Fire Archival information to generate Forest Fire Danger Ratings (ISFR 2021). Early warning technologies have advanced significantly thanks to improvements in communication, information technology, and our understanding of natural hazards. However, significant gaps, especially in developing countries, hinder establishing a global multi-hazard system. We must integrate improved modelling and prediction into disaster management to leverage these advancements fully. Bridging the gap between research and decision-making is crucial for maximizing early warning capacities. The major challenge remains ensuring that early warnings lead to swift and effective responses from governments and international organizations.

# 2 Understanding Fire Behaviour

Weather and climate significantly influence the potential for wildland fires to ignite, as well as their behavior and severity. Large-scale wind patterns across hemispheres affect local weather and climate, making wild and forest land susceptible to fire calamity. The characteristics of woodland fuels, particularly the amount of water, are crucial in studying the nature and behaviour of fire and their resulting influence on the forest ecosystem. Long-term prevailing weather elements greatly influence the extent and the spread of forest fires. Sometimes, seasons may also affect the rampant spreading of fires. Processes such as combustion and the onset of fire in a forest are affected by weather factors.

## *2.1 Factors that Influence Fire Behaviour*

### 2.1.1 Weather

Interaction between meteorological parameters like weather, fuel, fire, and the environment impacts the nature of forest fire spreading on a micro-scale (Singh 2024c). Researchers reported that daily weather fluctuations within 24 h can affect wind speed, air temperature, and the atmosphere's moisture content (relative humidity), which increases the risk of catching fires (Hirsch and Flannigan 1990; Kumar et al. 2020a). Fluctuations in the atmosphere and the surrounding air, such as instability in the air and prolonged dry weather, could lead to outbreaks of fires (Johnson and Miyanishi 2001; Singh 2024d). Electrical lightening due to clouds is likely to speed up the rate of spread of forest hazards (Latham and Williams 2001; MacGorman and Rust 1998).

### 2.1.2 Fuel Moisture

The wetness of the vegetation is crucial in determining the burning ability of the fuels. It is widely used in fire calamity/risk management and early warning systems (Fujioka et al. 2008; Roads et al. 2005; Sett et al. 2024; Kaushik and Singh 2024) and has been collaterally used to predict wildfire spread areas across continents (Nelson 2001). The concept identifies the spatial variability of water and oxygen capacity of forest lands, water retention capacity, and plant root conditions so that the right causes are identified, studied, and explained.

### 2.1.3 Climate Change

Climate change is the non-normal variations in the prevailing climate of a region over a prolonged period (Singh et al. 2024; Malik et al. 2024). These changes bring about a lot of variations in the vegetation, growing of plants, and changing temperature/rainfall patterns across countries, which significantly leads to the spread of drier spells and causes fire hazards. A rise in urbanization, deforestation, and changes in land use patterns over decades has led to the emission of greenhouse gases. Research suggests that global warming leads to the increased repetitive occurrence of fires and their harmful activity across regions (Flannigan et al. 2005; IPCC 1996; Tymstra et al. 2007; Kumar et al. 2020b). Other research suggests uncertainties about future fire occurrence and behaviour due to climate change. According to Beer and Williams (1995), fire calamity in Australia is projected to gain a boost, although climate models may not rely on humidity parameters, leading to an overestimation.

### 2.1.4 Lightning

Lightning is a global contributor to wildfires, but it only starts when the moisture content and fuel type are right. Whether positively charged lightning strikes result in more rapid burning than negatively charged ones is up for debate and constructive research. It has been studied that light fuel types are more common in igniting fire spreads in forests. In a 7-year study, Latham and Williams (2001) discovered that some fire-propellants perform better and show greater impacts in starting ignition (caused by lightning). The study extensively used geographic information systems (GIS) to correlate the type of combustible source with lightning occurrence and ignition. The study also found that trees, both coniferous and deciduous (0.03–0.05 fires/flash), burn faster than plants of the grass family (including bushes) (0.003–0.02 fires/flash). The rate of lightning and its swiftness in causing the burning of lands are influenced by various factors, including short- and long-term weather conditions, source and type of combustible sources, frequency and intensity of electric storms, and most significantly, the amount of moisture content of fuels (Johnson and Miyanishi 2001).

### 2.1.5 Fires and Wind

The wind is the most significant meteorological and climatic component influencing a fire's behaviour (Taylor et al. 2004). Research has shown different types of wind directly associated with wildland fires: winds resulting from aerial activity, winds originating due to fire's buoyancy (also known as entrainment), and local prevailing winds resulting from disproportionate heating of land and water. Depending on the topography, wind direction, and fuel supply, most forestland fires travel in varied directions. A fire may not successfully spread if the fuels are discontinuous, as in desert regions, unless the wind blows in the right direction and speed to allow the fire to "leap" the gap between the fuels.

## 2.2 *Importance of Understanding Fire Behaviour in Predicting and Responding to Fires*

The management of forest fires has long been studied and aided by modern technologies such as using definite geospatial data and satellite images. The information can be concisely presented as a map so that researchers/scientists can focus on these intrinsic issues and bring required changes in the areas of improvement. Remote sensing (RS) and geographic information systems (GIS) have been imparted with aerial photography some years ago. Georeferencing is done to link terrain data with other analyses, and the report is prepared and studied. As new types of sensors, such as spaceborne and UAV, have become available, their use has increased as high spatial and spectral resolution imageries are available. With high-resolution satellite imagery, these technologies can study small-scale environments, tree canopy, and adjoining land environments. Remote sensing data and statistics aid the GIS and other decision support systems.

## 2.3 *Susceptible Fire Conditions and Timely Management*

A forest fire is generally an outcome of the intricate synergy between the bio-ecological, physical, and meteorological factors. The probability/likelihood of the spread of fires, the nature of its extent, and the damage it could cause all depend on when these parameters interact singly or in combination. However, the environment and surrounding communities are on the receiving end of such outbreaks, especially when there is no awareness, early forecast, or prediction. Monitoring of forest fires could be enhanced with better technologies and their right application using models and statistics data to alert the public in general. The irrecoverable loss of biodiversity from the fire-struck area may take years to recover, and, in some cases, it is impossible to realign to pre-fire scenarios (Kurvits et al. 2022; Hardy 2005; Barwa and Singh 2024). The early warning systems were based on meteorological data, and predictions were made based on the available data (Finney 2005; Hernandez et al. 2015).

Weather data (temperature, rainfall, humidity) and the prevailing vegetation must be carefully studied to make the break of fires alarming. However, there is a need for a greater degree of an extended network of weather stations, and their active functioning and participation are necessary. Also, the need for the introduction and generation of fire calamity maps could be an excellent way to study the outbreak of forest fires. Since geo-informatics technologies can deduce uninterrupted and accurate data for a particular geographical area, the ability to scientifically analyze the obtained data should be of priority to create fire risk maps (Adab et al. 2013; Yu et al. 2017).

## *2.4 Fire Detection and Monitoring Fire Conditions*

Fire danger signaling or smoke detection is the earliest and the most critical step of forest fire management, aimed at conscientiously surveying the fire outbreak. Preliminary detection and monitoring systems using aerial surveys, geospatial data, and satellite images could be considered seriously (Leblon et al. 2012). The current stage of technological advancement incorporates the use of GIS, GPS, Remote sensing, and the like, which has helped scientists use multi-date high-resolution satellite and meteorological data to predict forest fire risks. Sensors' present spatial and temporal data status should be enhanced with better options and technologies to aid the fire-fighting forest managers. Detecting damage and monitoring the overall vegetation requires high-resolution imagery and multispectral imaging capabilities. A critical aspect of this is a quick turnaround time from data interpolation to the distribution of information. Understanding/studying an image that reflects the forest fire conditions of a week earlier would not help in natural-time forest risk management or risk mitigations. Vegetation conditions can also be inferred with hyperspectral vegetation indices. These can even measure delicate changes, and the targeted areas can be forewarned.

# 3 Advanced Technologies in Fire Prediction

The advent of advanced technologies has significantly transformed the landscape of wildfire prediction, offering new horizons in the accuracy and timeliness of forecasts. In the face of escalating wildfire incidents exacerbated by climate change and human activities, integrating Artificial Intelligence (AI) and remote sensing has emerged as a pivotal development in predictive methodologies (WFCA 2024). These technologies harness vast datasets, enabling the synthesis of predictive models that account for complex variables such as topography, vegetation moisture content, and meteorological conditions. AI, in particular, has been instrumental in refining fire behavior prediction models, analyzing historical fire data alongside real-time inputs to generate precise forecasts.

Moreover, Earth System Digital Twins and streaming data techniques have furthered the capability to simulate and predict fire spread with remarkable spatial resolution, enhancing the decision-making tools available to wildfire managers and first responders (Halem et al. 2022). The shift from traditional empirical methods to these sophisticated, data-driven approaches marks a significant milestone in wildfire management, promising a future where the impact of wildfires can be mitigated more effectively through technology.

## 3.1 Machine Learning (ML) and Deep Learning (DL)

Machine learning (ML) has significantly contributed to the field of fire prediction by providing sophisticated tools for analyzing complex datasets. ML models are adept at identifying intricate patterns indicating an increased fire risk. These models can process various types of data, including satellite imagery, weather conditions, and historical fire occurrences, to predict potential fire events with remarkable accuracy (Wang et al. 2022). One of the critical strengths of ML in fire prediction is its ability to adapt and improve over time. As more data becomes available, ML algorithms can be trained to refine their predictions, making them more accurate and reliable. This continuous learning process is vital for developing robust fire prediction systems to assist in early warning and mitigation efforts (Chen et al. 2023).

Machine learning algorithms have revolutionized fire prediction by processing vast datasets to identify patterns and predict potential fire outbreaks. By employing models such as Convolutional Neural Networks (CNNs), Random Forest (RF), and Deep Learning-based Learning Without Forgetting (LWF), these algorithms can analyze historical and real-time data, including weather conditions, topography, and drought indices, to forecast the likelihood and trajectory of wildfires (Boroujeni et al. 2024). Integrating machine learning with remote sensing data further refines the predictive accuracy, enabling a more nuanced understanding of fire dynamics (Abid 2021).

Machine learning (ML) and Deep learning (DL) have made quite notable progress in the field of fire prediction (Jain et al. 2020). The usage of computer techniques has greatly transformed the process of early detection and prediction of forest fires. As a result, these technological advancements have played a crucial role in minimizing the devastating consequences on the ecosystem and the economy (Abid 2021; Table 1).

ML and DL algorithms have been used to analyze different patterns in forest fires, employing datasets containing meteorological and Fire Weather Index (FWI) data. The application of several models, including decision trees, random forests, and neural networks, has facilitated the development of predictive systems that exhibit enhanced accuracy and recall metrics (Sathishkumar et al. 2023). These technologies are applied to the user interfaces of websites, where users can enter certain features to calculate the likelihood of forest fires. The practical implementation highlights ML and DL's capacity to offer practical and effective insights for managing and preventing forest fires (Bhatt and Chouhan 2024). The continuous evolution of fire prediction technologies promises further precision and response time improvements. Incorporating more comprehensive datasets and exploring novel deep learning techniques are anticipated to refine the predictive models, offering more robust solutions for global forest fire challenges.

**Table 1** The ongoing progress in the application of ML and DL techniques to enhance forest fire prevention strategies

| Year | Advancement | Description | References |
|---|---|---|---|
| 2020 | Transfer learning | Utilisation of pre-trained models like VGG16, Inception V3, and Xception for forest fire detection with improved accuracy and reduced computational complexity | Sathishkumar et al. (2023) |
| 2021 | Vision-based system | Using optimal convolution neutral networks (OPCNN) for accurate and efficient detection of forest fires | Jayasingh et al. (2024) |
| 2022 | Intelligent systems | Wildforest fires could be detection saw improvements with the introduction of (the IWFFDADL) model, which provided real-time monitoring and detection | Almasoud (2023) |
| 2023 | Extraction feature enhancement | New deep-learning approaches for detecting fires, which use pre-trained ResNet-50 and Xception for feature extraction, provide detailed comparisons with long-standing traditional models | Davis and Shekaramiz (2023) |
| 2024 | Analysis of risk and resilience | Using machine learning to look at complicated data about the risk and resistance of forest fires helps create models to predict how bad the fire will be and how likely it will happen | Varma et al. (2024) |

## 3.2 *Remote Sensing in Fire Prediction*

Remote sensing technology has become a cornerstone in fire prediction, leveraging satellite imagery and aerial photography to monitor environmental conditions. The synthesis of data from these high-resolution images enables the detection of potential fire hazards by analyzing vegetation dryness, land surface temperatures, and changes in landscape patterns. Applying thermal imaging cameras and sensor networks facilitates the real-time surveillance of vast terrains, thereby providing critical data for fire behavior modeling. These technological advancements have been instrumental in the early detection of fire incidents, allowing for prompt response and containment efforts (Shreya et al. 2022).

Remote sensing has become an indispensable tool in fire prediction, offering a multi-temporal perspective crucial for monitoring fire-prone areas. Using satellites and aerial sensors, remote sensing technologies can detect changes in vegetation moisture content, a key indicator of fire risk. This data, when analyzed, can predict potential fire outbreaks with significant accuracy (Szpakowski and Jensen 2019). Further enhancing fire prediction, remote sensing facilitates the mapping of fire hazards by integrating meteorological data and geographic information systems (GIS). This comprehensive approach allows for assessing fire danger over large areas, providing valuable insights for fire prevention strategies (Zhang et al. 2024).

### 3.3 Case Studies of Successful Fire Predictions

The efficacy of these technologies is substantiated by numerous case studies that demonstrate successful fire predictions. For instance, a study conducted in North Algeria utilized machine learning models to analyze data from NASA's fire information system and climate data API, achieving a prediction accuracy of up to 86% (Alkhatib et al. 2024). Another notable example is the Stanford University research team's model, which utilized satellite imagery and weather data to accurately predict fire occurrences in California (Miao et al. 2023). These instances underscore the potential of remote sensing and machine learning in enhancing fire prediction capabilities. The amalgamation of remote sensing and machine learning technologies represents a paradigm shift in fire prediction methodologies. The continuous evolution of these technologies promises further advancements in the accuracy and timeliness of fire predictions, thereby playing a crucial role in disaster management and environmental conservation. Future research directions may focus on integrating these technologies with innovative approaches, such as the Internet of Things (IoT) and big data analytics, to develop more comprehensive fire prediction systems (Ghibeche et al. 2024).

## 4 Early Warning Systems in Fire Prediction

The Early Warning System (EWS) is an important tool that enables the efficient use of countermeasures that aid in avoiding or reducing human and environmental losses. It gathers and disseminates warning information about hazards and risks to vulnerable people. EWS must diligently engage at-risk communities, promote public awareness, efficiently disseminate warnings and messages, and maintain a constant state of preparedness. Early Warning (EW) was defined as "the set of capacities needed to generate and disseminate timely and meaningful warning information to enable individuals, communities, and organizations threatened by a hazard to prepare and to act appropriately and in sufficient time to reduce the possibility of harm or loss" (UNISDR 2009). IPCC (2012) has also endorsed this definition. EWS delivers information using fire weather forecasts, enhanced by remotely sensed fuel data and fire activity. These systems can operate locally and globally depending on the fire management needs.

### 4.1 Elements of a Warning System

An effectual end-to-end warning system has four interrelated vital elements: (1) Risk knowledge, (2) Monitoring, (3) Response capability, (4) Warning communication (IFRC 2012; Fig. 2).

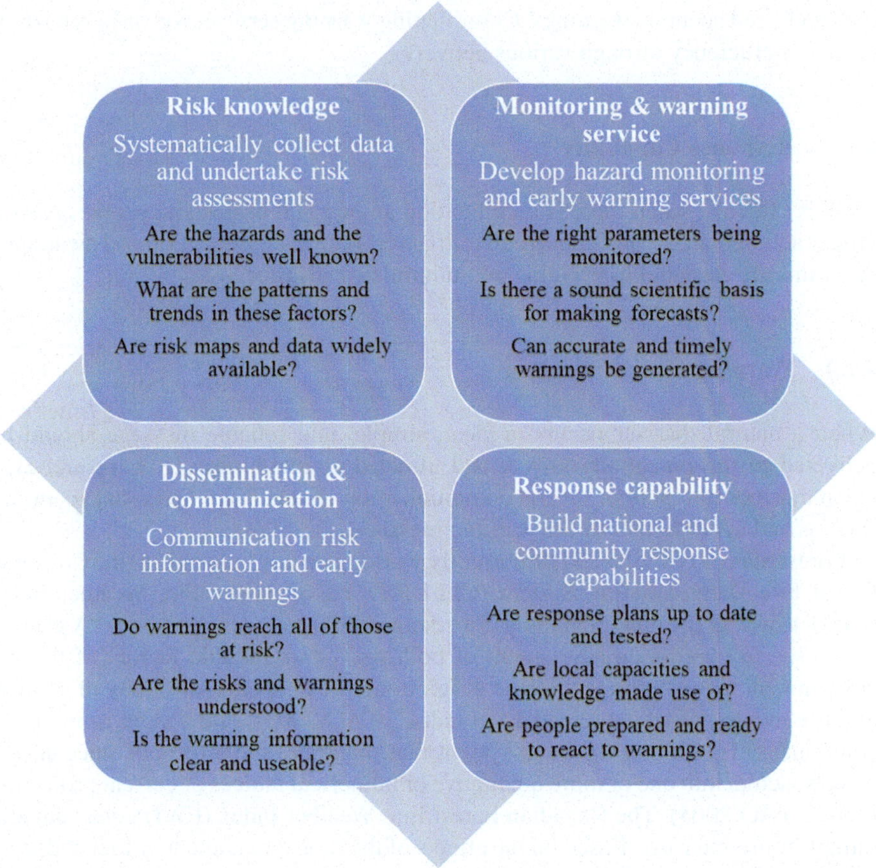


**Fig. 2** Four elements of EWS. *Source* UN/ISDR Platform for the Promotion of Early Warning

### 4.1.1 Risk Knowledge

With the help of warning systems, injury, damage, and loss of lives and property can be avoided or minimized to a great extent. Risk assessments done after systematically collecting data and analysis can guide people, design effective early warning systems, and prioritize prevention and mitigation strategies.

### 4.1.2 Monitoring

Warning services are central to the system. Monitoring, a key EWS element, involves comprehensive information gathering. Precise hazard prediction requires a robust scientific foundation, a reliable 24/7 forecasting system, and continuous monitoring for timely and accurate warnings. Coordinating warning services for various hazards enhances efficiency through various networks.

### 4.1.3 Response Capability

EWS is effective when there is coordination among authorities and people. Also, a proper action plan for prevention and mitigation is critical. Education and awareness programs are essential in avoiding and minimizing disaster loss.

### 4.1.4 Warning Communication

When a natural disaster occurs, a clear, simple, and reliable message should be delivered to the potentially at-risk and affected communities. To warn and alert the authorities and public, several communication channels are necessary to ensure success and avoid failure of the warning message.

Forest Survey of India (FSI) is currently working on a system akin to the Canadian Forest Fire Danger Rating System (CFFDRS), based on the Fire Weather Index (FWI), to assess fire danger in India. Fire danger (FD) has been defined as "A general term used to express an assessment of both fixed and variable factors of the fire environment that determine the ease of ignition, rate of spread, difficulty of control, and fire impact; often expressed as an index" (FAO 2005). Fire danger rating is "A component of a fire management system that integrates the effects of selected fire danger factors into one or more qualitative or numerical indices of current protection needs" (FAO 2005). The Canadian Forest Fire Weather Index (FWI) gives numeric ratings of fire intensity. Based on the effects of the moisture content of fuel and wind on fire behavior, the FWI comprises six components. The calculations are based on consecutive daily observations of temperature, relative humidity, wind speed, and precipitation over the past 24 h (Jain et al. 2017). The available data shows fire danger categorization based on the FWI index in Table 2. Continuous year-round data collection is crucial, as deep soil dryness accumulates over time (Asplund et al. 2013). The weather station provides real-time measurements needed to estimate the FWI.

**Table 2** Potential FD versus FWI value

| FWI class | Value range | Type of fire | Potential danger |
|---|---|---|---|
| Extreme | 30+ | Progressing active fire | Instant and aggressive measures are crucial |
| Very high | 20–30 | Profound surface fire (SF) | Tough to manage |
| High | 10–20 | Moderate vigorous SF | Requires hoses and power pumps |
| Moderate | 5–10 | Low vigorous SF | Easily extinguishable with hand tools |
| Low | 0–5 | Groveling SF | Fire will extinguish itself |

Creating and deploying an effective early warning system necessitates the contributions and coordination of various individuals and organizations.

## 5 Enhancing Response to Fires

In the critical domain of fire management, enhancing response strategies is paramount to safeguarding both human life and environmental resources. The integration of advanced communication systems plays a pivotal role in this enhancement, facilitating rapid dissemination of information and coordination among emergency services (Carta et al. 2023). When effectively implemented, these systems ensure that critical information reaches the relevant stakeholders promptly, enabling swift evacuation procedures and deployment of firefighting resources. Evacuation plans, meticulously crafted and regularly rehearsed, form the backbone of any fire response strategy. They provide a clear roadmap for safe egress and incorporate contingencies for unexpected developments during fire incidents.

The role of early warning systems (EWS) in augmenting these response strategies cannot be overstated. EWS serves as the harbinger of potential fire outbreaks, allowing preemptive actions to be taken before a fire can escalate (Li et al. 2022). By leveraging a multitude of detection technologies, including terrestrial, aerial, and satellite monitoring, EWS can identify the early indicators of a fire, such as abnormal heat signatures or smoke patterns, thereby initiating the alarm protocols that activate response mechanisms (Barmpoutis et al. 2020). The effectiveness of EWS is further enhanced by their integration with artificial intelligence and machine learning algorithms, which can analyze vast amounts of data to predict fire behavior and potential spread, thus optimizing the response efforts (Šakić Trogrlić et al. 2022).

Moreover, the strategic implementation of EWS within the broader framework of disaster risk reduction contributes significantly to the resilience of communities against fire hazards. By timely alerts, EWS enable authorities to mobilize resources, execute evacuation plans, and effectively engage community members in risk mitigation practices (Šakić Trogrlić et al. 2022). However, these systems' success is contingent upon the seamless collaboration between technological solutions and human agency. The synergy between sophisticated early detection technologies and the trained personnel who interpret and act upon the data ultimately determines the fire response's efficacy (Twigg 2003).

Enhancing fire response strategies through improved communication systems and robust evacuation plans is crucial. Integrating EWS within these strategies elevates the effectiveness of fire management efforts. As we continue to witness the exacerbation of fire incidents due to climate change and other anthropogenic factors, refining these systems and strategies becomes not just a matter of efficiency but of imperative necessity for preserving life and property.

# 6 Case Studies of Successful Fire Predictions and Responses

## 6.1 Rationale

Fire is one of the essential elements necessary for life on Earth, playing a crucial role in various ecosystems. The 'Fire Triangle' comprising fuel, oxygen, and a heat source explains how fire is sustained. In forests, ignition sources can include lightning or human activities such as discarded cigarettes or deliberate burning. Forest fires impact biodiversity, contribute to soil erosion, and release greenhouse gases like $CO_2$ (Urbanski 2014). However, they also replenish soil nutrients and maintain plant diversity. Human actions, such as clearing land or accidental fires from agricultural activities, often exacerbate fire risks, especially in areas like Uttarakhand, which is rich in biodiversity and crucial watersheds (Dobriyal and Bijalwan 2017; Ashutosh et al. 2019). Anthropogenic activities cause about 90% of fire ignitions, but the prevailing climatic and environmental factors intensify the fires (Bar et al. 2021; Taylor 2010). Effective disaster management is crucial, given the significant hazards that uncontrolled fires pose to communities and natural habitats. Remote sensing and Geographic Information Systems (GIS) play a vital role in all phases of disaster management: mitigation, preparedness, response, and recovery. Recent advances in Deep Learning, a subset of machine learning, have enhanced fire-related studies' analysis and prediction capabilities (Christin et al. 2019). These technologies facilitate the classification of high-resolution images, the detection of built-up and slum areas, and the prediction of ecological phenomena like forest fires (Bergado et al. 2018, 2021; Persello and Stein 2017; Kondylatos et al. 2022). This study aims to utilize deep learning models to generate fire forecast maps for Uttarakhand by analyzing spatiotemporal data on factors influencing fires.

## 6.2 Research Objectives

This research aims to generate fire forecast maps for Uttarakhand by analyzing spatiotemporal datasets using deep learning models. The key questions addressed include identifying the factors influencing fires, assessing the impact of multi-temporal data on forecast accuracy, determining the predictive timeframe, and integrating various datasets with the developed deep learning model for practical fire forecasting applications.

## *6.3 Study Area and Dataset*

Uttarakhand is located in the northwestern part of India within the Central Himalayas and borders Tibet, Nepal, Himachal Pradesh, and Uttar Pradesh. The state features diverse landscapes, ranging from the greater Himalayan Mountains and glaciers in the north to plains in the south. It is divided into 13 districts with elevations ranging from 300 to 7000 m above sea level (Chauhan 2010). Uttarakhand receives an average annual rainfall of 1631 mm, with 70% of it occurring during the southwest monsoon from June to September (Uttarakhand at a Glance 2017–18, 2019; Southwest Monsoon-2022 End of Season Report for Uttarakhand State 2022).

## *6.4 Methodology*

The research methodology comprised six sections: preliminary analysis, data preparation, model building, hyperparameter tuning, model validation, and creating a user interface for fire forecasting. Initially, examining fire hotspot distribution across various terrain features in Uttarakhand was conducted to understand the spatial patterns and influencing factors of forest fires. Terrain features were analyzed using the Shuttle Radar Topography Mission (SRTM) Digital Elevation Model (DEM) from Google Earth Engine (GEE). Essential meteorological data were sourced from ERA5 Land reanalysis data and ground meteorological stations, with ERA5 primarily used for climatological studies. Spectral indices were calculated using visible and infrared satellite images from the VIIRS sensor, and meteorological variables such as wind speed, wind direction, and relative humidity were derived from available ERA5 data. The data was split into training, validation, and testing sets based on different periods and fire seasons, ensuring comprehensive coverage of fire-prone periods. Deep learning models were then employed for fire forecasting. A modified, shallower U-Net model architecture was used for active fire forecasting with non-sequential fire-influencing variables. Fire forecasting was implemented for sequential data using a series of convolutional Long Short-Term Memory (LSTM) and convolutional layers. Hyperparameter tuning was performed to optimize the deep learning models, adjusting the neural network structure to fit the input and corresponding label features. Model validation was conducted using the testing dataset, with various metrics calculated to evaluate the model's efficiency in forecasting fires. Due to the imbalanced distribution of fire and non-fire classes, metrics beyond overall accuracy were used to avoid bias toward non-fire samples. To disseminate fire forecasts effectively to the forest department officials of Uttarakhand, a user-friendly interface was developed using Greppo, a Python package for publishing web applications (Krishnan and Tangirala 2023).

## 6.5 Outcome

An initial assessment was conducted to analyze the distribution of active fire points in Uttarakhand relative to terrain features from 2016 to 2023. This involved mapping fire points and correlating them with slope and aspect characteristics. Slope analysis determined fire distribution across varying slope categories, indicating susceptibility based on slope steepness. Aspect analysis evaluated orientation's impact on fire occurrence, distinguishing between north-facing and south-facing slopes. Meteorological data sourced from ground stations and ERA5 Reanalysis via Google Earth Engine ensured consistent data availability. A correlation analysis between the ground station and ERA5 data for 2021 assessed their relationship, focusing on maximum temperature. Fire forecasting models were developed using both non-sequential and sequential fire-influencing variables. Non-sequential data from a single day were utilized across different scenarios, training on 2016–2020 data and tuning on 2021 data. Models predicted fire events for the next day, two days, and five days. Label data were aggregated and matched with feature datasets for the latter forecasts. Hyperparameters were optimized, and models were evaluated quantitatively and qualitatively. Results demonstrated model performance, presented alongside elevation, land use, and reference plots. Additionally, a Convolutional LSTM model was employed to forecast fires using temporal data for the next day. Hyperparameters were tuned, and model performance was assessed, emphasizing prediction accuracy and output analysis.

## 6.6 Summary

The study demonstrates the potential of deep learning models for fire forecasting, encompassing various influencing factors and their interrelationships. These models exhibit higher prediction accuracy for forest fires and anthropogenic agricultural residue burning. They effectively distinguish between fire and non-fire seasons and accurately predict fires on different dates. However, extending the forecasting period beyond two days increases false positives, limiting operational use. The LSTM model, tailored for high temporal dependency, struggles to forecast fires when the temporal range exceeds the problem's context. Incorporating datasets on fire ignition could enhance model precision and understanding of the complex interplay between fire triangle elements, thus improving forecasting accuracy.

## 6.7 Case Study of Australian Bushfires

Australia's Early Warning System (EWS) was a cornerstone in strategically managing and controlling bushfires, particularly highlighted by the "Black Summer" bushfires of 2019–2020. Integrating advanced technologies and coordinated communication strategies significantly enhanced the country's ability to predict, prepare for, and respond to bushfire threats. As part of the EWS, satellite monitoring employed high-resolution imagery to detect hotspots indicative of potential fires. This real-time surveillance allowed for rapidly deploying firefighting resources to critical areas, mitigating the spread of fires and reducing potential damage.

Meteorological forecasting played a pivotal role in the EWS by providing accurate and timely predictions of weather conditions conducive to bushfires. These forecasts were crucial for pre-emptive measures, such as issuing fire bans and mobilizing emergency services in anticipation of fire outbreaks. The alert systems, which formed an integral part of the EWS, were instrumental in saving lives and properties. They disseminated warnings through multiple channels, ensuring residents in high-risk zones received prompt notifications to take necessary precautions or evacuate if needed.

Public information campaigns complemented these technological approaches by fostering a culture of awareness and preparedness among the community. These campaigns educated the public on interpreting warnings and responding effectively, empowering individuals to safeguard themselves and their properties proactively. The Australian Warning System (AWS), established post-Black Summer, exemplified the evolution of the EWS. The AWS provided a unified framework for emergency warnings, with clear levels of alerts and associated actions, ensuring consistency and clarity in communication across the nation.

The effectiveness of the EWS was measured by the immediate response to fires and the long-term resilience it built within communities. By integrating traditional knowledge, such as Indigenous fire management practices, with modern science, the EWS fostered a holistic approach to bushfire management. This synergy of knowledge enhanced the overall effectiveness of the EWS, making it a model for bushfire-prone regions worldwide.

## 7 Challenges and Future Directions in Fire Prediction and Response

The power to anticipate and address a fire, particularly wildfires, integrates various complexities, including technology, the environment, and human collaboration. From a technological point of view, effectively predicting fires can be a very challenging task despite breakthroughs such as machine learning. The main reason for this is primarily due to worries regarding data's substandard quality and the volatile characteristics of fires caused by human actions. While satellite photography is beneficial, it occasionally fails to detect minor fires promptly. Climate change is exacerbating the occurrence and intensity of wildfires, making their prediction more challenging from an environmental standpoint. The inequitable distribution of resources and breakdowns in communication among different groups frequently limit the efficacy of human response.

In order to prepare for the future, it is of utmost importance to create more advanced early warning systems that are user-friendly and understandable, integrate the most up-to-date technology, and facilitate smooth collaboration among all stakeholders, ranging from local communities to multinational organizations. Prioritizing education and fostering community engagement will guarantee that individuals can promptly and efficiently respond to warnings to safeguard lives and property. Although there have been significant developments in fire prediction technologies, some hurdles still need to be addressed to improve the effectiveness of fire prediction and response systems.

## *7.1 Technological Challenges*

One of the primary technological challenges in fire prediction is the accuracy of current models. While machine learning (ML) and artificial intelligence (AI) have made significant strides in predicting fire outbreaks, the reliability of these predictions can be compromised by data quality issues, such as class imbalance and misclassification (Sathishkumar et al. 2023). Additionally, the unpredictability of certain fires, especially man-made ones, adds another layer of complexity to accurate prediction (Awatade et al. 2024). Remote sensing, a key tool in fire detection, also faces limitations. The resolution of satellite imagery may not be sufficient to detect small-scale fires or those obscured by canopy cover. Moreover, the time lag between satellite passes can result in delayed detection, critical in the early stages of fire development (Sathishkumar et al. 2023).

## *7.2 Environmental and Human Factors*

Environmental and human factors significantly impact the ability to predict and respond to wildfires. Climate change is a major environmental factor, as rising temperatures, prolonged droughts, and shifting weather patterns create conditions that make wildfires more frequent and intense. These changes can outpace the current models used for prediction, making it harder to foresee and prepare for such events. Additionally, the health of our forests and lands plays a critical role. Overgrown forests, accumulated dry vegetation, and land use changes can all increase the risk and spread of wildfires.

Human factors are equally crucial in this context. Effective fire response relies heavily on coordinating and allocating resources, such as firefighting equipment, personnel, and emergency services. However, these resources are often unevenly distributed, particularly between urban and rural areas, which can delay responses in more vulnerable regions. Communication is another vital aspect; breakdowns in communication among local communities, government agencies, and emergency responders can lead to confusion and inefficient responses during critical moments. Furthermore, public awareness and education about fire risks and safety measures are often lacking, leading to inadequate preparedness and slow reactions when fires occur.

Addressing these factors requires a concerted effort. Environmentally, this means adapting our models to account for the changing climate and managing our forests and lands more sustainably. From a human perspective, it involves improving resource distribution, enhancing communication channels, and investing in public education to ensure communities are better prepared and respond more quickly when fires threaten.

### *7.3 Emerging Technologies and Strategies*

In response to these challenges, emerging technologies and strategies are being developed to improve early warning systems. Advances in ML algorithms, such as deep learning and transfer learning, enhance fire detection systems' accuracy (Kale et al. 2022). Integrating AI with geospatial technologies, the Internet of Things (IoT), and improved connectivity is paving the way for more sophisticated and responsive fire prediction models (Diaconu 2023).

### *7.4 Strategies for Improvement*

To improve early warning systems, a holistic approach is necessary. This includes developing institutional and legal frameworks, deploying advanced technology, engaging communities and the private sector, and fostering international cooperation and data sharing. Effective communication strategies are also vital to ensure that warnings reach all stakeholders promptly and are understood (Šakić Trogrlić et al. 2022).

Looking ahead, the focus should be on creating integrated, multi-hazard early warning systems that are people-centered and capable of addressing the full disaster management cycle 89. Investment in research and development is essential to overcome current technological limitations and to adapt to the changing environmental landscape. Additionally, enhancing the social components of warning systems, such as public education and community involvement, will be vital to ensuring that early warnings lead to early action 8. While the challenges in fire prediction and response are significant, the potential of emerging technologies and strategies offers a promising path forward. By addressing the current shortcomings and capitalizing on new developments, early warning systems can become more robust, inclusive, and effective in safeguarding lives, property, and ecosystems against the threat of fires.

## 8 Conclusion

The future of early warning systems for fire prediction and response lies in continuously advancing and integrating cutting-edge technologies. The recommendations for utilizing higher-resolution datasets and addressing data imbalances with Explainable AI tools underscore the importance of improving model reliability and operational applicability. Incorporating diverse data sources, such as CubeSat satellites, and comparing global and regional reanalysis datasets will refine our understanding of meteorological variables crucial for accurate forecasting. Experimentation with historical data periods and model architectures like Convolutional LSTM will further enhance predictive capabilities. The rapid progress in technology and communication has revolutionized forest fire management. Sophisticated systems leveraging sensor networks, GPS, IoT, and real-time alert mechanisms are pivotal in reducing response times and minimizing human intervention. Effective implementation of these systems demands collaboration among stakeholders and robust communication protocols, particularly in regions vulnerable to climate change and population pressures. Advanced automated detection systems powered by machine learning, computer vision, and remote sensing provide cost-effective, low-maintenance solutions. Future innovations, such as automatic fire extinguishers and wind sensors, promise to elevate these systems further. The active involvement of local governments, institutions, and communities in policymaking is essential for a comprehensive fire management strategy, ensuring preparedness through coordinated prevention, early warning, and rapid response. These collective efforts can significantly mitigate the impact of forest fires and safeguard our natural ecosystems.

## References

Abid F (2021) A survey of machine learning algorithms based forest fires prediction and detection systems. Fire Technol 57(2):559–590

Adab H, Kanniah KD, Solaimani K (2013) Modeling forest fire risk in the north east of Iran using remote sensing and GIS techniques. Natura 65:1723–1743

Alkhatib A, Jaber K, Alzo'by H, Abdallah M, Salah M (2024) Exploring progress in forest fire detection, prediction, and behavior: an in-depth survey. Int J Comput Digit Syst 15(1):789–800

Almasoud AS (2023) Intelligent deep learning enabled wild forest fire detection system. Comput Syst Sci Eng 44(2)

Ashutosh S, Kumar S, Chaudhary A, Biswas T, Ghosh S (2019) Technical information series

Asplund G, Intorf E, Droeg (2013) The osborne fire finder the basic lookout tools: firemans guide California region. U.S. Department of Agriculture Forest Service

Awatade A, Pawar P, Lakshmi D (2024) The power of machine learning in forest fire risk analysis and resilience: navigating best practices, challenges, and opportunities. In: Geospatial technology to support communities and policy: pathways to resiliency, 149–170

Bar S, Parida BR, Roberts G, Pandey AC, Acharya P, Dash J (2021) Spatio-temporal characterization of landscape fire in relation to anthropogenic activity and climatic variability over the Western Himalaya, India. GISci Remote Sens 58(2):281–299. https://doi.org/10.1080/15481603.2021.1879495

Barmpoutis P, Papaioannou P, Dimitropoulos K, Grammalidis N (2020) A review on early forest fire detection systems using optical remote sensing. Sensors 20:6442

Barwa S, Singh H (2024) Climate-induced changes in forest ecosystems of Indian Himalayan region: a meta-analysis. In: Singh H (ed) Urban forests, climate change and environmental pollution. Springer, Cham. https://doi.org/10.1007/978-3-031-67837-0_39

Beer T, Williams A (1995) Estimating Australian forest fire danger under conditions of doubled carbon dioxide concentrations. Climatic Change 29:169–188

Bergado JR, Persello C, Stein A (2018) Recurrent multiresolution convolutional networks for VHR image classification. IEEE Trans Geosci Remote Sens 56(11):6361–6374. https://doi.org/10.1109/TGRS.2018.2837357

Bergado JR, Persello C, Reinke K, Stein A (2021) Predicting wildfire burns from big geodata using deep learning. Saf Sci 140:105276. https://doi.org/10.1016/j.ssci.2021.105276

Bhatt S, Chouhan U (2024) An enhanced method for predicting and analysing forest fires using an attention-based CNN model. J For Res 35(1):1–13

Boroujeni SPH, Razi A, Khoshdel S, Afghah F, Coen JL, O'Neill L, Fule P, Watts A, Kokolakis NMT, Vamvoudakis KG (2024) A comprehensive survey of research towards AI-enabled unmanned aerial systems in pre-, active-, and post-wildfire management. Inf Fusion 102369

Carta et al (2023). Communication systems and their impact on fire emergency responses. J Emerg Manag 21(2):15–29

Chauhan M (2010) A perspective on watershed development in the central Himalayan state of Uttarakhand, India. Int J Ecol Environ Sci 36:253–269

Chen Z, Zhang C, Li W, Gao L, Liu L, Fang L, Zhang C (2023) Fire danger forecasting using machine learning-based models and meteorological observation: a case study in Northeastern China. Multimed Tools Appl 1–21

Christin S, Hervet É, Lecomte N (2019) Applications for deep learning in ecology. Methods Ecol Evol 10(10):1632–1644. https://doi.org/10.1111/2041-210X.13256

Davis M, Shekaramiz M (2023) Desert/forest fire detection using machine/deep learning techniques. Fire 6(11):418

Diaconu BM (2023) Recent advances and emerging directions in fire detection systems based on machine learning algorithms. Fire 6(11):441

Dobriyal MJ, Bijalwan A (2017) Forest fire in western Himalayas of India: a review. N Y Sci J 10(6):39–46

FAO (2005) Wildland fire management terminology. Food and Agriculture Organization, United Nations, Rome

Finney MA (2005) The challenge of quantitative risk analysis for wild land fire. For Ecol Manage 211:97–108

Flannigan MD, Logan KA, Amiro BD, Skinner WR, Stocks BJ (2005) Future area burned in Canada. Clim Change 72:1–16

Fujioka FM, Gill AM, Viegas DX, Wotton BM (2008) Fire danger and fire behavior modeling systems in Australia, Europe, and North America. In: Bytnerowicz A, Arbaugh M, Riebau A, Andersen C (eds) Wildland fires and air pollution. Elsevier, Oxford, UK, pp 469–497

Gaur A, Singh A, Kumar A, Kulkarni KS, Lala S, Kapoor K, Srivastava V, Kumar A, Mukhopadhyay SC (2019) Fire sensing technologies: a review. IEEE Sens J 19:3191–3202

Ghibeche Y, Sellam A, Nouri N, Khaldi A, Harrane A, Ghibeche I (2024) Machine learning for forest fire prediction: a case study in north Algeria. Ing Syst d'Inf 29(1)

Global Forest Resources Assessment (2020) Main report. Report by Food and Agriculture Organization of the United Nations

Grasso VF (2016) Early warning systems: state-of-art analysis and future directions. Draft report. United Nations Environment Programme (UNEP)

Halem M, Kochanski AK, Mandel J, Sleeman J, Demoz B, Bargteil A et al (2022) Towards a dynamic data driven wildfire digital twin (WDT): impacts on deforestation, air quality and cardiopulmonary disease. In: International conference on dynamic data driven applications systems, Oct 2022. Springer Nature Switzerland, Cham, pp 403–410

Hardy CC (2005) Wildland fire hazard and risk: problems, definitions, and context. For Ecol Manage 211:73–82

Hernandez C, Drobinski P, Turquety S (2015) How much does weather control fire size and intensity in the Mediterranean region. Annal Geophys 33:931–939

Hirsch KG, Flannigan MD (1990) Meteorological and fire behavior characteristics of the 1989 fire season in Manitoba, Canada. In: International conference on forest fire research, Coimbra. Springer, New York, pp B.06-1–B.06-16

India State of Forest Report, Forest Survey of India (2019)

India State of Forest Report, Forest Survey of India (2021)

Intergovernmental Panel on Climate Change (IPCC) (2012)

Intergovernmental Panel on Climate Change (1996) Climate change 1995 impacts, adaptations, and mitigation of climate change: scientific-technical analyses. Cambridge University Press, Cambridge

International Federation of Red Cross and Red Crescent Societies (IFRC) (2012) Community early warning systems: guiding principles. Geneva

International Strategy for Disaster Reduction (2006) Developing early warning systems: a checklist. In: EWC III third international conference on early warning

Jain P, Wang X, Flannigan MD (2017) Trend analysis of fire season length and extreme fire weather in North America between 1979 and 2015. Int J Wildl Fire 26(12):1009–1020

Jain P, Coogan SC, Subramanian SG, Crowley M, Taylor S, Flannigan MD (2020) A review of machine learning applications in wildfire science and management. Environ Rev 28(4):478–505

Jayasingh SK, Swain S, Patra KJ, Gountia D (2024) An experimental approach to detect forest fire using machine learning mathematical models and IoT. SN Comput Sci 5(1):148

Johnson EA, Miyanishi K (2001) Forest fires: behavior and ecological effects. Academic Press, San Diego, CA, p 594

Kale MP, Mishra A, Pardeshi S, Ghosh S, Pai DS, Roy PS (2022) Forecasting wildfires in major forest types of India. Front For Glob Change 5:882685

Kaushik P, Singh H (2024) Climate change-induced forest ecosystem functioning and alteration in water–atmosphere–soil–plant (WASP) relationship. In: Singh H (ed) Urban forests, climate change and environmental pollution. Springer, Cham. https://doi.org/10.1007/978-3-031-67837-0_38

Kondylatos S, Prapas I, Ronco M, Papoutsis I, Camps-Valls G, Piles M, Fernández-Torres M, Carvalhais N (2022) Wildfire danger prediction and understanding with deep learning. Geophys Res Lett 49(17). https://doi.org/10.1029/2022GL099368

Krishnan A, Tangirala S (2023) Welcome to Greppo's documentation. Greppo. https://docs.greppo.io/#

Kumar A, Dwivedi GK, Tewari S, Paul J, Anand R, Kumar N, Kumar P, Singh H, Kaushal R (2020a) Carbon mineralization and inorganic nitrogen pools under *Terminalia chebula* Retz.-based agroforestry system in Himalayan foothills, India. For Sci 66(5):634–643

Kumar A, Kumar P, Singh H, Kumar N (2020b) Adaptation and mitigation potential of roadside trees with bio-extraction of heavy metals under vehicular emissions and their impact on physiological traits during seasonal regimes. Urb For Urb Green 58:126900

Kumar N, Singh H, Kumar A, Singh AK, Pandey PK, Kumar A (2024) Forest-fire-induced land degradation. In: Mishra G, Giri K, Singh S, Kumar M (eds) Sustainable land management in India. Springer Nature, Singapore. https://doi.org/10.1007/978-981-97-5223-2_4

Kurvits T, Popescu A, Paulson A, Sullivan A, Ganz D, Burton C (2022) Spreading like wildfire: the rising threat of extraordinary landscape fires. United Nations Environment Programme, Nairobi

Latham D, Williams E (2001) Lightning and forest fires. In: Johnson EA, Miyanishi K (eds) Forest fires: behavior and ecological effects. Academic Press, San Diego, CA, pp 375–418 (594 pp)

Leblon B, Bourgeau-Chavez L, San-Miguel-Ayanz J (2012) Use of remote sensing in wildfire management. Current international perspectives on wild land fires. In: Curkovic S (ed) Sustainable development—authoritative and leading-edge content for environmental management. Intech Open

Li X, Vázquez-López A, del Río S, Sáez J, Wang DY (2022) Recent advances on early-stage fire-warning systems: mechanism, performance, and perspective. Nano-Micro Lett 14(1):197
MacGorman DR, Rust WD (1998) The electrical nature of storms. Oxford University Press
Malik A, Kumar K, Kumari G, Kumar N, Singh H (2024) Application of functional traits in modelling productivity and resilience under climate change. Springer, pp 77–96. ISBN: 978-981-97-1509-1
Miao X, Li J, Mu Y, He C, Ma Y, Chen J et al (2023) Time series forest fire prediction based on improved transformer. Forests 14(8):1596
Nelson RM Jr (2001) Water relations of forest fuels. In: Johnson EA, Miyanishi K (eds) Forest fires: behavior and ecological effects. Academic Press, San Diego, CA
Persello C, Stein A (2017) Deep fully convolutional networks for the detection of informal settlements in VHR images. IEEE Geosci Remote Sens Lett 14(12):2325–2329. https://doi.org/10.1109/LGRS.2017.2763738
Roads J, Fujioka F, Chen S, Burgan R (2005) Seasonal fire danger forecasts for the USA. Int J Wildl Fire Spec Issue Fire For Meteorol 14:1–18
Roohi UKM (2015) Comparison and application of forest fire detection system based on a ZigBee wireless sensor network. Int J Eng Res Gen Sci 3(1):427–430
Šakić Trogrlić R, van den Homberg M, Budimir M, McQuistan C, Sneddon A, Golding B (2022) Early warning systems and their role in disaster risk reduction. Towards the "perfect" weather warning: bridging disciplinary gaps through partnership and communication. Springer International Publishing, Cham, pp 11–46
Sathishkumar VE, Cho J, Subramanian M, Naren OS (2023) Forest fire and smoke detection using deep learning-based learning without forgetting. Fire Ecol 19(1):9
Sett T, Nikam BR, Singh H, Purohit S (2024) Hydrological variability in Indian forest ecosystem: analysis of drought resilience, recovery and water use efficiency in moist and dry deciduous forests. In: Singh H (ed) Urban forests, climate change and environmental pollution. Springer, Cham. https://doi.org/10.1007/978-3-031-67837-0_37
Shreya M, Rai R, Shukla S (2022) Forest fire prediction using machine learning and deep learning techniques. In: Computer networks and inventive communication technologies: proceedings of fifth ICCNCT 2022. Springer Nature Singapore, Singapore, pp 683–694
Singh H (2024a) Climate change and forests: biological perspectives. In: Singh H (ed) Forests and climate change. Springer Nature, Singapore. https://doi.org/10.1007/978-981-97-3905-9_1
Singh H (ed) (2024b) Forests and climate change: biological perspectives on impact, adaptation, and mitigation strategies. Springer Nature, Singapore. https://doi.org/10.1007/978-981-97-3905-9
Singh H (2024c) Climate change and environmental pollution: biological consequences on urban forests. In: Singh H (ed) Urban forests, climate change and environmental pollution. Springer Nature, Cham, Switzerland. https://doi.org/10.1007/978-3-031-67837-0_1
Singh H (ed) (2024d) Urban forests, climate change and environmental pollution: physio-biochemical and molecular perspectives to enhance urban resilience. Springer Nature, Cham, Switzerland. https://doi.org/10.1007/978-3-031-67837-0
Singh H, Kumar N, Kumar A (2024) Enhancing resource use efficiency in crops through plant functional traits. Springer, pp 97–117. ISBN: 978-981-97-1509-1
Szpakowski DM, Jensen JL (2019) A review of the applications of remote sensing in fire ecology. Remote Sens 11(22):2638
Taylor D (2010) Biomass burning, humans and climate change in Southeast Asia. Biodivers Conserv 19(4):1025–1042. https://doi.org/10.1007/s10531-009-9756-6
Taylor SW, Wotton BM, Alexander ME, Dalrymple GN (2004) Variation in wind and crown fire behaviour in a northern jack pine—black spruce forest. Can J For Res 34:1561–1576
Twigg J (2003) The human factor in early warnings: risk perception and appropriate communications. In: Early warning systems for natural disaster reduction, pp 19–26. Springer Berlin Heidelberg, Berlin, Heidelberg
Tymstra C, Flannigan MD, Armitage OB, Logan K (2007) Impact of climate change on area burned in Alberta's boreal forest. Int J Wildl Fire 16:153–160

UNISDR (2009) Terminology on DDR. The United Nations Office for Disaster and Risk Reduction, Geneva

Urbanski S (2014) Wildland fire emissions, carbon, and climate: emission factors. For Ecol Manage 317:51–60. https://doi.org/10.1016/J.FORECO.2013.05.045

Varma S, Roy SS, Rai PK (2024) Machine learning for forest fire risk and resilience. In: Geospatial technology to support communities and policy: pathways to resiliency. Springer Nature Switzerland, Cham, pp 171–184

Wang SSC, Qian Y, Leung LR, Zhang Y (2022) Interpreting machine learning prediction of fire emissions and comparison with FireMIP process-based models. Atmos Chem Phys 22(5):3445–3468

WFCA (2024) How AI is aiding in wildfire prediction and prevention. Western Fire Chiefs Association. https://wfca.com/wildfire-articles/ai-to-predict-and-prevent-fires/

World Wide Fund for Nature (WWF), Switzerland and Boston Consulting Group (BCG) (2020) Fires, forests and the future: a crisis raging out of control. Joint report by the World Wide Fund for Nature (WWF), Switzerland and Boston Consulting Group (BCG)

Yu B, Chen F, Li B, Wang L, Wu M (2017) Fire risk prediction using remote sensed products: a case of Cambodia. Photogramm Eng Remote Sens 83:19–25

Zhang Z, Tian Y, Wang G, Zheng C, Zhao F (2024) A forest fire prediction method for lightning stroke based on remote sensing data. Forests 15(4):647

# Ecosystem Recovery After Forest Fire: Methodologies and Technologies

**Sucharita Panda, Auindrila Biswas, Dristi Majumdar, Sourakanti Sarkar, Sumana Balo, Unshani Daryal, and Aditya Pratap Singh**

**Abstract** Forest fires pose a significant threat to ecosystems, causing economic losses, deteriorated real estate, burned wood, and harm to recreational, environmental, and ecological values. Anthropogenic factors, such as careless human activities, and natural factors like lightning, volcanic eruptions, and friction from rolling stones are the prime causes of forest-fire. Forest fires have immediate and long-term impacts on flora and fauna, influencing ecosystem function and regulation. Intense fires can alter the climate by reducing the carbon storage capacity of forests, potentially leading to more frequent and severe fires in the future. Early detection and real-time monitoring of fire activities are crucial for effective management and conservation efforts. Remote sensing and geographic information systems (GIS) are cost-effective for monitoring fire damage across vast geographical regions. Various types of sensors, such as thermal imaging, LiDAR, and multispectral sensors,

S. Panda
Department of Agroforestry, School of Agriculture, GIET University, Gunupur, Rayagada, Odisha, India

A. Biswas
Department of Agriculture, Jharkhand Rai University, Ranchi, Jharkhand, India

D. Majumdar
Department of Biotechnology, Siksha-Bhavana, Viswa-Bharati, Santiniketan, West Bengal, India

S. Sarkar
Department of Agricultural Economics, School of Agriculture, Seacom Skills University, Kendradangal, Bolpur, West Bengal, India

S. Balo
Department of Soil Science and Agricultural Chemistry, School of Agriculture, GIET University, Gunupur, Rayagada, Odisha, India

U. Daryal
Department of Silviculture and Agroforestry, Sher-e-Kashmir University of Agricultural Sciences and Technology, Chatha, Jammu and Kashmir, India

A. P. Singh (✉)
Department of Genetics and Plant Breeding, Bidhan Chandra Krishi Viswavidyalaya, Mohanpur, West Bengal, India
e-mail: adityapratapbckv@gmail.com

 
H. Singh and N. Kumar (eds.), *Forest Fire and Climate Change*,
https://doi.org/10.1007/978-3-031-89967-6_26

are used for fire detection, fire risk mapping, burn damage intensity assessment, damage extension, and damage intensity distribution. Recent technologies, including satellite-based remote sensing platforms and advanced GIS tools, are essential for understanding the impact of fires on forest ecosystems and ensuring the success of recovery efforts. Over half of the world's tropical forests have been transformed for other land use, leading to habitat degradation, biodiversity decline, and loss of essential ecosystem services. Forest and landscape restoration (FLR) is crucial for species conservation, climate change mitigation, and the sustainability of forest ecosystems. Seed dispersal, soil stabilization techniques, post-fire management strategies, precision forest restoration (PFR), and ecosystem restoration are all essential to mitigating ongoing wildfires and preventing future fires. More research is needed to understand the natural fire regime, differentiate between natural and anthropogenic fires, and address current challenges in forest fire management and restoration methodologies.

**Keywords** Biodiversity decline · Ecosystem restoration · Forest and landscape restoration (FLR) · Forest fire management · Remote sensing · Soil stabilization techniques

## 1 Introduction

Fire has a significant impact on the development and management of forests. Some forest ecosystems have evolved in response to frequent fires from natural causes as a secondary succession. At the same time, the effects of wild fire have been degrading and detrimental to flora and fauna in an ecosystem (Kumar et al. 2024). Forest fire is a combustion that naturally or artificially consumes inflammable materials, impacting biodiversity. Forest fires have spontaneously started and spread throughout the forest throughout recorded history. However, not all woods are affected equally by fire. Since the dawn of humanity, fire and forests have been closely related. There has been an intrinsic relation between forest and forest fire. While controlled fire enhances the productivity of the forest by encouraging emerging shoots and affecting nutrient cycling, uncontrolled fire or wildfire converts luscious vegetation into barren land. That is why there is a saying, "Fire is a good servant but a bad Master." Forests are one of the crucial natural entity play a remarkable role in balancing the environment (Barwa and Singh 2024). The forest's health could predict an area's ecological condition (Kaushik and Singh 2024; Sett et al. 2024). Due to climate change and global warming, fires are increasing yearly, destroying a major part of the forest. The frequency of fires in different years in India is shown in Fig. 1.

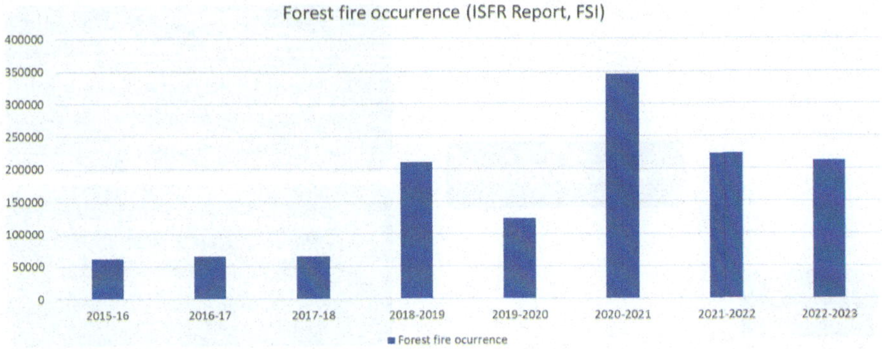


**Fig. 1** Forest fire occurrence in India

## *1.1 Types and Causes of Forest Fire*

Forest fires are not always the same; they may differ depending on their nature, size, spreading speed, behavior, etc. Forest fires can be grouped into four types depending on their nature and size. The different types of forest fires have been depicted in Fig. 2.

Mostly, the causes of forest fires are attributed to human negligence. Forest fire causes can be categorized as natural or manmade based on the heat source. The human causes of forest fires may be separated into two categories: purposeful causes and unintentional or accidental causes. Natural causes of forest fires include lightning, volcanic eruptions, friction from rolling stones, etc. (Fig. 3).

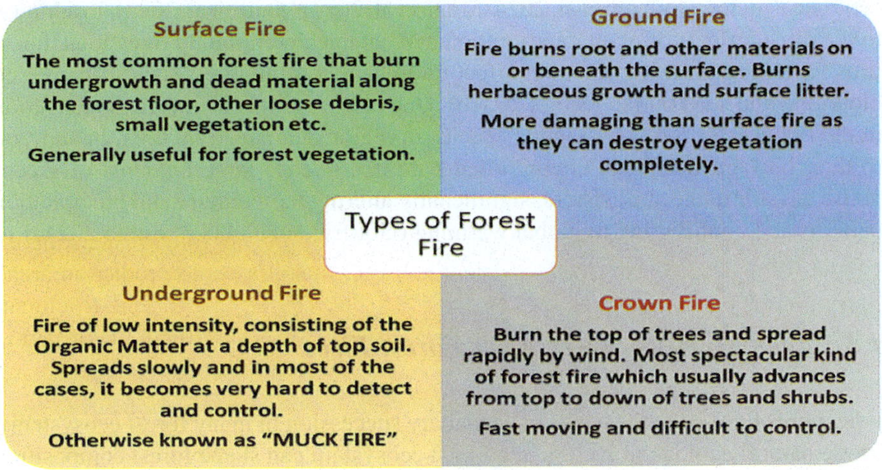


**Fig. 2** Different types of forest fires

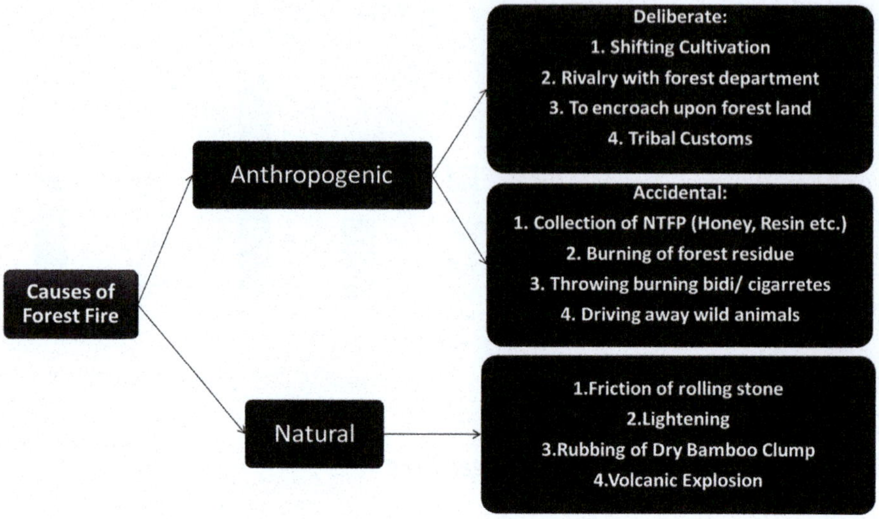


**Fig. 3** Various causes of forest fires

## 1.2 Forest Fire and Ecosystem

Forest Fire significantly impacts ecosystem regulation, including changes in floral composition, threats to biodiversity, alterations in species and habitats, and emissions of greenhouse gases leading to climate change (Darshni and Jamwal 2021; Singh 2024a). The influence of wildfires on animal habitats is remarkable as they can lead to significant changes in native environments, which plays a crucial role in shaping ecological niches (Jodhani et al. 2024; Singh et al. 2024; Malik et al. 2024). In addition to the direct effects of plant deterioration and changed community dynamics, forest fires have long-term impacts that affect biological communities, such as increased blooming and seed output (Singh 2024b). One of the main concerns is the reduction in the richness of forest biological species in regions damaged by fires. In addition to their biological effects, uncontrolled wildfires release large amounts of greenhouse gases into the atmosphere, significantly altering temperature and precipitation patterns and contributing to global warming (Gottfried et al. 2003; Singh 2024c).

## 1.3 Ecosystem Recovery After Forest Fires

Forest fires have been the cause of secondary succession in many forest ecosystems. These natural ecological factors in a forest ecosystem can shape forest composition and structure. Vegetation recovery after a fire event depends upon the resprouting ability of thick bark establishment from viable seeds buried in soil or canopy. High-intensity fire harms forest soil and water. All ecosystems frequently experience forest fires, which are crucial to soil, microorganisms, biodiversity, and ecosystem dynamics (Singh 2024d). The low-intensity fire burns organic debris in the soil and litter, making more nutrients available to promote the formation of the post-fire community and the regeneration of plants (Chandra and Bhardwaj 2015). Higher-intensity fires destroy soil organic matter and cause nitrogen, phosphorus, and potassium to volatilize. However, Mn, Mg, Cu, and other micronutrients must burn completely at extremely high temperatures. High temperatures impact most soil microorganisms and are very sensitive to them. The frequency of forest fires raises the bacterial richness while decreasing the number of actinomycetes, fungal species, and arbuscular mycorrhizal fungi (Kumar et al. 2017, 2019; Jodhani et al. 2024).

## 2 Understanding the Impact of Forest Fire

Forest fire has both immediate and longstanding impacts on the flora and fauna, eventually influencing the ecosystem function and regulation (Fig. 4).

a. ***Short-Term Impacts on Flora and Fauna***

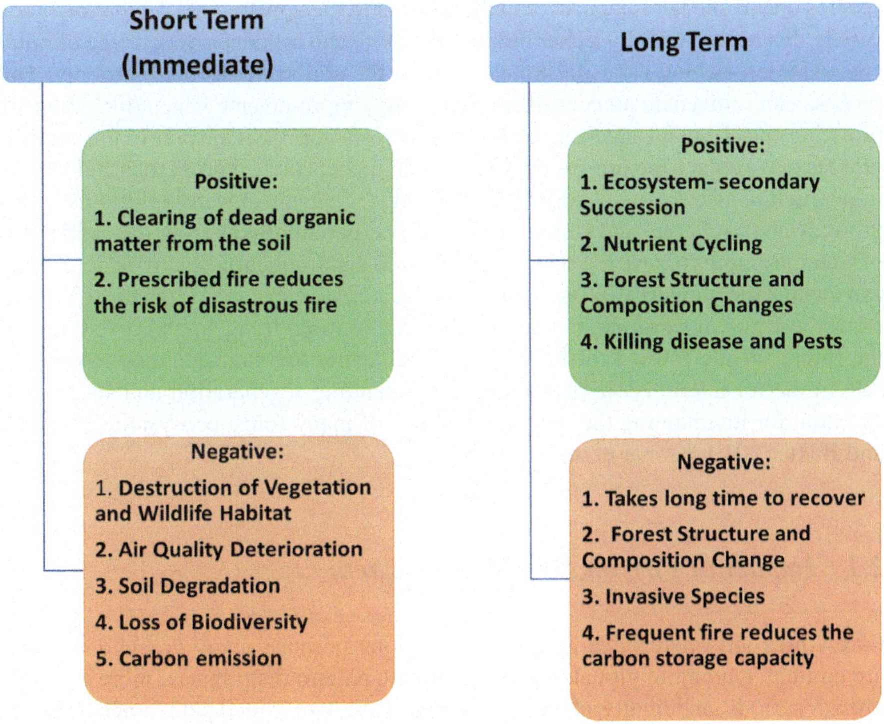


**Fig. 4** Immediate and long-term effects of forest fires

Forest fires have significant immediate impacts on ecosystems, which can be destructive and beneficial to various degrees depending on the specific circumstances and the natural adaptations of the ecosystem to fire. The rapid destruction of vegetation results in the immediate loss of habitat and food for fauna species, because of which the survivors are forced to migrate or face death (Arshad et al. 2022). Through biomass combustion, large quantities of smoke, carbon monoxide, volatile organic compounds, and particulate matter are released that degrade air quality, pose health risks to humans and animals, and contribute to climate change by increasing the atmospheric concentration of greenhouse gases. Fires make soil hydrophobic (water-repellent) by reducing the ability of water absorption, leading to increased runoff and erosion. The fire strips the soil of essential nutrients, degrading the land quality (Chandra and Bhardwaj 2015). The death of mammals, reptiles, and insects, which cannot escape the rapidly advancing flames, disrupt local food chains and ecological balance. Forest fires release significant amounts of carbon dioxide and other greenhouse gases stored in vegetation. This exacerbates climate change and global warming (Kumar et al. 2021a).

b. ***Long-Term Impacts on Flora and Fauna***

The burned areas undergo ecological secondary succession, where different plants and animals gradually recolonize the area on time. Initially, fire-adapted and pioneer species invade the landscape, eventually leading to the appearance of a mature biodiversity that has potentially richer biodiversity. Fire can influence the cycling of nutrients. The left behind ashes by fires replenish the soil minerals and nutrients. This process can rejuvenate an ecosystem, fostering a more diverse vegetation composition (Stevens-Rumann and Morgan 2019). Fire resistants like Chirpine or fire-adapted grasses may replace the fire-prone. Long-term impacts on soil properties and vegetation structure affect watershed hydrology, that is, changes in water infiltration rates, groundwater recharge, and stream flow patterns, potentially altering the availability of water resources (Gottfried et al. 2003). Intense and frequently occurring forest fires can change the climate by reducing the carbon storage capacity of forests, which can exacerbate global warming and potentially lead to more frequent and severe fires in the future. While forest fires have immediate destructive impacts, they also play a role in the long-term ecological processes, including regeneration and succession, essential for maintaining the dynamic balance of many forest ecosystems (Malison and Baxter 2010; Kumar et al. 2020a, b).

## *2.1 Impact of Fire on Overall Ecosystem*

Forest fires anywhere have global ramifications because many greenhouse gases are emitted, leading to global warming, climate pattern disturbance, more extended growth periods, and finally, change in biodiversity. An inclined percentage of smoke in the environment retard the photosynthetic activity, thereby affecting the health of human beings and animals (Jodhani et al. 2024).

At the regional level, the percentage of biomass stock, water cycle, nutrient cycles, coral reefs, and functioning of plant and animal species are unfavorably affected, leading to detrimental effects on the regional ecosystem. The trees falling due to fires act as fuel for the upcoming years and hence favoring the frequency of forest fires. It may also trigger the growth of fire-resistant species in large numbers, e.g. pyrophytic grasses. Insect colonization and infections followed by a fire occurrence degrade vegetation quality and disturb the ecological balance. One of the most negative ecological impacts of fire is the replacement of vast areas of forest with pyrophytic grasslands in tropical rain forests, which has already been observed in parts of Indonesia and Amazonia.

### 2.2 *Impact of Forest Fires on Plant Diversity*

Severe fires have a dramatically detrimental impact on floral composition and diversity. Stubble burning in agriculture is one of the pressing causes of forest fires in Tropical forests. Forests with excess human activities and fires caused by deforestation or shifting cultivation sometimes lead to complete combustion of the forest vegetation, resulting in barren soil (Malison and Baxter 2010). Forests have no fire adaptation mechanism, such as thick bark, loss of seedlings sprouts, lianas, climbers, and young regenerations. Damaged seeds, seedlings, and saplings act as an obstacle to recovering original species.

### 2.3 *Adaptations of Plants to Forest Fire*

The vegetations of tropical forests, where the frequency of fire occurrence is abundant in almost every dry season, have certain adaptations like thick bark (*Pinus roxburghii* having 2.5 cm bark), faster scar healing properties, and vegetative growth of plants from roots or stems, e.g. Mountain ash (*Eucalyptus regrans*). Some dicots have adaptations of protecting their seeds in their modified flower-like structure (cones) during fire occurrence and disseminating them after the fire period, e.g. Jack pine (*Pinus banksiana*) and Lodgepole pine (*Pinus contorta*) (Souza-Alonso et al. 2022). The positive implication of forest fire is that it vigorously promotes the invasion and survival of fire-tolerant species.

### 2.4 *Impacts of Fire on Forest Fauna*

Forest fires not only kill and injure the animals but also have long-standing impacts such as climatic stress, fragmentation of habitat, loss of territories, and decline in the amount of shelter and food. Forest recovery rates depend on certain vital organisms, such as pollinators and invertebrates. Hence, the process slows down or gets disrupted with the loss of their population. Small mammal species such as lemur, bats, or other cavity-nesting bird species are affected due to the destruction of snags in the forest, which act as their homes. Hence, many birds and mammals are displaced, resulting in loss of wildlife and alteration of biodiversity (Moretti et al. 2010; Kumar et al. 2021b, c). Due to the loss of fruit trees, many animal and bird species have declined as they depend on these trees for food. Many animals and bird species depend upon certain fruit-bearing trees, such as fruit-eating birds. The decline in their population affects them adversely. The hornbill population has declined significantly in tropical forests due to forest fires (Zhong et al. 2023). There has also been a reduction in the population of some small carnivores due to their running away from their prey, such as rodents. The community of arthropods, being the food source for omnivores and carnivores, is also destroyed by fire in leaf litter (Kinnaird and O'Brien 1998).

### *2.5 Adaptations of Animals to Forest Fire*

In the savannas of Australia, Grass layer beetles are fire-resistant. For the population, the growth of moose fire is essential. In the United States, while the moose and deer are resistant to fire, fire adversely affects them as their feed plant species are sensitive to fire. Hence, it affects the predators of moose and deer, e.g., Gray wolf (Orgeas and Andersen 2001).

## 3 Extent of Fire Damage and Various Methods for Measurement

Early detection and real-time monitoring of fire activities are crucial for fire damage assessment and management. The extent of fire damage to organisms, communities, and ecosystems can be estimated by comparing the severity of the fire to the severity of the burn. The importance of environmental change or socio-economic losses after a forest fire can be explained through fire severity. Assessment of damage by fire can be evaluated via following methods:

### *3.1 Traditional Methods*

Traditional methods for assessing forest fire damage primarily involve field surveys through direct observations using equipment or indirect observations based on rough sketches overlaid onto topography and cadastral maps (Lee 2005; Lee et al. 2006). In 2002, the categorization of post-burn ground characteristics was compiled with unburned, scorched, light, moderate, and deep forest fire damage (Ryan 2002). Based on the extent of canopy damage, fire severity can be classified into six categories based on assigned values of the canopy severity index (Table 1; Han et al. 2021). This method is still in use, but the limitation is that it only considers the damage above the ground level.

In 2021, Han et al. generalized and tweaked the categorization of fire damage based on the study of Keeley (2009) and Ryan (2002) (Table 2).

**Table 1** Extent of fire damage based on canopy-severity index

| Canopy-severity index values | Extent of fire damage | Characteristics |
|---|---|---|
| 0 | Unburned | No direct damage from fire, the canopy color is green, and no change in appearance |
| 1 | Low | Slight alteration in appearance, slightly less number of leaves |
| 2 | Low to moderate | Significantly less number of leaves, but the majority of twigs are present |
| 3 | Moderate | Significantly less number of twigs but majority of branches are present |
| 4 | Moderate to high | Upper branches are consumed, a small portion of lower branches are present |
| 5 | High | Majority of branches and cones are destroyed |

**Table 2** Modified classification of fire damage severity with characteristics

| The extent of fire damage | Characteristics |
|---|---|
| Unburned | Plant parts remain unchanged and retain their natural green colour. They are not directly affected by the fire heat |
| Scorched | Plants are not burned but experience leaf loss due to exposure to radiated or convicted heat. Mosses and leaves exhibit brown or yellow hues. However, their distinctive species traits remain discernible |
| Light | Partial death of canopy cover but the presence of green foliage despite the charred stems. The organic layer of soil remains unchanged with destroyed twigs, mosses, and herbs |
| Moderate | Canopy trees are perished, and all understory vegetation is burnt, but the foliage is not consumed. Earth's surface is covered with decomposed twigs, and logs are also burned. Surface litter, mosses, and herbaceous plants are destroyed. The organic layer of soil is also predominantly absorbed |
| Heavy | Nearly all organic layers in the soil, including mosses, shrubs, herbaceous plants, twigs, and small branches, are destroyed. Due to oxidation, the uppermost layer of mineral soil turns reddish to yellow. All organic matter is deeply burned to several centimeters in depth |

These methods often require significant time and energy to visit sites and visually assess the damage. Moreover, in many cases, it has been observed that it is not physically possible to reach severely damaged areas due to complex terrain morphology, large scenes, and bad weather conditions, ultimately resulting in complete omission from the assessment (Liu et al. 2020).

## 3.2 *Remote Sensing*

A unique and cost-effective alternative to the time and energy-consuming field surveys is geospatial technologies such as remote sensing (RS) and geographic information systems (GIS) for monitoring fire damage across vast and distant geographical regions (Hua and Shao 2017; MacDonald and McKenney 2020). They complement each other, in which remote sensing provides the raw data, and GIS uses it to generate valuable insights and visualizations. The release of energy in the form of light and heat during combustion can be easily detected in the electromagnetic spectrum (EMR). Various aerial and satellite sensors have been used to monitor the differences in the reflectance spectrum between the standing plant and the resulting ash and char after the burning (Joseph et al. 2009). Since the 1980s, there have been several types of remote sensing platforms and sensors, such as satellites, UAVs, active sensors (SAR, LiDAR), and passive sensors (optical, thermal infrared, multispectral). Satellite images acquired through remote sensing technologies are utilized to identify burned regions and evaluate the extent of destruction following a forest fire, even without going to the actual field. One of the most important aspects of remote sensing is the spatial resolution of the images, which determines the level of detail in an image based on the satellite's spatial resolution (Solanky and Katiyar 2016). Three types of spatial resolution are present in various remote sensing sensors, which are low (less than 5 m), medium (between 5 and 30 m), and high (more than 30 m) (Kurbanov et al. 2022). When there is a significant forest fire, multispectral analysis is a helpful tool for identifying burned 0 areas (Oncu and Corumluoglu 2023). Multispectral Landsat systems of medium resolution, such as Multispectral Scanner System (MSS), Thematic Mapper (TM), Enhanced Thematic Mapper Plus (ETM+), and the Operational Land Imager (OLI) are the most frequently used optical sensor (Masek et al. 2013). The Sentinel Multispectral Instrument (MSI), multiple sensors of the SPOT satellite, Linear Imaging Self-Scanning Sensors (LISS) of the Indian ResourceSat family satellites, the Advanced Spaceborne Thermal Emission and Reflection Radiometer (ASTER) of the Terra satellite, and the Airborne Visible InfraRed Imaging Spectrometer (AVIRIS) are the primary sources of medium resolution images for burnt area studies. In the case of low-resolution remote sensing imagery, Envisat Medium Resolution Imaging Spectrometer (MERIS), Visible Infrared Imaging Radiometer Suite (VIIRS), Advanced Very-High-Resolution Radiometer (AVHRR), and Moderate Resolution Imaging Spectroradiometer (MODIS) are popular. Whereas high spatial resolution systems like WorldView, Ikonos, -2,3, Quickbird, GeoEye-1, Pleiades-1A, and Resurs are also utilized but only for a small number of investigations (Kurbanov et al. 2022). Utilizing all these types of sensors (Table 3), remote sensing is used in many areas, such as fire detection, fire risk mapping in forest fires, burn damage intensity assessment, damage extension, and damage intensity distribution (Oncu and Corumluoglu 2023).

Considering both precision and effectiveness, spectral index-based approaches are often regarded as the most popular and practical methods for retrieving burned areas (Veraverbeke et al. 2011). Many remote sensing indices are in the literature for quantifying fire severity, making it difficult to extract a unified index. No single index can be applied universally in all cases (Han et al. 2021; Filipponi 2018). Indices for extracting burned areas included both a single-date approach and a bitemporal approach (Epting et al. 2005). Measurement of fire damage through burn severity in the field is typically done by using the composite burn index (CBI), which entails visually evaluating burned regions to estimate the ecological effects of the fire. This tool enables visual estimation by assigning a grade to the extent of fire damage and the anticipated vegetation regeneration for a given region, using a scale ranging from 0 to 3 (Kasischke et al. 2008). The major limitation of CBI is time dependency and the requirement of physical visits to the affected areas to conduct the evaluation (Szpakowski and Jensen 2019). Normalized difference vegetation Index (NDVI) and Normalized burned ratio (NBR) are the two most prevalent and frequently employed spectral indices for evaluating burn severity and estimating burned area at a local level (Fernández-Manso et al. 2016; Mallinis et al. 2018). Although NDVI is still utilized in present-day studies, NBR has mostly overtaken as the prevailing index for evaluating burn severity. NBR has been extensively used to estimate damage extent by analyzing Near Infrared (NIR) and Shortwave Infrared (SWIR) data that typically falls between 2.08 and 2.35 μm. NIR wavelengths are responsive to the leaf structure of live vegetation, whereas SWIR is responsive to soil moisture content and conditions (Slaton et al. 2001). NBR reflects the remote sensing spectral index at a specific moment. While it is positively correlated with fire severity, it cannot accurately pinpoint the burned area or account for vegetation heterogeneity (Han et al. 2021). That is why the difference in the normalized burned ratio (dNBR) is calculated by taking the difference before and after a fire.

Nonetheless, dNBR is unsuitable for large-scale severity measurement due to the heterogeneous nature of forest landscapes. Further, there is no consensus on the classification of fire severity under dNBR, leading to subjective judgments. Various other indices that can be applied to assess fire damage and severity have been represented with their formulas in this section (Table 4).

**Table 3** Various types of sensors are used for fire damage assessment on a global and local scale

| Sensor | Spatial resolution | Pros | Cons | Use |
|---|---|---|---|---|
| BIRD | 185 m | Night-time data is possible to collect, specialized | Unclear | Detecting burn areas along with burn area assessment |
| MODIS | 250 $m^{-1}$ km | Free of charge, readily available, covers a large area, temporal resolution is high | Difficulty penetration through the canopy, limitation to smaller areas | Detecting burn areas along with burn area assessment |
| AVHRR | 1.1 km | Temporal frequency is high | 325 K saturation in channel 3 | Detection of active fire |
| DMSP-OLS | 0.56 km | Temporal frequency is high, and nighttime data | Limitation to day time data | Detection of active fire |
| GOES-10 and 12 | 1 km | Temporal resolution is very high | Spatial resolution is coarse, with 335 K saturation in channel 2 | Active fire identification |
| ATSER | 15–90 m | Free of charge, readily available, hyper spectral sensor, short-wave infrared (SWIR) band | Temporal resolution is low, difficulty in penetration through the canopy | Detecting burn areas along with burn area assessment |
| IRS-AWiFS | 56 m | Spatial resolution is high, short-wave infrared (SWIR) band | Unclear | Identification of active fire, estimation of burn area, and validation |
| Landsat MSS, TM and ETM+, OLI | 15–30 m | Free of charge, readily available, high spatial resolution, medium-infrared (MIR) band | Low temporal resolution, difficulty in penetration through the canopy | Identification of active fire, estimation of burn area, and validation |
| Sentinel-2 | 10–60 m | Free of charge, both spatial and temporal resolution are comparatively high, multiple near-infrared (NIR) bands | Difficulty penetration through the canopy | Fire hazard mapping |

(continued)

**Table 3** (continued)

| Sensor | Spatial resolution | Pros | Cons | Use |
|---|---|---|---|---|
| Pleiades Neo | 0.3–1.2 m | Spatial resolution is high | The cost is too high | Fuel mapping |
| AVIRIS | 4–20 m | Spatial resolution is high, sensor is hyper spectral | Cost is high, data processing is complicated | Fire hazard mapping |
| AVIRIS-NG | 0.3–4 m | Spatial resolution is high, sensor is hyper spectral | Cost is high, data processing is complicated | Fire hazard mapping |
| Hyperion | 10–30 m | Free of charge and easy to access, the sensor is hyper-spectral | Inactivated since 2017 | Fire hazard mapping |
| UAV | High centimetre level | Spatial resolution is high, hyper spectral sensor can be used, LiDAR sensor can be used | Cost depends on the sensor, low tolerance | Fire hazard mapping |

*Source* Joseph et al. (2009), Szpakowski and Jensen (2019) and Zagalikis (2023)

## 4 Recent Technologies for Monitoring Forest Ecosystem Improvement After Fire

Advancements in technology, such as remote sensing, offer a promising solution (Sasaki et al. 2018). Generally, it involves different passive sensors like optical and active sensors like LiDAR. By employing these sensors on platforms ranging from satellites to drones, data on tree structure, entire forest stands, and recovery of forest trees can be efficiently gathered in a time-efficient and cost-effective manner over large areas (Gouveia et al. 2010). This technology, including Orbital Multispectral Sensors, LiDAR, and UAS technology, not only aligns well with ground-truth data for traditional attributes but also facilitates the assessment of attributes that are difficult to measure using conventional methods.

### *4.1 Different Remote Sensing Tools for Monitoring Eco-system Recovery*

#### 4.1.1 Orbital Multispectral Sensors

Three categories may be used to classify vegetation recovery estimation techniques: picture categorization or image classification, vegetation indices (VIs), and spectral mixture analysis (SMA) (Szpakowski and Jensen 2019).

**Table 4** Various indices used for measuring the fire damage

| Category | Specific name of the index | How to calculate? |
|---|---|---|
| Vegetation index | Normalized difference vegetation index (NDVI) | $\frac{NIR-R}{NIR+R}$ |
| | Normalized difference vegetation index thermal (NDVIT) | $\frac{NIR-R\times TIR}{NIR+R\times TIR}$ |
| | Enhanced vegetation index (EVI) | $\frac{2.5(NIR-R)}{NIR-6R-7.5B+1}$ |
| | Global environment monitoring index (GEMI) | $\frac{2((NIR\times NIR)-(R\times R))+1.5NIR+0.5R}{NIR+R+0.5}$ |
| Fire index | Normalized burned ratio (NBR) | $\frac{NIR-LSWIR}{NIR+LSWIR}$ |
| | Normalized burned ratio thermal (NBRT) | $\frac{NIR-LSWIR\times TIR}{NIR+LSWIR\times TIR}$ |
| | Difference normalized burned ratio (dNBR) | (NBRprefire − NBRpostfire) − dNBRoffset |
| | Relativized burn ratio (RBR) | $\frac{dNBR}{NBRpre-fire+1.001}$ |
| | Burned area index (BAI) | $\frac{1}{(0.1+R)(0.1+R)+(0.06+NIR)(0.06+NIR)}$ |
| | Char soil index (CSI) | $\frac{NIR}{LSWIR}$ |
| | Char soil index thermal (CSIT) | $\frac{NIR}{LSWIR\times TIR}$ |
| | Normalized difference SWIR (NDSWIR) | $\frac{NIR-SSWIR}{NIR+SSWIR}$ |
| | Mid-infrared burn index (MIRBI) | 10 LSWIR − 9.8 SSWIR |
| | Differenced mid-infrared burn index (DMIRBI) | MIRBI pre-fire – MIRBI post-fire |
| | NIR-SWIR-temperature version 1 (NSTv1) | $\frac{NIR-LSWIR}{NIR+LSWIR}\times Ts$ |
| | NIR-SWIR-temperature version 2 (NSTv2) | $\frac{NIR-(LSWIR+Ts)}{NIR+(LSWIR+Ts)}$ |
| | NIR-SWIR-emissivity version 1 (NSEv1) | $\frac{NIR-LSWIR}{NIR+LSWIR}\times E$ |
| | NIR-SWIR-emissivity version 2 (NSEv2) | $\frac{NIR-(LSWIR+E)}{NIR+(LSWIR+E)}$ |

*Note* R (red, 0.6–0.7 μm); NIR (near infrared, 0.7–1.3 μm); B (blue, 0.45–0.5 μm); SSWIR (shorter short-wave infrared, 1.3–1.9 μm); LSWIR (longer short wave infrared, 1.9–2.5 μm); MIR (mid-infrared, 3–5.5 μm); TIR (thermal infrared, 8–13 μm); E (emissivity) and Ts (surface temperature)
*Source* Liu et al. (2020) and Han et al. (2021)

Image classification leverages spectral feedback to identify healthy vegetation at the pixel level. A study used a LiDAR-derived Digital Elevation Model (DEM), a Digital Orthophoto Quarter Quadrangle (DOQQ) picture, and high-resolution data from KH-4B photos from the CORONA mission. An alternative approach to image classification is Geographic Object-Based Image Analysis (GEOBIA), which uses geographic objects rather than individual pixels as the units of study (Addink et al.

2012). GEOBIA initiates the process with image segmentation, grouping similar pixels into objects.

Vegetation indices (VIs) are prevalent for monitoring vegetation recovery, as Gitas et al. (2012) noted. Among these, VIs based on near-infrared (NIR) reflectance are particularly favored. Generally, healthy vegetation reflects significantly more NIR than other landscape features, which makes this wavelength (NIR) highly effective for detecting and monitoring vegetation growth. Spectral indices are also utilized to estimate additional biological parameters connected to the rehabilitation of vegetation, such as leaf area index (LAI), fractional vegetation cover (FVC) (van Leeuwen 2008; Chu et al. 2016), and net primary productivity.

Because pixel sizes often surpass the dimensions of individual plants, techniques like Spectral Mixture Analysis (SMA), which addresses the issue of mixed pixels, are utilized in monitoring vegetation recovery (Veraverbeke et al. 2012). SMA calculates the proportion of a pixel that corresponds to a specific endmember, which is considered to represent the identified cover types in the image. Research has consistently validated the effectiveness of linear SMA models (Röder et al. 2008). However, a drawback of this method lies in the design of endmembers; they may not adequately capture the natural variability within a scene (Röder et al. 2008).

### 4.1.2 LiDAR

LiDAR, an active remote sensing technique, uses high-frequency laser impulses to distinguish objects by measuring the intensity and time-of-flight between the pulse's emission and reception. Using an inertial measurement unit (IMU) installed on the platform and a global navigation satellite system receiver, each return can be recorded with the appropriate observation angle, proper location, and sensor orientation. 3D point clouds are produced by LiDAR sensors, with each point representing the location in space of the item that was hit by the laser impulse (Price and Gordon 2016; Van Leeuwen and Nieuwenhuis 2010). Different numbers of return points can be captured from each pulse, depending on the type of sensor. Multiple returns give more information on items inside the laser footprint, indicating various structural features of the ground cover. In contrast, the initial return offers information on the first thing impacted by the laser pulse. Full waveform LiDAR captures the continuous distribution of the return radiation, providing a highly detailed explanation of flora foundations (Lim et al. 2010).

LiDAR's capability to deliver detailed information on crucial structure parameters of the flora makes it an ideal tool for watching vegetation repossession. Gordon et al. (2017) demonstrated that data generated from LiDAR effectively measures post-fire vegetation regrowth, with metrics aligning well with field-derived data. It also offers a reliable presentation of post-fire flora coverage. Similarly, Bolton et al. (2015) evaluated post-fire forest features in Canada's boreal north by combining LiDAR data with Landsat time-series analysis. They found noteworthy differences in forest flora structure between scorched and unburnt positions over 20–25 years of post-fire. This combination is a valuable method for analyzing post-fire vegetation characteristics. Despite the increasing research and availability of LiDAR data for vegetation recovery analysis, its high cost and the scarcity of repeat surveys limit its widespread use.

### 4.1.3 UAS and UAV

Unmanned aerial systems, often known as unmanned aerial vehicles or UAVs, offer quick, portable, and affordable sensor systems for monitoring and detecting fires (Ambrosia et al. 2011). The use of UASs in proactive fire control, post-fire damage assessment, and forest recovery monitoring is growing. According to Yuan et al. (2015), "fire monitoring" is the proactive search for possible fire incidents, whereas "fire detection" entails the identification of fires that are already underway. Unmanned Aerial Systems (UASs) equipped with visible and thermal infrared wavelengths operated by different sensors get high-resolution images, which is then subjected to image segmentation analysis to identify possible fires, which are then geolocated (Valavanis and Vachtsevanos 2015).

Post-fire, UASs capture valuable data on vegetation regrowth and ecosystem dynamics. A low-cost UAV with a camera offers a cost-effective tool for monitoring wildfire-affected forests (Larrinaga and Brotons 2019). They can acquire multispectral and RGB imagery at various resolutions, which is especially useful in boreal ecosystems where field campaigns are limited and satellite data is often compromised by short growing seasons and frequent cloud cover (Pádua et al. 2019). However, using UAVs to obtain multispectral ortho mosaics at ultra-high resolution posses challenges, but it offers a viable alternative to satellite imagery for evaluating post-fire forest vegetation (Fernández-Guisuraga et al. 2018).

With its comprehensive imaging capabilities, UAV technology offers quick, high-resolution data for assessing regions damaged by fires. Unmanned Aerial Vehicles (UAVs) fitted with computer-vision Structure-from-Motion (SfM) algorithms, or UAV-SfM, gather and analyze multi-spectral data to track the effects of wildfires (Reilly et al. 2021). Airborne laser scanners (ALS) and UAV-SfM comparisons highlight different remote sensing techniques for post-fire evaluation (McKenna et al. 2017; Carvajal-Ramírez et al. 2019). Using UAV footage, supervised classification mapping of land cover type and fire severity, and UAV LiDAR-derived variables, the intensity of the fire is determined (Hillman et al. 2021).

AI-enabled unmanned aerial systems (UASs) provide early warning systems through real-time aerial surveillance, offering a proactive approach to post-fire management. Prediction models that use UAS data facilitate identifying susceptible regions and applying focused mitigation techniques, enhancing post-fire evaluations and the distribution of resources for reforestation and erosion control (Boroujeni et al. 2024). Even though unmanned aerial systems (UASs) have great promise, their current application is restricted to small regions because of cost and battery life issues. However, UASs signify a change toward educated, flexible, and adaptable environmental management and post-fire disaster recovery plans.

## 5 Methods for Facilitating Ecosystem Recovery

More than half of the world's tropical forests have been transformed for other land use, leading to the degradation of habitats, a decline in biodiversity, and a loss of vital ecosystem services (Lewis et al. 2015). To address this, the conservation and management of remaining old-growth forests are crucial but insufficient (Grace et al. 2014). However, monoculture reforestation rarely delivers the benefits needed for landscape functionality, livelihoods, and biodiversity. Natural regeneration, less costly than tree planting, offers a cost-effective way to achieve large-scale restoration, as evidenced by the potential cost savings (Rozendaal and Chazdon 2015). Prioritizing natural regeneration in suitable areas allows funds to be allocated to more intensive interventions elsewhere, enhancing carbon mitigation potential (Chazdon et al. 2016). Natural regeneration faces economic, social, and legal challenges despite its benefits. Addressing these issues is critical for utilizing natural regeneration as a viable restoration tool at both national and global levels (Ceccon et al. 2015). Enabling natural regeneration through supportive policies and incentives would help to undertake large-scale tropical forest restoration more effectively planned and implemented (Fig. 5).

**Fig. 5** Methods for forest eco-system recovery

## *5.1 Supporting Key Wildlife for Seed Dispersal and Soil Regeneration*

Seed dispersal is crucial for successful restoration, especially in tropical ecosystems where over 65% of plant species rely on animals for dispersal (Correa et al. 2015). Early-successional species are dispersed by small birds, bats, and wind, forming canopies that attract larger birds and mammals for further dispersal. In deforested areas, seed movement depends on wind and small animals, while heterogeneous landscapes recover faster due to larger birds and mammals. Monitoring these interactions is often overlooked but crucial for project success. Initially, small birds, bats, and abiotic dispersal dominate, but as forests mature, large birds and flightless mammals become more critical, enhancing redundancy in dispersal modes and supporting forest regeneration (Marshall et al. 2023).

## *5.2 Soil Stabilization Techniques*

The rapid degradation following a fire necessitates swift implementation of mitigation strategies to counter increased runoff, erosion rates, and nutrient export (Ferreira et al. 2015). Management techniques such as prescribed fire and preventive forestry and ex-post methods like mulching, seeding, and hillslope barriers are vital for mitigating post-fire degradation. Mulching, particularly with straw, effectively reduces post-fire soil erosion. Integrating multiple techniques at various scales is essential to minimize water, sediment, and nutrient loss and preserve local ecosystems (Prats et al. 2012).

## *5.3 Repairing and Rehabilitating Fire-Affected Streams to Improve Water Quality and Habitat Conditions*

Post-fire management strategies, which can mitigate and exacerbate soil and vegetation impacts, are classified into emergency stabilization, restoration and rehabilitation. It can be applied at channel or hillslope scales. Emergency stabilization, implemented immediately after fires, aims to reduce runoff and related processes like sediment delivery and floods. Common techniques include log erosion barriers, mulching, grass seeding, and check-dams (Fernández et al. 2019). Rehabilitation and restoration practices, including afforestation and introducing shrubs to enhance biodiversity, have been widely implemented, particularly in the Mediterranean basin (Vallejo and Alloza 2012). Post-fire management can enhance hydrological dis-connectivity by increasing detention and infiltration of overland flow, thereby lowering runoff coefficients and erosion rates (Lucas-Borja and Delgado-Baquerizo 2019). The effectiveness of these treatments depends on their design, implementation timing, and wildfire severity (Badía et al. 2015). Check-dams are controversial due to their high economic cost, need for heavy machinery, and creation of new trails, which can lead to channel geomorphic alterations and concentrated overland flow (Martínez-Murillo and López-Vicente 2018). Hillslope bioengineering techniques comprise of different methods, such as log erosion barriers, contour-felled log debris, and afforestation (Albert-Belda et al. 2019). Afforestation reduces runoff and soil erosion due to forest litter (Porto et al. 2009). Effective planning and zoning, supported by tools like sediment connectivity indices, are crucial for optimizing resource use and understanding and measuring the impact of burned catchments (Martini et al. 2020). Increased wildfires affect water quality, elevating sediment, temperature, nutrients, chemicals, and metals in burned areas. Biological assemblages decline post-fire, with effects lasting up to 15 years. Few studies address pollutants from urban wildfires, highlighting benzene in drinking water and elevated metals in ash. Understanding these impacts is crucial for effective management (Paul et al. 2022).

### *5.4 Precision Restoration: For Promoting Forest Recovery in the Twenty-First Century*

"Precision forest restoration" (PFR) emphasizes individual plant success to foster resilient, self-regulating forests while minimizing the negative impacts of traditional reforestation. PFR leverages ecological knowledge and various technologies, from low-tech to high-tech, at both landscape and individual plant scales (Silva et al. 2020). Although initially more expensive per plant, PFR (Restore Key Plant—Animal Interactions) is now a viable and transformative approach to ecosystem restoration through different types of advance technology.

### *5.5 Comprehensive Burn Planning: Strategies for Effective and Safe Prescribed Burns*

Forests covering nearly a billion hectares across Latin America and the Caribbean (LAC) play a vital role in global biodiversity and climate regulation. Human-driven factors, particularly anthropogenic wildfires, contribute significantly to forest degradation in LAC. These fires, often exacerbated by deforestation, agriculture, invasive species, and climate change, threaten forest ecosystems substantially. Even in fire-adapted environments, human fire suppression efforts can disrupt ecosystem functioning (Souza-Alonso et al. 2022). Replanting trees is crucial for combating desertification, soil erosion, flooding, and zoonotic disease outbreaks while providing ecosystem services and livelihoods. However, current tree planting methods face limitations in cost-effectiveness and scalability. Unmanned aerial vehicle-supported seed sowing (UAVsSS) offers a promising solution by enabling rapid and environmentally friendly afforestation and reforestation, especially in inaccessible terrain. Recent literature on UAVsSS highlighting applications in post-wildfire reforestation, mangrove restoration, desert greening and their challenges including low seed survival rates, diversity concerns, weather limitations, and financial constraints. Mohan et al. (2021) provided twelve recommendations, such as publishing germination results and linking UAVsSS with carbon offset markets, to advance UAVsSS applications.

# 6 Post-forest Fire Ecosystem Restoration Approaches

## 6.1 *Active and Passive Restoration Approach*

Forest fires are a natural part of many ecosystems and can significantly shape landscapes and promote ecological diversity. Fire regime includes four major factors, i.e., fire season, fire intensity, severity of fie, and interval period, based on which a fire-prone ecosystem has a natural recovery pattern. The natural flora of a fire-prone ecosystem are adapted to the local climate; for instance, various obligate seeder species of plants (e.g., Pine) use the wildfire to break seed dormancy (Tangney et al. 2022). Humid tropical forest ecosystems are not as adapted to fire as dry deciduous forests, and triggers for wildfire, both anthropogenic and natural, tend to show differences in each of these cases. Hence, ecosystem recovery post-wildfire varies significantly in each case and requires extensive research (Souza-Alonso et al. 2022).

Ecosystem restoration can be done by either active or passive restoration procedure, where the active restoration includes plantation of indigenous seeds and seedlings along with management of resource utilization, and passive restoration relies on natural recovery of the system while ensuring the absence of environmental stressors, such as agriculture, grazing, and foraging. Passive restoration, although cost-effective, is slower than active restoration. Hence, an assisted natural recovery system, which meets both the restoration approaches in the middle, can answer both the economical and environmental concerns (Noulèkoun et al. 2021).

## 6.2 *Case Studies of Successful Ecosystem Recovery*

### 6.2.1 Yellowstone National Park Forest-Fire, 1988: An Example of Passive Restoration

In Yellowstone national park, located in the western part of North America, the majority (80%) of the natural flora of the ecosystem consists of Lodgepole Pines

(*Pinus coontorta*); the cones of said pine trees are serotinous, i.e., use the heat-shock method to break dormancy. Douglas fir, willow, and birch are other plants in the natural flora. These plants are all protected by thick bark and are deep-rooted, increasing their chances of fire survival and possible resprout after the mega fire incident. Herbaceous plants of Yellowstone germinate using seed dispersal and resprouting methods.

In the summer of 1988, Yellowstone National Park endured the severest case of forest fire in its history, burning over 36% of its forest cover. The severity of the fire was fueled by extreme climate conditions, draught, lightening, and human activity. Despite suppression efforts, the fire only stopped after the arrival of natural precipitation in mid-September (Turner et al. 2003).

#### 6.2.1.1 Fire-Mapping

To understand the nature of fire, Monica G. Turner and colleagues used satellite images to classify the burn patterns into four distinctive categories i.e. crown fire, severe surface burn, light surface burn, and unburned.

#### 6.2.1.2 Findings

The total burned area was estimated to be more than 3000 $km^2$ (more than 36% of the total park area), and the burn pattern was mosaic. Since most of the fire was controlled by airflow, there were several patches of 'crown fire' amid 'unburned' forest land (Turner et al. 1994). Although the fire could be classified as a megafire, wildlife loss was minimal: 345 elks out of an estimated 50,000 population were the largest number of animals lost from one species; 36 mules, 12 mooses, and 9 bison were also lost by the fatality.

#### 6.2.1.3 Approach

The passive restoration approach was primarily adopted to revive vegetation in Yellowstone. Although grass and seeds were dispersed in a few cases, no official reforestation approach was taken.

#### 6.2.1.4 Results

Regeneration of forest flora was immediate and unconstrained. The following year, substantial vegetative growth, including abundant seedling growth, was observed. Most regeneration was triggered by resprouting rather than seed dispersal. Invasive species were overpowered by native species in the recolonization of forest cover

(Turner et al. 2003). Another exciting change was the growth of Aspen over Lodgepole Pines. However, initially, Aspen's regeneration faced some challenges and exhibited slower growth. Eventually, some regions' growth rate surpassed that of Lodgepole Pines (Romme et al. 2011). Although the pine population was restored, the introduction of Aspen, a secondary population of plants, was an unexpected change in the ecosystem of Yellowstone (Turner et al. 2003). Above all, the wildlife species of Yellowstone did not face any long-term threat due to the heterogeneous mosaic spread of the fire. Various herbivore and omnivore species (e.g., Elk, Bear) benefitted from the abundance of food following the fire (Turner et al. 2003).

#### 6.2.1.5 Lessons

Based on the seemingly rapid recovery of forest cover, Yellowstone's passive recovery approach can be considered the ideal approach. Researchers claim introducing new saplings, exotic species, or seed-spreading is not ideal for a forest cover that is well-equipped with a fire regime. Instead, it might be counterproductive and harm the natural order of succession (Turner et al. 2003). However, researchers also agree that owing to anthropogenic activity and subsequent climate change, shifting of forest fire can harm the natural ecosystem; therefore, more severe forest fire occurrences are not unlikely, and a proper disaster management system should be employed to combat any such future incident (Romme et al. 2011). Moreover, the mere success of Yellowstone's pine forest does not ensure the same for others. In one case-study, part of the Ponderosa Pine forest in Arizona was reported to be transformed into permanent grassland after a forest fire (Savage and Mast 2005). Hence, more research is needed to understand a forest cover's natural fire regime and differentiate a natural fire from an anthropogenic one. Besides appropriate research in this field, monitoring forests and documenting bushfires and wildfires should be the primary objective. Above all, the climate crisis is a worldwide issue. Educating people on forest fires and appropriate actions in times of need is necessary to combat future disasters (Romme et al. 2011).

### 6.2.2 Mediterranean Basin Restoration Approach: Example of Active Restoration Initiative of Ecosystem

The Mediterranean forest ecosystem is the world's second-largest biodiversity hotspot, marked as the hotspot for climate change and environmental degradation by the UNEP (2024). The natural fire regime of this region has shifted rapidly, causing a massive threat to the ecosystem of the Mediterranean region. For the past 20 years, various researchers have developed scientific approaches to actively restore the ecosystem of this biodiversity hotspot (Vallejo and Alloza 2012). The local people of said region have also been practicing various restoration approaches that have been studied by another group of researchers (Nunes et al. 2016).

#### 6.2.2.1 Fire-Monitoring and Approach

As the restoration of this region is not caused by one particular forest fire incident, instead of fire mapping, the approach is shifted to forest cover monitoring with the help of Landsat, mainly depending on their NDVI values to understand the ecological urgency ideally (Pesaresi et al. 2020). Vallejo and colleagues successfully evaluated their restoration strategy in multiple regions of the Mediterranean basin, which was initially developed in Spain. The approach is discussed below:

(1) Establishment of management authorities to avoid short-term disasters with larger impacts, such as flash-floods, erosion, etc. The authority also monitors soil degradation, biodiversity welfare checks, fire resilience, prevention initiatives, etc. (Noce and Santini 2018).
(2) Surveying and monitoring remote sensing data to identify forest areas at risk of soil erosion, run-off, vulnerability to fire, etc. The goal is to identify key plant species for such vulnerable regions that can regenerate biodiversity and reduce the risk of vulnerability in general. The critical factor is using remote sensing, mapping, and surveying data to predict and prevent (Vallejo and Alloza 2015).
(3) In disasters, timely application of mitigation and restoration approaches is necessary (Vallejo and Alloza 2015).
(4) Short-term recovery plan: This is applicable for less than one year post-incident. The main agenda is to minimize the damage by stabilizing soil and following emergency fire rehabilitation protocols (Napper 2006).
(5) Long-term recovery plan: This plan is applicable for 2–5 years post-incident. The plan includes the planting of native plants to assist natural regeneration. The researchers recommend Woody resprouters with fire resilience (Vallejo and Alloza 2015).

At each step, the most crucial part is monitoring, surveying, and better understanding the shifted fire regime.

#### 6.2.2.2 Current Situation

The ‘Restoring Mediterranean Forest’ initiative group has officially revived a forest cover of 2 million ha since they started in 2017 (UNEP 2024). A study suggests the existence of more than 36 groups of ecosystem-restoration practitioners in the Mediterranean region actively working for the plantation and regeneration of the area’s ecological diversity (Nunes et al. 2016).

The Mediterranean ecosystem is not out of danger yet; Greece has faced the largest wildfire ever suffered by the European Union in 2023. The initiatives taken by various authorities need to be implemented more diligently. The nations worldwide have pledged to restore 1 billion ha of forest land to salvage the ecological damage (UNEP 2024). The researcher used a statistical analysis with the help of questionnaires to understand the involvement of scientific approaches in such initiatives. Results showed that 10% of European Union groups used non-native species and 40% non-European Union groups used the same (Nunes et al. 2016). The statistical significance of the data only shows the higher need for public awareness.

## 7 Conclusion

Effective mitigation of forest-fire-related disasters hinges upon a comprehensive understanding of the natural fire regime and its shifts. Recognizing the historical patterns of fire occurrence and the factors driving deviations from these norms is crucial for adopting appropriate approaches to mitigate wildfire risks. Leveraging remote sensing technologies is pivotal in monitoring fire dynamics and facilitating ecosystem restoration efforts. Remote sensing data enables timely detection of fire events and informs targeted interventions to restore fire-affected landscapes. Additionally, active and passive restoration strategies involving human-mediated and natural recovery processes are suggested to be employed based on the specific needs of the ecosystem. Research and continuous monitoring are essential for evaluating the effectiveness of restoration efforts and refining management strategies over time. Promoting the propagation of native species is central to restoration initiatives, contributing to ecosystem recovery and enhancing the region's natural biodiversity. Integrating these principles into fire management and restoration practices can foster resilient ecosystems capable of withstanding future fire disturbances while preserving biodiversity and ecosystem services. Combining historical knowledge, advanced technologies, and adaptive management can mitigate wildfire risks and support sustainable ecosystem recovery.

## References

Addink EA, Van Coillie FM, De Jong SM (2012) Introduction to the GEOBIA 2010 special issue: from pixels to geographic objects in remote sensing image analysis. Int J Appl Earth Obs Geoinf 15:1–6

Albert-Belda E, Bermejo-Fernández A, Cerdà A, Taguas EV (2019) The use of Easy-Barriers to control soil and water losses in fire-affected land in Quesada, Andalusia, Spain. Sci Total Environ 690:480–491. https://doi.org/10.1016/j.scitotenv.2019.06.303

Ambrosia VG, Wegener S, Zajkowski T, Sullivan DV, Buechel S, Enomoto F, Lobitz B, Johan S, Brass J, Hinkley E (2011) The Ikhana unmanned airborne system (UAS) western states fire imaging missions: from concept to reality (2006–2010). Geocarto Int 26(2):85–101. https://doi.org/10.1080/10106049.2010.539302

Arshad A, Azhar Ali A, Anjali KS (2022) Impact of forest fire on forest ecosystem. J Agric Technol 9(1&2):18–29

Badía D, Sánchez C, Aznar JM, Martí C (2015) Post-fire hillslope log debris dams for runoff and erosion mitigation in the semiarid Ebro Basin. Geoderma 237:298–307. https://doi.org/10.1016/j.geoderma.2014.09.004

Barwa S, Singh H (2024) Climate-induced changes in forest ecosystems of Indian Himalayan region: a meta-analysis. In: Singh H (ed) Urban forests, climate change and environmental pollution. Springer, Cham. https://doi.org/10.1007/978-3-031-67837-0_39

Bolton DK, Coops NC, Wulder MA (2015) Characterizing residual structure and forest recovery following high-severity fire in the western boreal of Canada using Landsat time-series and airborne LiDAR data. Remote Sens Environ 163:48–60

Boroujeni SPH, Razi A, Khoshdel S, Afghah F, Coen JL, O'Neill L et al (2024) A comprehensive survey of research towards AI-enabled unmanned aerial systems in pre-, active-, and post-wildfire management. Inf Fusion 102369. https://doi.org/10.1016/j.inffus.2024.102369

Camarretta N, Harrison PA, Bailey T, Potts B, Lucieer A, Davidson N, Hunt M (2020) Monitoring forest structure to guide adaptive management of forest restoration: a review of remote sensing approaches. New Forest 51(4):573–596. https://doi.org/10.1007/s11056-019-09754-5

Carvajal-Ramírez F, Marques da Silva JR, Agüera-Vega F, Martínez-Carricondo P, Serrano J, Moral FJ (2019) Evaluation of fire severity indices based on pre-and post-fire multispectral imagery sensed from UAV. Remote Sens 11(9):993. https://doi.org/10.3390/rs11090993

Ceccon E, Barrera-Cataño JI, Aronson J, Martínez-Garza C (2015) The socioecological complexity of ecological restoration in Mexico. Restor Ecol 23(4):331–336. https://doi.org/10.1111/rec.12228

Chandra KK, Bhardwaj AK (2015) Incidence of forest fire in India and its effect on terrestrial ecosystem dynamics, nutrient and microbial status of soil. Int J Agric For 5(2):69–78

Chazdon RL, Broadbent EN, Rozendaal DM, Bongers F, Zambrano AMA, Aide TM et al (2016) Carbon sequestration potential of second-growth forest regeneration in the Latin American tropics. Sci Adv 2(5):e1501639. https://doi.org/10.1126/sciadv.1501639

Chu T, Guo X, Takeda K (2016) Remote sensing approach to detect post-fire vegetation regrowth in Siberian boreal larch forest. Ecol Ind 62:32–46. https://doi.org/10.1016/j.ecolind.2015.11.026

Correa DF, Álvarez E, Stevenson PR (2015) Plant dispersal systems in Neotropical forests: availability of dispersal agents or availability of resources for constructing zoochorous fruits? Glob Ecol Biogeogr 24(2):203–214. https://doi.org/10.1111/geb.12248

Darshni P, Jamwal A (2021) Forest fires in India: a case study of Sunderbani forest range in foothills of Himalayas. Nveo-Nat Volat Essential Oils J Nveo 16080–16086

Epting J, Verbyla D, Sorbel B (2005) Evaluation of remotely sensed indices for assessing burn severity in interior Alaska using Landsat TM and ETM+. Remote Sens Environ 96(3–4):328–339. https://doi.org/10.1016/j.rse.2005.03.002

Fernandes PM, Guiomar N, Rossa CG (2019) Analysing eucalypt expansion in Portugal as a fire-regime modifier. Sci Total Environ 666:79–88. https://doi.org/10.1016/j.scitotenv.2019.02.237no

Fernández C, Fontúrbel T, Vega JA (2019) Effects of pre-fire site preparation and post-fire erosion barriers on soil erosion after a wildfire in NW Spain. CATENA 172:691–698. https://doi.org/10.1016/j.catena.2018.09.038

Fernández-Guisuraga JM, Sanz-Ablanedo E, Suárez-Seoane S, Calvo L (2018) Using unmanned aerial vehicles in postfire vegetation survey campaigns through large and heterogeneous areas: opportunities and challenges. Sensors 18(2):586. https://doi.org/10.3390/s18020586

Fernández-Manso A, Fernández-Manso O, Quintano C (2016) SENTINEL-2A red-edge spectral indices suitability for discriminating burn severity. Int J Appl Earth Obs Geoinf 50:170–175. https://doi.org/10.1016/j.jag.2016.03.005

Ferreira AJD, Alegre SP, Coelho COA, Shakesby RA, Páscoa FM, Ferreira CSS et al (2015) Strategies to prevent forest fires and techniques to reverse degradation processes in burned areas. CATENA 128:224–237. https://doi.org/10.1016/j.catena.2014.09.002

Filipponi F (2018) BAIS2: burned area index for Sentinel-2. In: The 2nd international electronic conference on remote sensing, vol 2, p 364

Flannigan MD, Krawchuk MA, de Groot WJ, Wotton BM, Gowman LM (2009) Implications of changing climate for global wildland fire. Int J Wildland Fire 18(5):483–507

Fultz LM, Moore-Kucera J, Dathe J, Davinic M, Perry G, Wester D et al (2016) Forest wildfire and grassland prescribed fire effects on soil biogeochemical processes and microbial communities: two case studies in the semi-arid southwest. Appl Soil Ecol 99:118–128

Gitas I, Mitri G, Veraverbeke S, Polychronaki A (2012) Advances in remote sensing of post-fire vegetation recovery monitoring—a review. In: Remote sensing of biomass—principles and applications, vol 1, p 334

Gordon CE, Price OF, Tasker EM (2017) Mapping and exploring variation in post-fire recovery following mixed severity wildfire using airborne LiDAR. Ecol Appl 27(5):1618–1632

Gottfried GJ, Neary DG, Baker Jr MB, Ffolliott PF (2003) Impacts of wildfires on hydrologic processes in forest ecosystems: two case studies. In: Proceedings, first interagency conference on research in the watersheds, pp 27–30

Gouveia C, DaCamara CC, Trigo RM (2010) Post-fire vegetation recovery in Portugal based on spot/vegetation data. Nat Hazards Earth Syst Sci 10(4):673–684. https://doi.org/10.5194/nhess-10-673-2010

Grace J, Mitchard E, Gloor E (2014) Perturbations in the carbon budget of the tropics. Glob Change Biol 20(10):3238–3255. https://doi.org/10.1111/gcb.12600

Han D, Di X, Yang G, Sun L, Weng Y (2021) Quantifying fire severity: a brief review and recommendations for improvement. Ecosyst Health Sustain 7(1):1973346. https://doi.org/10.1080/20964129.2021.1973346

Hillman S, Hally B, Wallace L, Turner D, Lucieer A, Reinke K, Jones S (2021) High-resolution estimates of fire severity—an evaluation of UAS image and LiDAR mapping approaches on a sedgeland forest boundary in Tasmania, Australia. Fire 4(1):14. https://doi.org/10.3390/fire4010014

Hua L, Shao G (2017) The progress of operational forest fire monitoring with infrared remote sensing. J For Res 28(2):215–229. https://doi.org/10.1007/s11676-016-0361-8

Jodhani KH, Patel H, Soni U, Patel R, Valodara B, Gupta N et al (2024) Assessment of forest fire severity and land surface temperature using Google Earth Engine: a case study of Gujarat State, India. Fire Ecol 20(1):23

Joseph S, Anitha K, Murthy MSR (2009) Forest fire in India: a review of the knowledge base. J For Res 14(3):127–134. https://doi.org/10.1007/s10310-009-0116-x

Kasischke ES, Turetsky MR, Ottmar RD, French NH, Hoy EE, Kane ES (2008) Evaluation of the composite burn index for assessing fire severity in Alaskan black spruce forests. Int J Wildland Fire 17(4):515–526. https://doi.org/10.1071/WF08002

Kaushik P, Singh H (2024) Climate change-induced forest ecosystem functioning and alteration in water–atmosphere–soil–plant (WASP) relationship. In: Singh H (ed) Urban forests, climate change and environmental pollution. Springer, Cham. https://doi.org/10.1007/978-3-031-67837-0_38

Keeley JE (2009) Fire intensity, fire severity and burn severity: a brief review and suggested usage. Int J Wildland Fire 18(1):116–126. https://doi.org/10.1071/WF07049

Kinnaird MF, O'Brien TG (1998) Ecological effects of wildfire on lowland rainforest in Sumatra. Conserv Biol 12(5):954–956. https://www.jstor.org/stable/2387568

Kumar N, Suyal DC, Sharma IP, Verma A, Singh H (2017) Elucidating stress proteins in rice (*Oryza sativa* L.) genotype under elevated temperature: a proteomic approach to understand heat stress response. 3 Biotech 7:205

Kumar N, Jeena N, Singh H (2019) Elevated temperature modulates rice pollen structure: a study from foothill Himalayan agro-ecosystem in India. 3 Biotech 9:175

Kumar A, Dwivedi G K, Tewari S, Paul J, Anand R, Kumar N, Kumar P, Singh H, Kaushal R (2020a) Carbon mineralization and inorganic nitrogen pools under *Terminalia chebula* Retz.-based agroforestry system in Himalayan foothills, India. For Sci 66(5):634–643

Kumar A, Kumar P, Singh H, Kumar N (2020b) Adaptation and mitigation potential of roadside trees with bio-extraction of heavy metals under vehicular emissions and their impact on physiological traits during seasonal regimes. Urb For Urb Green 58:126900

Kumar N, Jeena N, Kumar A, Khairakpam R, Singh H (2021a) Comparative response of rice cultivars to elevated air temperature in Bhabar region of Indian Himalaya: status on yield attributes. Heliyon 7:e07474

Kumar A, Kumar P, Singh H, Bisht S, Kumar N (2021b) Relationship of physiological plant functional traits with soil carbon stock in temperate forest of Garhwal Himalaya. Curr Sci 120(8):1368–1373

Kumar A, Kumar P, Singh H, Kumar N (2021c) Modulation of plant functional traits under essential plant nutrients during seasonal regime in natural forests of Garhwal Himalayas. Plant Soil 465:197–212

Kumar N, Singh H, Kumar A, Singh AK, Pandey PK, Kumar A (2024) Forest-fire-induced land degradation. In: Mishra G, Giri K, Singh S, Kumar M (eds) Sustainable land management in India. Springer Nature, Singapore. https://doi.org/10.1007/978-981-97-5223-2_4

Kurbanov E, Vorobev O, Lezhnin S, Sha J, Wang J, Li X, Cole J, Dergunov D, Wang Y (2022) Remote sensing of forest burnt area, burn severity, and post-fire recovery: a review. Remote Sens 14(19):4714. https://doi.org/10.3390/rs14194714

Larrinaga AR, Brotons L (2019) Greenness indices from a low-cost UAV imagery as tools for monitoring postfire forest recovery. Drones 3(1):6. https://doi.org/10.3390/drones3010006

Lee SY (2005) Distributional characteristics of wildfires at the eastern seashore and damage reduction methods. Reports for lessons from large-scale wildfires at the eastern seashore. Kangwon Development Research Institute (KDRI), pp 5–28

Lee KH, Kim JE, Kim YJ, Kim J, von Hoyningen-Huene W (2005) Impact of the smoke aerosol from Russian forest fires on the atmospheric environment over Korea during May 2003. Atmos Environ 39(1):85–99

Lee KW, We GJ, Kim EY, Oh JM, Lee JW (2006) Analysis of forest fire damages using airborne multi sensor data. In: Proceedings of 2006 Korean society of hazard mitigation conference, Suwon, Korea, 23 Feb 2006, pp 399–404

Lewis SL, Edwards DP, Galbraith D (2015) Increasing human dominance of tropical forests. Science 349(6250):827–832

Lim K, Treitz P, Woods M, Etheridge D (2010) Operationalizing the use of LiDAR in forest resource inventories: what is the optimal point density? In: ASPRS 2010 annual conference

Liu S, Zheng Y, Dalponte M, Tong X (2020) A novel fire index-based burned area change detection approach using Landsat-8 OLI data. Eur J Remote Sens 53(1):104–112. https://doi.org/10.1080/22797254.2020.1738900

Lucas-Borja ME, Delgado-Baquerizo M (2019) Plant diversity and soil stoichiometry regulates the changes in multifunctionality during pine temperate forest secondary succession. Sci Total Environ 697:134204. https://doi.org/10.1016/j.scitotenv.2019.134204

MacDonald H, McKenney D (2020) Envisioning a global forest transition: status, role, and implications. Land Use Policy 99(C):104808. https://doi.org/10.1016/j.landusepol.2020.104808

Machke ES, Turetsky MR, Ottmar RD, French NHF, Hoy EE, Kane ES (2008) Evaluation of the composite burn index for assessing fire severity in Alaskan black spruce forests. Int J Wildland Fire 17(4):515–526. https://doi.org/10.1071/WF08002

Malik A, Kumar K, Kumari G, Kumar N, Singh H (2024) Application of functional traits in modelling productivity and resilience under climate change. Springer, pp 77–96. ISBN: 978-981-97-1509-1

Malison RL, Baxter CV (2010) The fire pulse: wildfire stimulates flux of aquatic prey to terrestrial habitats driving increases in riparian consumers. Can J Fish Aquat Sci 67(3):570–579

Mallinis G, Mitsopoulos I, Chrysafi I (2018) Evaluating and comparing Sentinel 2A and Landsat-8 Operational Land Imager (OLI) spectral indices for estimating fire severity in a Mediterranean pine ecosystem of Greece. GISci Remote Sens 55(1):1–18. https://doi.org/10.1080/15481603.2017.1354803

Marshall AR, Waite CE, Pfeifer M, Banin LF, Rakotonarivo S, Chomba S et al (2023) Fifteen essential science advances needed for effective restoration of the world's forest landscapes. Philos Trans R Soc B 378(1867):20210065. https://doi.org/10.1098/rstb.2021.0065

Martínez-Murillo JF, López-Vicente M (2018) Effect of salvage logging and check dams on simulated hydrological connectivity in a burned area. Land Degrad Dev 29(3):701–712. https://doi.org/10.1002/ldr.2735

Martini L, Faes L, Picco L, Iroumé A, Lingua E, Garbarino M, Cavalli M (2020) Assessing the effect of fire severity on sediment connectivity in central Chile. Sci Total Environ 728:139006. https://doi.org/10.1016/j.scitotenv.2020.139006

Masek JG, Goward SN, Kennedy RE, Cohen WB, Moisen GG, Schleeweis K, Huang C (2013) United States forest disturbance trends observed using Landsat time series. Ecosystems 16:1087–1104. https://doi.org/10.1007/s10021-013-9669-9/

McKenna P, Erskine PD, Lechner AM, Phinn S (2017) Measuring fire severity using UAV imagery in semi-arid central Queensland, Australia. Int J Remote Sens 38(14):4244–4264. https://doi.org/10.1080/01431161.2017.1317942

Mohan M, Richardson G, Gopan G, Aghai MM, Bajaj S, Galgamuwa GP et al (2021) UAV-supported forest regeneration: current trends, challenges and implications. Remote Sens 13(13):2596. https://doi.org/10.3390/rs13132596

Montoya D, Rogers L, Memmott J (2012) Emerging perspectives in the restoration of biodiversity-based ecosystem services. Trends Ecol Evol 27(12):666–672. https://doi.org/10.1016/j.tree.2012.07.004

Moretti M, De Cáceres M, Pradella C, Obrist MK, Wermelinger B, Legendre P, Duelli P (2010) Fire-induced taxonomic and functional changes in saproxylic beetle communities in fire sensitive regions. Ecography 33(4):760–771

Napper C (2006) BAER-burn area emergency response. U.S. Department of Agriculture, Forest Service. http://www.fs.fed.us/eng/pubs/pdf/BAERCAT/lo_res/06251801L

Noce S, Santini M (2018) Mediterranean forest ecosystem services and their vulnerability. https://doi.org/10.13140/RG.2.2.30330.13760

Noulèkoun F, Mensah S, Birhane E, Son Y, Khamzina A (2021) Forest landscape restoration under global environmental change: challenges and a future roadmap. Forests 12(3):276. https://doi.org/10.3390/f12030276

Nunes A, Oliveira G, Mexia T, Valdecantos A, Zucca C, Costantini EAC, Abraham EM, Kyriazopoulos AP, Salah A, Prasse R, Correia O, Milliken S, Kotzen B, Branquinho C (2016) Ecological restoration across the Mediterranean Basin as viewed by practitioners. Sci Total Environ 566–567:722–732. https://doi.org/10.1016/j.scitotenv.2016.05.136

Oncu G, Corumluoglu O (2023) Determination of spatial distribution of damage intensity of Tınazlı-Izmir forest fire using remote sensing indexing techniques. Int J Environ Geoinform 10(2):151–158. https://doi.org/10.30897/ijegeo.1089014

Orgeas J, Andersen AN (2001) Fire and biodiversity: responses of grass-layer beetles to experimental fire regimes in an Australian tropical savanna. J Appl Ecol 49–62

Pádua L, Adão T, Guimarães N, Sousa A, Peres E, Sousa JJ (2019) Post-fire forestry recovery monitoring using high-resolution multispectral imagery from unmanned aerial vehicles. Int Arch Photogramm Remote Sens Spat Inf Sci 42:301–305. https://isprs-archives.copernicus.org/articles/XLII-3W8/301/2019/#:~:text=https%3A//doi.org/10.5194/isprs%2Darchives%2DXLII%2D3%2DW8%2D301%2D2019%2C%202019

Pan Y, Birdsey RA, Fang J, Houghton R, Kauppi PE, Kurz WA et al (2011) A large and persistent carbon sink in the world's forests. Science 333(6045):988–993. https://doi.org/10.1126/science.1201609

Paul MJ, LeDuc SD, Lassiter MG, Moorhead LC, Noyes PD, Leibowitz SG (2022) Wildfire induces changes in receiving waters: a review with considerations for water quality management. Water Resour Res 58(9):e2021WR030699. https://doi.org/10.1029/2021WR030699

Pesaresi S, Mancini A, Quattrini G, Casavecchia S (2020) Mapping Mediterranean forest plant associations and habitats with functional principal component analysis using Landsat 8 NDVI time series. Remote Sens 12(7):1132. https://doi.org/10.3390/rs12071132

Porto P, Walling DE, Callegari G (2009) Investigating the effects of afforestation on soil erosion and sediment mobilisation in two small catchments in Southern Italy. CATENA 79(3):181–188. https://doi.org/10.1016/j.catena.2009.01.007

Prats M, Perez J, Fernández JJ, Sanz PJ (2012) An open source tool for simulation and supervision of underwater intervention missions. In: 2012 IEEE/RSJ international conference on intelligent robots and systems, Oct 2012. IEEE, pp 2577–2582. https://doi.org/10.1109/IROS.2012.6385788

Price OF, Gordon CE (2016) The potential for LiDAR technology to map fire fuel hazard over large areas of Australian forest. J Environ Manage 181:663–673

Reilly S, Clark ML, Bentley LP, Matley C, Piazza E, Oliveras Menor I (2021) The potential of multispectral imagery and 3D point clouds from unoccupied aerial systems (UAS) for monitoring forest structure and the impacts of wildfire in Mediterranean-climate forests. Remote Sens 13(19):3810. https://doi.org/10.3390/rs13193810

Röder A, Hill J, Duguy B, Alloza JA, Vallejo R (2008) Using long time series of Landsat data to monitor fire events and post-fire dynamics and identify driving factors. A case study in the Ayora region (eastern Spain). Remote Sens Environ 112(1):259–273. https://doi.org/10.1016/j.rse.2007.05.001

Romme WH, Boyce MS, Gresswell R, Merrill EH, Minshall GW, Whitlock C, Turner MG (2011) Twenty years after the 1988 Yellowstone fires: lessons about disturbance and ecosystems. Ecosystems 14(7):1196–1215. https://doi.org/10.1007/s10021-011-9470-6

Rozendaal DM, Chazdon RL (2015) Demographic drivers of tree biomass change during secondary succession in northeastern Costa Rica. Ecol Appl 25(2):506–516. https://doi.org/10.1890/14-0054.1

Ryan KC (2002) Dynamic interactions between forest structure and fire behavior in boreal ecosystems. Silva Fenn 36(1):13–39. https://doi.org/10.14214/sf.548

Sasaki T, Ishii H, Morimoto Y (2018) Evaluating restoration success of a 40-year-old urban forest in reference to mature natural forest. Urb For Urb Green 32:123–132. https://doi.org/10.1016/j.ufug.2018.04.008

Savage M, Mast JN (2005) How resilient are southwestern ponderosa pine forests after crown fires? Can J For Res 35(4):967–977. https://doi.org/10.1139/x05-028

Sett T, Nikam BR, Singh H, Purohit S (2024) Hydrological variability in Indian forest ecosystem: analysis of drought resilience, recovery and water use efficiency in moist and dry deciduous forests. In: Singh H (ed) Urban forests, climate change and environmental pollution. Springer, Cham. https://doi.org/10.1007/978-3-031-67837-0_37

Sewak R, Vashisth M, Gupta L (2021) Forest fires in India: a review. J Univ Shanghai Sci Technol 23(7):247–259

Silva WR, Zaniratto CP, Ferreira JO, Rigacci ED, Oliveira JF, Morandi ME et al (2020) Inducing seed dispersal by generalist frugivores: a new technique to overcome dispersal limitation in restoration. J Appl Ecol 57(12):2340–2348. https://doi.org/10.1111/1365-2664.13731

Singh H (2024a) Climate change and forests: biological perspectives. In: Singh H (ed) Forests and climate change. Springer Nature, Singapore. https://doi.org/10.1007/978-981-97-3905-9_1

Singh H (ed) (2024b) Forests and climate change: biological perspectives on impact, adaptation, and mitigation strategies. Springer Nature, Singapore. https://doi.org/10.1007/978-981-97-3905-9

Singh H (2024c) Climate change and environmental pollution: biological consequences on urban forests. In: Singh H (ed) Urban forests, climate change and environmental pollution. Springer Nature, Cham, Switzerland. https://doi.org/10.1007/978-3-031-67837-0_1

Singh H (ed) (2024d) Urban forests, climate change and environmental pollution: physio-biochemical and molecular perspectives to enhance urban resilience. Springer Nature, Cham, Switzerland. https://doi.org/10.1007/978-3-031-67837-0

Singh H, Kumar N, Kumar A (2024) Enhancing resource use efficiency in crops through plant functional traits. Springer, pp 97–117. ISBN: 978-981-97-1509-1

Slaton MR, Hunt ER, Smith WK (2001) Estimating near-infrared leaf reflectance from leaf structural characteristics. Am J Bot 88(2):278–284. https://doi.org/10.2307/2657019

Solanky V, Katiyar SK (2016) Pixel-level image fusion techniques in remote sensing: a review. Spat Inf Res 24:475–483. https://doi.org/10.1007/s41324-016-0046-6

Souza-Alonso P, Saiz G, García RA, Pauchard A, Ferreira A, Merino A (2022) Post-fire ecological restoration in Latin American forest ecosystems: insights and lessons from the last two decades. For Ecol Manage 509:120083. https://doi.org/10.1016/j.foreco.2022.120083

Stephens SL, Boerner RE, Moghaddas JJ, Moghaddas EE, Collins BM, Dow CB et al (2012) Fuel treatment impacts on estimated wildfire carbon loss from forests in Montana, Oregon, California, and Arizona. Ecosphere 3(5):1–17. https://doi.org/10.1890/ES11-00289.1

Stevens-Rumann CS, Morgan P (2019) Tree regeneration following wildfires in the western US: a review. Fire Ecol 15(1):1–17

Stueve KM, Cerney DL, Rochefort RM, Kurth LL (2009) Post-fire tree establishment patterns at the alpine treeline ecotone: Mount Rainier National Park, Washington, USA. J Veg Sci 20(1):107–120

Szpakowski DM, Jensen JL (2019) A review of the applications of remote sensing in fire ecology. Remote Sens 11(22):2638. https://doi.org/10.3390/rs11222638

Tangney R, Paroissien R, Le Breton TD, Thomsen A, Doyle CAT, Ondik M, Miller RG, Miller BP, Ooi MKJ (2022) Success of post-fire plant recovery strategies varies with shifting fire seasonality. Commun Earth Environ 3(1):126

Turner MG, Hargrove WW, Gardner RH, Romme WH (1994) Effects of fire on landscape heterogeneity in Yellowstone National Park, Wyoming. J Veg Sci 5(5):731–742

Turner MG, Romme WH, Tinker DB (2003) Surprises and lessons from the 1988 Yellowstone fires. Front Ecol Environ 1(7):351–358. https://doi.org/10.1890/1540-9295(2003)001[0351:SALFTY]2.0.CO;2

UN General Assembly (2019) United Nations decade on ecosystem restoration (2021–2030) A/RES/73/284

UNEP, Celine Barakat (2024) Fighting fire with forests across the Mediterranean, 19 Mar 20124. UNEP. https://www.unep.org/news-and-stories/story/fighting-fire-forests-acrossmediterranean#:~:text=Scientists%20point%20to%20the%20Mediterranean,for%20local%20ecosystems%20and%20communities

Valavanis KP, Vachtsevanos GJ (eds) (2015) Handbook of unmanned aerial vehicles, vol 1. Springer Netherlands, Dordrecht

Vallejo VR, Alloza JA (2012) Post-fire management in the Mediterranean Basin. Isr J Ecol Evol 58(2–3):251–264. https://doi.org/10.1560/IJEE.58.2-3.251

Vallejo VR, Alloza JA (2015) Postfire ecosystem restoration. In: Wildfire hazards, risks and disasters. Elsevier, pp 229–246. https://doi.org/10.1016/B978-0-12-410434-1.00012-9

van Leeuwen WJ (2008) Monitoring the effects of forest restoration treatments on post-fire vegetation recovery with MODIS multitemporal data. Sensors 8(3):2017–2042. https://doi.org/10.3390/s8032017

van Leeuwen M, Nieuwenhuis M (2010) Retrieval of forest structural parameters using LiDAR remote sensing. Eur J For Res 129:749–770. https://doi.org/10.1007/s10342-010-0381-4

Veraverbeke S, Harris S, Hook S (2011) Evaluating spectral indices for burned area discrimination using MODIS/-ASTER (MASTER) airborne simulator data. Remote Sens Environ 115(10):2702–2709. https://doi.org/10.1016/j.rse.2011.06.010

Veraverbeke S, Gitas I, Katagis T, Polychronaki A, Somers B, Goossens R (2012) Assessing post-fire vegetation recovery using red–near infrared vegetation indices: accounting for background and vegetation variability. ISPRS J Photogramm Remote Sens 68:28–39. https://doi.org/10.1016/j.isprsjprs.2011.12.007

Yuan C, Zhang Y, Liu Z (2015) A survey on technologies for automatic forest fire monitoring, detection, and fighting using unmanned aerial vehicles and remote sensing techniques. Can J For Res 45(7):783–792. https://doi.org/10.1139/cjfr-2014-0347

Zagalikis G (2023) Remote sensing and GIS applications in wildfires. https://doi.org/10.5772/intechopen.111616

Zhong C, Guo M, Zhou F, Li J, Yu F, Guo F, Li W (2023) Forest succession trajectories after fires in valleys and on slopes in the Greater Khingan Mountains, China. J For Res 34(3):623–640

# Analyzing Forest Fire Trends, Influencing Factors, and Strategies for Effective Management: A Spatial Examination in Madhesh Provine of Nepal

**Abhilasha Rajbhandari, Sewak Bhatta, Rajeev Joshi, and Santosh Ghimire**

**Abstract** Human-involved activities such as arson, deforestation, negligence, etc., have escalated forest fires. Such forest fires pose a risk to the ecosystem and biodiversity. This study focused on determining the pattern of forest fires, their drivers and effects on plant biodiversity, and management options to reduce the onset of fires. The study used MODIS data to determine forest fires' distribution pattern and frequency from 2010 to 2021. Field and 30 general household surveys and five key informant interviews were conducted in each of the three community forests to gain more insight into the impacts of forest fires. The collected data were then analyzed using spatial and statistical tools such as QGIS, Microsoft Excel, and RStudio. The research showed that Mahottari district experienced a total of 288 counts of fires in the last twelve years, i.e., from 2010 to 2021. Out of total fire incident detection, about 16 incidents were detected with greater than 30% confidence. The highest number of fire incidents was recorded in 2013, with 39 incidents; the lowest was recorded in 2020, with 16 fire incidents. It was observed that the fire season peaked in March and April, and the highest fire occurrence was observed in March 2019, with 27 fire incidents. It was also noted that 2013 had the driest days and the highest fire occurrences in the last twelve years, indicating a link between fire occurrence and dry days. The R square value of the correlation between fire days and fire occurrence was 0.255. Further, analysis showed that dry season and intentional fire were the leading causes of forest fires. Similarly, the construction of fire lines and clearing

---

A. Rajbhandari · S. Bhatta
Faculty of Science and Technology, School of Environmental Science and Management (SchEMS), Pokhara University, Pokhara, Nepal

R. Joshi (✉)
Faculty of Forestry, College of Natural Resource Management, Agriculture and Forestry University, 56310 Katari, Udayapur, Nepal
e-mail: joshi.rajeev20@gmail.com

College of Economics and Management, Northwest Agriculture and Forestry University, Yangling, Shaanxi-712100 Xianyang, China

S. Ghimire
Faculty of Forestry, Agriculture and Forestry University, Hetauda, Nepal

 
H. Singh and N. Kumar (eds.), *Forest Fire and Climate Change*,
https://doi.org/10.1007/978-3-031-89967-6_27

and the appointment of forest watchers were some of the local practices to reduce and manage forest fires.

**Keywords** Causes · Forest fire · Mahottari · MODIS · Monitoring

## 1 Introduction

Forests are home to over three-quarters of the world's life on land. Several species of flora and fauna inhabit and thrive in the forest landscape. Wild animals, along with several indigenous tribes worldwide, depend on forests and their products for their survival. Around 1.6 billion people, about 25% of the world's population, depend on forests for their livelihood (IUCN, n.d.). Forests are also the second largest storehouse for carbon after oceans (Chand et al. 2018). Approximately 2.6 billion tons of carbon dioxide, i.e., one-third of the $CO_2$ released from burning fossil fuels, is absorbed by forests yearly (Kumar et al. 2024; Thakur et al. 2024). Forests play a vital role in conserving water, soil, and other natural resources, along with maintaining an array of ecosystem services such as food, water supply, climate regulation, and calamities control (Singh 2024a, b; Barwa and Singh 2024). Globally, forests cover 31% of the total land area, i.e., approximately 4.06 billion ha (FAO 2020). However, these forests are not equally distributed around the world. According to FAO (2020), more than half of the world's forests are found in only five countries—the Russian Federation, Brazil, Canada, the United States of America, and China, and two-thirds (66%) of forests are found in ten countries. Although the forests improve environmental quality and economic wealth, the loss of forest areas and a wide range of biodiversity persist (Singh 2024c, d; Karki et al. 2025). In 2020, the world lost around 12.2 million ha of tree cover, mainly in the tropics (WWF, n.d.). Forests around the globe are facing degradation at an accelerated rate for various reasons. There are several natural and anthropogenic factors affecting the state of forests (Kaushik and Singh 2024; Sett et al. 2024; Bhatta et al. 2025). Natural factors include floods, landslides, forest fires, volcanic eruptions, invasive species, pests and insects, etc.

In contrast, anthropogenic factors include illegal logging and felling, deforestation, pollution, etc. These factors tend to affect and degrade the forest environment severely. Wildfires have been considered a severe issue in recent years and are responsible for altering the quality of forests. Forest fires occur naturally due to lightning, drought, wind, or volcanic activities. However, human-induced causes, such as deforestation, burning debris, arson, etc., have recently escalated and can be linked to 90% of wildfires (Paton et al. 2014). According to the European Space Agency, fires annually affect an estimated four million km$^2$ of Earth's land. Despite forest fires destroying the standing trees and other vegetation, in some cases, they do help in the germination of new vegetation. In some forest types, frequent surface fires reduce the possibility of large, high-intensity wildfires that can be catastrophic to the ecosystem and human population; further wildfires maintain diversity and genetic variability, thereby also contributing to the creation of a range of natural products (Pausas and

Keeley 2019). However, the adverse effects of fire outweigh the positive aspects in most cases. Fires burn down the vegetation and disrupt the whole ecosystem of floral and faunal diversity in forests. It further reduces food resources, pollutes the air and water resources with emissions, and risks the area from flooding, landslides, and other hazards.

# 2 Materials and Methods

## 2.1 *Study Area*

Mahottari district, a part of Madhesh province, has been selected as the study area for the research purpose. The district has ordained Jaleshwar as its headquarters and covers an area of 1002 $km^2$. The district is located at latitude 26° 52′ 22.08″ North and longitude 85° 49′ 42.96″ East. An outer Terai region, this district shares its border with Dhanusha in the east, Sarlahi and India in the west, Sindhuli in the north, and again India in the south. According to the 2021 census, the total population of this region is about 705,838. Most of the people in this region are dependent on agriculture. The people in this district are predominantly Hindu and Muslim, along with a community of at least 92 castes, and the majority of the population speaks Maithali in this region. Approximately 22,240 ha or 22.23% of the land is covered by forest area in this district (DFRS 2018). The district is majorly situated in the plains of the Terai in the southern region and the Chure region in the northern region. The fieldwork for this study was conducted in three community forest areas: Urramarka Community Forest, Kalikhola Community Forest, and Bahunijhora Community Forest (Fig. 1).

## 2.2 *Data Collection*

The study was conducted according to the research design. The study area was selected based on the problems that were published in various articles, websites, and journals. Then, the objectives were established, which guided the survey and questionnaire preparation. Primary data collected from the field and secondary data obtained from MODISa and DHM were analyzed. Further, necessary charts, tables, and maps were prepared to demonstrate the findings. Literature reviews relevant to the study were also studied.

### 2.2.1 Primary Data Collection

A pre-coded structured questionnaire was developed for the primary data collection for the general household survey. Two types of questionnaire surveys were

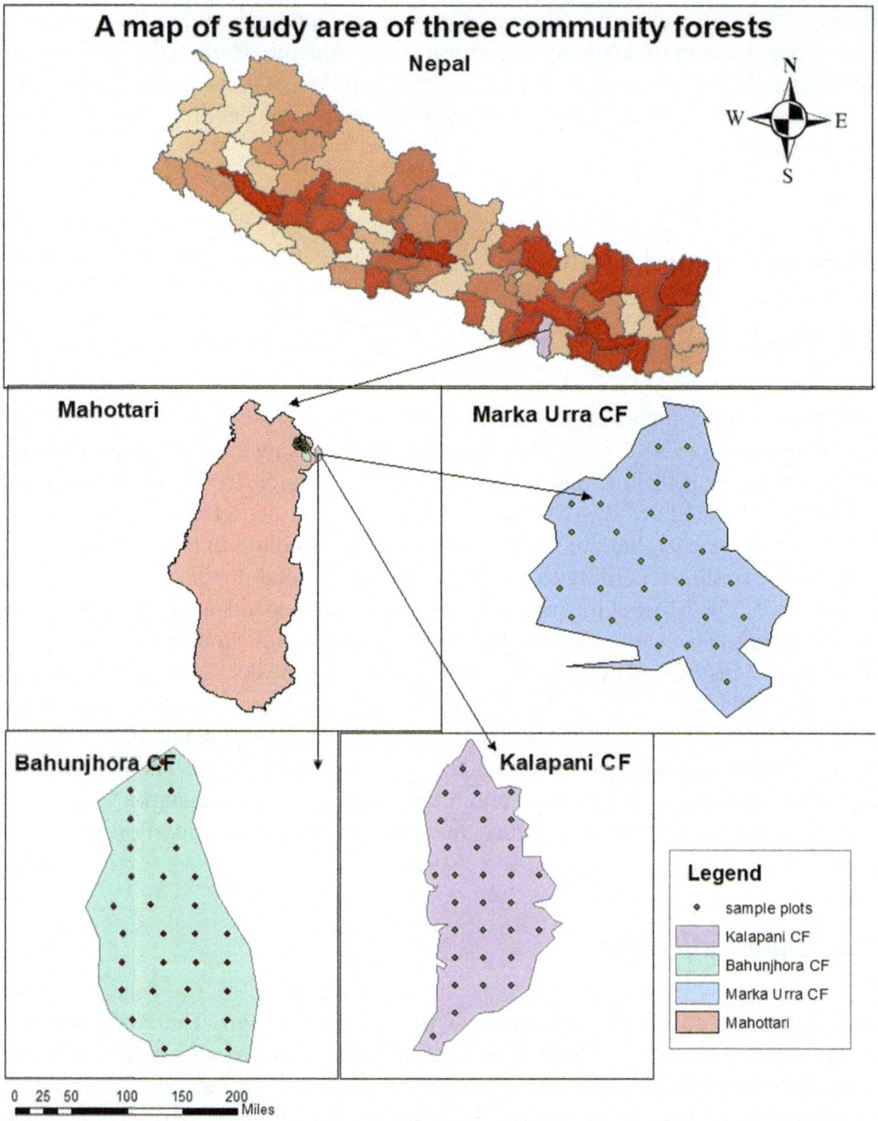


**Fig. 1** Map of the study area

conducted: a household survey/interview for the local villagers (elderly, women, farmers, herders) that helped to understand their perception about the issues that arises with the onset of forest fire. This questionnaire was conducted on 90 households, with 30 households in each community forest. Another type of survey, i.e., key informant interview, is aimed towards people such as the chairperson, secretary, and Community Forest Consumer Group members. Five critical informants in each community forest were interviewed with profound knowledge of the related subject. Furthermore, a Focus Group Discussion (FGD) was conducted with 10–12 participants. These questionnaires focused on information regarding forest fires, the effects on their lives, and their approaches to preventing them. Besides this, three community forests, namely Markaurra, Kalikhola, and Bahunijhora community forests, were chosen to compare the biodiversity of fire-affected and fire-non-affected forest areas. A total of 90 plots were established, with 30 plots in each community forest. The quadrats of $25 \times 20\ m^2$ for trees, $10 \times 10\ m^2$ for poles, $5 \times 5\ m^2$ for saplings, and $2 \times 1\ m^2$ for seedlings were laid out in all plots. The diameter at breast height (DBH) and height of trees, poles, and saplings were measured, and species-wise frequency was noted on each plot. These plots were created to determine the status of plant species diversity and species richness in the forest areas.

In order to check the accuracy of the fire data, a field visit was conducted, and random fire points from three community forests were recorded and compared with the fire occurrence year, month, and day provided by NASA.

### 2.2.2 Secondary Data Collection

#### MODIS Data

Since Nepal does not have reliable tools and technology for the detection and monitoring, the secondary information about fire data was obtained from the NASA website. The data from MODIS sensors onboard Terra and Aqua satellites were used for the study. The data helped to compare the occurrence, frequency, and intensity of forest fires in Mahottari district for the last 12 years. The forest fire incidence data was extracted from NASA FIRMS in shapefile format from 2010 to 2021. The shapefiles were further converted to csv (comma-separated value) format for better analysis. This study analyzed the forest fire pattern between the 12 years, i.e., 2010 and 2021, in the Mahottari district, which was compared using active fire records from the Moderate Resolution Imaging Spectroradiometer (MODIS) sensor. Apart from this, information regarding the research was gathered from various websites, journal articles, reports, books, etc.

## *2.3 Data Analysis*

#### Statistical Analysis

Data regarding the fire was interpreted using various statistical tools. The forest fire pattern was evaluated using time series analysis. Similarly, descriptive analysis such as mean (central tendency), standard error, standard deviation, minimum, and maximum was calculated for fire and biodiversity. Temperature data for the fire months was categorized into average minimum and maximum. Microsoft Excel evaluated the number of dry days greater than days with the average temperature for each fire month. Lastly, Principal Component Analysis (PCA) was used to determine the causes of forest fire and pre-fire monitoring measures.

The quantitative analysis of the forest biodiversity was calculated using indices such as:

(i) Shannon Weiner Index (H) (Nolan and Callahan 2006) was calculated using the formula:

$$H = -\sum_{i=1}^{s} p_i \ln p_i$$

where, H = Shannon Weiner Diversity index, s = number of species, $p_i$ = portion of individuals of each species belonging to the ith species of the total number of individuals.

(ii) Species richness (D)

$$D = s/\sqrt{N}$$

where, S = number of different species, N = total number of individuals.

(iii) Evenness Index (Zhang et al. 2012) was calculated with the following formula:

$$E = H'/\ln(S)$$

where, H′ = Shannon Weiner Index, ln (S) = species richness.

**Spatial Analysis**

The necessary data were obtained from various websites, organizations, and institutions. MODIS data product regarding the forest fire incidence was obtained from NASA FIRMS (https://firms.modaps.eosdis.nasa.gov/data/). Similarly, data regarding the temperature of Mahottari district were acquired from the Department of Hydrology and Meteorology (DHM). The software QGIS 3.10 was used to interpret fire incident data, along with its trend analysis of 12 years. Also, general information about fire incidents in three community forests was retrieved from the locals during the general household survey.

## 3 Result and Discussion

### *3.1 Temporal and Spatial Distribution of Forest Fires*

The spatial and temporal distribution of forest fire occurrence was analyzed from 2010 to 2021. According to the MODIS data, there were 288 counts of forest fire occurrences in Mahottari district from the year 2010 to 2021. Out of total fire incident detection, about 16 incidents were detected with greater than 30% confidence. The overall fire distribution showed that most forest fires occurred on the northern side of Mahottari district. Moreover, fire incidents were higher near the highway and close to the boundary of the Sindhuli district, which is marked by a circle in the figure. The fire distribution and occurrence of the last twelve years is demonstrated in Figs. 2 and 3.

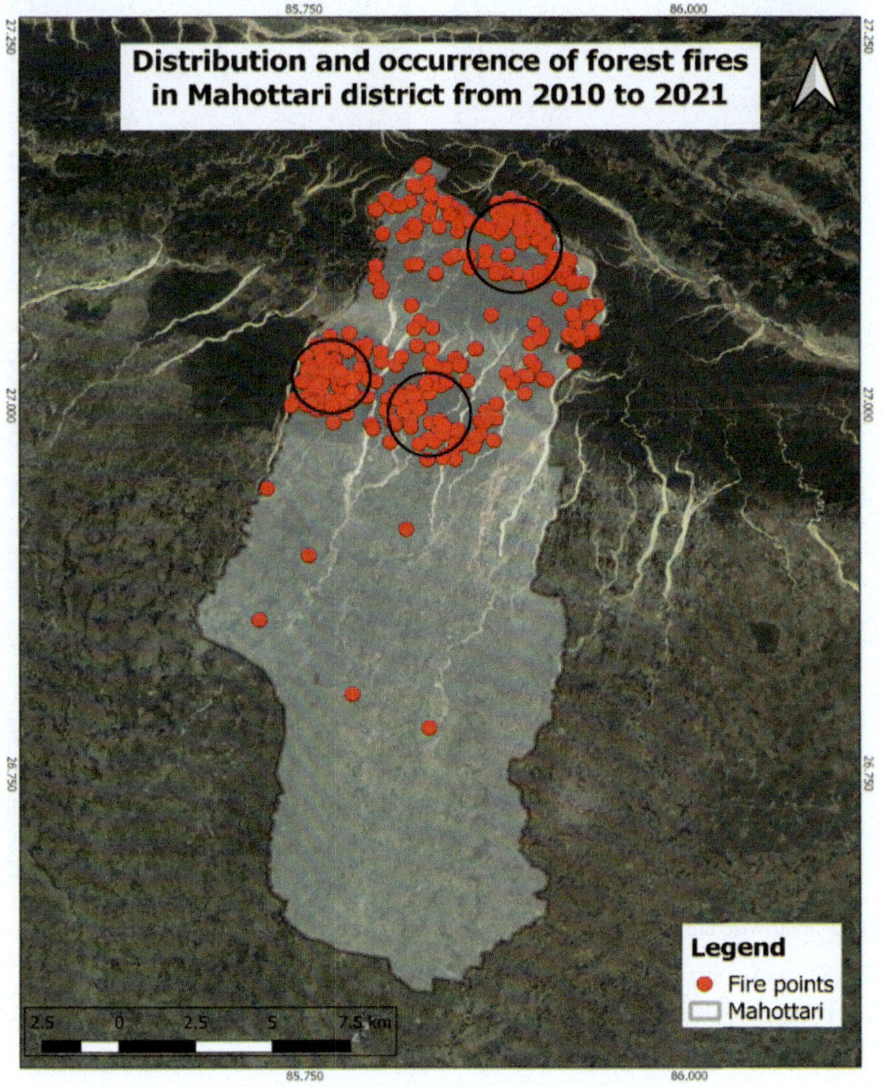


**Fig. 2** Distribution of fire events in Mahottari district from 2010 to 2021

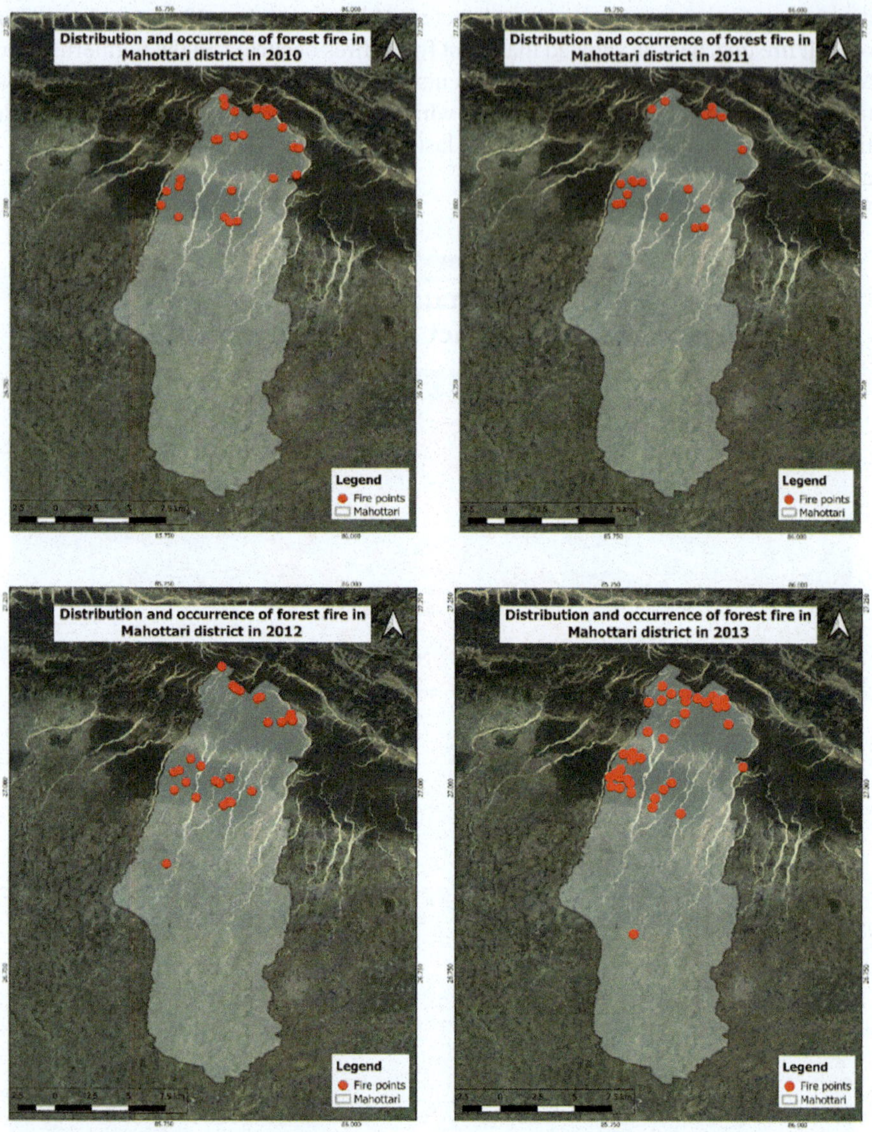


**Fig. 3** Year-wise fire distribution in Mahottari district from 2010 to 2021

The figures demonstrate the occurrence and distribution of forest fires of each year since 2010–2021 in Mahottari district.

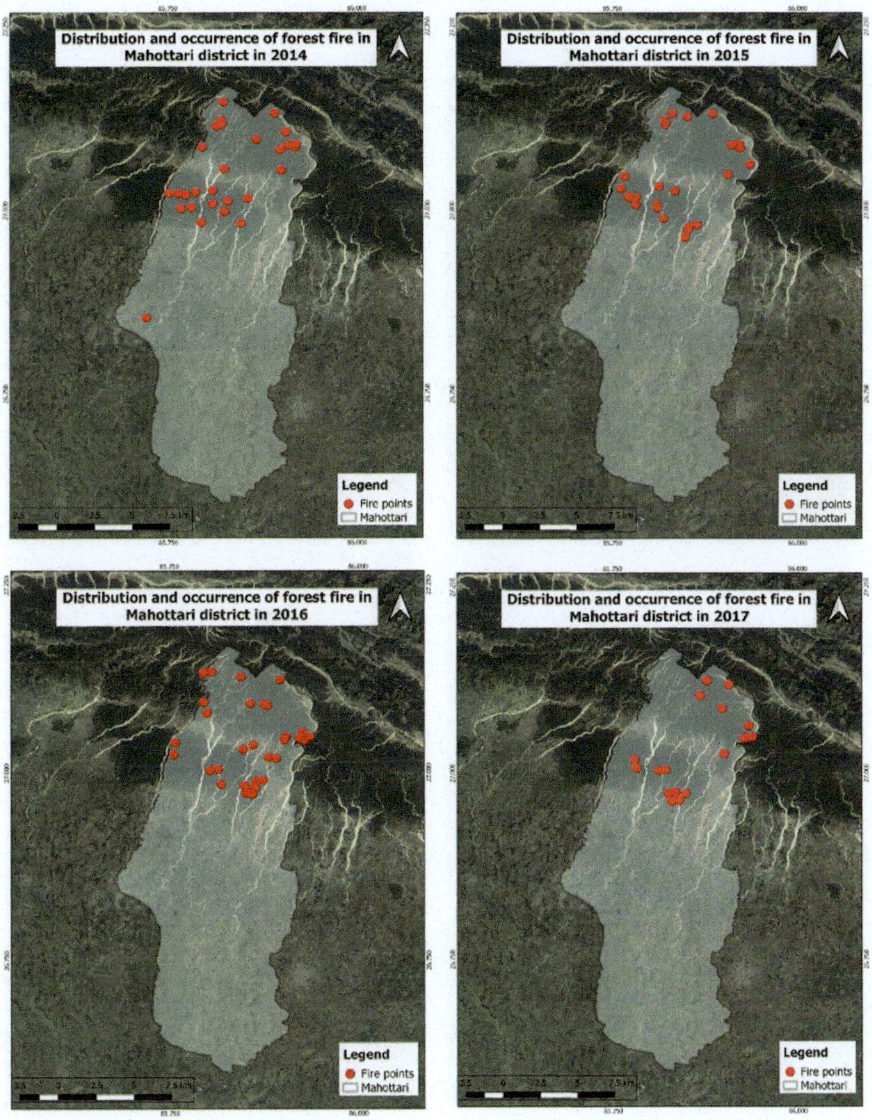


**Fig. 3** (continued)

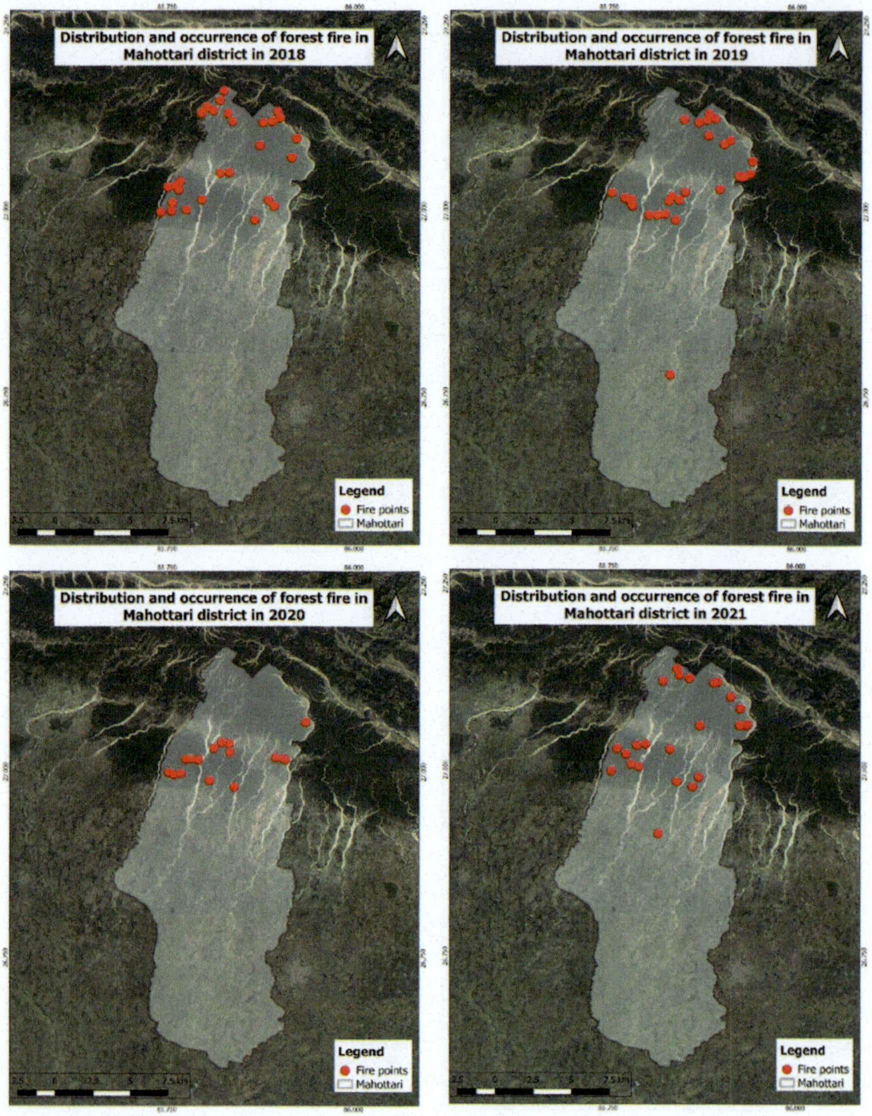


**Fig. 3** (continued)

### 3.1.1 Year-Wise Temporal and Spatial Distribution of Fire Occurrence

The graph in Fig. 4 shows the year-wise representation of forest fire occurrences from 2010 to 2021. The result showed that, out of the 288 fire counts, 26 fire events occurred in 2010, 19 in 2011, 27 in 2012, and 39 in 2013. Likewise, 28 fire incidents took place in 2014, 25, incidents in 2015, 32 incidents in 2016, 17 incidents in 2017,

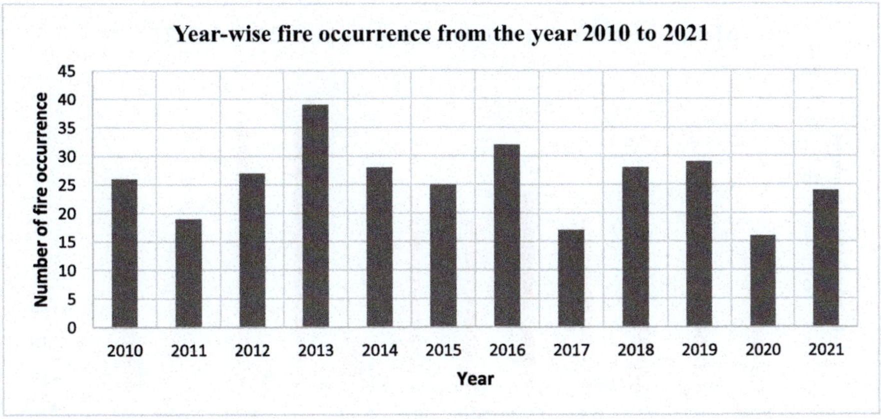


**Fig. 4** Forest fire occurrences from the year 2010 to 2021

**Table 1** Descriptive analysis of fire occurrence from 2010 to 2021

| *Descriptive analysis of number of fire occurrence* | |
|---|---|
| Mean | 25.83 |
| Standard error | 1.86 |
| Standard deviation | 6.45 |
| Minimum | 16 |
| Maximum | 39 |

28 incidents in 2018, 29 incidents in 2019, 16 incidents in 2020 and 24 incidents in 2021. It was found that 2013 had the highest forest fire occurrence, with 39 incidents. The lowest number of fire incidents was recorded in 2020, with 16. Moreover, the descriptive analysis showed that the central tendency was 25.83, with a standard error of 1.86 and a standard deviation of 6.45 (Table 1).

### 3.1.2 Month-Wise Temporal and Spatial Distribution of Fire Occurrence

The month-wise distribution of fire incidents from 2010 to 2021 is portrayed in Figs. 5 and 6. It was found that fire activities in forests usually start in January and end in May. The result showed that the fire activities reached the highest during March and April, while the least fire events occurred during May.

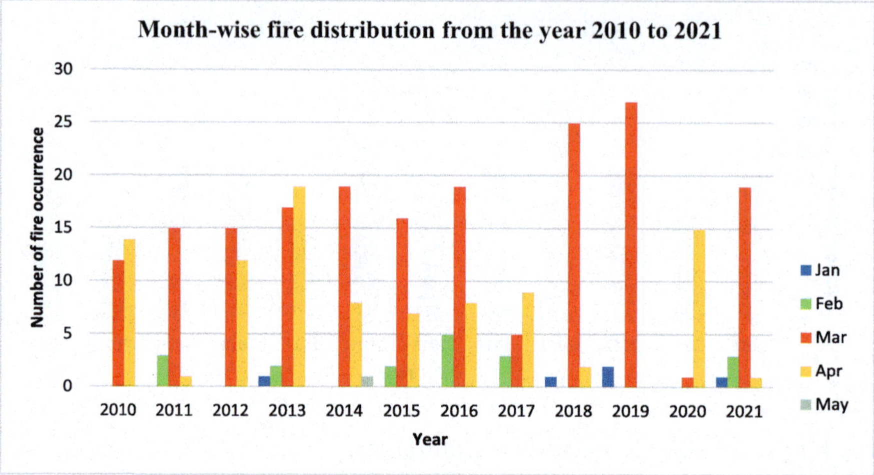


**Fig. 5** Distribution of fire incidents in different months from 2010 to 2021

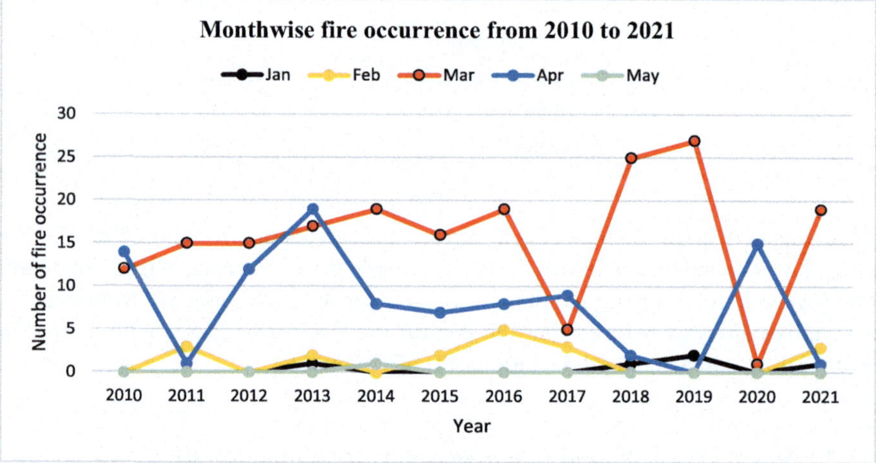


**Fig. 6** Graph showing month-wise fire incidents from 2010 to 2021

### 3.1.3 Fire Calendar

Table 2 demonstrates the fire calendar from 2010 to 2021. The table shows that fire season lasts five months, generally starting in January and ending in May. The fire incidents are the least in January and May, while peaks during March and April. In 2013 and 2021, the fire events began in January and continued till April. Similarly, in 2018 and 2019, the fire events started in January. However, no fire incidents were observed in February. In 2015, 2016 and 2017, fire incidents began in February and

**Table 2** Distribution of fire in different months from 2010 to 2021

| | Jan | Feb | Mar | Apr | May | Jun | Jul | Aug | Sep | Oct | Nov | Dec |
|---|---|---|---|---|---|---|---|---|---|---|---|---|
| 2010 | | | | | | | | | | | | |
| 2011 | | | | | | | | | | | | |
| 2012 | | | | | | | | | | | | |
| 2013 | | | | | | | | | | | | |
| 2014 | | | | | | | | | | | | |
| 2015 | | | | | | | | | | | | |
| 2016 | | | | | | | | | | | | |
| 2017 | | | | | | | | | | | | |
| 2018 | | | | | | | | | | | | |
| 2019 | | | | | | | | | | | | |
| 2020 | | | | | | | | | | | | |
| 2021 | | | | | | | | | | | | |

ended in April. It was observed that fire events commenced in March and continued until May. Further, in other years, fire incidents only occurred during two months, i.e., in March and April.

Table 3 shows the number of fire incidents that occurred in each fire month from 2010 to 2021. It was found that the year 2013 experienced the highest number of fire incidents, with a total of 39 fire incidents, ensuing from January to April, with 19 incidents occurring in March, followed by April with 17 incidents, while January and February suffered from a single and two fire incidents respectively. As for the highest number of fire incidents occurring in a month in twelve years, it was found that March of 2019 had the highest number of fire incidents, with a total of 27 incidents. It was followed by March of 2018, with 25 fire incidents. The forests were exposed to 19 fire incidents during March 2014, 2016 and 2021.

Similarly, 2010, 2012 and 2020 saw the occurrence of 14, 12 and 15 fire events in April. Meanwhile, fire occurrences were low during January, February, and May. The mean $\pm$ SE for 2010 was found to be $5.2 \pm 3.2$, and the standard deviation was 7.16. Likewise, the mean $\pm$ SE for 2011 was found to be $3.8 \pm 2.9$, and its standard deviation was found to be 6.38. The mean $\pm$ SE and standard deviation values were also similar for other years.

### 3.1.4 Correlation Between Fire Occurrence and Number of Dry days

The figure portrays the correlation between fire occurrence and number of dry days. The R square value of the correlation between fire days and fire occurrence was 0.255. The study showed a direct link between fire occurrence and dry days. The graph plot shows that the number of fires increased with the number of dry days. It was found that 2013, which had experienced the greatest number of fire incidents in twelve years, i.e., 39 fire incidents, was also the year with the driest days, with

**Table 3** Number of fire occurrences in fire months from 2010 to 2021

| | Jan | Feb | Mar | Apr | May | Total no. | Mean ± SE | SD |
|---|---|---|---|---|---|---|---|---|
| 2010 | 0 | 0 | 12 | 14 | 0 | 26 | 5.2 ± 3.2 | 7.16 |
| 2011 | 0 | 3 | 15 | 1 | 0 | 19 | 3.8 ± 2.9 | 6.38 |
| 2012 | 0 | 0 | 15 | 12 | 0 | 27 | 5.4 ± 3.3 | 7.47 |
| 2013 | 1 | 2 | 17 | 19 | 0 | 39 | 7.8 ± 4.2 | 9.36 |
| 2014 | 0 | 0 | 19 | 8 | 1 | 28 | 5.6 ± 3.7 | 8.20 |
| 2015 | 0 | 2 | 16 | 7 | 0 | 25 | 5 ± 3 | 6.78 |
| 2016 | 0 | 5 | 19 | 8 | 0 | 32 | 6.4 ± 3.5 | 7.83 |
| 2017 | 0 | 3 | 5 | 9 | 0 | 17 | 3.4 ± 1.7 | 3.78 |
| 2018 | 1 | 0 | 25 | 2 | 0 | 28 | 5.6 ± 4.9 | 10.88 |
| 2019 | 2 | 0 | 27 | 0 | 0 | 29 | 5.8 ± 5.3 | 11.88 |
| 2020 | 0 | 0 | 1 | 15 | 0 | 16 | 3.2 ± 3 | 6.61 |
| 2021 | 1 | 3 | 19 | 1 | 0 | 24 | 4.8 ± 3.6 | 8.01 |

313 dry days altogether. Meanwhile, 2016 had 187 dry days and experienced 32 fire incidents (Fig. 7).

The table depicts the fire months of twelve years along with the average monthly temperature of the fire months. These dry days are greater than days with average temperature, fire days in each fire month, and number of fire occurrences yearly. The table shows that in January 2013, a single fire incident occurred, and the average monthly temperature was 16.6 °C. Two fire incidents were observed in February, with an average temperature of 20.1 °C. Likewise, 17 fire incidents were observed

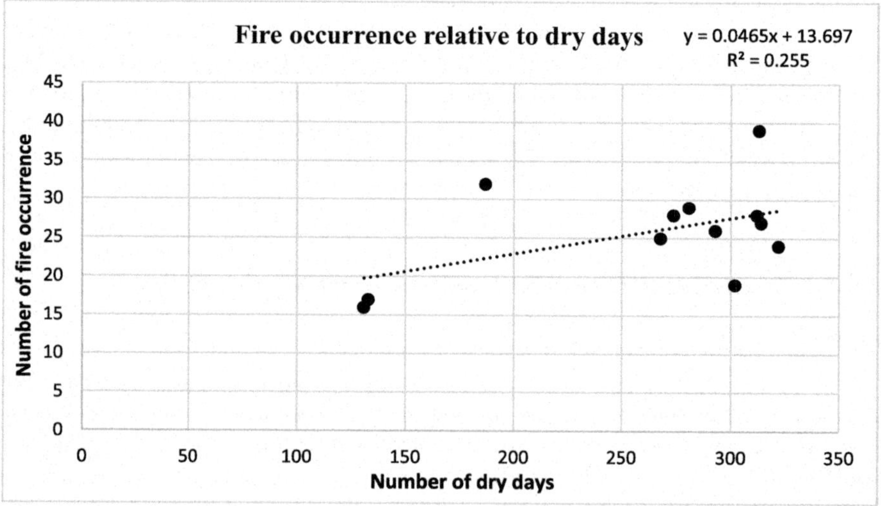


**Fig. 7** Relation between dry days and the number of fire events

in March, and the average monthly temperature was 23.2 °C. In contrast, the month of April experienced a total of 19 fire incidents with a monthly average temperature of 27.8 °C. Thus, the monthly average temperature of fire months was also similar in other years (Table 4).

### 3.1.5 Trend of Temperature in Fire Season

The maximum, minimum, and average fire season temperatures are plotted in the graph.

**Maximum Temperature During Fire Season**

Figure 8 shows the frequency of maximum temperatures during the fire season from 2010 to 2021. It was found that in 2014, March had a higher frequency with a temperature of 31.9 °C. Similarly, March 2018 had a frequency with a temperature of 33 °C. Likewise 2014, April had a higher frequency with a temperature of 37 °C.

Figure 9 shows the frequency of the maximum temperature and the fire occurrence during the fire months. It was found that as the temperature in the fire months increases, the fire occurrence also tends to increase. The graph shows the increase in the temperature during March and April. During these months, the number of fires was also observed to increase.

**Minimum Temperature During Fire Season**

Figure 10 shows the frequency of minimum temperatures during the fire season from 2010 to 2021. It was found that in 2014, March had a higher frequency with a temperature of 22.7 °C. In the same year, the temperature frequency in April was found to be 27.9 °C.

**Average Temperature During Fire Season**

Figure 11 shows the frequency of average temperatures during the fire season from 2010 to 2021. It shows that in 2014, March, April, and May had higher frequency with temperatures of 27.3 °C, 32.5 °C, and 31.7 °C, respectively.

### 3.1.6 Distribution of Forest Fires in the Community Forests

MODIS data provided information on fire incidents in the community forests from 2010 to 2021. It was found that there have been 12 fire incidents in Markaurra community forest in the last 12 years. The fire incidents occurred in 2010, 2011, 2012, 2014, 2015, 2018 and 2021. The highest fire incidents occurred in 2012, with four fire events. Two fire events were observed in 2010 and 2014, whereas only one was recorded in other years. Kalikhola Community Forest observed two fire events, with a single fire event occurring in 2012 and 2019. Overall, seven fire incidents were recorded in the Bahunijhora community forest. Three fire events were observed in

**Table 4** Number of dry days and fire incidents in the fire months

| Year | Jan | | | | Feb | | | | Mar | |
|---|---|---|---|---|---|---|---|---|---|---|
| | Avg. monthly temp (°C) | No. of dry days > Avg. temp | Fire days | No. of fire occurrence | Avg. monthly temp (°C) | No. of dry days > Avg. temp | Fire days | No. of fire occurrence | Avg. monthly temp (°C) | No. of dry days > Avg. temp |
| 2010 | 13.8 | 17 | 0 | 0 | 18.8 | 15 | 0 | 0 | 25.3 | 14 |
| 2011 | 13.8 | 17 | 0 | 0 | 20.2 | 17 | 2 | 3 | 23.5 | 14 |
| 2012 | 16.9 | 22 | 0 | 0 | 20.1 | 13 | 0 | 0 | 23.4 | 15 |
| 2013 | 16.6 | 20 | 1 | 1 | 20.1 | 13 | 2 | 2 | 23.2 | 14 |
| 2014 | 15.9 | 13 | 0 | 0 | 18.2 | 15 | 0 | 0 | 27.3 | 17 |
| 2015 | 16.1 | 10 | 0 | 0 | 15.6 | 11 | 2 | 2 | 22.7 | 15 |
| 2016 | 16.3 | 19 | 0 | 0 | 16.4 | 12 | 4 | 5 | 21.7 | 15 |
| 2017 | 16 | 17 | 0 | 0 | 15.3 | 11 | 2 | 3 | 23.5 | 14 |
| 2018 | 12.9 | 17 | 1 | 1 | 19.4 | 13 | 0 | 0 | 25.5 | 15 |
| 2019 | 16.4 | 14 | 2 | 2 | 19.1 | 16 | 0 | 0 | 23.5 | 20 |
| 2020 | 16.1 | 14 | 0 | 0 | 16.6 | 15 | 0 | 0 | 24.4 | 15 |
| 2021 | 16.1 | 14 | 1 | 1 | 18.6 | 15 | 3 | 3 | 24.4 | 16 |

(continued)

**Table 4** (continued)

| Year | Mar | | Apr | | | | May | | | |
|---|---|---|---|---|---|---|---|---|---|---|
| | Fire days | No. of fire occurrence | Avg. monthly temp (°C) | No. of dry days > Avg. temp | Fire days | No. of fire occurrence | Avg. monthly temp (°C) | No. of dry days > Avg. temp | Fire days | No. of fire occurrence |
| 2010 | 7 | 12 | 30 | 19 | 7 | 14 | 30.1 | 18 | 0 | 0 |
| 2011 | 6 | 15 | 26.5 | 17 | 1 | 1 | 28.8 | 17 | 0 | 0 |
| 2012 | 7 | 15 | 27.9 | 12 | 5 | 12 | 29.7 | 16 | 0 | 0 |
| 2013 | 8 | 17 | 27.8 | 17 | 12 | 19 | 29.5 | 27 | 0 | 0 |
| 2014 | 7 | 19 | 32.5 | 14 | 5 | 8 | 31.7 | 18 | 1 | 1 |
| 2015 | 8 | 16 | 26.3 | 16 | 5 | 7 | 28.6 | 17 | 0 | 0 |
| 2016 | 10 | 19 | 25.2 | 17 | 4 | 8 | 28.2 | 17 | 0 | 0 |
| 2017 | 3 | 5 | 30 | 16 | 5 | 9 | 30.1 | 18 | 0 | 0 |
| 2018 | 15 | 25 | 27.9 | 16 | 2 | 2 | 29.1 | 19 | 0 | 0 |
| 2019 | 10 | 27 | 27.7 | 15 | 0 | 0 | 30.6 | 19 | 0 | 0 |
| 2020 | 1 | 1 | 27.6 | 14 | 5 | 15 | 26.2 | 18 | 0 | 0 |
| 2021 | 9 | 19 | 27.7 | 13 | 1 | 1 | 26.7 | 17 | 0 | 0 |

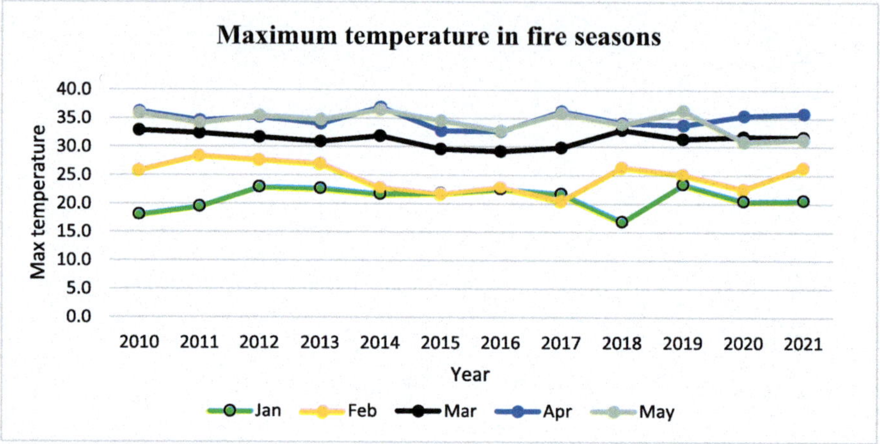


Fig. 8 Maximum temperature during fire seasons

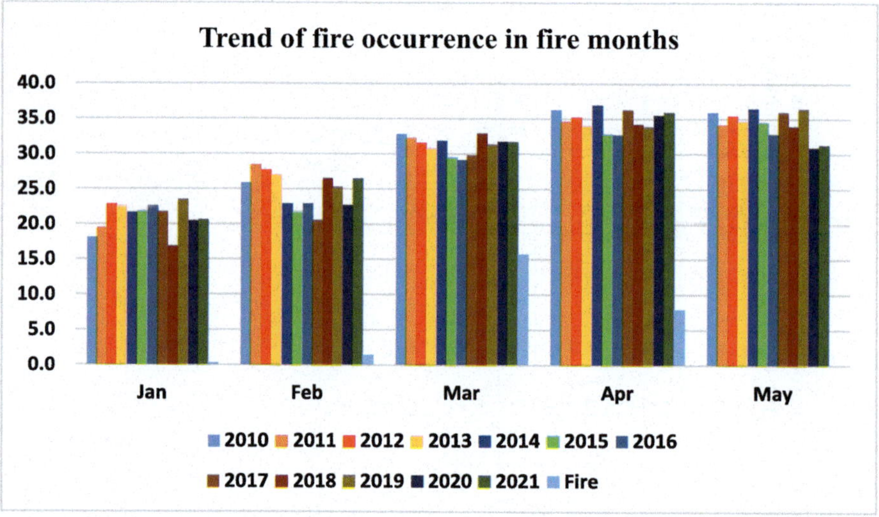


Fig. 9 Maximum temperature during fire season and fire occurrence

2021, while single fire events were observed in 2010, 2011, 2017 and 2020 (Figs. 12, 13 and 14).

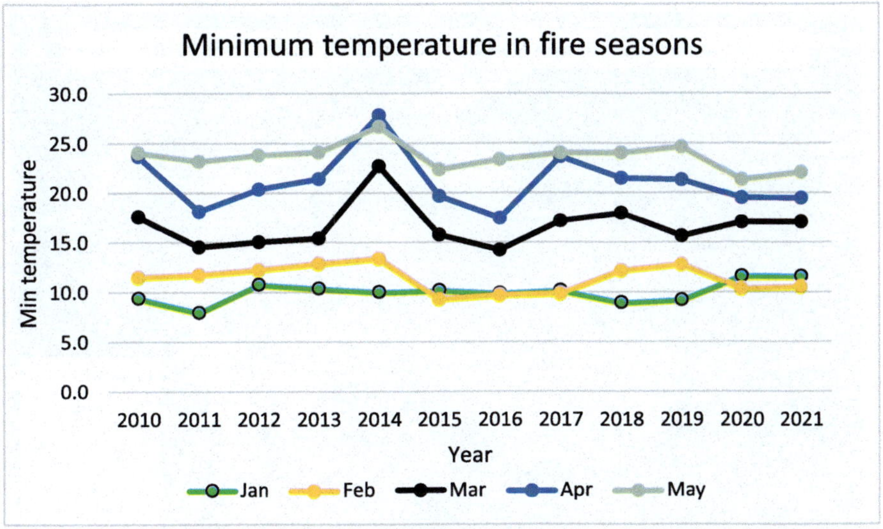


**Fig. 10** Minimum temperature during fire seasons

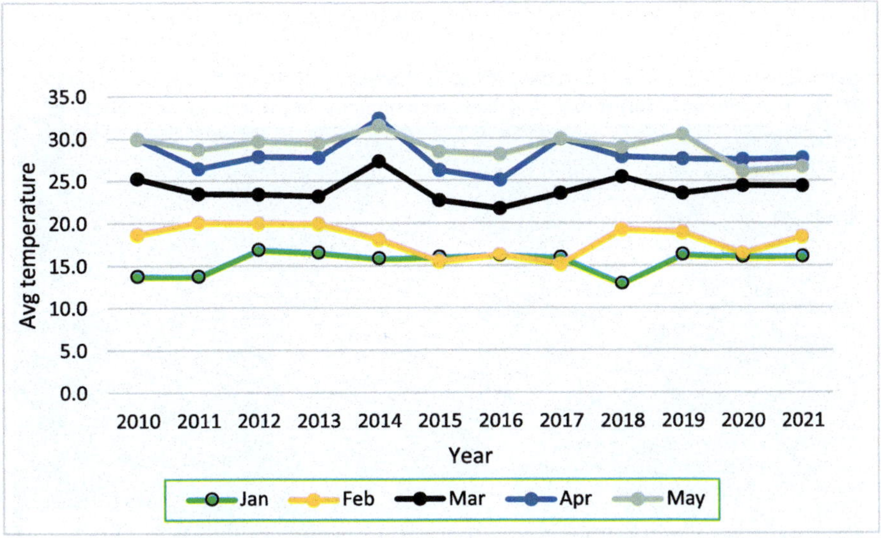


**Fig. 11** Average temperature during fire seasons

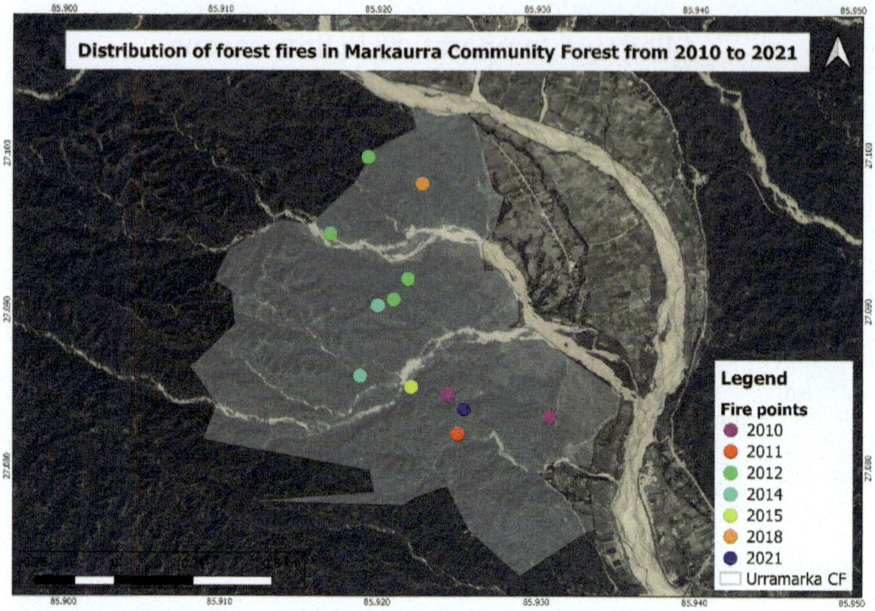


**Fig. 12** Distribution of forest fire in Markaurra community forest from 2010 to 2021

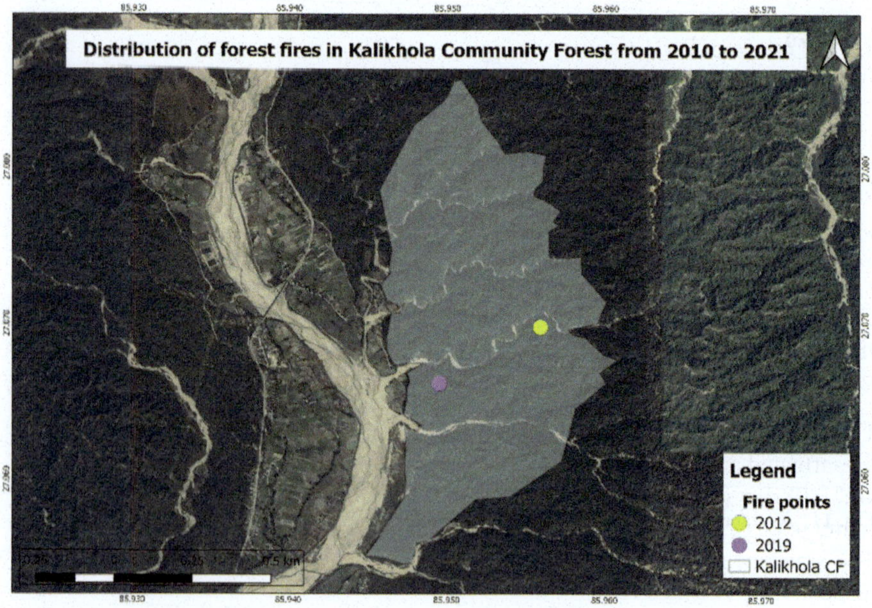


**Fig. 13** Distribution of forest fire in Kalikhola community forest from 2010 to 2021

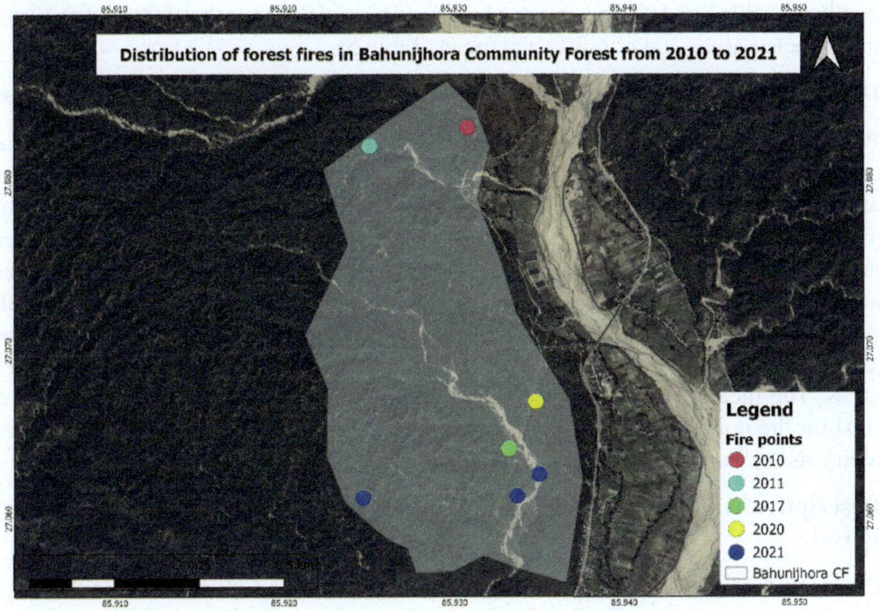


**Fig. 14** Forest fire distribution in Bahunijhora community forests from 2010 to 2021

## 3.2 *Descriptive Statistics of Species Diversity in Fire Affected and Non-affected Plots*

The descriptive analysis of fire affected and non-affected areas in three community forests were calculated.

### 3.2.1 Descriptive Analysis of the Shannon Weiner Index

**Descriptive Analysis of Shannon Weiner Index in Markaurra Community Forest**

The mean ± SE for trees in fire-affected areas was found to be 0.46 ± 0.12; in fire-affected areas, the value was 0.63 ± 0.04. The standard deviation (SD) was 0.20 and 0.08 for fire-affected and non-affected areas. For poles, the mean ± SE was found to be 0.52 ± 0.04 in fire-affected areas and 0.69 ± 0.07 in non-affected areas. The standard deviation in fire-affected and non-affected areas was 0.08 and 0.14, respectively. The mean ± SE for saplings in fire-affected and non-affected areas was determined to be 0.75 ± 0.24 and 0.77 ± 0.27, respectively. The standard deviation for the affected area was 0.34, whereas for the non-affected area, it was 0.55. For seedlings, fire affected had mean ± SE of 0.55 ± 0.05 and non-affected had mean

± SE of 0.97 ± 0.10. The standard deviation was found to be 0.09 and 0.32 for fire-affected areas and non-affected areas (Table 5).

**Descriptive Analysis of Shannon Weiner Index in Kalikhola Community Forest**

For tees, the mean ± SE in Kalikhola community forest for the affected area was calculated to be 0.68 ± 0.02, and that of the non-affected area was 0.66 ± 0.08. Fire-affected and non-affected areas had standard deviations of 0.04 and 0.17, respectively. The poles in this community forest had mean ± SE of 0.40 ± 0.29 in fire-affected areas and 0.52 ± 0.08 in non-affected areas. The standard deviation for fire-affected and non-affected areas was 0.41 and 0.18, respectively. Similarly, the mean ± SE for sapling was 0.64 ± 0.06 for fire-affected areas and 0.67 ± 0.06 for non-affected areas. The standard deviation was 0.11 and 0.16 for fire-affected and non-affected areas. The mean ± SE for the seedlings in the areas was 0.82 ± 0.09 and 0.75 ± 0.04 for the non-affected area. In comparison, the standard deviation of the affected area was 0.19 and 0.12 for the non-affected area.

**Descriptive Analysis of Shannon Weiner Index in Bahunijhora Community Forest**

For trees, the mean ± SE for fire-affected and non-affected areas in Bahunijhora CF was found to be 0.65 ± 0.02 and 0.63 ± 0.04, respectively. The standard deviation for the fire-affected area was determined to be 0.03 and 0.09 for the non-affected area. As for poles, the mean ± SE of fire-affected areas was 0.56 ± 0.03; for non-affected areas, it was found to be 0.54 ± 0.04. The standard deviation of poles in the fire-affected and non-affected areas was calculated as 0.05 and 0.10, respectively. The mean ± SE for sapling in fire-affected areas was 0.49 ± 0.10; in non-affected areas, the value was 0.68 ± 0.06. The standard deviation was found to be 0.17 in both areas. As for seedlings, the affected area had mean ± SE 0.76 ± 0.11, and for the non-affected area the mean ± SE was 0.78 ± 0.13. Fire-affected and non-affected areas had standard deviations of 1.27 and 1.33, respectively.

### 3.2.2 Descriptive Analysis of Species Evenness

**Descriptive Analysis of Species Evenness in Markaurra Community Forest**

The mean ± SE of trees in fire-affected and non-affected areas was determined to be 0.23 ± 0.06 and 0.31 ± 0.02, respectively. The standard deviation of trees in the areas was 0.10, and in non-affected areas it was 0.04. The poles in the affected area had a mean ± SE of 0.26 ± 0.02, and in the non-affected area the mean ± SE was 0.31 ± 0.01. The standard deviation for fire-affected and non-affected areas was 0.04 and 0.02, respectively. Meanwhile, for a sapling, the mean ± SE in the affected area was calculated to be 0.29 ± 0.04, and for the non-affected area, mean ± SE was found to be 0.25 ± 0.02. The standard deviation was 0.05 and 0.03 for saplings in fire-affected and non-affected areas, respectively. For seedlings, the mean ± SE was 0.27 ± 0.02 in the affected area and 0.26 ± 0.02 in the non-affected area. The

**Table 5** Shannon Weiner Index in fire-affected and non-affected areas in three community forests

| Shannon Weiner Index | | | | | | | | | | | | | | | | | |
|---|---|---|---|---|---|---|---|---|---|---|---|---|---|---|---|---|---|
| CFUG | Area | Tree | | | | Pole | | | | Seedling | | | | Sapling | | | |
| | | Mean ± SE | Min | Max | SD | Mean ± SE | Min | Max | SD | Mean ± SE | Min | Max | SD | Mean ± SE | Min | Max | SD |
| Markaurra | Fire non affected area | 0.63 ± 0.04 | 0.56 | 0.69 | 0.08 | 0.69 ± 0.07 | 0.57 | 0.90 | 0.14 | 0.97 ± 0.10 | 0.62 | 1.57 | 0.32 | 0.77 ± 0.27 | 0.41 | 1.58 | 0.55 |
| | Fire affected area | 0.46 ± 0.12 | 0.35 | 0.69 | 0.20 | 0.52 ± 0.04 | 0.45 | 0.60 | 0.08 | 0.55 ± 0.05 | 0.40 | 0.60 | 0.09 | 0.75 ± 0.24 | 0.51 | 0.99 | 0.34 |
| Kalikhola | Fire non affected area | 0.66 ± 0.08 | 0.55 | 0.95 | 0.17 | 0.52 ± 0.08 | 0.32 | 0.69 | 0.18 | 0.75 ± 0.04 | 0.57 | 0.94 | 0.12 | 0.67 ± 0.06 | 0.52 | 0.98 | 0.16 |
| | Fire affected area | 0.68 ± 0.02 | 0.64 | 0.73 | 0.04 | 0.40 ± 0.29 | 0.10 | 0.69 | 0.41 | 0.82 ± 0.09 | 0.62 | 1.06 | 0.19 | 0.64 ± 0.06 | 0.52 | 0.71 | 0.11 |
| Bahunijhora | Fire non affected area | 0.63 ± 0.04 | 0.50 | 0.69 | 0.09 | 0.54 ± 0.04 | 0.38 | 0.63 | 0.10 | 0.78 ± 0.13 | 0.43 | 1.33 | 0.34 | 0.68 ± 0.06 | 0.33 | 0.81 | 0.17 |
| | Fire affected area | 0.65 ± 0.02 | 0.63 | 0.69 | 0.03 | 0.56 ± 0.03 | 0.52 | 0.61 | 0.05 | 0.76 ± 0.11 | 0.51 | 1.27 | 0.26 | 0.49 ± 0.10 | 0.33 | 0.67 | 0.17 |

standard deviation was found to be 0.05 for both fire-affected areas and non-affected areas (Table 6).

**Descriptive Analysis of Species Evenness in Kalikhola Community Forest**

The mean ± SE of trees in fire-affected and non-affected areas in Kalikhola CF was determined to be 0.34 ± 0.01 and 0.29 ± 0.01, respectively. The standard deviation of trees was 0.02 in both areas. The poles in fire-affected areas had a mean ± SE of 0.20 ± 0.15, and in non-affected areas, its value was 0.24 ± 0.04. The standard deviation in fire-affected and non-affected areas was determined to be 0.21 and 0.09, respectively. For saplings, the mean ± SE value of 0.24 ± 0.01 was found for fire-fire-affected areas, and the value 0.30 ± 0.01 was found for non-affected areas. The values 0.01 and 0.03 were obtained respectively for the standard deviation for fire-affected and non-affected areas. As for seedlings, the mean ± SE in both fire-affected and non-affected areas was found to be 0.28 ± 0.01. Also, the standard deviation for fire-affected and non-affected areas was determined to be 0.03.

**Descriptive Analysis of Species Evenness in Bahunijhora Community Forest**

The mean ± SE of trees in fire areas was found to be 0.33 ± 0.0; for non-affected areas, the value was 0.31 ± 0.02. The standard deviation in fire-affected areas was found to be 0.02, while in non-affected areas, the standard deviation was found to be 0.04. The poles in fire-affected areas had mean ± SE of 0.23 ± 0.03 whereas mean ± SE was found to be 0.27 ± 0.02 in non-affected areas. The standard deviation was calculated as 0.06 and 0.05 in fire-affected and non-affected areas, respectively. Likewise, the mean ± SE value of saplings in fire-affected and non-affected areas was 0.21 ± 0.02 and 0.23 ± 0.02, respectively. The standard deviation for fire-affected and non-affected areas was 0.04. For seedlings, the mean ± SE value for fire-affected areas was 0.25 ± 0.02, while for non-affected areas, the value was 0.22 ± 0.01. The standard deviation for both areas was found to be 0.04.

### 3.2.3 Descriptive Analysis of Species Richness Index

**Descriptive Analysis of Species Richness in Markaurra Community Forest**

The mean ± SE of trees in fire-affected areas and non-affected areas was found to be 2. For the pole, the mean ± SE for the fire-affected area was 2, while for non-affected areas, the value was 2.25 ± 0.25. The standard deviation for non-affected areas was found to be 0.50. For sapling, the mean ± SE was found to be 2.5 ± 0.5 for the affected area, while the value was found to be 3 ± 1. The standard deviation for fire-affected and non-affected areas was 0.71 and 2, respectively. For seedlings, the mean ± SE in the affected area was determined to be 2; for the non-affected area, the value was 3.70 ± 0.26. The standard deviation for non-affected areas was found to be 0.82 (Table 7).

 

**Table 6** Species evenness in fire-affected and non-affected areas in three community forests

| Species evenness | | | | | | | | | | | | | | | | | |
|---|---|---|---|---|---|---|---|---|---|---|---|---|---|---|---|---|---|
| CFUG | Area | Tree | | | | Pole | | | | Seedling | | | | Sapling | | | |
| | | Mean ± SE | Min | Max | SD | Mean ± SE | Min | Max | SD | Mean ± SE | Min | Max | SD | Mean ± SE | Min | Max | SD |
| Markaurra | Fire non affected area | 0.31 ± 0.02 | 0.28 | 0.35 | 0.04 | 0.31 ± 0.01 | 0.28 | 0.33 | 0.02 | 0.26 ± 0.02 | 0.18 | 0.33 | 0.05 | 0.25 ± 0.02 | 0.21 | 0.29 | 0.03 |
| | Fire affected area | 0.23 ± 0.06 | 0.17 | 0.35 | 0.10 | 0.26 ± 0.02 | 0.23 | 0.30 | 0.04 | 0.27 ± 0.02 | 0.20 | 0.30 | 0.05 | 0.29 ± 0.04 | 0.25 | 0.33 | 0.05 |
| Kalikhola | Fire non affected area | 0.29 ± 0.01 | 0.27 | 0.32 | 0.02 | 0.24 ± 0.04 | 0.16 | 0.34 | 0.09 | 0.28 ± 0.01 | 0.24 | 0.33 | 0.03 | 0.30 ± 0.01 | 0.26 | 0.34 | 0.03 |
| | Fire affected area | 0.34 ± 0.01 | 0.32 | 0.37 | 0.02 | 0.20 ± 0.15 | 0.05 | 0.34 | 0.21 | 0.28 ± 0.01 | 0.25 | 0.31 | 0.03 | 0.24 ± 0.01 | 0.24 | 0.26 | 0.01 |
| Bahunijhora | Fire non affected area | 0.31 ± 0.02 | 0.25 | 0.34 | 0.04 | 0.27 ± 0.02 | 0.19 | 0.31 | 0.05 | 0.22 ± 0.01 | 0.18 | 0.29 | 0.04 | 0.23 ± 0.02 | 0.17 | 0.27 | 0.04 |
| | Fire affected area | 0.33 ± 0.01 | 0.31 | 0.34 | 0.02 | 0.23 ± 0.03 | 0.17 | 0.30 | 0.06 | 0.25 ± 0.02 | 0.22 | 0.32 | 0.04 | 0.21 ± 0.02 | 0.17 | 0.24 | 0.04 |

**Table 7** Species richness in fire-affected and non-affected areas in three community forests

| Species Richness Index | | | | | | | | | | | | | | | | | |
|---|---|---|---|---|---|---|---|---|---|---|---|---|---|---|---|---|---|
| CFUG | Area | Tree | | | | Pole | | | | Seedling | | | | Sapling | | | |
| | | Mean ± SE | Min | Max | SD | Mean ± SE | Min | Max | SD | Mean ± SE | Min | Max | SD | Mean ± SE | Min | Max | SD |
| Markaurra | Fire non affected area | 2 | 2 | 2 | 0 | 2.25 ± 0.5 | 2 | 3 | 0.50 | 3.70 ± 0.26 | 3 | 5 | 0.82 | 3 ± 1 | 2 | 6 | 2.00 |
| | Fire affected area | 2 | 2 | 2 | 0 | 2 | 2 | 2 | 0.00 | 2 | 2 | 2 | 0.00 | 2.5 ± 0.5 | 2 | 3 | 0.71 |
| Kalikhola | Fire non affected area | 2.2 ± 0.2 | 2 | 3 | 0.45 | 2.2 ± 0.2 | 2 | 3 | 0.45 | 2.75 ± 0.16 | 2 | 3 | 0.46 | 2.25 ± 0.16 | 2 | 3 | 0.46 |
| | Fire affected area | 2 | 2 | 2 | 0 | 2 | 2 | 2 | 0.00 | 3 ± 0.41 | 2 | 4 | 0.82 | 2.67 ± 0.33 | 2 | 3 | 0.58 |
| Bahunijhora | Fire non affected area | 2 | 2 | 2 | 0 | 2 | 2 | 2 | 0 | 3.57 ± 0.57 | 2 | 6 | 1.51 | 2.86 ± 0.14 | 2 | 3 | 0.38 |
| | Fire affected area | 2 | 2 | 2 | 0 | 2.5 ± 0.29 | 2 | 3 | 0.58 | 3 ± 0.26 | 2 | 4 | 0.63 | 2.33 ± 0.33 | 2 | 3 | 0.58 |

**Descriptive Analysis of Species Richness in Kalikhola Community Forest**

For trees and poles, the mean ± SE in fire fire-affected areas was 2, whereas the value was 2.2 ± 0.2 for non-affected areas. The standard deviation in the non-affected area was determined to be 0.45. The mean ± SE for saplings in fire fire-affected areas was 2.67 ± 0.33 while the value was 2.25 ± 0.16 for non-affected areas. The standard deviation for fire-affected and non-affected areas was calculated to be 0.58 and 0.46, respectively. For seedlings, the fire-affected area had a mean ± SE of 3 ± 0.41, and the non-affected area had a value of 2.75 ± 0.16. The standard deviation for fire-affected and non-affected areas was determined to be 0.82 and 0.46, respectively.

**Descriptive Analysis of Species Richness in Bahunijhora Community Forest**

For trees, fire-affected and non-affected areas had a mean ± SE value of 2. The poles in the affected area had a mean of 2.5 ± 0.29 and had a value of 2 in non-affected areas. The standard deviation in fire-affected areas was found to be 0.58. For saplings, the mean ± SE in fire-affected and non-affected areas was 2.33 ± 0.33 and 2.86 ± 0.14 respectively. The fire-affected area had a standard deviation 0.58, and the non-affected area had a standard deviation of 0.38. Further, the seedlings had a mean ± SE of 3 ± 0.26 in the affected area, while the value was 3.57 ± 0.57 in the non-affected area. The standard deviation was 0.63 in the affected area and 1.51 in the non-affected area.

## *3.3 Drivers of Forest Fires and Management Options Measures*

The study showed varied causes and monitoring measures of forest fires in Mahottari district. The causes and management practices of forest fires were analyzed using Principal Component Analysis.

### 3.3.1 Principal Component Analysis Showing Drivers of Forest Fire

The household survey provided crucial information regarding the causes of forest fires. According to the study, there are five main drivers for forest fires. These drivers of fires in the community forests were deliberate fire, natural occurrence, dry season, and carelessness by the smokers.

Principal Component Analysis of the Markaurra community forest showed that deliberate fire and natural occurrence were positively correlated since they fell in the same quadrant (Fig. 15). Principal Component Analysis of the causes in the Kalikhola community (Fig. 16) showed that carelessness by smokers and natural occurrence were positively correlated as they fell in the same quadrant. Similarly, deliberate fire and dry season also fell in the same quadrant, meaning they were positively correlated.

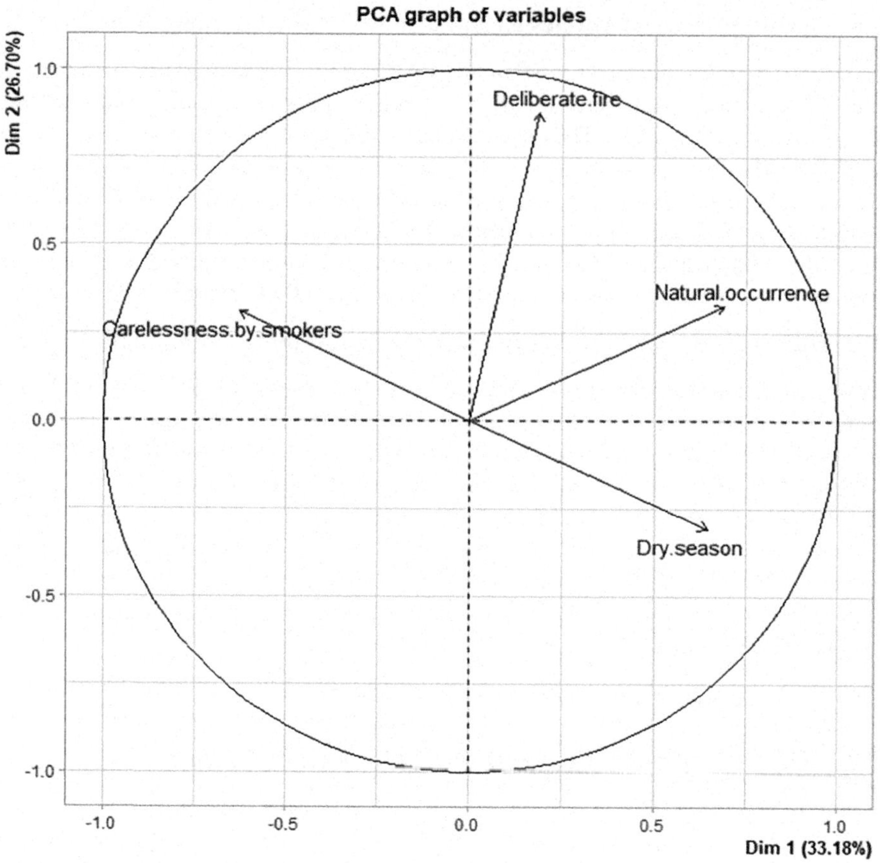


**Fig. 15** PCA demonstrating the causes of fire in Markaurra community forest

Principal Component Analysis of Bahunijhora community forest (Fig. 17) showed that the causes were not positively correlated.

### 3.3.2 Principal Component Demonstrating Management Options

Principal Component Analysis presented information about the measures adopted by the locals to mitigate the fire in the community forests. According to the survey, mainly five measures have been adopted they are Fire Line Construction (FC), Fire Line Cleaning (FLC), Construction of Watch Tower (CWT), Appointment of Forest Watchers (AFW) and Awareness (AW) that includes training and workshops.

Principal Component Analysis of the Markaurra community forest showed that Awareness (AW) and Appointment of Forest Watchers (AFW) positively correlated since they both fell in the same quadrant. Likewise, fire line construction (FC) and

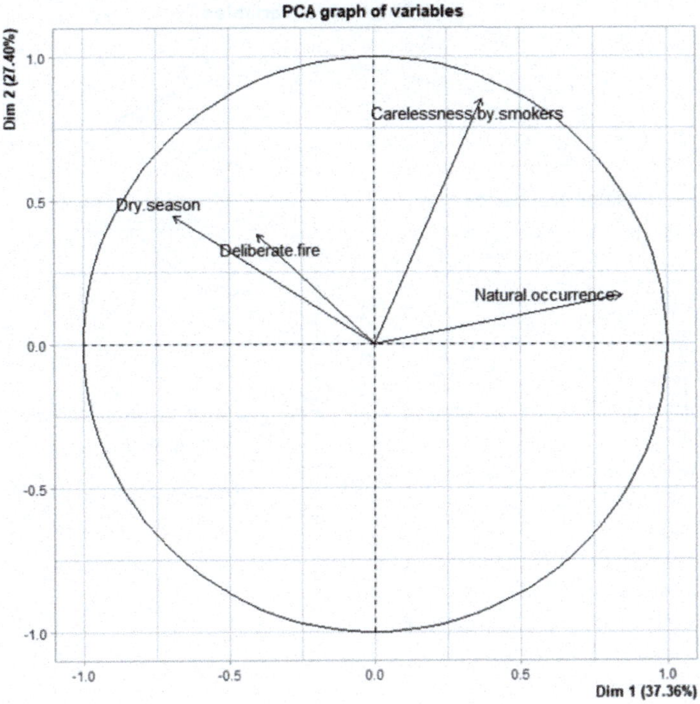


**Fig. 16** PCA demonstrating the causes of fire in Kalikhola community forest

clearing (FLC) were positively correlated as they fell in the same quadrant. Principal Component Analysis of the Kalikhola community forest showed that the construction of the watch tower (CWT) and the appointment of forest watcher (AFW) were positively correlated since they fell on the same quadrant. The results from the PCA from the responses provided by the locals of the Bahunijhora community forest showed that Appointment of Forest Watchers (AFW) and Awareness (AW) were positively correlated with each other (Figs. 18, 19 and 20).

## 3.4 Discussion

### 3.4.1 Spatial and Temporal Distribution of Forest Fire

The spatial and temporal distribution of forest fire in Mahottari district from 2010 to 2021 was analyzed. According to MODIS fire data, the result showed 288 fire events in the Mahottari district in the last twelve years. Among the total fire incidents, there were 26 fire events in 2010, 19 in 2011, and 27 in 2012. Similarly, 39 fire incidents were observed in 2013, 28 incidents were observed in 2014, 25 incidents

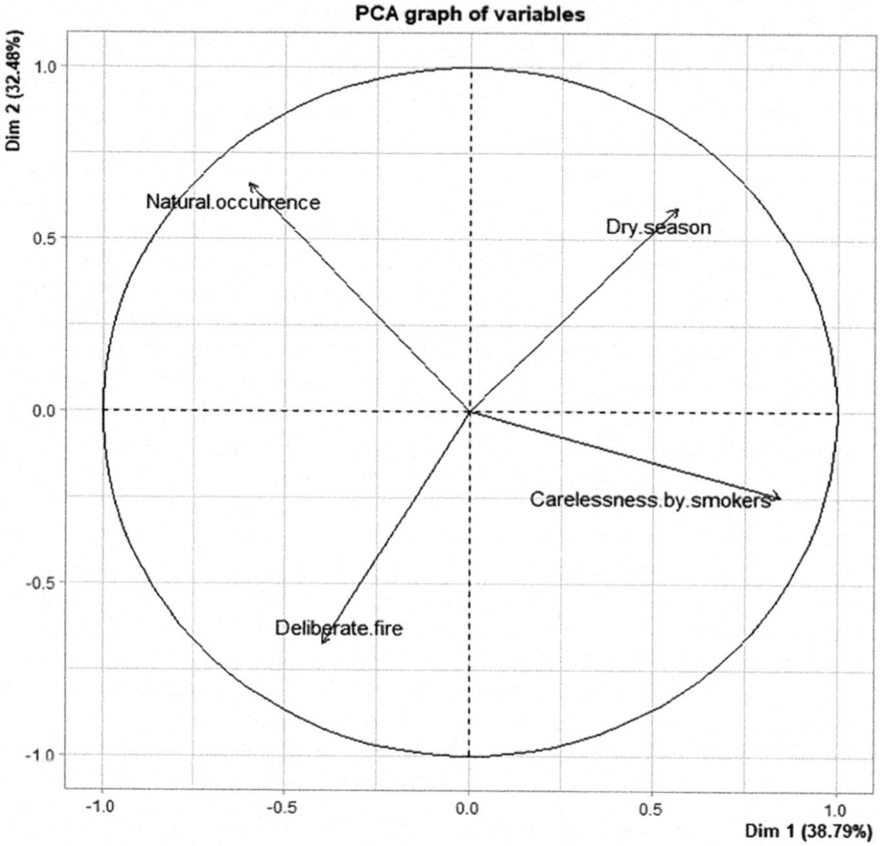


**Fig. 17** PCA demonstrating the causes of fire in Bahunijhora community forest

were recorded in 2015, 32 incidents were observed in 2016, and 17 fire incidents were recorded in 2017. Further, 28 fire incidents were followed in 2018, 29 fire incidents were observed in 2019, 16 fire incidents in 2020, and 24 fire incidents in 2021. The highest number of fire events occurred in 2013, with 39 fire events that year.

In comparison, the lowest number of fires occurred in 2020, with 16 fire incidents. Markaurra community forest experienced 12 fire incidents, Kalikhola community forest experienced two counts, and Bahunijhora community forest endured seven fire incidents. During the last twelve years, Mahottari district has experienced fire incidents from January to May. The maximum number of fires occurred during March and April, while the lowest number of fire incidents occurred in May. March 2019 experienced the highest number, i.e., 27 counts of fire events in twelve years. This result has been consistent with the previous studies conducted by Kunwar and Khaling (2006), Matin et al. (2017) as well as by Badal and Mandal (2021). The reason for the culmination of fire during these months is that the tree species in this region shed their leaves during these months. The accumulation of large amounts of

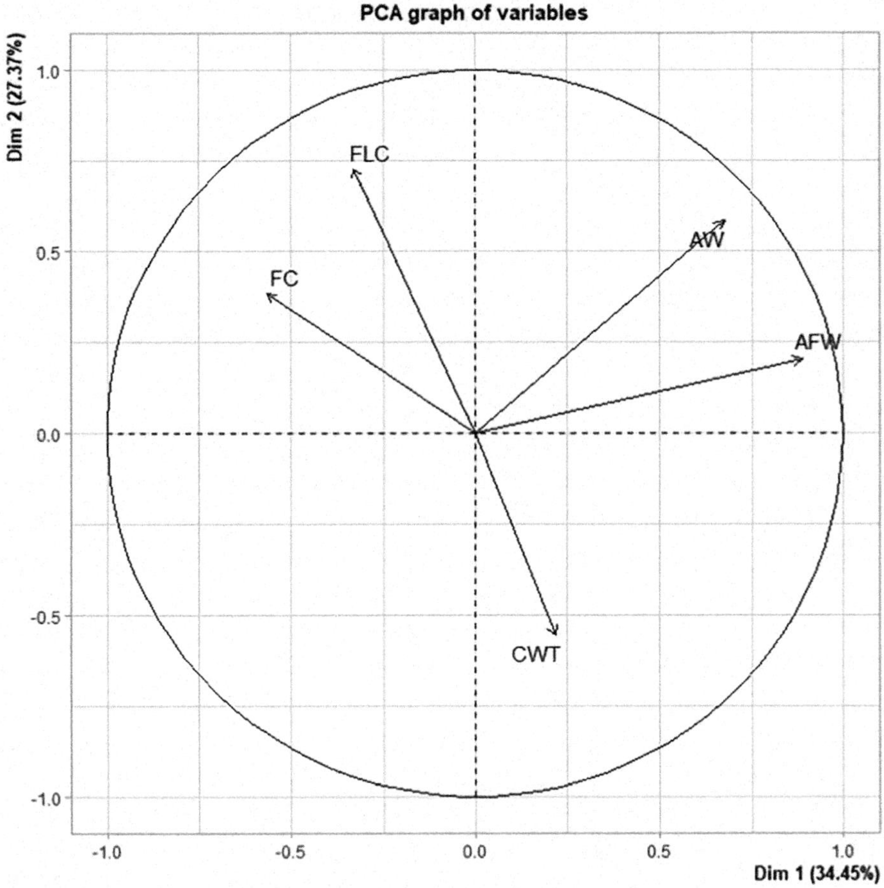


**Fig. 18** PCA showing mitigation measures in Markaurra community forest

dry leaves, twigs, and litter on the forest floor, accompanied by undergrowth species, grasses, and weeds, fuels the outbreak of forest fires (Kunwar and Khaling 2006). Additional causes for forest fires might be due to the increase in drought in the area, prolonged dry season, negligence in the disposal of cigarettes after smoking, mischief during playtime by children, and accumulation of dry litter such as dry leaves, branches, bushes, etc. on the forest ground.

### 3.4.2 Fire Occurrence Relative to Dry days

The result showed that the number of dry days directly affects the commencement of forest fires. It was found that the fire months with days greater than the days with average temperature were more prone to fire occurrence. The more dry days, the

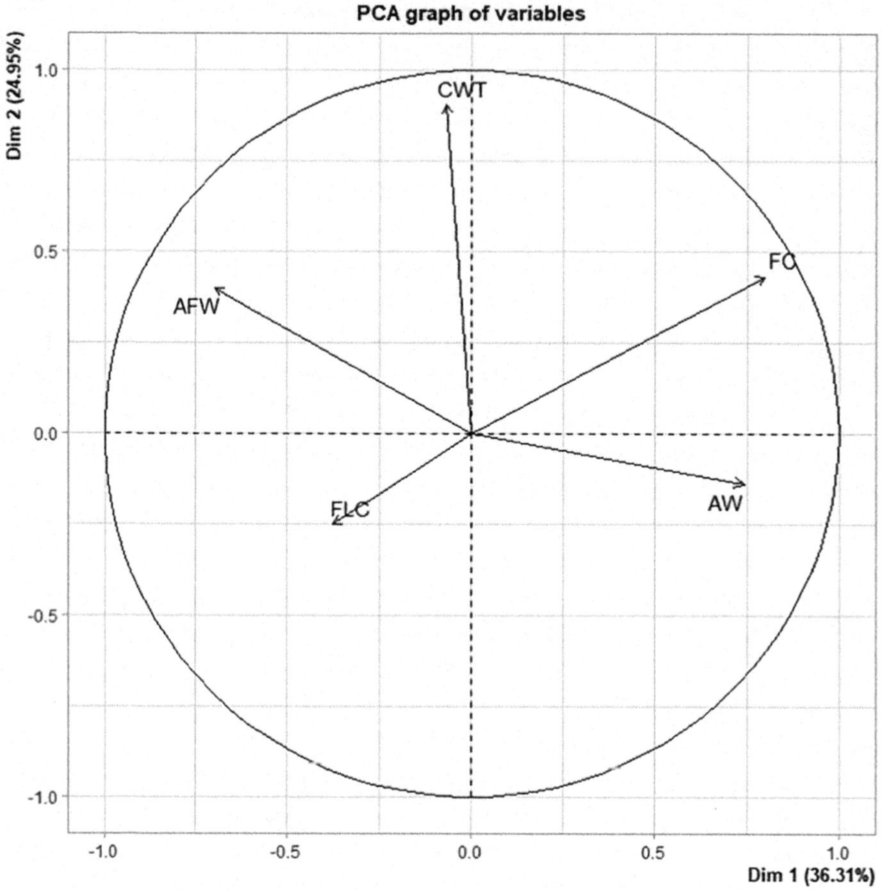


**Fig. 19** PCA showing mitigation measures in Kalikhola community forest

more the chance of starting fires in the forest. According to the result, the highest number of fires was recorded in 2013, with 39 fire incidents, which was also the same year with the highest number of dry days with a total of 313. This might be because the rise in temperature leads to moisture loss from the surroundings. This creates dry conditions in the forest, which act as a fueling agent and thus give rise to fires.

### 3.4.3 Temperature Trend and Fire Season

The result showed the frequency of average, minimum, and maximum temperatures during the fire season and that the frequency was higher during the peak fire months. It was observed that the fire occurrences increased as the temperature increased during the fire months.

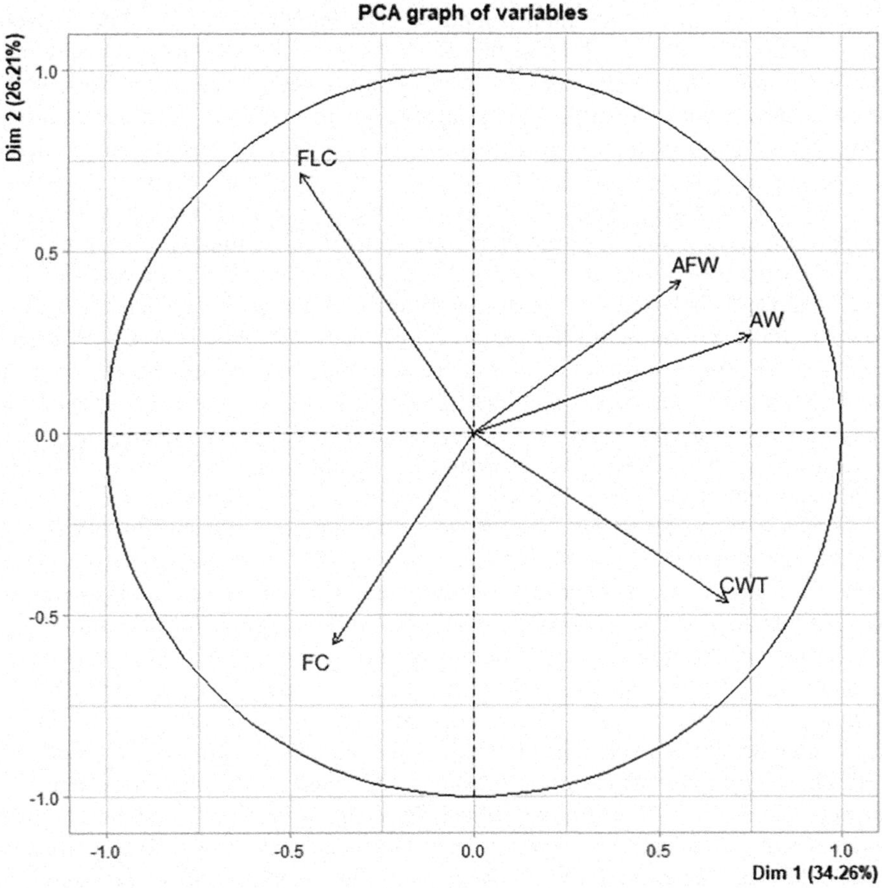


**Fig. 20** PCA showing mitigation measures in Bahunijhora community forest

### 3.4.4 Occurrence of Forest Fires in the Community Forests

The descriptive analysis of fire-affected and non-affected plots showed that seedlings and saplings were comparatively more affected than trees and poles, which had different diversity and regeneration in most places.

### 3.4.5 Drivers of Forest Fire and Management Options to Control Forest Fire

The drivers of forest fire and the measures adopted by the locals to control the spread of fire were analyzed using Principal Component Analysis. According to the study, four major factors are the causes of forest fires. It was found that most forest fires in the Mahottari district are human-induced. The main causes were a dry season, carelessness by smokers, deliberate fire, and natural occurrence. Fires generally start due to natural reasons in the forest. However, the dry season increase in recent years from January to May has further contributed to the onset of fires. Negligence during the disposal of cigarettes and *bidis* by smokers in the forest area has also increased the chances of spreading fire. Moreover, cattle grazers' arson or mischievous behaviour during playtime has also been found to be the leading cause of forest fires. Another main cause of forest fire has been deliberate fire by herders to get rid of insects and snakes during the collection of grass and fodders, as well as cattle grazing.

The study established that the locals followed five major management options to control and reduce the further onset of fires. They were Fire Line Construction, Fire Line Clearing, Construction of Watch Tower, Appointment of a Forest Watcher, and Awareness. It has been found that following two or more than two monitoring measures and management practices have been more effective in reducing the spread of forest fires. The PCA in the Markaurra community forest showed that Forest Line Construction and Forest Line Clearing were positively correlated, meaning that these measures effectively reduced forest fires when used together. Similarly, for the Kalikhola community forest, the Appointment of a Forest Watcher and the Construction of a Watch Tower were positively correlated, which means these two measures were effective when used together. CDFAN has contributed to the community-level approach to minimizing the effects of forest fire by organizing workshops, plantation programs, and construction of conservation ponds. According to IUCN, community management is a highly effective form of conservation. Community-level monitoring and management practices are more successful since the communities have profound insight into fire history, cause and effects, traditional management practices, etc. There are four critical elements for effective forest management strategies in a community: prevention, preparedness, suppression, and response and relief. Pre-fire management (fuel management), fire detection and suppression, and post-fire management (mopping up) are the primary aspects of effective forest fire management.

## 4 Conclusion

The study revealed fluctuations in annual fire incidents, with the highest occurrence observed in 2013 and the lowest in 2020. Additionally, it was determined that the fire season typically extends from January to May, with March and April registering the highest number of fire incidents, contrasting with May having the fewest. Furthermore, the research demonstrated a direct correlation between the number of dry days and fire incidents over the past twelve years, with years featuring more dry days experiencing a corresponding increase in fire incidents. Principal Component Analysis indicated that in Markaurra, Kalikhola, and Bahunijhora community forests, fires primarily resulted from dry conditions, deliberate actions, natural causes, and negligence by herders. Notably, the predominant local management practices encompassed constructing fire lines, clearing fire lines, appointing forest watchers, erecting watchtowers, and raising awareness. Since the fires typically ignite in January, it is essential to consistently undertake pre-fire precautions, including routine fire line maintenance and proper clearing and disposal of dry leaves and debris. Notably, the highest frequency of fire incidents occurs in March and April, emphasizing the need for enhanced fire suppression techniques and reduced anthropogenic activities within the forest regions. It is imperative to organize comprehensive awareness programs, training sessions, and workshops to educate local communities about the detrimental impacts of fires and equip them with various strategies for fire prevention. Furthermore, empowering Community Forest User Groups (CFUGs) by providing necessary tools for capacity building in fire risk management is crucial. The local government should actively support a community-based management approach, allotting the appropriate budget for fire line construction, conservation ponds, reforestation efforts, and the erection of watchtowers.

To address the issue of forest fires comprehensively, it's essential to implement regular pre-fire measures like cleaning firebreaks and ensuring proper clearance and disposal of dry leaves and litter, as these fires typically begin in January. March and April witnessed the highest incidence of fires, emphasizing the need for increased attention to fire suppression techniques and a reduction in human activities in forested areas. Moreover, conducting awareness programs, including training sessions and workshops, is crucial to educate local communities about the detrimental effects of fires and teach them various methods to mitigate such incidents. Furthermore, it is vital to empower Community Forest User Groups (CFUGs) by providing the necessary tools for capacity building in fire risk management. Encouraging a community-based management approach, the local government should allocate an appropriate budget for building firebreaks, conservation ponds, reforestation, and watchtower development. Additionally, there's a need for ongoing research on forest fires, examining factors like wind velocity, rainfall patterns, humidity data, and more to gain a deeper understanding of these events.

## References

Andela N, Morton DC, Giglio L, Paugam R, Chen Y, Hantson S et al (2019) The Global Fire Atlas of individual fire size, duration, speed and direction. Earth Syst Sci Data 11(2):529–552. https://doi.org/10.5194/essd-11-529-2019

Badal D, Mandal RA (2021) Spatial analysis of forest fire distribution, occurrence and dynamics in province-2, Nepal

Barwa S, Singh H (2024) Climate-induced changes in forest ecosystems of Indian Himalayan region: a meta-analysis. In: Singh H (ed) Urban forests, climate change and environmental pollution. Springer, Cham. https://doi.org/10.1007/978-3-031-67837-0_39

Bhatta S, Joshi R, Ghimire S (2025) Assessing tree structure and biodiversity in community forests and spatial distribution of local disturbance. Iran For Mag. https://doi.org/10.22034/ijf.2024.413621.1942

Chand HB, Singh H, Chhetri R (2018) Carbon sequestration potential in Sahid Smriti community forest: A case study of Terai region of Nepal. In International Conference on Agriculture and Allied Sciences. The Prod Food Secur and Ecol (Vol. 2018, pp 108–113)

DFRS (2018) Forest cover maps of local levels (753) of Nepal. Department of Forest Research and Survey, Kathmandu

FAO (2020) The state of the world's forests. Food and Agriculture Organization of the United Nations

IUCN (n.d.) Forests and climate. Retrieved from IUCN National Committee of the Netherlands: https://www.iucn.nl/en/our-work/forests-and-climate/#:~:text=Approximately%202.6%20billion%20tonnes%20of,weather%20events%2C%20floods%20and%20droughts

Karki R, Bhatta S, Pandey S (2025) Impact of forest cover and land use change on tree species diversity and carbon stocks in Shivapuri Nagarjun National Park, Nepal. Discov Environ 3(1). http://dx.doi.org/10.1007/s44274-025-00205-x

Kaushik P, Singh H (2024) Climate change-induced forest ecosystem functioning and alteration in water–atmosphere–soil–plant (WASP) relationship. In: Singh H (ed) Urban forests, climate change and environmental pollution. Springer, Cham. https://doi.org/10.1007/978-3-031-67837-0_38

Kumar N, Singh H, Kumar A, Singh AK, Pandey PK, Kumar A (2024) Forest-fire-induced land degradation. In: Mishra G, Giri K, Singh S, Kumar M (eds) Sustainable land management in India. Springer Nature, Singapore. https://doi.org/10.1007/978-981-97-5223-2_4

Kunwar RM, Khaling S (2006) Forest fire in the Terai, Nepal causes and community management interventions. IFFN 46–54

Matin MA, Chitale VS, Murthy MS, Uddin K, Bajracharya B, Pradhan S (2017) Understanding forest fire patterns and risk in Nepal using remote sensing, geographic information system and historical fire data. Int J Wildland Fire 276–286. https://doi.org/10.1071/WF16056

Nolan KA, Callahan JE (2006) Beachcomber biology: the Shannon-Weiner species diversity index. In: Tested studies for laboratory teaching, pp 334–338

Paton D, Buergelt PT, McCaffrey S, Tedim F (2014) Wildfire hazards, risks, and disasters. Elsevier

Pausas JG, Keeley JE (2019) Wildfires as an ecosystem service. Front Ecol Environ 17(5):289–295

Randerson JT, Chen Y, van Der Werf GR, Rogers BM, Morton DC (2012) Global burned area and biomass burning emissions from small fires. J Geophys Res Biogeo-Sci 117:1–23. https://doi.org/10.1029/2012jg002128

Sett T, Nikam BR, Singh H, Purohit S (2024) Hydrological variability in Indian forest ecosystem: analysis of drought resilience, recovery and water use efficiency in moist and dry deciduous forests. In: Singh H (ed) Urban forests, climate change and environmental pollution. Springer, Cham. https://doi.org/10.1007/978-3-031-67837-0_37

Singh H (2024a) Climate change and forests: biological perspectives. In: Singh H (ed) Forests and climate change. Springer Nature, Singapore. https://doi.org/10.1007/978-981-97-3905-9_1

Singh H (ed) (2024b) Forests and climate change: biological perspectives on impact, adaptation, and mitigation strategies. Springer Nature, Singapore. https://doi.org/10.1007/978-981-97-3905-9

Singh H (2024c) Climate change and environmental pollution: biological consequences on urban forests. In: Singh H (ed) Urban forests, climate change and environmental pollution. Springer Nature, Cham, Switzerland. https://doi.org/10.1007/978-3-031-67837-0_1

Singh H (ed) (2024d) Urban forests, climate change and environmental pollution: physio-biochemical and molecular perspectives to enhance urban resilience. Springer Nature, Cham, Switzerland. https://doi.org/10.1007/978-3-031-67837-0

Thakur P, Joshi R, Bhatta S, Ghimire S, Silwal R (2024) Computing aboveground carbon stocks using ground-based and sentinel imagery approach. In: Forests and Climate Change, pp 771–787. http://dx.doi.org/10.1007/978-981-97-3905-9_37

WWF (n.d.) Responsible forestry. Retrieved from www.worldwildlife.org, https://www.worldwildlife.org/industries/responsible-forestry

WWF (n.d.) Retrieved from www.worldwildlife.org: https://www.worldwildlife.org/initiatives/forests

Zhang Y, Chen H, Reich PB (2012) Forest productivity increases with evenness, species richness and trait variation: a global meta-analysis. J Ecol 742–749. https://doi.org/10.1111/j.1365-2745.2011.01944.x

The manufacturer's authorised representative in the EU is Springer Nature Customer Service Centre GmbH, Europaplatz 3, 69115 Heidelberg, Germany. If you have any concerns regarding our products, please contact ProductSafety@springernature.com

Printed and bound by CPI Group (UK) Ltd, Croydon, CR0 4YY
01/10/2026
02225871-0019